Lecture Notes in Mathematics

Edited by A. Dold, F. Takens and B. Teissier

Editorial Policy
for the publication of monographs

1. Lecture Notes aim to report new developments in all areas of mathematics – quickly, informally and at a high level. Monograph manuscripts should be reasonably self-contained and rounded off. Thus they may, and often will, present not only results of the author but also related work by other people. They may be based on specialized lecture courses. Furthermore, the manuscripts should provide sufficient motivation, examples and applications. This clearly distinguishes Lecture Notes from journal articles or technical reports which normally are very concise. Articles intended for a journal but too long to be accepted by most journals, usually do not have this "lecture notes" character. For similar reasons it is unusual for doctoral theses to be accepted for the Lecture Notes series.

2. Manuscripts should be submitted (preferably in duplicate) either to one of the series editors or to Springer-Verlag, Heidelberg. In general, manuscripts will be sent out to 2 external referees for evaluation. If a decision cannot yet be reached on the basis of the first 2 reports, further referees may be contacted: the author will be informed of this. A final decision to publish can be made only on the basis of the complete manuscript, however a refereeing process leading to a preliminary decision can be based on a pre-final or incomplete manuscript. The strict minimum amount of material that will be considered should include a detailed outline describing the planned contents of each chapter, a bibliography and several sample chapters.
Authors should be aware that incomplete or insufficiently close to final manuscripts almost always result in longer refereeing times and nevertheless unclear referees' recommendations, making further refereeing of a final draft necessary.
Authors should also be aware that parallel submission of their manuscript to another publisher while under consideration for LNM will in general lead to immediate rejection.

3. Manuscripts should in general be submitted in English.
Final manuscripts should contain at least 100 pages of mathematical text and should include
– a table of contents;
– an informative introduction, with adequate motivation and perhaps some historical remarks: it should be accessible to a reader not intimately familiar with the topic treated;
– a subject index: as a rule this is genuinely helpful for the reader.

Lecture Notes in Mathematics 1721

Lecture Notes in Mathematics

Editors:
A. Dold, Heidelberg
F. Takens, Groningen
B. Teissier, Paris

1724

Springer
Berlin
Heidelberg
New York
Barcelona
Hong Kong
London
Milan
Paris
Singapore
Tokyo

Anthony Iarrobino   Vassil Kanev

Power Sums, Gorenstein Algebras, and Determinantal Loci

With an Appendix
The Gotzmann Theorems and the Hilbert Scheme
by Anthony Iarrobino and Steven L. Kleiman

 Springer

Authors

Anthony Iarrobino
Mathematics Department
Northeastern University
Boston, MA 02115, USA

E-mail: iarrobin@neu.edu

Vassil Kanev
Institute of Mathematics
Bulgarian Academy of Sciences
1113 Sofia, Bulgaria

E-mail: kanev@math.bas.bg

Cataloging-in-Publication Data applied for

Die Deutsche Bibliothek - CIP-Einheitsaufnahme

Iarrobino, Anthony A.:
Power sums, Gorenstein algebras, and determinantal loci / Anthony
Iarrobino ; Vassil Kanev. With an appendix The Gotzmann theorems
and the Hilbert scheme / by Anthony Iarrobino and Steven L.
Kleiman. - Berlin ; Heidelberg ; New York ; Barcelona ; Hong Kong
; London ; Milan ; Paris ; Singapore ; Tokyo : Springer, 1999
    (Lecture notes in mathematics ; 1721)
    ISBN 3-540-66766-0

Mathematics Subject Classification (1991):
14M12, 14C05, 13C40, 14N99, 13H10

ISSN 0075-8434
ISBN 3-540-66766-0 Springer-Verlag Berlin Heidelberg New York

Typesetting: Camera-ready T$_E$X output by the authors
Printed on acid-free paper   SPIN: 10700327   41/3143-543210

To our parents
Elizabeth and Anthony Iarrobino
Givka i Ivan Kanevi

Preface

This book is devoted to a classical problem with a long history — that of representing a homogeneous polynomial as a sum of powers of linear forms. This problem is closely related to another interesting topic — the study of the loci which parametrize homogeneous polynomials with a given sequence of dimensions for the spaces spanned by their order-i higher partial derivatives. Here a convenient tool to work with are the catalecticant matrices associated to a homogeneous polynomial, whose columns are the coefficients of its partial derivatives in appropriate monomial bases — the above dimensions are then the ranks of the catalecticant matrices, and the above parametric varieties are their determinantal loci.

In the Introduction we define all basic notions in an informal way, in the classical setting of characteristic zero. We hope this will facilitate reading the book, where setting of arbitrary characteristic is adopted. Our experience has been that with a little more effort almost all results valid in characteristic zero can be extended to arbitrary characteristic, replacing the ring of polynomials by the ring of divided powers; the two rings are isomorphic when the characteristic is zero.

The first two chapters are mainly expository and are intended to give an account of what was already known about catalecticant matrices, especially those associated with the first partial derivatives or with homogeneous polynomials in two variables. We aimed to make this part of the book as self-contained as possible, and include full proofs of some material scattered in the literature or contained in some hardly available old books. Chapters 3 – 8 as well as Section 2.2 of Chapter 2 contain our new results on the subject. We also included Sections 4.4 and 6.4 in which some recent development due to various authors is surveyed — a development partially inspired by earlier preliminary versions of this memoir circulated in 1995 – 1996 [IK].

The expert already familiar with the basic notions and notation may wish to skip to the Brief Summary of Chapters at the end of the Introduction, then skim the expository part of the first two chapters, noting especially the Detailed Summary (Section 1.4), before looking for topics of particular interest.

ACKNOWLEDGMENT. We thank Mats Boij, Joel Briançon, Young Hyun Cho, Steve Kleiman, Philippe Maisonobe, Michel Merle, Juan Migliore, Yves Pitteloud, Richard Porter, Bernard Teissier, Junzo Watanabe, Jerzy Weyman, and Joachim Yameogo for their comments. We particularly thank Anthony Geramita, Andy Kustin, Jan Kleppe, and Giuseppe Valla for detailed answers to our questions, and Young Hyun Cho, Bae Eun Jung, Hal Schenck, and Junzo Watanabe for their careful reading of certain sections, and corrections. We are most appreciative to Steve Kleiman for joining with us in writing Appendix C.

We are grateful to Richard Porter whose advice and LaTeX expertise guided us in improving the appearance of the book — and as well for access to an early version of his LaTeX guidebook [**Por**]; and we thank Tania Parhomenko for essential help in compiling the book in LaTex. The first author thanks the Laboratory of Mathematics at the University of Nice and its members for their hospitality; and his spouse Gail Charpentier for her support of the project. The authors were supported in part by the Bulgarian foundation *Scientific Research*, and by the National Science Foundation under the US-Bulgarian project *Algebra and Algebraic Geometry*.

We thank Anthony and Elizabeth Iarrobino, parents of the first author, for designing the frontispiece, based on a drawing by Anthony Iarrobino.

The authors
September 1999

Contents

Introduction: Informal History and Brief Outline

0.1. Canonical forms, and catalecticant matrices of higher partial derivatives of a form

A standard fact from linear algebra is that if $f(x) = xAx^t$, $x \in k^r$ is a quadratic form in r variables over a field k of characteristic not 2, then f can be represented as a sum of s squares of linear forms if and only if the rank of A satisfies $\text{rk}(A) \leq s$. For homogeneous forms of higher degree one can ask a similar question.

PROBLEM 0.1. What are the conditions on a homogeneous polynomial (shortly a form) of degree j in r variables, so that it can be represented as a sum

$$f = L_1^j + \cdots + L_s^j, \qquad (0.1.1)$$

where L_i is a linear form and s is fixed? When is such a representation unique? If $PS(s,j;r)$ (for *power sum*) denotes the set of such forms in the space $\mathcal{R}_j$ of all degree-j forms, what are the generators of the ideal of the affine variety $\overline{PS(s,j;r)}$?

A particular case is Waring's problem for general forms

WARING'S PROBLEM. What is the minimum integer s such that a general form of degree j in r variables can be represented as sum of powers as in (0.1.1)?

Waring's problem was only recently solved by J. Alexander and A. Hirschowitz [AlH3]. Their result also yields (via Terracini's Lemma) the dimension of $PS(s,j;r)$ (see Section 2.1 for details).

These problems attracted much attention among geometers and algebraists who worked in the field of theory of invariants in the second half of the nineteenth and first decades of the twentieth century. It suffices to mention the names of A. Clebsch, J. Lüroth, T. Reye, J. J. Sylvester, G. Scorza, E. Lasker, H. W. Richmond, A. Dixon,

A. Terracini, and J. Bronowski, who made important contributions to the subject (see [**Bro, Cle, Dix, Las, Lur, Rey, Ri, Sc1, Sc2, Sy1, Sy2, Sy3, Ter1, Ter2**]). When a representation of the type (0.1.1) is unique it is called a *canonical form* of f and a great deal of the above cited research was devoted to finding canonical forms of homogeneous polynomials. In fact even more general canonical forms close to sums of powers were studied, as e.g. $f = X^3 + Y^3 + Z^3 + mXYZ$ for cubic polynomials in 3 variables (see R. Ehrenborg and G.-C. Rota's [**EhR**] for a modern account and amplification).

The case of two variables (binary forms) is not difficult and was essentially solved by J. Sylvester [**Sy1, Sy2, Sy3**] who proved that a general binary form of odd degree $j = 2t - 1$ has a canonical form $f = L_1^j + \cdots + L_t^j$. He also introduced a *catalecticant invariant* in the case of even degrees and proved that among the binary forms of degree $j = 2t$ with vanishing catalecticant invariant every sufficiently general one has a canonical form $f = L_1^j + \cdots + L_t^j$. [1] An extension of Sylvester's arguments is easy and the outcome is that the algebraic closure $\overline{PS(s, j; 2)}$ is the rank $\leq s$ determinantal locus in the space of $(s+1) \times (j-s+1)$ catalecticant matrices (to be introduced below); the canonical forms are determined by the solutions of linear homogeneous systems whose matrices of coefficients are catalecticant matrices [2].

If one wants to obtain canonical forms of arbitrary homogeneous polynomials in 2 variables a new feature appears: one has to consider representations more general than (0.1.1), namely

$$f = G_1 L_1^{j-d_1+1} + \cdots + G_m L_m^{j-d_m+1} \qquad (0.1.2)$$

where $\deg G_i = d_i - 1$ and $\sum_{i=1}^m d_i = s$. We refer the reader to Section 1.3 for an exposition of this subject. A different solution to the problem of representing a binary form as sum of powers of s linear forms was obtained by S. Gundelfinger [**Gu1, Gu2**] (see also [**GrY, Ku1, Ku2**]), who expressed the condition $f \in \overline{PS(s, j; 2)}$ in terms of the vanishing of certain covariants of f.

The problem of finding the generators of the ideal of $\overline{PS(s, j; r)}$ is more difficult. For instance in the case of quadrics ($j = 2$) this is

---

[1]Sylvester introduced the name "catalecticant" from prosody, where a catalectic line of verse means one missing a foot, or beat: the catalectic binary forms have t summands, instead of the $t+1$ needed in general for binary forms of degree $j = 2t$ (see B. Reznick's [**Rez1**, p. 49] for a historical remark).

[2]This should have been well-known to Sylvester and later invariant theorists, although maybe not written explicitly. At those times the aim was to express such conditions in terms of invariants and covariants.

the content of the Second Fundamental Theorem of invariant theory for the orthogonal group $O(s)$ (see discussion in Section 1.2). In the binary case ($r = 2$) the generators are the $(s + 1) \times (s + 1)$ minors of certain catalecticant matrices — the Hankel matrices. We refer to the papers [**GruP, Ei1, Wa4**] as well as to Section 1.3 about this result.

Much less is known if the number of variables r is greater than two. One of the primary goals of this book is to study the locus $PS(s, j; r)$ ($= P_s$ for short) of homogeneous forms of degree j in r variables which have a representation of the form (0.1.1). We call this a length-s additive decomposition of f. We aim to generalize the above mentioned cases $j = 2$ or $r = 2$ and to relate the closure $\overline{P_s}$ to certain determinantal loci of catalecticant matrices, which we now introduce informally. We refer the reader to Section 1.1 for precise definitions.

We will suppose that k is an algebraically closed field. Let $\mathcal{R} = k[X_1, \ldots, X_r]$ denote the polynomial ring, and let $\mathcal{R}_j$ be the space of homogeneous polynomials of degree j. Let us assume for the moment that char(k) = 0. If a form $f \in \mathcal{R}_j$ has a length-s additive decomposition as in (0.1.1) (so $f \in PS(s, j; r)$) then it is clear that for every v, $1 \leq v \leq j - 1$ the partial derivatives $\partial^v f / \partial X^V$ with $|V| = v$ span a subspace $\langle L_1^{j-v}, \ldots, L_s^{j-v} \rangle$ in $\mathcal{R}_u$, $u = j - v$ of dimension no greater than s.

This can be equivalently stated in the following way. Consider the polynomial ring $R = k[x_1, \ldots, x_r]$ and for fixed f, taking derivatives, consider the linear operator $C_f(u, v) : R_v \to \mathcal{R}_u$ which transforms $\phi \in R_v$ to $\phi \circ f = \phi(\partial_1, \ldots, \partial_r) f$. The above condition implies that for every v, $1 \leq v \leq j - 1$ one has $\mathrm{rk}\, C_f(u, v) \leq s$. In order to express this in matrix form take a basis $x_1^{v_1} \cdots x_r^{v_r}$, $v_1 + \cdots + v_r = v$ for R_v and a basis $\frac{(j-v)!}{U!} X^U := \frac{(j-v)!}{u_1! \cdots u_r!} X_1^{u_1} \cdots X_r^{u_r}$, $u_1 + \cdots + u_r = j - v = u$ for $\mathcal{R}_u$. Then an easy calculation shows that the matrix of $C_f(u, v)$ for the polynomial

$$f = \sum_{w_1 + \cdots + w_r = j} \frac{j!}{w_1! \cdots w_r!} a_{w_1, \ldots, w_r} X_1^{w_1} \cdots X_r^{w_r}$$

is equal to $j!/(j - v)!$ times the *catalecticant matrix*

$$Cat_f(u, v; r) = (b_{U,V} = a_{U+V})_{|U|=u, |V|=v}. \tag{0.1.3}$$

So, the above rank conditions can be equivalently stated by saying that the variety $PS(s, j; r)$ is contained in the rank $\leq s$ determinantal loci of the catalecticant matrices (0.1.3) for each u, $1 \leq u \leq j - 1$. Notice that ${}^t Cat_F(u, v; r) = Cat_F(v, u; r)$, so half of the values of u suffice, $1 \leq u \leq j/2$.

The catalecticant matrices in the binary case $r = 2$, which we discussed above have the following form. If

$$f = a_0 X_1^j + ja_1 X_1^{j-1} X_2 + \ldots + \binom{j}{v} a_v X_1^{j-v} X_2^v + \ldots + a_j X_2^j,$$

one has with $j = u + v$

$$Cat_f(u, v; 2) = \begin{pmatrix} a_0 & a_1 & \ldots & a_v \\ a_1 & a_2 & \ldots & a_{v+1} \\ \cdots\cdots\cdots\cdots\cdots\cdots \\ a_u & a_{u+1} & \ldots & a_j \end{pmatrix}, \tag{0.1.4}$$

which is also known as the u-th Hankel matrix of the binary form f. The Hankel matrices have been extensively studied because of applications to control theory, to data-processing, or to discretization of differential equations (see, e.g. [**HeiR, Io, HMP, GoS**]). Most of these studies have been in a non-homogeneous context; the homogeneous context that we adopt may lead to more simply stated results.

In the general case we ask the following questions.

PROBLEM 0.2. What are the possible sequences of ranks of the catalecticant matrices of a form $f \in \overline{PS(s, j; r)}$? What kind of forms belong to the boundary $\overline{PS(s, j; r)} - PS(s, j; r)$?

In the binary case the second question is easy and requires considering the decompositions (0.1.2) (this is attributed to J. H. Grace in [**EhR**]); the first question was solved by Macaulay [**Mac1**] (see also Section 1.3). We study this problem and in the case $r = 3$ we answer the first question , and we give partial results for the second in Chapter 5 (see Section 5.6).

In general this problem as well as the question of the explicit forms of the homogeneous polynomials belonging to the boundary $\overline{PS(s, j; r)} - PS(s, j; r)$ (a generalization of the decomposition (0.1.2)) are open problems. One can prove (see Lemma 1.17) that when f is general — lies outside a certain proper closed subvariety of $\overline{PS(s, j; r)}$ — these ranks $t_i = \text{rk}\,Cat_f(i, j - i; r)$ are equal to the following sequence of positive integers $T = (t_0, \ldots, t_j)$, $t_i = H(s, j, r)_i$, where

$$H(s, j, r)_i = \min(s, \dim_k R_i, \dim R_{j-i}), \quad 0 \le i \le j. \tag{0.1.5}$$

We conclude that $\overline{PS(s, j; r)}$ is contained in the algebraic closure of the quasiaffine set $Gor(T)$ consisting of polynomials for which the ranks of the catalecticant matrices are equal to the sequence $T = H(s, j, r)$ given by (0.1.5). Clearly $Gor(T)$ is an open subset of the determinantal locus given by the vanishing of all $(s + 1) \times (s + 1)$ minors of all the catalecticant matrices $Cat_f(i, j - i; r)$.

One of the main results in this book is that for $r \geq 3, j = 2t$ or $2t+1$, and $s \leq \dim_k R_{t-1}$ the affine variety $\overline{PS(s, j; r)}$ is an irreducible component of $\overline{Gor(T)}$ (Theorem 4.10A). Moreover, in case $r = 3$ one has in fact equality (Theorems 4.1A and 4.5A). This result is a step toward a solution of Problem 0.1 when $r = 3$. It can be reformulated as follows. If $s \leq \binom{t+1}{2}$ and a ternary form f of degree $j = 2t$ or $2t+1$ has the same sequence of ranks of its catalecticant matrices as that of a general form of $PS(s, j; 3)$, i.e. equal to (0.1.5), then f is a sum of s powers of linear forms (0.1.1) or is a degeneration of such a sum (i.e. belongs to $\overline{PS(s, j; 3)}$).

The sequence $H(s, j, 3)$ is only one of the possible sequences $T = H_f$ of ranks of catalecticant matrices for ternary forms f. In Chapter 5 we consider a more general and difficult problem – representing a ternary form with arbitrary sequence T of ranks of catalecticant matrices as a sum of s powers of linear forms, or as a limit of such sums, where $s = \max\{T_i\}$, the minimum possible number of summands. With the exception of some sequences occuring when the degree $j \geq 4$ is even, we find criteria for such a representation to exist, in terms of the catalecticant matrices. While the proofs are complicated and use the Buchsbaum-Eisenbud structure theorem for height three Gorenstein ideals, from commutative algebra, the criteria are simple and are analogous to the rank conditions for Hankel matrices in the binary case. In particular, if the form f is such that the sequence T of ranks of catalecticant matrices contains a constant subsequence (s, s, s), then f is in the closure $\overline{PS(s, j, 3)}$. The reader may find a further discussion and a precise statement of our results in this direction in the summarizing Section 5.4, which is written in the language of matrices and polynomials, so as to be accessible to a general reader.

The geometrically minded reader might look at this subject from another point of view. The projectivization $\mathbb{P}\overline{PS(s, j; r)}$ is the s-secant variety of the Veronese variety $v_j(\mathbb{P}^{r-1})$. This is the closure in $\mathbb{P}^{r-1}$ of the variety traced out by the projective subspaces spanned by the s-tuples of points of $v_j(\mathbb{P}^{r-1})$. So the above results can be restated in the following way. Provided $s \leq \dim_k R_{t-1}$ as above, the s-secant variety to the Veronese variety $v_j(\mathbb{P}^{r-1})$ is an irreducible component of the determinantal locus given by the vanishing of the $(s+1) \times (s+1)$ minors of all catalecticant matrices associated to forms of degree j. A warning: $\mathbb{P}\overline{Gor(T)}$ need not be equal to the latter determinantal variety (see Example 7.11). In some cases ($j = 2$, or $r = 2$, or $s \leq 2$) it is known that the $(s+1) \times (s+1)$ minors of the catalecticant matrices generate the homogeneous ideal of the s-secant variety to the Veronese variety (equal to the ideal of $\overline{PS(s,j;r)}$, cf. Problem 0.1). We refer to Sections 1.2 and 1.3 for a discussion.

0.2. Apolarity and Artinian Gorenstein algebras

The notation $Gor(T)$ comes from another important connection, the study of graded Artinian Gorenstein algebras. Setting $x_i = \partial_i$ we may think of R as the ring of linear differential operators with constant coefficients acting on the ring $\mathcal{R}$ of polynomials. Let $f \in \mathcal{R}_j, f \neq 0$, and let $I \subset R$ be the ideal of all differential operators which annihilate f; these are also called polynomials apolar to f if considering $R = k[x_1, \ldots, x_r]$ as before. We use the notation $I = \mathrm{Ann}(f)$. These ideals were called *principal systems* by F. H. S. Macaulay and the quotients $A_f = R/I$ were studied in [**Mac2**]. In fact Macaulay worked over arbitrary characteristic of the base field, which requires a slightly different setting, which is equivalent (if $\mathrm{char}(k) = 0$ or $\mathrm{char}(k) > j$) to the one we have so far considered (see page 266).

It is well known that the quotients A_f are exactly the graded Artin algebras A whose socle,

$$Soc(A) = 0 : m = \{h \in A \mid m \cdot h = 0, \text{ where } m = R_{\geq 1} = R_1 + \cdots \}$$

satisfies $\dim_k Soc(A) = 1$ (see e.g. [**Ei2**, p.526], or also Lemma 2.14 below). These algebras $A_f = R/\mathrm{Ann}(f)$, known classically and studied extensively by Macaulay and others, became known as graded Artinian *Gorenstein* algebras, after the influential article of H. Bass [**Bas**][3]: D. Gorenstein had studied a self-duality property of semigroup rings for certain singularities of curves [**Gor**]. We will follow usual practice in calling $I = \mathrm{Ann}(f)$ a Gorenstein ideal — although it is the quotient algebra A_f that is Gorenstein and that in the Artinian case has finite dimension as a k-vector space.

Recall that the Hilbert function of a graded algebra $A = \oplus A_i$ is the sequence $H = (h_0, \ldots, h_i, \ldots)$, where $h_i = \dim_k A_i$. The dimensions of the spaces of partial derivatives of f (see (0.1.3)) form a sequence H_f equal to the Hilbert function $H(A_f)$. From this point of view it is natural to consider arbitrary sequences of positive integers $T = (t_0, \ldots, t_j)$ with $t_0 = 1, t_1 \leq r, t_i = t_{j-i}$ and let $Gor(T)$ denote the subset of the affine space $\mathcal{A}_j = \mathbb{A}(\mathcal{R}_j)$ consisting of the polynomials f with $T = H(A_f)$. Then if $Gor(T)$ is nonempty, $\mathbb{P}Gor(T)$ parametrizes all graded Artinian Gorenstein algebras A_f with Hilbert function $H_f =$

---

[3]H. Bass reports that the name was originally given by A. Grothendieck

$H(A_f) = T$. We call such a sequence T a Gorenstein sequence, and we let $\mathcal{H}(j,r)$ be the set of all such Gorenstein sequences.

When $r = 3$ a powerful tool that we and other authors have used for studying the varieties of sums of powers $PS(s, j; 3)$ and the varieties $Gor(T)$ is the Buchsbaum-Eisenbud structure theorem of height three Gorenstein ideals [**BE2**]. It gives in particular the minimal resolution of the ideal $\mathrm{Ann}(f)$, for $f \in k[X, Y, Z]$ in terms of a skew-symmetric matrix and its Pfaffians. From this it is straightforward to determine the set of height three Gorenstein sequences $\mathcal{H}(j, 3)$ (see [**St1**], and Section 4.4 below.)

We consider several related problems concerning the decomposition of $\mathcal{A}_j = \mathbb{A}(\mathcal{R}_j)$ as a disjoint union

$$\mathcal{A}_j = \bigcup_{T \in \mathcal{H}(j,r)} Gor(T).$$

The questions we study or survey are

PROBLEM (A). Determine the minimum length s of an additive decomposition $f = L_1^j + \ldots + L_s^j$ of a general degree-j homogeneous polynomial f in r variables as sums of s powers of linear forms (Waring's problem). When is there a unique additive decomposition of minimum length? How is the Zariski closure $\overline{PS(s,j,r)}$ in $\mathbb{A}(\mathcal{R}_j)$ of the variety $PS(s, j; r)$ parametrizing forms having an additive decomposition of length s, related to the determinantal loci of catalecticant matrices? What is the dimension of $PS(s, j; r)$?

PROBLEM (B). Describe the algebraic set $Gor(T) \subset \mathbb{A}(\mathcal{R}_j)$. Is $Gor(T)$ irreducible? What is its dimension? What is its Zariski closure?

PROBLEM (C). If T is a given sequence, and $I = \mathrm{Ann}(f)$ is a Gorenstein ideal in R such that $A_f = R/I$ has Hilbert function $H(A_f) = T$, determine the possible Hilbert functions and minimal resolutions for the ideal I^2.

PROBLEM (D). Given a set Z of s general points of $\mathbb{P}^{r-1}$, consider the square $\mathcal{I}_Z^2$ and the symbolic square $\mathcal{I}_Z^{(2)}$ of the graded ideal $\mathcal{I}_Z$ in $R = k[x_1, \ldots, x_r]$ of polynomials vanishing at Z. What are the Hilbert functions of $R/\mathcal{I}_Z^2$ and $R/\mathcal{I}_Z^{(2)}$?

PROBLEM (E). Generalizing the binary forms case $r = 2$, understand the relation between the Gorenstein ideal $I = \mathrm{Ann}(f), f \in Gor(T)$, and the ideal $J = (I_{\le c})$ generated by the elements of I having degree no greater than c, especially when $c \approx j/2$ and J is the defining ideal of a zero-dimensional (or punctual) scheme in $\mathbb{P}^{r-1}$. What is the connection between $Gor(T)$ and the punctual Hilbert scheme?

Problems (A) and (B) we discussed above. As we show in Sections 3.2 and 4.1 Problem (C) is related to the study of the tangent spaces to the determinantal scheme $\mathbf{Gor}(T)$ whose associated reduced scheme is the locus $Gor(T)$ considered above. As for Problem (D), why do ideals $\mathcal{I}_Z$ of finite sets of points appear in the problem of representing a homogeneous form f as sums of powers of linear forms? This is the core of the classical apolarity method, which we use throughout in the memoir, and connecting I^2 with $\mathcal{I}_Z^2$ and eventually with $\mathcal{I}_Z^{(2)}$ permits us to obtain some results on Problems (A) and (B). We introduce the reader to it by sketching the beautiful solution of Clebsch [Cle] to the problem of representing a quartic form in 3 variables as a sum of 4th powers of 5 linear forms. If one tries to represent such a form as sum of powers

$$f = L_1^4 + \cdots + L_5^4 \qquad\qquad (0.2.1)$$

the first thought is that this should always be possible for a generic $f \in \mathcal{R}_4$ since on both sides we have the same number of parameters, 15. Now, let $L_i = \sum_{j=1}^3 p_{ij} X_j$, let $p_i = (p_{i1} : p_{i2} : p_{i3}) \in \check{\mathbb{P}}^2$. Let $R = k[x_1, x_2, x_3]$. We may think of R as the homogeneous coordinate ring of the dual projective space $\check{\mathbb{P}}^2$, where $\mathbb{P}^2$ has homogeneous coordinate ring $\mathcal{R} = k[X_1, X_2, X_3]$. This is natural by the differentiation pairing. Now, consider the five points $p_1, \ldots, p_5 \in \check{\mathbb{P}}^2$ which correspond to the five lines $L_i = 0$ in (0.2.1). There is a quadratic form $\phi \in R_2$ which vanishes simultaneously on all p_i. Differentiate f by $\phi(\partial_1, \partial_2, \partial_3)$. Then an easy calculation (see e.g. Lemma 1.15) shows that

$$\phi \circ f = const \cdot (\phi(p_1)L_1^2 + \cdots + \phi(p_5)L_5^2) = 0.$$

Equivalently, the linear map $C_f(2,2) : R_2 \to \mathcal{R}_2$ whose matrix is the catalecticant matrix $Cat_f(2,2;3)$ has a nonzero element ϕ in the kernel, hence the catalecticant determinant $\det(Cat_f(2,2;3)) = 0$. This is a polynomial relation of degree 6 on the coefficients of f in order that f has a representation of the form (0.2.1). Why is this a nontrivial relation? Take $g = L_1^4 + \cdots + L_6^4$ with 6 general forms. Then the same argument as above coupled with the fact that no conic in $\mathbb{P}^2$ can pass through 6 general points, shows that $Cat_g(2,2;3)$ is nondegenerate. Thus $\det(Cat_g(2,2;3)) \neq 0$. In fact, one can show that the hypersurface of degree 6 in $\mathbb{A}^{15}$ given by the vanishing of the catalecticant determinant is equal to the variety $\overline{PS(5, 4; 3)}$ (see e.g. [**DK**, §6] or Corollary 2.3).

Remark. All problems and discussion so far were stated for sums of powers. In fact we work mainly with another type of additive decomposition $f = L_1^{[j]} + \cdots + L_s^{[j]}$ which permits us to obtain new results, and extend many old ones to arbitrary characteristic of the base field k. Namely, f belongs to the ring of divided powers $\mathcal{D}$ and $L^{[j]}$ is the divided power of a linear form L. We refer to Appendix A for definitions and details. The reader who is interested in the char(k) = 0 case or characteristic sufficiently large (char(k) > j, when working with polynomials of degree $\leq j$) may replace the ring of divided powers $\mathcal{D}$ by the ring of polynomials $\mathcal{R}$, and the divided powers $L^{[j]}$ by ordinary powers L^j in all statements of the book — since under these assumptions $L^{[j]} = \frac{1}{j!} L^j$ (see Proposition A.12).

0.3. Families of sets of points

A representation of f as a sum of powers of linear forms as in (0.1.1) determines s distinct points of $\mathbb{P}^n$, or a point of the s-th symmetric product $Sym^s(\mathbb{P}^n)$, $n = r - 1$. Likewise, a form f in the closure of $PS(s, j; r)$, may determine a degree-s zero-dimensional subscheme Z_f of $\mathbb{P}^n$, that is *smoothable* — lies in a flat family of such schemes, the general member of which consists of s distinct points, so is smooth.

A degree-s, zero-dimensional subscheme Z of $\mathbb{P}^n$ consists of s points, counting multiplicities, with the further structure of an Artin ring concentrated at each distinct point - these are sometimes called thick points. Such a scheme Z determines a point of the punctual Hilbert scheme $\mathbf{Hilb}^s(\mathbb{P}^n)$, introduced by Grothendieck, and studied further by the Nice school of J. Briançon, M. Granger, J. Yaméogo, as well as J. Fogarty, S. Kleiman, and others. Recently the punctual Hilbert scheme, particularly for surfaces, has seen striking applications to symmetric functions and physics (see, for example [**Hai, Nak1, Nak2**]). When $n \leq 2$, $\mathbf{Hilb}^s(\mathbb{P}^n)$ is smooth. This fact has been recently used by G. Ellingsrud and A. Strømme to determine the degree of the secant variety to the j-th Veronese embedding of $\mathbb{P}^2$ in certain cases [**ElS**].

How does a zero-dimensional scheme $Z \subset \mathbb{P}^n$ connect to a form f? Given Z, we can consider the set of forms f of degree-j, such that $\mathcal{I}_Z \subset I = \text{Ann}(f)$. If we are very fortunate, this process determines a vector bundle of forms f over $\mathbf{Hilb}^s(\mathbb{P}^n)$. However, this is not always the case, and the study of when there might be a fibration $Gor(T)$ to $\mathbf{Hilb}^s(\mathbb{P}^n)$ is one of the more technically difficult problems that we explore in Part II.

0.4. Brief summary of chapters

We now briefly describe the contents of the book, and postpone a more detailed summary of the main results until Section 1.4. For simplicity we state the results in the framework of usual power sums, which are valid for $\operatorname{char}(k) = 0$ or $\operatorname{char} k > j$ (sufficiently large), although the theorems in the paper are stated and proved for divided powers and arbitrary characteristic (see the remark above).

Chapter 1 contains preliminary material, examples and definitions, a discussion of determinantal varieties of catalecticant matrices in the simplest cases $Cat(1, j-1; r)$ and $Cat(i, j-i; 2)$ (Sections 1.2, 1.3), and the detailed summary (Section 1.4). In Section 1.1 we define the determinantal scheme $\mathbf{V}_s(u, v; r)$ of catalecticant matrices, whose ideal is generated by the $(s+1) \times (s+1)$ minors of the generic catalecticant matrix $Cat_F(u, v; r)$. We also define the scheme $\mathbf{Gor}(T)$ whose associated reduced subscheme $Gor(T)$ we discussed above. We introduce the basic definitions of apolarity, prove the Apolarity Lemma 1.15 and give some corollaries of it (cf. [**DK**, §2,4] for this classical material).

Section 1.2 is a survey of the known results about the determinantal locus $V_s(1, j-1; r)$, where $s < r$. Besides the classical case of quadrics ($j = 2$) we describe a recent work of O. Porras [**Po**] about this variety and report on a recent result of the second author [**Ka**] describing the structure of $\overline{PS(2, j; r)}$, as well as that of $\overline{Gor(T)}$ when $t_1 \leq 2$. Section 1.3 is a self-contained exposition with full proofs of the theory in the binary forms case ($r = 2$). It may serve as an introduction to the subject, and as a model for what we would like to achieve for $r \geq 3$. The detailed summary, Section 1.4, also states results from the literature that we use in the sequel.

Chapter 2 contains a discussion of the Waring problem (Section 2.1), a new result concerning the uniqueness of the representation of a form f as sum of powers (Section 2.2), and an introduction to Artinian Gorenstein algebras (Section 2.3). Section 2.1 contains a translation of the J. Alexander-A.Hirschowitz vanishing theorem to a solution of the Waring problem, in all cases except when $\operatorname{char} k | j$; and Section 2.2 contains Theorem 2.6, which proves that a representation $f = L_1^j + \cdots + L_s^j$ is unique provided $f \in PS(s, j; r)$ is general enough and $s \leq \dim_k R_t - r$ for $j = 2t$, or $s \leq \dim_k R_t$ for $j = 2t + 1$.

Section 2.3 is an introduction to Artinian Gorenstein algebras: it includes all the basic results needed for the sequel. It also includes an application of the Minimal Resolution Theorem for ideals of s general enough points of $\mathbb{P}^2$, to determine the minimal resolution of $\operatorname{Ann}(f)$, when $r = 3$ (Proposition 2.19): this is an easy example of the relation between the study of points in $\mathbb{P}^{r-1}$ and the study of $PS(s, j; r)$, that is deepened in the later chapters of the book.

Chapter 3 contains two simple but important results, Theorems 3.2 and 3.9 which reduce the calculation of the tangent spaces to the schemes $\mathbf{V}_s(u, v; r)$ and $\mathbf{Gor}(T)$ at a point f, to determining the degree-j graded piece of the square I^2 of the annihilating ideal $I = \mathrm{Ann}(f)$. We give also several examples and corollaries, including a proof of the irreducibility of the catalecticant determinant $\det(Cat_F(t, t; r))$ in Proposition 3.13.

More generally, in Theorem 3.14 and Remark 3.15 we give a result which describes the structure of the corank-one determinantal loci $V_s(u, v; r)$ with $u \leq v$, and $s = \dim_k R_u - 1$. Lemma 3.17 provides the connection between I^2 and the symbolic square $\mathcal{I}_Z^{(2)}$, where $f = L_1^j + \cdots + L_s^j$ and $p_i \in \check{\mathbb{P}}^{r-1}$ are the points in the dual space which correspond to L_i, and $Z = \{p_1, \ldots p_s\}$. This lemma is the basic tool for the calculation in Chapter 4 of the tangent space to $\mathbf{Gor}(T)$ for $T = H(s, j; r)$ and s in the range $s \leq \dim_k R_{t-1}$, $j = 2t$ or $2t + 1$. Chapter 3 concludes with three conjectures. Conjecture 3.25 concerns the Hilbert function of $R/\mathcal{I}_Z^2$. It is related to Conjectures 3.20 and 3.23 about $Gor(T)$, $T = H(s, j; r)$, when s is in the range $\dim_k R_{t-1} < s < \dim_k R_t$, $j = 2t$ or $j = 2t + 1$.

Chapter 4 contains some of the main results of the book. In Theorems 4.10A and 4.10B we prove that, provided $s \leq \dim_k R_{t-1}$ with $j = 2t$ or $2t + 1$, then the closure $\overline{PS(s, j; r)}$ is an irreducible component of both $\overline{Gor(T)}$, $T = H(s, j, r)$ and of the determinantal locus $V_s(t, j - t; r)$. When $r = 3$, by S. J. Diesel's result showing the irreducibility of $\overline{Gor(T)}$ for any Gorenstein sequence [Di], it follows that in fact $\overline{PS(s, j; r)} = \overline{Gor(T)}$.

Examples in Chapter 7 show that $V_s(t, j - t; r)$ may be reducible when $r = 3$ and $Gor(T)$ can be reducible for certain T when $r \geq 4$, so an equality $\overline{PS(s, j; r)} = \overline{Gor(T)}$, with $T = H(s, j, r)$ seems to hold only for $r \leq 3$. For s in the range $\dim_k R_{t-1} < s < \dim_k R_t$ the dimension of $Gor(T)$ is larger than the dimension of $PS(s, j; r)$. In some cases we are able to prove that the scheme $\mathbf{Gor}(T)$ is generically smooth along $PS(s, j; r)$, and that the unique irreducible component of $Gor(T)$ which contains $PS(s, j; r)$ is of expected dimension (Theorem 4.1B for $r = 3, j = 2t$; Theorem 4.13 for $r \geq 3, s > \dim_k(R_t) - r$).

In Section 4.3, we study families of Gorenstein algebras $R/\operatorname{Ann}(f)$ such that the initial generators of $\operatorname{Ann}(f)$ form a complete intersection: we show that they form an irreducible component of $Gor(T)$ (Theorem 4.19). This work is related to results of J. Watanabe in [**Wa2**].

Our results in this chapter were obtained in 1994–1995. After distributing the preliminary draft of the early version of this memoir [**IK**], our Theorems 4.1A, 4.1B, 4.5A, 4.5B for $r = 3$ were reproved and substantially generalized by various authors. They use our Theorem 3.9 and some techniques specific to the $r = 3$ case, in particular the Buchsbaum-Eisenbud structure theorem for height 3 Gorenstein ideals. At present, it is proved that for $r = 3$, the schemes $\mathbf{Gor}(T)$ are smooth for every Gorenstein sequence T (J. Kleppe, [**Kl2**]) and there are several general formulas for $\dim Gor(T)$ in terms of T. We survey these developments in Section 4.4 (see also [**I9**] for a broader survey through 1997). We think it is worthwhile to include our original arguments since they focus more on apolarity and the ideals associated to finite sets of points, so they seem more promising for generalization to arbitrary r.

In Chapter 5 we consider mainly forms in 3 variables. We focus on the problem of finding criteria for representing a ternary form belonging to $Gor(T)$ as a sum of s powers of linear forms, where $s = \max\{T_i\}$ is the minimum possible number of summands. We also study the problem of explicitly describing the boundary $\overline{PS(s, j; r)} - PS(s, j; r)$, obtaining some partial results when $r = 3$. The case of binary forms suggests that in order to obtain satisfactory results in these and other related problems it is desirable to extend the classical apolarity relation between finite sets of points Z in $\check{\mathbb{P}}^{r-1}$ and forms, to one between zero-dimensional subschemes of $\check{\mathbb{P}}^{r-1}$ and forms. This is done in Section 5.1 where we introduce and give some basic properties of annihilating schemes of a form. These are zero-dimensional schemes, whose graded ideal consists of polynomials ϕ which annihilate f : $\phi \circ f = 0$. For a $f \in Gor(T)$ such a scheme is called tight if $\deg Z = s = \max\{T_i\}$. Section 5.2 contains some material about limits of punctual schemes and limits of ideals which is used later in connection with studying the closure $\overline{PS(s, j; r)}$. We also give the definition and some properies of the "postulation" Hilbert scheme $\mathbf{Hilb}^H \mathbb{P}^{r-1}$ whose closed points correspond to zero-dimensional subschemes of $\mathbb{P}^{r-1}$ with fixed Hilbert function.

Section 5.3 contains our results on power sum representations when $r = 3$. In particular we show that if a ternary form f of degree $j \geq 4$

is such that the sequence $T = H_f$ of ranks of catalecticant matrices (recall that $(H_f)_i$ is the dimension of the vector space of order $j - i$ higher partial derivatives of f) contains a subsequence (s, s, s), then f has an annihilating scheme Z_f of degree s. Moreover this scheme is unique, and its graded ideal is generated by the forms ϕ of degree $\leq \tau + 1$ apolar to f, $\phi \circ f = 0$, where $\tau = \min\{i \mid T_i = s\}$; so Z_f can be recovered explicitly from f. When $f \in Gor(T)$ is general enough the annihilating scheme is smooth – a set of distinct points in $\check{\mathbb{P}}^2$ – and one obtains by apolarity a length-s power sum decomposition $f = \sum_{u=1}^{s} L_u^j$. Notice the complete analogy of this result with the binary case, where a similar statement holds when $T \supset (s, s)$. If T does not contain (s, s, s) it is no longer true that a general element $f \in Gor(T)$ has a length-s power sum representation.

When $j = \deg(f)$ is odd (so $T \supset (s, s)$), we give necessary and sufficient conditions for the existence of annihilating schemes of degree s, and again they are unique and can be explicitly recovered from f, as above. Our results in the even case $T \subset (s - a, s, s - a), a > 0$ are less satisfactory, and our methods do not cover some classes of sequences T: however, these are exactly the cases when even if an annihilating scheme of degree s existed, its ideal could not be generated by apolar forms of degree $\leq j/2$, so the above procedure to recover Z_f explicitly from f could not be applied.

The canonical way an annihilating scheme is associated with a form permits us to construct, when $T \supset (s, s, s)$ a morphism from $Gor(T)$ to the postulation stratum $Hilb^H \mathbb{P}^2 = (\mathbf{Hilb}^H \mathbb{P}^2)_{red}$ (Theorem 5.31). The morphism is dominant and fibered by open subsets in $\mathbb{A}^s$. As an application one obtains a dimension formula for $Gor(T)$ when $T \supset (s, s, s)$, derived from Gotzmann's formula [**Got3, Got5**] for $\dim Hilb^H \mathbb{P}^2$ (Corollary 5.50). Conversely, a simple formula for $\dim Hilb^H \mathbb{P}^2$, can be derived from the recent Conca-Valla formula [**CoV1**] for $\dim Gor(T)$ (Corollary 5.34). In Corollary 5.49 it is proved that the Gorenstein ideals $\mathrm{Ann}(f)$ satisfy a certain weak Lefshetz property, when $T \supset (s, s, s)$.

Section 5.4 summarizes some of the results from the previous sections. Here we avoid the language of commutative algebra used in Section 5.3, and formulate our results in terms of matrices and polynomials. We focus on how to explicitly calculate a power sum representation of a given form. An algorithm and some examples are given. In Section 5.5 some further applications of the morphism $p : Gor(T) \rightarrow Hilb^H \mathbb{P}^2$ are given. Using it we derive some results about the Betti strata of $Hilb^H \mathbb{P}^2$ – the locally closed subsets of $Hilb^H(\mathbb{P}^2)$

parametrizing schemes Z whose ideal sheaves have fixed generator degrees.

In Section 5.6 we apply our work towards answering Problem 0.2 above, for $r = 3$: we determine the set of Hilbert functions that occur for forms f in the closure $\overline{PS(s, j; 3)}$ (Proposition 5.70). Also, it follows from the results of Section 5.3, that the sets $Gor(T)$, $T \supset (s, s, s)$ are completely inside the closure of $PS(s, j; 3)$ (Theorem 5.71). We give other criteria, that test either $f \in \overline{PS(s, j; 3)}$ or the contrary. Nevertheless when $r \geq 3$ the problems of describing explicitly the forms in the boundary $\overline{PS(s, j; r)} - PS(s, j; r)$, and of finding effective, simple criteria in terms of the form f itself, to decide whether each f is in $\overline{PS(s, j; r)}$, remain open, even when $r = 3$.

In Section 5.7 we discuss a generalization to higher dimensions of the apolarity relation between punctual schemes and graded Artinian Gorenstein algebras.

In Chapter 6 we study the component structure of $Gor(T)$, when $r \geq 4$, extending our work on the annihilating scheme. We work out the example where the annihilating scheme consists of one point, and its defining ideal in the local ring of the point is equal to the annihilating ideal of some homogeneous polynomial in $r - 1$ variables. We call this a "conic Gorenstein scheme" concentrated at a single point. In Lemma 6.1 we prove several statements that give a parallel between the case Z is smooth (this is the usual apolarity), and the conic Gorenstein case.

A main application of our tight annihilating schemes — those whose degree equals $s = \max\{T_i\}$ — is the construction of families $Gor(T)$ having two irreducible components for $r \geq 5, r \neq 8$ variables. This contrasts with the irreducibility of $Gor(T)$ for $r = 3$, proven by S. J. Diesel [Di]. The cases $r = 5, 6$ are treated in Corollaries 6.28 and 6.29 and require j relatively high. For $r \geq 7, r \neq 8$ we consider the sequence $T = T(j, r)$ where $T(j, r) = (1, r, 2r - 1, 2r, \ldots, 2r, 2r - 1, r, 1)$. In Theorem 6.26 we prove under the assumption $j \geq 8, r \neq 8$ that $Gor(T), T = T(j, r)$ has at least two irreducible components. One of them, $C_1(T)$ has a Zariski open subset which consists of polynomials $f = L_1^j + \cdots + L_{2r}^j$, where the points $p_1, \ldots, p_{2r} \in \check{\mathbb{P}}^{r-1}$ which correspond to the linear forms L_i, are a self-associated set Z (see [Co2, Co1, DO]). The other component $C_2(T)$ consists of polynomials whose tight annihilating scheme is a conic Gorenstein scheme concentrated in one point, and whose ideal at the point is the annihilating ideal of a general cubic form g in $r - 1$ variables.

For any r, we show that the forms having a smoothable tight annihilating scheme - of degree $s = \max_i\{(H_f)_i\}$ - are in the closure of $PS(s,j;r)$; but we prove a weaker statement than the converse (Proposition 6.7). Nevertheless, this approach suffices to give examples of reducible $Gor(T)$, for $r \geq 5, r \neq 8$.

We should notice that the cases $r = 4, r = 8$ were settled by M. Boij [**Bo2**], who constructed two irreducible components of certain $Gor(T)$; his construction works for arbitrary $r \geq 4$ and sufficiently large degrees of the forms. In Section 6.4 we discuss this and other recent developments, in the construction of components of $Gor(T)$.

In Chapter 7 we first show that the projectivization $\mathbb{P}V_s(u,v;r)$ is connected: in fact, each component contains $\mathbb{P}PS(1,j;r)$ (Theorem 7.6). We then in Section 7.2 give several criteria for deciding if $f \in \overline{Gor(T)}$; and we use the criteria to show that even when $r = 3$, $V_s(t,t;3)$ may have a large number $\approx \lfloor t/4 \rfloor$ irreducible components.

In Section 7.3 we restate some of the results obtained for $\overline{PS(s,j;r)}$ in terms of the s-secant variety to the Veronese variety $Sec_s(v_j(\mathbb{P}^{r-1}))$; one has $\mathbb{P}(\overline{PS(s,j;r)}) = Sec_s(v_j(\mathbb{P}^{r-1}))$. We also report a result by Ellingsrud-Strømme which gives the degree of $Sec_s(\mathbb{P}^2)$ when $s \leq 8$.

Chapter 8 compares the structure and closure of two natural embeddings of $\mathbb{P}Gor(T)$: the first, into $\mathbb{P}(\mathcal{D}_j)$, is the one that we use throughout; the second one sends an ideal into the ordered set of its graded components, in a product of Grassmanians.

In Chapter 9 we discuss open problems.

Appendix A contains the definition of the ring of divided powers and gives an exposition of the few properties of these rings we use in the book. In particular Proposition A.12 shows that when $\operatorname{char}(k) = 0$ or $\operatorname{char}(k)$ is sufficiently large one may interchange the ring of divided powers $\mathcal{D}$ and the polynomial ring $\mathcal{R}$.

Appendix B contains several results, mostly concerning height three Gorenstein algebras, that we use. Section B.1 contains basic definitions concerning Pfaffians, and as well a result concering the Pfaffian of a matrix with a large block of zeroes; and a decomposition formula of H. Srinivasan for Pfaffians. Section B.2 states for reference the Buchsbaum-Eisenbud structure theorem for height three Gorenstein ideals, in the graded case we use; and a result of Kustin-Ulrich concerning the minimal resolution of the square I^2 when I is a height three Gorenstein ideal. Section B.3 states several useful results of M. Boij connecting the minimal resolutions of the defining ideal for punctual schemes of $\mathbb{P}^n$, and the minimal resolutions for related Artinian Gorenstein algebras. In Section B.4 we state and prove a nice formula of A. Conca and G. Valla determining the maximal set of generator degrees $D_{\max}(T)$ possible for a height three Gorenstein sequence T, directly from the second difference of T. We also show a result of S. J. Diesel enumerating height three Gorenstein sequences T of given order and socle degree, by showing a one-to-one correspondence between the maximal sets of generator degrees, and certain partitions $P(T)$.

Appendix C was written by the first author and Steven L. Kleiman, and it holds a particular place in the book. Section C.1 contains what is used in the body: Macaulay's characterization of Hilbert functions of graded quotients of the polynomial ring, using O-sequences, and the particular case of graded quotients of $k[x_1, x_2]$ (Corollary C.6). The rest of the appendix treats various aspects of the theory of the Hilbert scheme of $\mathrm{Hilb}^P(\mathbb{P}^n)$, which parametrizes subschemes of $\mathbb{P}^n$ with Hilbert polynomial P. The last Section C.5 proves some new results about annihilating schemes of homogeneous polynomials and the varieties $Gor(T)$.

In Section C.2 we define Macaulay and Gotzmann polynomials, and give some of their properties; in Section C.3 we state the Persistence and Regularity Theorems of G. Gotzmann [Got1]. Next, in Section C.4 we report on the relation of these results to G. Gotzmann's identification of the Hilbert scheme $\mathrm{Hilb}^P(\mathbb{P}^n)$ as a simply described subscheme in a product of two Grassmanians. Gotzmann's description is based on Grothendieck construction and on his own improvement of Mumford's effective bound for the Castelnuovo–Mumford regularity degree. Here we also establish D. Bayer's related conjecture that the Hilbert scheme is a determinantal subscheme of a Grassmannian.

The last Section C.5 of the Appendix applies the Gotzmann Hilbert Scheme Theorem and a related result [Got4] to certain schemes $Gor(T)$, in a manner similar to some applications in an article of A. Bigatti, A. Geramita, and J. Migliore [BGM]. Thus we obtain another example, in addition to those of M. Boij [Bo2], of a scheme $Gor(T)$ for $r = 4$ having several irreducible components.

In Appendix D we give some examples of the Macaulay computer algebra scripts used [BaS1].

In Appendix E we give a comparison with the earlier manuscript of this book of May, 1996 [IK]. We do this in particular because some published articles refer to this earlier version, which was circulated.

Throughout the book we assume, unless otherwise specified, that the base field k is algebraically closed of arbitrary characteristic. We mean by variety a separated, reduced scheme of finite type over k (i.e.

not necessarily irreducible). Unless otherwise stated, by a point of a scheme we mean a closed point of a scheme. We will use as a rule math bold font to denote schemes and math slanted font to denote the corresponding varieties — the associated reduced subschemes. For instance we frequently work with the schemes $\mathbf{Gor}(T)$, $\mathbf{V}_s(u, v; r)$ etc. and the corresponding reduced subschemes (varieties) $Gor(T)$, $V_s(u, v; r)$ etc. Expressions like "Property '$\mathcal{P}$' holds for sufficiently general (or general enough, or general) element f" mean that from the context it is clear that f belongs to an irreducible variety and Property '$\mathcal{P}$' holds for every f in a certain Zariski open dense subset of this irreducible variety. The expression '...for generic F...' means F is the scheme-generic point of the irreducible scheme considered.

Dependence of the chapters/sections. A reader new to this material or to our viewpoint on it might wish to start with this Introduction and Sections 1.1–1.3, to get a flavor of the book, then look through 1.4 and Chapter 2, before perusing later sections of interest.

Prerequisites:

Each Chapter may be read independently, after reading Section 1.1. Although Chapter 4 uses Chapter 3, it is not necessary to read Chapter 3 first, just look at what one uses. We do feel that Chapters 1 and 2 give a desirable foundation for further reading. Here are some further suggestions.

Chapter 3: 1.1, 2.3; Section 3.2: 3.1.

Section 4.2: 4.1

Chapter 5: 1.1, 2.3; Section 5.3: 5.1, 1.3; Section 5.4: 1.1 only;
      Section 5.5: 5.2;   Section 5.6: 5.1,5.2;   Section 5.7: 5.3

Chapter 6: 5.1,5.2; Section 6.3: 6.2

Chapter 7: 5.1; Section 7.2: 7.1

Chapter 8: 2.3.

Part I

Catalecticant Varieties

CHAPTER 1

Forms and Catalecticant Matrices

This Chapter contains basic definitions and examples (Section 1.1); a discussion of determinantal varieties of catalecticant matrices in the simplest cases of the Jacobian $Cat(1, j - 1; r)$ (Section 1.2), and of binary forms, $Cat(i, j - i; 2)$ (Section 1.3). The latter includes full proofs of the classical results in this case, related to the Hankel matrix. Section 1.4 is the detailed summary, which includes preparatory material and results from the literature that we will use, and is organized around five main goals of our work.

1.1. Apolarity and catalecticant varieties: the dimensions of the vector spaces of higher partials

We first introduce notation and summarize what we need concerning the contraction action which also defines the Macaulay or Matlis duality.

DEFINITION 1.1. CONTRACTION ACTION AND MACAULAY OR MATLIS DUALITY. Let $\mathcal{D}$ be the divided power ring in r variables $X_1, \ldots, X_r$ over an algebraically closed field k of arbitrary characteristic. We refer to Appendix A for the basic properties of divided power rings that we use below. If char(k) = 0 then $\mathcal{D}$ is isomorphic to the polynomial ring $\mathcal{R} = k[X_1, \ldots, X_r]$. The divided powers monomials (DP-monomials) $X^{[U]} = X_1^{[u_1]} \ldots X_r^{[u_r]}$ form a basis of the k-space $\mathcal{D}$. If char(k) = 0, then $X^{[U]} = \frac{1}{u_1! \ldots u_r!} X_1^{u_1} \ldots X_r^{u_r}$. The same holds if $|U| = u_1 + \cdots + u_r \leq j$ and char(k) > j. If $L = a_1 X_1 + \cdots + a_r X_r$ then we denote by $L^{[j]}$ the sum

$$L^{[j]} = \sum_{j_1 + \cdots + j_r = j} a_1^{j_1} \ldots a_r^{j_r} X_1^{[j_1]} \ldots X_r^{[j_r]}$$

in the divided power ring. Then $L^j = (j!)L^{[j]}$. We let $R = k[x] = k[x_1, \ldots, x_r]$ and define the *contraction action* of R on $\mathcal{D}$: if $J =$

$(j_1, ..., j_r)$ and $U = (u_1, ..., u_r)$ are non-negative multiindices, then

$$x^U \circ X^{[J]} = \begin{cases} X^{[J-U]} & \text{if } U \le J; \\ 0 & \text{otherwise.} \end{cases} \qquad (1.1.1)$$

which is extended bilinearly to an action of $k[x]$ on $\mathcal{D}$. In characteristic 0 this is equal to the differentiation action

$$\phi \in R, \; f \in k[X_1, \dots, X_r] \Rightarrow \phi \circ f = \phi(\frac{\partial}{\partial X_1}, \dots, \frac{\partial}{\partial X_r}) f \qquad (1.1.2)$$

The partial derivative action (1.1.2) is the one used in the 19th century works on apolarity, while the contraction action is that of Macaulay's inverse systems (see Proposition A.3(iv) from Appendix A). The two actions are equivalent when the characteristic of k is zero, or is larger than any degree j considered (Appendix A, Proposition A.12). The contraction action gives a duality isomorphism between the set of finite-dimensional R-submodules M of $\mathcal{D}$, and the set of Artin algebra quotients $A = R/\operatorname{Ann}(M)$ of R; the module M satisfies $M \cong \operatorname{Hom}_k(A, R) = A^\vee$ (see §60 of [**Mac2**], Lemma 1.2 of [**I6**], or, in the graded case, Lemma 2.12 below). For a more general discussion of Macaulay or Matlis duality see [**BrS**, Chapter 10] and the references cited there, also [**Ei2**, Chapter 21],[**No**],[**NoR**], and [**Mat**].

DEFINITION 1.2. CATALECTICANT HOMOMORPHISM. If $v \ge 0$, $j = u + v$ and $f \in \mathcal{D}_j$, we let $C_f(u, v) : R_v \to \mathcal{D}_u$ be the *catalecticant homomorphism*:

$$C_f(u, v)(g) = g \circ f. \qquad (1.1.3)$$

Let $f = \sum_{|J|=j} a_J X^{[J]}$. Let us calculate the matrix of $C_f(u, v)$ in the bases: $x^V = x_1^{v_1} \dots x_r^{v_r}$, $|V| = v$ for R_v and $X^{[U]} = X_1^{[u_1]} \dots X_r^{[u_r]}$, $|U| = u$ for $\mathcal{D}_v$. We have

$$C_f(u, v)(x^V) = x^V \circ \sum_{|J|=j} a_J X^{[J]} = \sum_{|J|=j} \sum_{U+V=J} a_J X^{[U]}$$

Thus the matrix of $C_f(u, v)$ equals $(b_{U,V} = a_{U+V})_{|U|=u, |V|=v}$. Here as usual the sum of multiindices is understood to be the sum by components.

The matrices obtained are called *catalecticant matrices*. Their determinantal loci are the main object of study in this memoir, and we now introduce them with more details.

DEFINITION 1.3. CATALECTICANT MATRICES. We first fix (r, j), and denote by $k[Z]$, $Z = \{Z_J \mid J = (j_1, ..., j_r), j_i \ge 0, |J| = j\}$ the polynomial ring of coefficients of a generic degree-j element $F = \sum_{|J|=j} Z_J X^{[J]}$ of $\mathcal{D}_j$. We let $\mathbb{A}_j = \mathbb{A}(\mathcal{D}_j)$, or $\mathbb{A}^N$, $N = \dim_k \mathcal{D}_j = \binom{r+j-1}{r-1}$, denote the affine space $\operatorname{Spec} k[Z]$, and we denote by $\mathbb{N}$ the nonnegative integers. If $u + v = j$, $r_u = \dim_k R_u$, the $r_u \times r_v$ generic *catalecticant matrix* has rows indexed by the degree-u divided powers monomials in $X_1, \ldots, X_r$, hence by elements $U = (u_1, \ldots, u_r)$ of $\mathbb{N}^r$ of length $|U| = \sum u_i = u$, in lexicographic order (see Appendix C.1); its columns are indexed similarly by the degree-v monomials in $x_1, \ldots, x_r$. Each entry $Z_{U,V} = Z_{U+V}$, a variable of the ring $k[Z]$. Thus, we have

$$Cat_F(u, v; r) = (Z_{U,V} = Z_{U+V}), \quad |U| = u, |V| = v. \qquad (1.1.4)$$

If $p : Z_J = a_J$ is a k-valued point of $\mathbb{A}^N$, we let $f_p \in \mathcal{D}_j$ denote the corresponding homogeneous element. Then we define $Cat_{f_p}(u, v; r)$ as the matrix $Cat_F(u, v; r)$ evaluated at the point $Z = p$. Usually we will more simply write $Cat_f(u, v; r)$ for $f \in \mathcal{D}_j$. As we saw in Definition 1.2, $Cat_f(u, v; r)$ is the matrix of the catalecticant homomorphism $C_f(u, v) : R_v \to \mathcal{D}_u$. Clearly the transpose ${}^t Cat_f(u, v; r)$ satisfies

$$ {}^t Cat_f(u, v; r) = Cat_f(v, u; r). \qquad (1.1.5)$$

When $u = v$, we will refer to the symmetric matrix $Cat_f(u, u; r)$ as the square catalecticant. Its determinant is the classically known catalecticant invariant of a form f of even degree $j = 2u$ when $\operatorname{char}(k) = 0$ (see e.g. [**DK, GrY**]).

DEFINITION 1.4. DETERMINANTAL LOCI OF CATALECTICANTS. We define the determinantal loci $U_s(u, v; r) \subset V_s(u, v; r) \subset \mathbb{A}(\mathcal{D}_j)$ as

$$U_s(u, v; r) = \{f \in \mathcal{D}_j \mid \operatorname{rk} Cat_f(u, v; r) = s\}$$
$$V_s(u, v; r) = \{f \in \mathcal{D}_j \mid \operatorname{rk} Cat_f(u, v; r)) \leq s\}.$$

We denote by $\mathbf{V}_s(u, v; r)$ the subscheme of $\mathbb{A}(\mathcal{D}_j)$ defined by the ideal $I_{s+1}(Cat_F(u, v; r))$ generated by the $(s+1) \times (s+1)$ minors of $Cat_F(u, v; r)$, in the ring $k[Z]$. We denote by $\mathbf{U}_s(u, v; r)$ its open subscheme whose closed points are those DP-forms f for which $Cat_f(u, v; r)$ has rank exactly s. The algebraic set $\mathbf{V}_s(u, v; r)_{red} = V_s(u, v; r)$ is the set of closed points of $\mathbf{V}_s(u, v; r)$ in $\mathbb{A}(\mathcal{D}_j)$, and it may have several irreducible components. We call the $V_s(u, v; r)$ *catalecticant varieties* and the $\mathbf{V}_s(u, v; r)$ *catalecticant schemes*.

In the following four examples we assume the ground field k is of characteristic 0. Then the ring $\mathcal{D}$ is the polynomial ring $\mathcal{R} =$

$k[X_1, \ldots, X_r]$. The divided powers case (when $\text{char}(k) > 0$) is similar and left to the reader.

EXAMPLE 1.5. QUADRATIC FORMS. Let $j = 2$. For a general quadratic form $F = {}^t X Z X$ with $X = {}^t(X_1, \ldots, X_r)$, $Z = (Z_{ij})$, ${}^t Z = Z$ one has $\text{Cat}_F(1, 1; r) = Z$. From linear algebra $V_s(1, 1; r) \subset \mathcal{R}_2$ consists of quadratic forms representable as sums of squares of $\leq s$ linear forms. We refer the reader to Section 1.2 for more details on this example.

EXAMPLE 1.6. GRADIENT OF A FORM. The catalecticant homomorphism $C_f(j - 1, 1)$ is given by the gradient $\left(\frac{\partial f}{\partial X_1}, \ldots, \frac{\partial f}{\partial X_r}\right)$. We discuss the catalecticant varieties in this case in Lemma 1.22.

EXAMPLE 1.7. BINARY FORMS. Let $r = 2$. For a general binary form

$$F = Z_0 X_1^j + \cdots + \binom{j}{d} Z_d X_1^{j-d} X_2^d + \cdots + Z_j X_2^j$$

$$= j! \sum_{d=0}^{j} Z_d X_1^{[j-d]} X_2^{[d]}$$

the u-th catalecticant matrix is

$$\text{Cat}_F(u, j - u; 2) = j! \begin{pmatrix} Z_0 & Z_1 & \ldots & Z_{j-u} \\ Z_1 & Z_2 & \ldots & Z_{j-u+1} \\ \vdots & \vdots & & \vdots \\ Z_u & Z_{u+1} & \ldots & Z_j \end{pmatrix}$$

known also as a Hankel matrix. We refer the reader to Section 1.3 for a detailed study of this case.

EXAMPLE 1.8. CLEBSCH'S CATALECTICANT MATRIX. We saw in the introduction that Clebsch's solution of the problem of representing a ternary quartic (i.e. the case $r = 3$, $j = 4$) as a sum of 5 fourth powers of linear forms requires considering the second partial derivatives of the quartic form f and the corresponding catalecticant matrix. Below we write this 6×6 matrix explicitly. Writing $X^{[i]}$ for $\frac{1}{i!} X^i$ and similarly for divided powers of Y and Z we consider the generic quartic form $F \in \mathcal{R} = k[X, Y, Z] : \quad F = aX^{[4]} + bX^{[3]}Y + cX^{[3]}Z + dX^{[2]}Y^{[2]} + eX^{[2]}YZ + fX^{[2]}Z^{[2]} + gXY^{[3]} + hXY^{[2]}Z + iXYZ^{[2]} + jXZ^{[3]} + kY^{[4]} + lY^{[3]}Z + mY^{[2]}Z^{[2]} + nYZ^{[3]} + oZ^{[4]} = \sum_{u+v+w=4} a_{uvw} X^{[u]} Y^{[v]} Z^{[w]}$ We notice that, multiplying by $4!$, one obtains the classical expression in the theory of invariants of a generic ternary quartic with binomial coefficients involved. Then the square catalecticant matrix $Cat_F(2,2;3)$ satisfies,

$$
Cat_F(2,2;3) = \begin{matrix} F_{xx} & F_{xy} & F_{xz} & F_{yy} & F_{yz} & F_{zz} \end{matrix} \\
\begin{pmatrix}
a & b & c & d & e & f \\
b & d & e & g & h & i \\
c & e & f & h & i & j \\
d & g & h & k & l & m \\
e & h & i & l & m & n \\
f & i & j & m & n & o
\end{pmatrix}
$$

The basis of the target space $\mathcal{R}_2$ is $(X^{[2]}, XY, XZ, Y^{[2]}, YZ, Z^{[2]})$ (ordered lexicographically). For instance the second partial derivative $F_{xz} = cX^{[2]} + eXY + fXZ + hY^{[2]} + iYZ + jZ^{[2]}$ yields the third column of the catalecticant matrix.

If one wants to work instead with the classical basis $\frac{4!}{u!v!w!}X^u Y^v Z^w$, $u+v+w = 4$ for $\mathcal{R}_4$ and the basis $\frac{2!}{m!n!p!}X^m Y^n Z^p$, $m+n+p = 2$ for $\mathcal{R}_2$ then one should multiply the above matrix by $\frac{4!}{2!}$. Similarly in the general case one should multiply the matrix (1.1.4) by $\frac{(u+v)!}{u!}$. The advantage of working with divided powers is twofold. First, one does not care about coefficients like $\frac{(u+v)!}{u!}$ and second the considerations are meaningful in char$(k) > 0$, when one replaces the polynomial ring $\mathcal{R}$ by the divided power ring $\mathcal{D}$.

DEFINITION 1.9. THE R-SUBMODULE OF HIGHER DERIVATIVES OF A FORM. If f is a degree-j homogeneous element of $\mathcal{D}$, we denote by $\mathcal{D}_{j-v}(f)$ the space $R_v \circ f = \{\phi \circ f | \phi \in R_v\}$, the span in $\mathcal{D}_{j-v}$ of the contractions $x^V \circ f, |V| = v$. By definition $\mathcal{D}_{j-v}(f)$ is the image of the catalecticant homomorphism $C_f(u,v)$. We see that if char$(k) = 0$ or char$(k) > j$ the determinantal locus $V_s(u,v;r)$ equals the set of forms of degree j in r variables, whose v-th partial derivatives span a subspace of dimension $\leq s$ in $\mathcal{R}_u$. A similar statement holds in arbitrary characteristic replacing forms by DP-forms in $\mathcal{D}_j$ and partial derivatives by contractions $x^V \circ f$ with $|V| = v$. We let $\mathcal{D}_f = \oplus_{0 \leq i \leq j}\mathcal{D}_i(f)$. This is an R-submodule of $\mathcal{D}$. Let $t_i(f) = \dim_k \mathcal{D}_i(f)$, and denote by H_f the sequence

$$H_f = (t_0(f), \ldots, t_i(f), \ldots, t_j(f)) \qquad (1.1.6)$$

It is clear from (1.1.5) that

$$t_0 = 1, t_1 \leq r \quad \text{and} \quad t_{j-i}(f) = t_i(f). \qquad (1.1.7)$$

DEFINITION 1.10. PARAMETER SCHEME $\mathbf{Gor}(T)$ FOR FORMS f WITH $H_f = T$. We fix r. Suppose that T is a symmetric nonnegative sequence of integers T satisfying (1.1.7). We let

$$\mathbf{Gor}_{\leq}(T) \subset \mathbb{A}(\mathcal{D}_j)$$

be the affine subscheme defined by the ideal

$$\textstyle\sum_{u=1}^{j-1} I_{t_u+1}(Cat_F(u, j - u; r))$$

generated by the $(t_u + 1) \times (t_u + 1)$ minors in $k[Z]$ of the matrix $Cat_F(u, j - u; r)$, (see Definition 1.3 ; note that $t_u = t_{j-u}$). We define the open subscheme $\mathbf{Gor}(T)$ of $\mathbf{Gor}_{\leq}(T)$ by the nonvanishing of some $t_u \times t_u$ minor for each u. We denote by $Gor(T)$ the corresponding reduced scheme. It is a quasiaffine algebraic set. Namely, it is the set of points $p : (Z_J = p_J) \in \mathbb{A}^N$ parameterizing degree-j homogeneous DP-forms f_p such that the rank of $Cat_{f_p}(u, j - u; r)$ is t_u for each $u \leq j$.

We will also refer to $\mathbf{Gor}(T)$ and to $Gor(T)$ as *catalecticant schemes*, *catalecticant varieties* respectively (cf. Definition 1.4). Evidently, $Gor(T)$ is an affine cone over the projective algebraic set $\mathbb{P}Gor(T) \subset \mathbb{P}^{N-1}$, parameterizing such forms f_p up to nonzero constant multiple. If for some sequence T the set $Gor(T)$ is nonempty we call T a *Gorenstein sequence*. We denote by $\mathcal{H}(j, r)$ the set of all such Gorenstein sequences obtained from DP-forms of degree j in r variables.

DEFINITION 1.11. APOLARITY AND GORENSTEIN ARTIN ALGE-BRAS. To each degree-j homogeneous element $f \in \mathcal{D}_j$ one associates the ideal $I = \text{Ann}(f)$ in $R = k[x_1, \ldots, x_r]$ consisting of polynomials ϕ such that $\phi \circ f = 0$ (see Definition 1.1). If $\phi \circ f = 0$ we call ϕ and f *apolar* to each other. One associates to f also the quotient algebra $A_f = R/I$. (Warning to the commutative algebraist: this is not a localization, and in this book we will never localize by the multiplicative system of powers of f)). F. H. S. Macaulay called such an ideal a *principal system* (see also Proposition A.3), but we now know them as *Gorenstein ideals*, since A_f is a *Gorenstein Artin algebra*[1]. We refer to Section 2.3 or [Ei2, p.527] for the equivalence between Macaulay's principal systems and Gorenstein Artin algebras.

---

[1]H. Bass reports that A. Grothendieck first used the term "Gorenstein ideal"; despite the earlier occurence in the work of F.H.S. Macaulay. See p. xviii. We used "Artinian Gorenstein algebra" in the Introduction, but henceforth use "Gorenstein Artin algebra", which is more current among specialists

Recall that if M is a graded R-module, and M_i denotes its degree-i portion, the Hilbert function $H(M)$ is the sequence $H(M) = (\ldots, h_i = \dim_k M_i, \ldots)$. Clearly there is an isomorphism of vector spaces $(A_f)_i \xrightarrow{\sim} (\mathcal{D}_f)_{j-i}$. Since the Hilbert function of $\mathcal{D}_f$ is symmetric with respect to $\frac{j}{2}$ (this is the equality $t_i(f) = t_{j-i}(f)$ from Definition 1.9) we conclude that the Hilbert functions of A_f and $\mathcal{D}_f$ coincide, and thus $H(A_f) = H(\mathcal{D}_f) = H_f$; hence, the Hilbert function of A_f is symmetric with respect to $\frac{j}{2}$.

Another way to see this property of $H(A_f)$ is to consider the bilinear map $B : R_i \times R_j \to k$ defined by

$$B(\phi, \psi) = \langle \phi\psi, f \rangle = \phi \circ (\psi \circ f) = \psi \circ (\phi \circ f).$$

Clearly the left and right kernels of B are I_i, I_{j-i} respectively. Hence B induces a duality $(A_f)_i \times (A_f)_{j-i} \to k$. We obtain another interpretation of the variety $Gor(T)$, which explains its name:

The projective algebraic set $\mathbb{P}Gor(T)$ parameterizes the graded Gorenstein Artin quotients $A = R/I$ of Hilbert function T.

As we mentioned in the introduction, the reader who is interested in the cases of $\mathrm{char}(k) = 0$ or characteristic sufficiently large ($\mathrm{char}(k) > j$) may replace everywhere below the divided power ring $\mathcal{D}$ by the polynomial ring $\mathcal{R} = k[x_1, \ldots, x_r]$, the divided power $L^{[j]}$ by $\frac{1}{j!}L^j$, and the contracting action $\phi \circ f$ by the differentiation action (see Appendix A).

If $a = (a_1, \ldots, a_r) \in k^r$, let L_a denote the linear form

$$L_a = a_1 X_1 + \cdots + a_r X_r \in \mathcal{D}_1 \tag{1.1.8}$$

Denote by $L_a^{[j]}$ the degree-j DP-form

$$L_a^{[j]} = \sum_{j_1 + \cdots + j_r = j} a_1^{j_1} \ldots a_r^{j_r} X_1^{[j_1]} \ldots X_r^{[j_r]},$$

$$= \frac{1}{j!}(L_a)^j \quad \text{if} \quad \mathrm{char}(k) = 0 \quad \text{or} \quad \mathrm{char}(k) > j. \tag{1.1.9}$$

where the power $(L_a)^j$ in (1.1.9) is taken in the divided power ring $\mathcal{D}$. We have the following useful equality: for any $\phi \in R_v, v \le j$ and any $L_a \in \mathcal{D}_1$

$$\phi \circ L_a^{[j]} = \phi(a) L_a^{[j-v]}. \tag{1.1.10}$$

Indeed, by linearity it suffices to check this formula for a monomial
$\phi = x_1^{v_1} \cdots x_r^{v_r}$, $v_1 + \cdots + v_r = v$. We have

$$x_1^{v_1} \cdots x_r^{v_r} \circ \sum_{j_1 + \cdots + j_r = j} a_1^{j_1} \cdots a_r^{j_r} X_1^{[j_1]} \cdots X_r^{[j_r]}$$

$$= \sum a_1^{j_1} \cdots a_r^{j_r} X_1^{[j_1 - v_1]} \cdots X_r^{[j_r - v_r]}$$

$$= a_1^{v_1} \cdots a_r^{v_r} \left(\sum_{u_1 + \cdots + u_r = j - v} a_1^{u_1} \cdots a_r^{u_r} X_1^{[u_1]} \cdots X_r^{[u_r]} \right)$$

$$= \phi(a) \, (L_a)^{[j-v]}.$$

Abusing notation we will write $\phi(L_a)$ for $\phi(a)$.

DEFINITION 1.12. LOCUS OF POWER SUMS $PS(s, j; r)$ AND MUL-
TISECANT VARIETIES OF THE VERONESE VARIETIES. Consider the
DP-forms $f \in \mathcal{D}_j$ that can be written as a sum

$$f = L_1^{[j]} + \cdots + L_s^{[j]} \tag{1.1.11}$$

for some choice of linear forms $L_1, \ldots, L_s \in \mathcal{D}_1$. They form the image
of the regular map

$$\mu : \underbrace{\mathcal{D}_1 \times \cdots \times \mathcal{D}_1}_{s \text{ times}} \longrightarrow \mathcal{D}_j \tag{1.1.12}$$

defined by $\mu(L_1, \ldots, L_s) = L_1^{[j]} + \cdots + L_s^{[j]}$. Let us denote by $PS(s, j; r)$
this image. It is a constructible set, (i.e a union of locally closed sets
in Zariski's topology) by the theorem of Chevalley (see e.g. [**Hum**,
§4.4]). Its algebraic closure $\overline{PS(s, j; r)}$ is an irreducible affine variety,
which is one of the main objects of study in this memoir. The variety
$\overline{PS(s, j; r)}$ is invariant under multiplication by elements of k^*. Its projectivization $\mathbb{P}\overline{PS(s, j; r)}$ is a classical object in projective geometry.
Namely by Corollary A.10 in Appendix A if we consider the Veronese
variety $v_j(\mathbb{P}^{r-1})$, $\mathbb{P}^{r-1} = \mathbb{P}(\mathcal{D}_1)$, and if $s < \dim_k R_j = \binom{j+r-1}{r-1}$, then
for general enough forms $L_1, \ldots, L_s$ the projectivization of the span
$\langle L_1^{[j]}, \ldots, L_s^{[j]} \rangle$ is an $(s-1)$-plane that intersects the Veronese variety in
the points $v_j(\langle L_i \rangle) = \langle L_i^{[j]} \rangle$, $i = 1, \ldots, s$. Thus $\mathbb{P}\overline{PS(s, j; r)}$ is exactly
the s-secant variety to the Veronese variety $Sec_s(v_j(\mathbb{P}^{r-1}))$.

A classical problem in the theory of invariants is to find polynomial
equations which determine $\overline{PS(s, j; r)}$ set-theoretically. A more difficult problem is to find generators for the ideal in R of this
variety. Some polynomials in this ideal are easily found. Namely,

if $f \in PS(s, j; r)$ and $f = L_1^{[j]} + \cdots + L_s^{[j]}$, then for every pair of positive integers u, v with $u + v = j$ and every $\phi \in R_v$ we have from (1.1.10)

$$\phi \circ f = \phi(L_1) L_1^{[u]} + \cdots + \phi(L_s) L_s^{[u]}.$$

This shows that $\mathcal{D}_u(f)$, the image of the catalecticant homomorphism $C_f(u, v)$ has dimension $\leq s$. Hence the $(s + 1) \times (s + 1)$ minors of the catalecticant matrices $Cat_F(u, v; r)$ vanish on $PS(s, j; r)$ for every $u, v \geq 1$ with $u + v = j$. This can be reformulated equivalently as follows. If $T = H(s, j, r)$ (see (0.1.5)) then

$$\overline{PS(s, j; r)} \subset Gor_{\leq}(T). \tag{1.1.13}$$

In order to obtain a stronger result one needs more information about the polynomials apolar to the elements in $PS(s, j; r)$. This is our next goal.

DEFINITION 1.13. POLAR POLYHEDRONS. If a homogeneous element $f \in \mathcal{D}_j$ can be written as a sum

$$f = L_1^{[j]} + \cdots + L_s^{[j]}$$

where $L_i \in \mathcal{D}_1$ and each pair L_i, L_j, $i \neq j$ is non-proportional we say that the hyperplanes $H_i \subset \mathbb{P}(R_1)$, defined by the equations $L_i(x) = 0$ form a *polar polyhedron*. The coordinates $p(i) \in k^r$ of L_i, $i = 1, \ldots, s$ modulo multiplication by k^* yield the projective coordinates of s points in $\mathbb{P}^{r-1} = \mathbb{P}(\mathcal{D}_1)$ which correspond to the polar polyhedron of f by projective duality. In fact these are the s points corresponding to the s lines $\{\langle L_i \rangle\} \subset \mathbb{P}(\mathcal{D}_1)$. Abusing notation we denote this set of s points in $\mathbb{P}^{r-1}$ again by $\{p(i)\}$. Conversely, given a subset $P = \{p(1), \ldots, p(s)\}$ of $\mathbb{P}^{r-1}$ we let

$$L_P^{[j]} = \langle L_{p(1)}^{[j]}, \ldots, L_{p(s)}^{[j]} \rangle \subset \mathcal{D}_j$$

denote the span of the j-th divided powers of the corresponding linear forms. The Zariski open dense subset of $L_P^{[j]}$ consisting of linear combinations with s non-zero coefficients, is the set of degree-j elements f in $\mathcal{D}_j$ having polar polyhedron P. We denote by $\mathcal{I}_P$ the graded ideal in R of polynomials vanishing on P.

EXAMPLE 1.14. If $r = 3$, $R = k[x, y, z]$, and $F = X^{[3]} + Y^{[3]} + Z^{[3]}$, then the corresponding point set is $P = \{(1, 0, 0), (0, 1, 0), (0, 0, 1)\}$. We shall prove later (Proposition 1.25) that F is not a sum of two cubes and P is the unique polar triangle of F. But F has many polar polyhedrons with four or more sides. For instance for each primitive cubic root of unity α the point set $P'_\alpha = \{(1,0,0),(0,1,1),(0,1,\alpha),(0,1,\alpha^2)\}$ corresponds by duality to a polar quadrangle of F, since

$$F = X^{[3]} + (1/3)\left((Y+Z)^{[3]} + (Y+\alpha\cdot Z)^{[3]} + (Y+\alpha^2\cdot Z)^{[3]}\right).$$

LEMMA 1.15. APOLARITY LEMMA. *Let* $p_1,\dots,p_s \in k^r$, *let* $L_i = L_{p_i}$, *let* $P = \{[p_1],\dots,[p_s]\} \subset \mathbb{P}(k^r) = \mathbb{P}^{r-1}$ *and let* $\mathcal{I}_P$ *be the homogeneous ideal in* R *of polynomials vanishing on* P. *Then*

i. *For every* $\phi \in R_i$

$$\phi\circ\left(L_1^{[j]} + \cdots + L_s^{[j]}\right) = \phi(p_1)\,L_1^{[j-i]} + \cdots + \phi(p_s)\,L_s^{[j-i]}. \quad (1.1.14)$$

ii. *With respect to the contraction pairing* $R_j \times \mathcal{D}_j \to k$ *one has*

$$((\mathcal{I}_P)_j)^\perp = \langle L_1^{[j]},\dots,L_s^{[j]}\rangle.$$

iii. *The points* $[p_1],\dots,[p_s] \in \mathbb{P}^{r-1}$ *impose independent conditions on the linear system* $|\mathcal{O}_{\mathbb{P}^{r-1}}(j)|$ *if and only if* $L_1^{[j]},\dots,L_s^{[j]}$ *are linearly independent.*

iv. *Suppose* $s \leq \dim_k R_{j-i}$ *and the linear forms* $L_1,\dots,L_s$ *have the property that the corresponding set* P *imposes independent conditions on the linear system* $|\mathcal{O}_{\mathbb{P}^{r-1}}(j-i)|$. *Let* $f = L_1^{[j]} + \cdots + L_s^{[j]}$. *Then we have for the apolar forms to* f *of degree* i *the equality* $\mathrm{Ann}(f)_i = (\mathcal{I}_P)_i$.

PROOF. (i). This follows by linearity from (1.1.10).
(ii). We have for every $\phi \in R_j$

$$\phi\circ\left(\sum_{n=1}^{s}\lambda_n L_n^{[j]}\right) = \sum_{n=1}^{s}\lambda_n\phi(p_n).$$

This shows that

$$\langle L_1^{[j]},\dots,L_s^{[j]}\rangle^\perp = \{\phi\mid\phi(p_i)=0\} = (\mathcal{I}_P)_j.$$

Now, (ii) follows from the fact that the contraction yields a nondegenerate pairing between R_j and $\mathcal{D}_j$.

(iii). This is immediate from (ii).

(iv). The inclusion $(\mathcal{I}_P)_i \subset \mathrm{Ann}(f)_i$ follows from (1.1.14) without any assumptions on $L_n, n = 1,\dots,s$. By Part (ii) the DP-forms $L_1^{[j-i]},\dots,L_s^{[j-i]}$ are linearly independent. This implies by (1.1.14) the opposite inclusion $\mathrm{Ann}(f)_i \subset (\mathcal{I}_P)_i$. $\square$

COROLLARY 1.16. EXISTENCE OF A DECOMPOSITION. *If k is algebraically closed, every degree-j DP-form $f \in \mathcal{D}$ has an additive decomposition of length s no greater then $\binom{j+r-1}{r-1}$.*

PROOF. Let $s = \binom{j+r-1}{r-1} = \dim_k R_j$. By Part (iii) of the preceding lemma if $L_1, \ldots, L_s$ are sufficiently general linear forms in $\mathcal{D}_1$, then $L_1^{[j]}, \ldots, L_s^{[j]}$ are linearly independent. Thus $\mathcal{D}_j = \langle L_1^{[j]}, \ldots, L_s^{[j]} \rangle$. □

LEMMA 1.17 (A. Iarrobino [I3, Prop. 4.7]). HILBERT FUNCTION OF A GENERAL SUM OF s POWERS. *If $L_1, \ldots, L_s$ are general enough linear forms, $L_i = \sum \beta_{ij} X_j$ in $\mathcal{D}_1$, then $f = L_1^{[j]} + \cdots + L_s^{[j]}$ satisfies, $H_f = H(s, j, r)$ (see (0.1.5)). If $s \leq \min(\dim_k R_i, \dim_k R_{j-i})$ we have*

$$R_i \circ f = \langle L_1^{[j-i]}, \ldots, L_s^{[j-i]} \rangle. \tag{1.1.15}$$

PROOF. We have $t_i(f) = t_{j-i}(f)$ (see Definition 1.9). So, it suffices to prove that $t_i(f) = \min(s, \dim_k R_i)$ if $i \leq j - i$. Suppose first that $s \leq r_{j-i} = \dim_k R_{j-i}$. Then from Lemma 1.15(iv) we have $\mathrm{Ann}(f)_i = (\mathcal{I}_P)_i$. If $s < r_i = \dim_k R_i$, then the s sufficiently general points of P impose independent conditions on the linear systems $|\mathcal{O}_{\mathbb{P}^{r-1}}(i)|$ and $|\mathcal{O}_{\mathbb{P}^{r-1}}(j - i)|$, so

$$t_i(f) = \dim_k R_i / \mathrm{Ann}(f)_i = \dim_k R_i/(\mathcal{I}_P)_i = s. \tag{1.1.16}$$

If $s \geq r_i$, then $(\mathcal{I}_P)_i = 0$, so $t_i(f) = r_i$.

Now, suppose $s > r_{j-i}$. Already when f is the sum of r_i powers of general linear forms, we have $\dim_k R_{j-i} \cdot f = r_i$, the maximum possible: we show that when f is the sum of a larger number of general enough powers, that dimension remains r_i. Consider the constructible set

$$X = \{ f = c_1 M_1^{[j]} + \cdots + c_s M_s^{[j]} \mid M_u \in \mathcal{D}_1, \, c_u \in k \},$$

which is the image of $\mathcal{D}_1 \times \cdots \times \mathcal{D}_1 \times k \times \cdots \times k$ (see Definition 1.12). Its closure is an irreducible affine variety and equals the closure $\overline{PS(s, j; r)}$.

We claim that the closed subvariety $\overline{X}_i$ defined by the condition, $\mathrm{rk} \, Cat_f(i, j - i; r) < r_i$ is a proper subvariety of $\overline{X}$. Indeed, let $s' = r_{j-i}$. Set $c_{s'+1} = \ldots = c_s = 0$. Then we obtain an irreducible closed subvariety of $\overline{X}$ whose sufficiently general points f' have catalecticant matrices $Cat_{f'}(i, j-i; r)$ of rank r_i by the already proved case $s \leq r_{j-i}$. This proves our claim, which in turn implies that for $f = L_1^{[j]} + \cdots + L_s^{[j]}$ with $L_1, \ldots, L_s$ general enough one has $\mathcal{D}_i(f) = R_{j-i} \circ f = \mathcal{D}_i$. Therefore $t_i(f) = r_i$.

For the second part of the lemma we observe that the condition $s \leq \min(r_i, r_{j-i})$ implies from Lemma 1.15(iv) that $\dim_k(R_i \circ f) =$

s. On the other hand by equality (1.1.14) this space is contained in $\langle L_1^{[j-i]}, \ldots, L_s^{[j-i]} \rangle$. Thus they are equal.   □

This lemma permits us to strengthen (1.1.13).

COROLLARY 1.18. *Let* $T = H(s, j, r)$ *(see* (0.1.5)*).   Then* $\overline{PS(s, j; r)} \subset \overline{Gor(T)} \subset Gor_{\leq}(T)$.

LEMMA 1.19. RELATION BETWEEN THE ANNIHILATING IDEAL OF A SUM OF POWERS, AND A VANISHING IDEAL AT POINTS. *Let* $L_1, \ldots, L_s$ *be general enough linear forms in* $\mathcal{D}_1$. *Let* $P = \{p_1, \ldots, p_s\} \subset \mathbb{P}(\mathcal{D}_1) = \mathbb{P}^{r-1}$ *be the points corresponding to the forms* $L_1, \ldots, L_s$, *and let* $\mathcal{I}_P$ *be the graded ideal in* R *of all forms vanishing at* P. *Let* $I = \text{Ann}(f)$ *be the ideal of polynomials apolar to* f. *Then if* i *satisfies* $s \leq \dim_k R_{j-i}$, *we have*

$$I_i = (\mathcal{I}_P)_i.$$

Furthermore $H(R/\mathcal{I}_P)_d = \min(\dim_k R_d, s)$ *for every* $d \geq 0$.

PROOF. Since $s \leq r_{j-i}$ and $L_1, \ldots, L_s$ are general enough, the corresponding set of points imposes independent conditions on $|\mathcal{O}_{\mathbb{P}^{r-1}}(j-i)|$. Lemma 1.15(iv) yields the equality $I_i = (\mathcal{I}_P)_i$.

For the proof of the second part fix an integer i with $r_i \geq s$ and let U be the open subset of points in $\mathbb{P}^{r-1} \times \cdots \times \mathbb{P}^{r-1}$ (s times) consisting of s-tuples P such that $H(R/\mathcal{I}_P)_d = \min(s, \dim_k R_d)$ for every $d \leq i$. We wish to prove that $H(R/\mathcal{I}_P)_d = s$ for every $d \geq i$. This equality is equivalent to the following restriction map being surjective:

$$H^0(\mathbb{P}^{r-1}, \mathcal{O}_{\mathbb{P}^{r-1}}(d)) \longrightarrow H^0(P, \mathcal{O}_P(d)). \qquad (1.1.17)$$

Let us choose a linear form ℓ which does not vanish on any of the points of P. We know (1.1.17) is epimorphic for $d = i$. Multiplying by ℓ^{d-i} we see it is epimorphic for every $d \geq i$ as well.   □

EXAMPLE 1.20. COMPARISON OF $\text{Ann}(f)$ AND $\mathcal{I}_P$. Let $\mathcal{R} = k[X, Y, Z]$, $s = 3$, $f = X^4 + Y^4 + Z^4$. Then $H_f = (1, 3, 3, 3, 1)$ and $I = \text{Ann}(f) = (xy, xz, yz, x^4 - y^4, x^4 - z^4)$. Since $P = \{(1, 0, 0), (0, 1, 0), (0, 0, 1)\}$, the ideal $\mathcal{I}_P = (y, z) \cap (x, z) \cap (x, y) = (xy, xz, yz)$, and we have $(\mathcal{I}_P)_i = I_i$ for $1 \leq i \leq 3$, the values of i for which $(H_f)_i = 3$. But $(\mathcal{I}_P)_4 \subsetneq I_4$.

REMARK 1.21. The proof in [I3] of Lemma 1.17 uses the classical result that the i-th divided powers of a general enough set of $s = \dim_k \mathcal{D}_i$ linear forms span $\mathcal{D}_i$: this suffices to show the existence of a choice of $L_1, \ldots, L_s$ such that the dimension of $\langle L_1^{[i]}, \ldots, L_s^{[i]} \rangle$

is $\min(s, \dim_k \mathcal{D}_i)$, for each i. This fact and the Apolarity Lemma 1.15 also show the last statement of Lemma 1.19, that $H(R/\mathcal{I}_P)_d = \min(\dim_k R_d, s)$.

1.2. Determinantal loci of the first catalecticant, the Jacobian

We assume that the number of variables $r \geq 2$. We illustrate the notions introduced in the previous section by the well-studied case $Cat_F(1, j - 1; r)$ and put together some known results about the catalecticant varieties $V_s(1, j - 1; r)$. The matrix $Cat_f(1, j - 1; r)$ is the transpose of $Cat_f(j - 1, 1; r)$. If we assume char $k = 0$ or char $k > j$ and take a homogeneous polynomial $f \in \mathcal{R}$ then $Cat_f(j - 1, 1; r)$ is up to a diagonal matrix factor, the matrix of first partials of f. Evidently, if f is in the subring of the variables $Y_1, \ldots, Y_s$ then the $\operatorname{rk} Cat_f(j - 1, 1; r) \leq s$. The following Lemma, stated as usual in terms of the divided power ring $\mathcal{D}$ gives a converse.

LEMMA 1.22. *Suppose $f \in \mathcal{D}_j$ and $\operatorname{rk} Cat_f(1, j - 1; r) = s < r$. Then there is a linear change of coordinates $Y_1, \ldots, Y_s, Y_{s+1}, \ldots, Y_r$ such that f is a divided powers polynomial*

$$f = \sum_{j_1 + \cdots + j_s = j} b_{j_1, \ldots, j_s} Y_1^{[j_1]} \ldots Y_s^{[j_s]}.$$

For fixed linearly independent linear forms $Y_1, \ldots, Y_s$ the space of DP-polynomials expresssible as $\sum c_{i_1, \ldots, i_s} Y_1^{[i_1]} \ldots Y_s^{[i_s]}$ forms a subring of $\mathcal{D}$ isomorphic to the ring of divided powers in s variables.

PROOF. Indeed, by (1.1.5) the catalecticant homomorphism

$$C_f(j - 1, 1) \; : \; R_1 \longrightarrow \mathcal{D}_{j-1}$$

(cf. Example 1.6) has kernel of dimension $r - s$. Choosing new variables $y_1, \ldots, y_s, y_{s+1}, \ldots, y_r$ so that $y_{s+1}, \ldots, y_r$ span this kernel, taking the dual coordinates $Y_1, \ldots, Y_r$ in $\mathcal{D}_1$ and using the GL_r-equivariant properties proved in Appendix A we obtain the first statement of the lemma. The second statement is obvious if $Y_i = X_i$ and the general case follows again by GL_r-equivariance. □

We now let $\mathbb{G} = Grass(r - s, R_1)$ be the Grassmanian parametrizing $(r - s)$-dimensional vector subspaces of R_1, and consider the algebraic variety $X \subset \mathbb{G} \times \mathcal{D}_1$ defined as follows:

$$X = \{(\Delta, f) \mid \Delta \circ f = 0\}.$$

the two projections yield the following diagram

$$X \xrightarrow{\quad q \quad} V_s(1, j-1; r) \qquad\qquad (1.2.1)$$
$$\pi \downarrow$$
$$\mathbb{G}$$

By Lemma 1.22 $\pi : X \to \mathbb{G}$ is a vector bundle of rank $\binom{j+s-1}{s-1}$.

PROPOSITION 1.23. *Suppose* $1 \le s < r$. *The variety* $V_s = V_s(1, j-1; r)$ *is irreducible of dimension*

$$\dim V_s = s(r-s) + \binom{j+s-1}{s-1}. \qquad\qquad (1.2.2)$$

The map q *is a birational morphism and* q^{-1} *is regular on* $V_s - V_{s-1}$. *The projective variety* $\mathbb{P}(V_s)$ *is rational.*

PROOF. The map q is surjective by definition, hence V_s is irreducible. The open set $V_s - V_{s-1}$ is nonempty, since it contains the elements $f = L_1^{[j]} + \cdots + L_s^{[j]}$ with sufficiently general $L_1, \ldots, L_s \in \mathcal{D}_1$ according to Lemma 1.17. Cramer's formulas applied to the homogeneous linear system with matrix the generic catalecticant $Cat_F(j-1, 1; r)$ shows that q has a regular inverse q^{-1} on $V_s - V_{s-1}$. The projective variety $\mathbb{P}(V_s)$ is birational to the projectivization of the vector bundle X over $\mathbb{G}$, hence it is rational. □

REMARK 1.24. Notice an elementary corollary from this proposition. If $f \in \mathcal{D}_j$, $f \ne 0$ and $\operatorname{rk} Cat_f(1, j-1; r) = 1$, then $f = L^{[j]}$ for some $L \in \mathcal{D}_1$. Indeed $\overline{PS(1, j; r)}$ is the cone over the Veronese variety $v_j(\mathbb{P}^{r-1})$ (see Corollary A.10), thus it has dimension r and is closed in $\mathbb{A}(\mathcal{D}_1)$. It is contained in $V_s(1, j-1; r)$ of dimension r by (1.2.2). Hence they are equal. One can prove the same fact in another way. Equating to zero the 2×2 minors of the generic catalecticant matrix $Cat_F(1, j-1; r)$ one obtains a system of quadratic equations which is easily seen to be equivalent to the larger system

$$Z_{u_1 \ldots u_j} Z_{v_1 \ldots v_j} = Z_{u'_1 \ldots u'_j} Z_{w'_1 \ldots w'_j}$$

for every $u_1 + v_1 = u'_1 + v'_1, \ldots, u_j + v_j = u'_j + v'_j$. The latter one is the standard system of equations of the Veronese variety [**Sha**, Ch.I§4]. In case $r = 2$ the variety $\overline{PS(1, j; 2)} = v_j(\mathbb{P}^1) = C$ is a rational normal curve $C = v_j(\mathbb{P}^1)$ and one obtains the standard determinantal description

$$C = \left\{ (Z_0 : \ldots : Z_j) \mid \mathrm{rk} \begin{pmatrix} Z_0 & \cdots & Z_{j-1} \\ Z_1 & \cdots & Z_j \end{pmatrix} \leq 1 \right\} \qquad (1.2.3)$$

(cf. Example 1.7).

Sums of s powers of linear forms, $s \leq r$. Suppose $j \geq 3$. Let $L_1, \ldots, L_s$ be s linearly independent linear forms, $s \leq r$. Consider the DP-form $f = L_1^{[j]} + \cdots + L_s^{[j]}$. Then the Hilbert function $H(A_f)$ (see Definition 1.11) has maximum value s, and equals

$$H(A_f) = (1, s, \ldots, s, 1)$$

This is immediate from Lemma 1.15 since $s \leq r$ linearly independent points in $\mathbb{P}^{r-1}$ impose independent conditions on $|\mathcal{O}_{\mathbb{P}^{r-1}}(i)|$ for every $i \geq 1$. If $T = (1, s, \ldots, s, 1) = H(s, j, r)$ we conclude that

$$\overline{PS(s, j; r)} \subset \overline{Gor(T)} \subset V_s(1, j-1; r)$$

The following proposition will be generalized later in Section 2.2 (Theorem 2.6).

PROPOSITION 1.25. *Let $j \geq 3$. Suppose $s \leq r$ and let $L_1, \ldots, L_s$ be linearly independent elements in $\mathcal{D}_1$. Then $f = L_1^{[j]} + \cdots + L_s^{[j]}$ is the unique additive decomposition of f by $\leq s$ linearly independent linear forms, up to multiplication of L_i, $i = 1, \ldots, s$ by j-th roots of unity and their permutation. The variety $\overline{PS(s, j; r)}$ has dimension rs.*

PROOF. Let $P = \mathbb{P}(\mathcal{D}_1) = \mathbb{P}^{r-1}$ be the points corresponding to $\{L_1, \ldots, L_s\}$. Lemma 1.15(iv) shows that

$$\mathrm{Ann}(f)_{j-1} = (\mathcal{I}_P)_{j-1} \qquad (1.2.4)$$

The uniqueness of additive decomposition follows obviously from the following claim:

The set P is the base locus of the linear system $|\mathrm{Ann}(f)_{j-1}|$.

By equality (1.2.4) this is equivalent to the following simple fact: for every $i \geq 2$ we have for the base locus

$$Bs \, |\mathcal{O}_{\mathbb{P}^{r-1}}(i)(-P)| = P. \qquad (1.2.5)$$

To prove this we observe that for every $q \notin P$ there is a partition $P = P_1 \cup P_2$ such that each of the sets $P_\alpha \cup \{q\}$, $\alpha = 1, 2$ is linearly independent. So, there exist linear forms H_α, $\alpha = 1, 2$ which vanish on P_α, but do not vanish on q. The degree-i form $H_1^{i-1}H_2$ vanishes on P but $H_1^{i-1}H_2(q) \neq 0$. This proves (1.2.5).

That $\dim \overline{PS(s, j; r)} = rs$ is immediate from the proved uniqueness of additive decomposition. $\qquad \square$

Until the end of this section we will assume unless otherwise specified that $\mathrm{char}(k) = 0$; then $\mathcal{D} = \mathcal{R} = k[X_1, \ldots, X_r]$.

The case $j = 2$. For a quadratic form $f = {}^tXAX$ where $X = {}^t(X_1, \ldots, X_r)$ we have $Cat_F(1, 1; r) = A$. From linear algebra the form f can be written as a sum of s squares of linear forms

$$f = L_1^2 + \cdots + L_s^2 \qquad (1.2.6)$$

where $L_1, \ldots, L_s$ are linearly independent if and only if $\mathrm{rk}(A) = s$. Thus

$$\overline{PS(s, 2; r)} = V_s(1, 1; r).$$

Notice that the representation (1.2.6) is not unique unless $s = 1$ because of the action of the orthogonal group $O(s)$. The only symmetric Hilbert functions possible are $T = (1, s, 1)$ for $s \leq r$. Then the algebraic set $Gor(T) = U_s(1, 1; r)$ is isomorphic via polarization ($f \mapsto A = Cat_f(1, 1; r)$) to the rank-$s$ locus of the space of symmetric $r \times r$ matrices and $V_s(1, 1; r) = \overline{U_s(1, 1; r)} = \overline{Gor(T)}$.

The varieties $V_s(1, 1; r)$ were much studied and well understood (see e.g. [**Ro, Kut, JPW**]). We state below some of their properties. Let $F = {}^tXZX$ be the generic quadratic form with $Z = (Z_{ij})$, where $Z_{ij} = Z_{ji}$, $i \leq j$ are indeterminates over k. The ideal $I_{s+1}(Z) = I_{s+1}(Cat_F(1, 1; r))$ generated by the $(s + 1) \times (s + 1)$ minors of Z determines the affine scheme $\mathbf{V}_s(1, 1; r) \subset \mathbb{A}(\mathcal{R}_2)$.

THEOREM 1.26. THE CATALECTICANT VARIETY $V_s(1, 1; r)$, $j = 2$. *Suppose that $1 \leq s < r$. Set $s^\vee = r - s$. Let $V_s = V_s(1, 1; r)$ and let $I(V_s)$ be the ideal of the irreducible affine variety $V_s \subset \mathbb{A}(\mathcal{R}_2)$*

i. *The dimension and codimension of V_s in $\mathbb{A}(\mathcal{R}_2)$ and the degree of $\mathbb{P}(V_s)$ in $\mathbb{P}(\mathcal{R}_2)$ are given by the following formulas*

$$\dim(V_s) = sr - \frac{s(s-1)}{2} = \frac{r(r+1)}{2} - \frac{s^\vee(s^\vee + 1)}{2},$$

$$\mathrm{cod}(V_s) = \frac{r(r+1)}{2} - s(s+1)/2 - s(r-s) = \frac{s^\vee(s^\vee + 1)}{2},$$

$$\deg(V_s) = \prod_{\alpha=0}^{r-s-1} \frac{\binom{r+\alpha}{r-s-\alpha}}{\binom{2\alpha+r}{\alpha}} \qquad (1.2.7)$$

ii. $I(V_s) = I_{s+1}(Z) = I_{s+1}(Cat_F(1, 1; r))$
iii. V_s *is normal, Cohen-Macaulay and has rational singularities. Its singular locus is V_{s-1}.*

PROOF. (i). The dimension is from Proposition 1.23. The degree is classical [**Ro**]. A modern proof may be found in [**HaT**, p.78]. Notice however the discussion of the validity of [**HaT**, Proposition 9] in [**Ful**] where the main result of [**HaT**] is generalized.

(ii). A symmetric $r \times r$ matrix A is of rank s if and only if it is the Gram matrix $((v_i, v_j))$ of r vectors $v_1, \ldots v_r$ in k^s which span k^s. Let us recall the proof of this fact. Suppose A is of rank s. One diagonlizes the quadratic form (AX, X) by a linear transformation $X = TY$. We have $a_{ij} = (Ae_i, e_j)$ and if $e_i = Tw_i$ we let $v_i \in k^s$ be the vectors whose first coordinates equal those of w_i, $i = 1, \ldots, r$. Then $a_{ij} = (v_i, v_j)$ as claimed. The converse is obvious. Using this fact Part (ii) is equivalent to the Second Fundamental Theorem of the theory of invariants for the group $O(s)$ (see [**Weyl, DP, VP**]): the key point is that the determinantal ideals $I_{s+1}(Z)$ are radical. We refer the reader also to Porras' Theorem 1.28 where a more general fact is proved.

(iii). That $Sing(V_s) = V_{s-1}$ follows easily from (ii). It has codimension $s^\vee + 1 \geq 2$ if $s \leq r-1$. Kutz proved that V_s is Cohen-Macaulay [**Kut**]; normality follows from Serre's Criterion (see e.g. [**Ei2**, pp.457–458] or [**ACGH**, p.98]). Another approach to proving these facts, which also yields that V_s has rational singularities is applying Kempf's method [**Ke2**]. We discuss this after the statement of Porras' Theorem 1.28.   □

The case $j = 3$. Here the only possible symmetric Hilbert functions are $T = (1, s, s, 1) = H(s, 3, r)$ with $s \leq r$. For such T we have $Gor(T) = U_s(1, 2; r)$ and according to Lemma 3.5

$$V_s(1, 2; r) = \overline{Gor(T)} = \bigcup_{s' \leq s} Gor(H(s', 3, r)).$$

THEOREM 1.27. THE CATALECTICANT VARIETY $V_s(1, 2; r)$. *Let $j = 3$, $s < r$, let $V_s = V_s(1, 2; r)$. Then*

   i. *V_s is irreducible of dimension $rs + \frac{s(s-1)(s-2)}{2}$.*

  ii. *$\overline{PS(s, j; r)}$ is properly contained in V_s unless $s = 1$ or $s = 2$ in which cases the two varieties are equal.*

 iii. *The variety V_s is normal, Cohen-Macaulay with rational singularities. The ideal $I(V_s)$ is generated by the $(s + 1) \times (s + 1)$ minors of the generic catalecticant matrix $Cat_F(1, 2; r)$.*

PROOF. Part (i) is a particular case of Proposition 1.23. Part (ii) follows from Proposition 1.25. The assumption $char(k) = 0$ is used

only in Part (iii), which is a particular case of Porras' Theorem 1.28.
□

As a corollary of this theorem one obtains that the ideal of the chordal variety to the Veronese variety $v_3(\mathbb{P}^{r-1}) = \mathbb{P}(PS(2,3;r))$ (see Definition 1.12) is generated by the 3×3 minors of the catalecticant matrix $Cat_F(1,2;r)$. This answers affirmatively Problem 10.7 in [**G1**, p.102].

The case $j > 3$. Suppose $j \geq 4$. Then (1.2.2) and Proposition 1.25 show that $\overline{PS(s,j;r)}$ is a proper subvariety of $V_s(1,j-1;r)$ unless $s = 1$ when the two varieties are equal (see Remark 1.24). Referring to some results to be proved later we can give more information about the relation between the varieties $\overline{PS(s,j;r)}$, $\overline{Gor(T)}$, $T = (1, s, \ldots, s, 1)$ and $V_s(1,j-1;r)$. If f is a sufficiently general element of $V_s(1,j-1;r)$, then from Lemma 1.22 and Lemma 1.17 or Proposition 3.12 it follows that the Hilbert sequence $H(A_f)$ satisfies $H(A_f)_i = \min(s_i, s_{j-i})$, or

$$H(A_f)_i = \min\left(\binom{i+s-1}{s-1}, \binom{j-i+s-1}{s-1}\right).$$

Unless $s = 1$ this sequence differs from $T = (1, s, \ldots, s, 1)$ which is the Hilbert sequence of A_f for sufficiently general $f \in PS(s,j;r)$. We see that $\overline{Gor(T)} \subset V_s(1, j-1;r)$, this inclusion being proper unless $s = 1$, when there is equality. In Theorem 4.10A it is proved that $\overline{PS(s,j;r)}$ is an irreducible component of $\overline{Gor(T)}$ and that one has equality $\overline{PS(s,j;r)} = \overline{Gor(T)}$ in the case of 3 variables, $r = 3$. The varieties $V_s(1, j-1;r)$ were studied in the paper [**Po**, Section 4] as a particular case of the rank varieties of tensors.

THEOREM 1.28 (O. Porras). *Supose* char$(k) = 0$. *Let* $j \geq 2$, $1 \leq s \leq r-1$. *The following properties hold.*

   i. *The variety $V_s(1, j-1;r)$ is normal, Cohen-Macaulay with rational singularities.*

   ii. *Its ideal is generated by the $(s+1) \times (s+1)$ minors of the catalecticant matrix $Cat_F(1, j-1;r)$.*

   iii. *The singular locus of $V_s(1, j-1;r)$ equals $V_{s-1}(1, j-1;r)$*

The proof of this theorem (see [**Po**] or [**FW**]) is based on the so-called geometric method of calculating syzygies. It works in characteristic 0 since uses Bott's theorem for cohomology of homogeneous vector bundles. It was invented by G. Kempf [**Ke1**, **Ke2**] and later developed and used successfully for calculating the ideal and the syzygies of various types of determinantal varieties (see [**La1**, **JPW**, **We2**,

Po, FW]). We refer the reader to the paper [**FW**] for the proof of the basic theorem of the geometric method. A simplified proof of Porras' Theorem, using our Theorem 3.2 may be found in [**Ka**, Section 2].

Porras' Theorem gives affirmative answers to the first two questions of Problem 11.6 in [**G1**]. The case $j = 2$ was previously considered in Theorem 1.26. The case $s = 1$ yields the char $k = 0$ case of the following corollary, proved independently by M. Pucci [**Puc**], who also showed that the ideals of 2×2 minors of $Cat_F(u, j - u; r), 1 \leq u < j$ are independent of u (see also Chapter 9, Problem Q).

COROLLARY 1.29. *The Veronese variety* $v_j(\mathbb{P}^{r-1})$ *is projectively normal, arithmetically Cohen-Macaulay, its affine cone has rational singularities and its ideal is generated by the* 2×2 *minors of the catalecticant matrix* $Cat_F(1, j - 1; r)$.

Another application of the geometric method may be found in [**Ka**], where it is proved that for $j \geq 3$ the ideal of the variety $\overline{PS(2, j; r)}$ is generated by the 3×3 minors of the catalecticant matrices $Cat_F(1, j - 1; r)$ and $Cat_F(2, j - 2; r)$, moreover the variety $\overline{PS(2, j; r)}$ is normal, Cohen-Macaulay with rational singularities. More generally, similar properties are proved for the varieties $Gor_{\leq}(T)$, where T is a Gorenstein sequence with $t_1 = 2$ ([**Ka**, Section 4]). The special case $r = 2$ is treated in the next Section 1.3.

When $t_1 \geq 3$ and $i \neq 1, j - 1$, then the rank $s < r$ locus of $Cat_F(i, j - i; r)$ is more complicated. See Chapter 7 below and Problems A and D of Chapter 9.

1.3. Binary forms and Hankel matrices

In this section we consider the case of binary forms ($r = 2$). We focus on the following topics.

- Canonical forms of homogeneous polynomials in 2 variables and their connection with catalecticant matrices (Theorems 1.43 and 1.53).
- The structure of the Gorenstein Artin algebra $k[x, y]/\operatorname{Ann}(f)$ (Theorems 1.44 and 1.54).
- The equations and geometric properties of the multisecant varieties of a rational normal curve (Theorems 1.45 and 1.56).

Part of this material is contained in some old books [**GrY, El**] as well as in the papers [**KuR, DK, EhR**]. Another part is based on more recent results from [**GruP, Ei1, Wa4**]. The subject is a kind of folklore which should be well-known to the specialists. Our exposition is self-contained with full proofs and may serve as an introduction and a good example to the problems treated in this memoir. Unless otherwise stated the base field k is algebraically closed. In order to keep a better connection with the older sources we work in this section mainly in characteristic 0, or char$(k) > j$ when dealing with polynomials of degree j (pages 22–35). At the end of the section (pages 36–39) we indicate what changes should be made in order to remove the restrictions on the characteristic of k.

We assume for now char$(k) = 0$ or char$(k) > j$ when working with binary forms of degree j.

DEFINITION 1.30. Let $f \in \mathcal{R}_j$ be a binary form of degree j. Let $L_1, \ldots, L_m$ be linear forms. A representation of f as a sum

$$f = G_1 L_1^{j-d_1+1} + \cdots + G_m L_m^{j-d_m+1} \tag{1.3.1}$$

where $G_i \in \mathcal{R}_{d_i-1}$ is called a *generalized additive decomposition* (GAD) of f. If all $d_i = 1$ we obtain the usual additive decomposition

$$f = c_1 L_1^j + \cdots + c_s L_s^j \tag{1.3.2}$$

with $c_i \in k$. A GAD is called *normalized* if no pair L_α, L_β is proportional to each other and none of the G_i is divisible by L_i. For the sake of simplicity we will not distinguish two GADs which are obtained from one another by multiplying the linear forms L_i by nonzero constants. In this sense we will speak of unique representations etc. If (1.3.1) is a normalized GAD its *length* is by definition $\sum_{i=1}^m d_i$. Every GAD can be reduced to a normalized GAD. The length of the latter is called the length of a GAD. So, the length of a GAD given by (1.3.1) is less than or equal to $s = \sum_{i=1}^m d_i$. Equality takes place if and only if a GAD is normalized.

LEMMA 1.31. *Let $\phi = \prod_{i=1}^m (b_i x - a_i y)^{d_i}$ be a prime decomposition of a nonzero form in R_s. Let $L_i = a_i X + b_i Y$. Then a form $f \in \mathcal{R}_j$ with $j \geq s$ has a generalized additive decomposition (1.3.1) if and only if ϕ is apolar to f. If all roots of ϕ are simple, then this is an additive decomposition (1.3.2).*

PROOF. Let $\phi_i = (b_i x - a_i y)^{d_i}$. Then $R\phi = \cap_{i=1}^m R\phi_i$. These are graded ideals in R, so for the j-th graded components we have $R_{j-s}\phi = \cap_{i=1}^m R_{j-d_i}\phi_i$. Now, ϕ is apolar to f if and only if f is apolar to $R_{j-s}\phi$, indeed the differentiation pairing is non-degenerate, so $\phi \circ f = 0$ iff $0 = R_{j-s} \circ (\phi \circ f) = (R_{j-s}\phi) \circ f$. We conclude that $\phi \circ f = 0$ iff $f \in (R_{j-s}\phi)^\perp = \sum_{i=1}^m (R_{j-d_i}\phi_i)^\perp$. It remains to prove that the forms of degree j apolar to $\phi_i = (b_i x - a_i y)^{d_i}$ are those that may be written

$G_i L_i^{j-d_i+1}$ with $G_i \in R_{d_i-1}$. Since the differentiation action is GL_2-equivariant this statement follows from the following obvious fact: the forms F of degree j apolar to x^n, i.e. whose derivative $\frac{\partial^n}{\partial X^n} F = 0$, may be written $F = G(X, Y) Y^{j-n+1}$ with $G(X, Y)$ a form of degree $n - 1$. $\qquad\square$

DEFINITION 1.32. The *length* of a binary form f is the minimun length of a generalized additive decomposition of f. We denote it by $\ell(f)$.

LEMMA 1.33. *Let $j = 2t$ or $2t + 1$, let $f \in R_j$. Then $\ell(f) \leq t + 1$. If $I = \mathrm{Ann}(f)$ is the ideal of forms apolar to f, then $\ell(f)$ equals the order d (i.e. the initial degree) of the graded ideal $I = I_d + I_{d+1} + \cdots$*

PROOF. The order of I is the minimal degree of a nonzero polynomial apolar to f, so the second statement follows from Lemma 1.31. In order to prove the first one we observe that I_{t+1} is the kernel of the linear map

$$C_f(j - t - 1, t + 1) \ : \ R_{t+1} \longrightarrow R_{j-t+1},$$

so $I_{t+1} \neq 0$ since $t + 1 > j - t - 1$. Thus $\ell(f) = d \leq t + 1$. $\qquad\square$

In the examples to follow in this section we give explicitly the generators of the annihilating ideal of certain binary forms. This can be calculated using the "Macaulay" or other algebra program (see Appendix D), or may be readily checked, using the well known result that $\mathrm{Ann}(f)$ is a complete intersection when $r = 2$ (Theorems 1.44, 1.54).

EXAMPLE 1.34. The form $f = X(X + 2Y)^2$ is written as a GAD of length two, with $G_1 = X$, $L_1 = (X + 2Y)$; the ideal $\mathrm{Ann}(f) = ((2x - y)^2, 4x^3 - 3yx^2)$. By Lemma 1.31, a form F can be written $F = G_1(X + 2Y)^2$ with G_1 a linear form iff $(2x - y)^2 \in \mathrm{Ann}(F)$. By Proposition 1.36 below, the length two GAD for f is unique. That $x^2(4x - 3y) \in \mathrm{Ann}(f)$ implies by Lemma 1.31 that there is a length three GAD that can be written $f = G_1 Y^2 + c(3X + 4Y)^3$, for suitable $G_1 \in R_1$ and $c \in k$: we have $f = \frac{1}{27} \left((-144X - 64Y) Y^2 + (3X + 4Y)^3 \right)$. There are many length three decompositions, each corresponding to the factorization of an element in the two dimensional vector space $\mathrm{Ann}(F)_3$.

Consider a second form $g = X^4 + Y^4 + (X + Y)^4$, whose ideal of apolar forms is $\mathrm{Ann}(g) = (xy(x - y), x^3 + y^3 - 3x^2y)$. Thus, g has no length two GAD's, but many length three GAD's, each corresponding to an element of the vector space $V = \operatorname{Ann}(g)_3$, and all but four of which are additive decompositions[2].

Finally, the form $h = X^4 + Y^4$ has apolar ideal $\operatorname{Ann}(h) = (xy, x^4 - y^4)$, thus, it has a unique decomposition of length two, and no further normalized GAD's of length ≤ 3 according to Proposition 1.36 below.

LEMMA 1.35 (Jordan's Lemma [**GrY**]). *Assuming the notation of Definition 1.30 suppose that the linear forms L_i, $i = 1, \ldots, m$ are not proportional to each other and*

$$0 = G_1 L_1^{j-d_1+1} + \cdots + G_m L_m^{j-d_m+1} \qquad (1.3.3)$$

with $\sum_{i=1}^m d_i \leq j + 1$. Then $G_i = 0$ for every i.

FIRST PROOF (ONLY FOR $d_i = 1$). Let $G_i = c_i \in k$, $m = s$. Changing the coordinates if necessary we can assume that none of the zeros of L_i, $i = 1, \ldots, s$ is ∞. Then passing to affine coordinates (1.3.3) is equivalent to

$$c_1(t - \beta_1)^j + \cdots + c_s(t - \beta_s)^j = 0$$

where $\beta_i \neq \beta_j$. Taking $(s - 1)$ derivatives of this equation we obtain a homogeneous system for $c_1, \ldots, c_s$ whose determinant is a multiple of the Vandermonde determinant, so is nonzero. Therefore $c_1 = \ldots = c_s = 0$. $\qquad \square$

SECOND PROOF. The lemma is equivalent to the statement that the vector space of the forms on the right-hand side of (1.3.3), i.e. $\sum_{i=1}^m \mathcal{R}_{d_i-1} L_i^{j-d_i+1}$ has dimension $s = \sum_{i=1}^m d_i$. According to Lemma 1.31 and its proof this space equals the annihilator $(R_{j-s}\phi)^\perp \subset \mathcal{R}_j$, where $\phi = \prod_{i=1}^m (b_i x - a_i y)^{d_i}$ and $L_i = a_i X + b_i Y$. Hence $\dim \sum_{i=1}^m R_{d_i-1} L_i^{j-d_i+1} = (j+1) - (j-s+1) = s$. $\qquad \square$

PROPOSITION 1.36. UNIQUENESS OF GAD. *Suppose $j = 2t$ or $2t + 1$. Let*

$$f = G_1 L_1^{j-d_1+1} + \cdots + G_m L_m^{j-d_m+1} \qquad (1.3.4)$$

be a normalized GAD of $f \in \mathcal{R}_j$ of length $s = \sum_{i=1}^m d_i \leq t + 1$. Then f has no other GAD of length $\leq j + 1 - s$ and $\ell(f) = s$. In particular if $s \leq t$ or if $s = t + 1$, $j = 2t + 1$ (equivalently $2s \leq j + 1$) then (1.3.4) is the unique normalized GAD of f having length $\leq t + 1$.

---

[2]The existence of only four exceptional GAD's can be seen by bounding the number of elements of V having multiple roots: these roots are given by the zeroes of a Wronskian determinant associated to V (see [**I2**, Appendix],[**IY**]).

PROOF. We use Jordan's Lemma. Another proof, which avoids this lemma and reconstructs the GAD from a catalecticant matrix via Lemma 1.31, is given in Proposition 1.50.

Suppose f has a second normalized GAD

$$f = H_1 M_1^{j-e_1+1} + \cdots + H_n M_n^{j-e_n+1}$$

of length $s' = \sum_{i=1}^n e_i \leq j + 1 - s$. Then we obtain an equality

$$\begin{aligned} 0 = {} & G_1 L_1^{j-d_1+1} + \cdots + G_m L_m^{j-d_m+1} \\ & - H_1 M_1^{j-e_1+1} - \cdots - H_n M_n^{j-e_n+1}. \end{aligned} \tag{1.3.5}$$

Suppose first that none of the L_i is proportional to some M_l. Then we obtain a relation as in Jordan's Lemma 1.35 with $s + s' \leq j + 1$ summands, so all $G_i = 0 = M_l$. Suppose some L_i is proportional to some M_l. Reordering $\{L_\alpha\}$ and $\{M_\beta\}$ we may assume this happens for $i = 1 = l$. By symmetry we may assume $d_1 \leq e_1$. Then (1.3.5) reduces to

$$0 = (G_1 - cL_1^{e_1-d_1})L_1^{j-d_1+1} + \cdots$$

for some constant $c \in k$. This sum has an even smaller number of summands than $s+s' \leq j+1$, so applying Jordan's Lemma we conclude that $G_1 - cL_1^{e_1-d_1} = 0$. If $e_1 > d_1$ we would obtain that L_1 divides G_1 which is excluded by the hypothesis that the GADs are normalized. Thus $e_1 = d_1$ and $G_1 L_1^{j-d_1+1} = H_1 M_1^{j-e_1+1}$. Canceling these terms and continuing in the same way we obtain the required uniqueness, in the sense of Defintion 1.30.

By Lemma 1.33 f has a normalized GAD of length $s \leq t + 1$. We have already proved that f has no other normalized GAD of length less or equal $t + 1$, unless $s = t + 1$ and $j = 2t$. Then $j + 1 - s = t$, so f has no GAD of length $\leq t$ as already proved, so whatever the case $\ell(f) = s$. $\qquad\square$

DEFINITION 1.37. Let $f \in \mathcal{R}_j$ and let $2\ell(f) \leq j + 1$. Then the unique normalized GAD of length $s = \ell(f)$ (according to Proposition 1.36) is called the *canonical form* of f.

Let $P_s \subset PS(s, j; 2)$ be the locus of binary forms of degree j which have an additive decomposition $f = L_1^j + \cdots + L_s^j$ of length s (i.e. each pair L_α, L_β is non-proportional).

LEMMA 1.38. *Suppose* $s \leq j$, $f \in \mathcal{R}_j$.

i. *The binary form f belongs to $\overline{P_s}$ if and only if it has a generalized additive decomposition of length $\leq s$.*

ii. *Suppose* $2s \leq j+2$. *Then* f *belongs to* $\overline{P_s} - \overline{P_{s-1}}$ *if and only if it has a normalized GAD of length* s. *This holds iff* $\ell(f) = s$.

iii. *Suppose* $2s \leq j+1$. *Then* $\dim \overline{P_s} = 2s$. *If* $2s = j+2$ *then* $\overline{P_s} = \mathcal{R}_j$ *and* $\dim \overline{P_s} = 2s - 1$.

Proof. The irreducible variety $\overline{P_s}$ is invariant under multiplication by elements of k^*, so we can work with the projectivization $\mathbb{P}(\overline{P_s}) \subset \mathbb{P}(\mathcal{R}_j)$. Consider the closed subvariety $X \subset \mathbb{P}(R_s) \times \mathbb{P}(\mathcal{R}_j)$ which consists of pairs

$$X = \{ ([\phi], [f]) \mid \phi \circ f = 0 \}$$

and the two projections

$$
\begin{array}{c}
X \xrightarrow{\ \pi_2\ } \mathbb{P}(\mathcal{R}_j) \\
{\scriptstyle \pi_1} \Big\downarrow \\
\mathbb{P}(R_s)
\end{array}
\qquad (1.3.6)
$$

Lemma 1.31 and its proof show that the fiber of the surjective projection $\pi_1 : X \to \mathbb{P}(R_s)$ over a point $[\phi]$ is the projective space $\mathbb{P}(R_{j-s}\phi)^\perp$ of dimension $s - 1$ and consists of points $[f]$ where f has a GAD of length $\leq s$. Thus X is a smooth irreducible projective variety of dimension $2s - 1$. The image of the second projection $\pi_2 : X \to \mathbb{P}(\mathcal{R}_j)$ is therefore an irreducible projective variety and by Lemma 1.31 the points of $\pi_2(X)$ are the forms of length $\leq s$.

Part (i). It suffices to verify that $\mathbb{P}(P_s)$ is a dense subset of $\pi_2(X)$. Let us consider the subset $X_0 \subset X$ consisting of pairs $([\phi], [f])$ such that ϕ has s distinct linear factors and f has an additive decomposition of length s. By definition $\mathbb{P}(P_s) = \pi_2(X_0)$, so it suffices to verify that X_0 contains an open subset of X. Let $U \subset \mathbb{P}(k^2 \times \cdots \times k^2)$ be the open subset consisting of s-tuples $((a_1, b_1), \ldots, (a_s, b_s))$ with $a_i b_l - a_l b_i \neq 0$ for any $i \neq l$. The morphism $\sigma : U \to X$

$$
\sigma((a_1, b_1), \ldots, (a_s, b_s)) = \left([\phi = \prod_{i=1}^{s} (b_i x - a_i y)],\ [f = \sum_{i=1}^{s} (a_i X + b_i Y)^j] \right)
$$

is dominant since it contains an open subset of any fiber of π_1 over the open subset of $\mathbb{P}(R_s)$ consisting of those $[\phi]$ with s distinct linear factors (the latter is the complement to the discriminant hypersurface). Thus $\sigma(U) = X_0$ contains an open subset of X.

Part (ii). This follows from Part (i) and Proposition 1.36 according to which if f has a normalized GAD of length $s \leq t+1$ then it has no GADs of smaller length.

Part (iii). The equality $\dim \overline{P_s} = 2s$ follows from the uniqueness statement of Proposition 1.36 if $2s \leq j+1$. The last assertion $\overline{P_{t+1}} = R_{2t}$ follows from Lemma 1.33. $\square$

EXAMPLE 1.39. CLOSURE OF $PS(2, 3; 2)$. We set $R = k[x, y]$ and $\mathcal{R} = k[X, Y]$. The form $f = X^2 Y$ is already written as a GAD of length two, since the coefficient of X^2 has degree one. By Proposition 1.36 this is its canonical form. By Lemma 1.38 f is in the closure of $PS(2, 3; 2)$. One can see this directly, since

$$f = \lim_{t \to 0} \frac{1}{3t} \left((X + tY)^3 - (X - tY)^3 \right).$$

Also f can be written in many ways as an element of $PS(3, 3; 2)$ — each corresponding to an element of $\mathrm{Ann}(f)_3 = (y^2, x^3)_3$, by Lemma 1.31. Thus, the decomposition

$$f = \frac{1}{6a} \left((X + aY)^3 - (X - aY)^3 - 2a^3 Y^3 \right),$$

corresponds to the element $(ax - y)(ax + y)x \in \mathrm{Ann}(f)$.

Sylvester's Theorem below shows that $\overline{PS(2, 3; 2)} = \mathbb{A}(\mathcal{R}_3) = \mathbb{A}^4$, the affine space parametrizing all degree-3 forms.

THEOREM 1.40 (J. Sylvester). *Every general binary form f of odd degree $j = 2t - 1 \geq 3$ has a unique canonical form of length t.*

$$f = L_1^j + \cdots + L_t^j.$$

Every binary form f of odd degree $j = 2t - 1$ has a unique canonical form

$$f = G_1 L_1^{j - d_1 + 1} + \cdots + G_m L_m^{j - d_m + 1}$$

of length $\ell(f) = \sum_{i=1}^{m} d_i \leq t$.

PROOF. This follows from $\dim \overline{P_t} = 2t = \dim R_{2t-1}$ and Proposition 1.36. $\square$

REMARK 1.41. In the even case $j = 2t$ one has $\dim \overline{P_t} = 2t = \dim R_{2t} - 1$, so $\overline{P_t}$ is a hypersurface which coincides with the catalecticant hypersurface $V_t(t, t; 2)$ as we prove below in Theorem 1.43. If $s = t + 1$, $j = 2t$ the map $\pi_2 : X \to \mathbb{P}(\mathcal{R}_j)$ is surjective and $\dim X = \dim \mathbb{P}(\mathcal{R}_j) + 1$, so every general form of even degree $2t$ has a one-dimensional family of GADs of length $t+1$ and no GAD of smaller length.

EXAMPLE 1.42. When $j = 4$, the form $G = X^4 + Y^4 + (X+Y)^4$ is general enough so it does not lie in $\overline{P}_2$: as we saw in Example 1.34 it has no GAD of length 2, and an infinite number of GAD's of length three. It is a consequence of Theorem 1.44 below that the open subvariety of $\mathbb{A}(\mathcal{R}_4) = \mathbb{A}^5$ parametrizing forms $f \in R_4$ *not* having a length two GAD, is isomorphic to the variety $CI(3,3)$ parametrizing complete intersections $(h_1, h_2) = \mathrm{Ann}(f)$, having two generators of degree 3. This illustrates the second case in Remark 1.41, and generalizes to even degrees $j = 2t$: the open subvariety in $\mathbb{A}(\mathcal{R}_j) = \mathbb{A}^{j+1}$ of forms f not having a length-t GAD, is isomorphic to $CI(t+1, t+1)$, parametrizing the ideals $\mathrm{Ann}(f)$. It is only these forms that fail to have a canonical GAD as described by Sylvester's Theorem when j is odd, or by Remark 1.41 for $f \in \overline{P}_t$ when j is even.

The forms f having no canonical GAD can be recognized, according to Theorem 1.44, among those of degree $j = 2t$, by their having the maximum possible Hilbert function $H_f = H(R/\mathrm{Ann}(f)) = (1, 2, \ldots, t, t+1, t, \ldots, 1)$, or more simply by $\ell(f) = t + 1$, which is equivalent to the non-vanishing of their catalecticant invariant $\det Cat_f(t, t; 2)$.

Now let $f \in \mathcal{R}_j$,

$$ f = a_0 X^j + \cdots + \binom{j}{d} a_d X^{j-d} Y^d + \cdots + a_j Y^j. \qquad (1.3.7) $$

Choose a basis of R_v to be $x^{v-i}y^i$, $i = 0, \ldots, v$ and a basis of $\mathcal{R}_j$ to be $\binom{j}{d} X^{j-d} Y^d$. In these bases the matrix of the linear operator

$$ C_f(j - v, v) : R_v \to \mathcal{R}_{j-v}, \qquad \phi \mapsto \phi \circ f \qquad (1.3.8) $$

equals $\frac{j!}{(j-v)!}$ times the Hankel matrix

$$ Cat_f(j - v, v; 2) = \begin{pmatrix} a_0 & a_1 & \cdots & a_v \\ a_1 & a_2 & \cdots & a_{v+1} \\ \vdots & \vdots & & \vdots \\ a_{j-v} & a_{j-v+1} & \cdots & a_j \end{pmatrix} \qquad (1.3.9) $$

We refer the reader to Example 1.7; notice however the difference by a constant coefficient multiple, which appears since we want to keep the classical setting in this section (see the discussion in Example 1.8), and which is irrelevant for our considerations here. Recall from Definition 1.4 that $V_s(j - v, v; 2)$ denotes the catalecticant variety

$$ V_s(j - v, v; 2) = \{ f \in \mathcal{R}_j \mid \mathrm{rk}\, Cat_f(j - v, v; 2) \leq s \}. \qquad (1.3.10) $$

Theorem 1.43. *Let $j = 2t$ or $2t + 1$, let $f \in \mathcal{R}_j$.*

i. *Let $s = \operatorname{rk} Cat_f(j - t, t; 2)$. Then $\ell(f) = s$. If $2s \leq j + 1$, then f has a unique generalized additive decomposition of length s and no other GADs of length $\leq t + 1$.*

ii. *For every pair of integers s, v with $1 \leq s \leq v \leq j - v + 1$ one has $\overline{P_s} = V_s(j - v, v; 2)$. If $\ell(f) = s$, then $\ell(f) = s = \operatorname{rk} Cat_f(j - v, v; 2)$.*

Proof. We first prove Part (ii). Assume $2s \leq j$. Fix v with $s \leq v \leq j - v$ and let $V_s = V_s(j - v, v; 2)$. First we prove that $\overline{P_s} \subset V_s$. Since V_s is closed it suffices to verify that $P_s \subset V_s$. Any form $g \in P_s \subset \mathcal{R}_j$ is apolar to some $\phi \in R_s$, thus it is apolar to the vector space $R_{v-s}\phi$ of dimension $v - s + 1$. This implies $\operatorname{rk} Cat_g(j - v, v; 2) \leq s$, hence $g \in V_s$.

Next we prove that $\overline{P_s} = V_s(j - s, s; 2)$. We proved above that $\overline{P_s} \subset V_s(j - s, s; 2)$. Conversely, if $g \in V_s(j - s, s; 2)$, then by the definition of the latter (1.3.10) there is a nonzero form $\phi \in R_s$ apolar to g. Hence g has a normalized GAD of length $\leq s$, thus $g \in \overline{P_s}$ according to Lemmas 1.31 and 1.38.

Let F be a generic binary form, i.e the coefficients $a_d = Z_d$, $d = 0, \ldots, j$ in (1.3.7) are indeterminates over k. We claim that the $(s + 1) \times (s + 1)$ minors of the catalecticant matrix $Cat_F(j - s, s; 2)$ are minors of any matrix $Cat_F(j - v, v; 2)$ with $s \leq v \leq j - v$. Indeed, the $(s + 1) \times (s + 1)$ minors of $Cat_F(j - v, v; 2)$ which are contained either in its first $s + 1$ columns or in its last $s + 1$ rows are exactly the $(s + 1) \times (s + 1)$ minors of $Cat_F(j - s, s; 2)$. This observation implies that

$$\overline{P_s} \subset V_s(j - v, v; 2) \subset V_s(j - s, s; 2) = \overline{P_s}$$

which proves the equality $\overline{P_s} = V_s(j - v, v; 2)$. The same equality holds for $s = t + 1$, since in both cases $j = 2t$ or $2t + 1$ the rank conditions in (1.3.10) are empty, so $V_s(j - t - 1, t + 1; 2) = \mathcal{R}_j$ and we can apply Lemma 1.38. One obtains that for $s \leq v \leq j - v$ as well as for $s = t + 1, v = t$

$$\overline{P_s} - \overline{P_{s-1}} = V_s(j - v, v; 2) - V_{s-1}(j - v, v; 2). \tag{1.3.11}$$

According to Lemma 1.38 the left-hand side consists of the forms f with $\ell(f) = s$. The forms of the right-hand side are by definition those with $\operatorname{rk} Cat_f(j - v, v; 2) = s$. Therefore if $\ell(f) = s$ then $s = \operatorname{rk} Cat_f(j - v, v; 2)$. This proves Part (ii).

Part (i). Setting $v = t$ in (1.3.11) we obtain the equality $\ell(f) = \operatorname{rk} Cat_f(j - t, t; 2)$. The remaining statement follows from Proposition 1.36. $\qquad\square$

The portion of the following Theorem, showing that $\operatorname{Ann}(f)$ is a complete intersection (CI), was shown by F. H. S. Macaulay in 1904, in the more general context of an element $f \in \mathcal{R}$ that is not necessarily homogeneous [**Mac1, Mac2**]; the homogeneous case is rather simpler, though often reproved. We show the case of arbitrary $\operatorname{char}(k)$ later in Theorem 1.54. There is a well known generalization due to Serre, "Gorenstein height two implies CI" (cf. Remark 1.55 below).

THEOREM 1.44. *Let f be a binary form of degree $j = 2t$ or $2t + 1$. Let $I = \operatorname{Ann}(f)$ be the ideal of forms apolar to f. Let $A_f = R/I$ be the associated Gorenstein Artin algebra. Let $s = \max\{\dim_k(A_f)_i\}$. Then*

   i. *$s = \ell(f)$ and the Hilbert function of A_f satisfies*

$$H(A_f) = (1, 2, \ldots, s-1, \overset{s-1}{s}, s, \ldots, \overset{j-s+1}{s}, s-1, \ldots, 2, 1) \quad (1.3.12)$$

   ii. *Let T be the sequence on the right-hand side of of the above equality ($T = H(s, j, 2)$). Then $\overline{P_s} = Gor(T)$.*

   iii. *Suppose $2s \leq j + 1$. Then $\dim I_s = 1$, $I_s = \langle \phi \rangle$ and for every integer v with $s \leq v \leq j - s + 1$ one has $I_v = R_{v-s}\phi$.*

   iv. *The ideal I is generated by two homogeneous polynomials $\phi \in I_s$ and $\psi \in I_{j+2-s}$. Equivalently, the ring A_f is a complete intersection of generator degrees $s, j + 2 - s$.*

PROOF. The Hilbert function $H(A_f)$ is symmetric, so it suffices to verify (1.3.12) for the first half of the terms $i \leq t$. One has (see Definitions 1.9, 1.11)

$$\dim_k(A_f)_i = \operatorname{rk} Cat_f(j - i, i; 2).$$

Let $s' = \operatorname{rk} Cat_f(j - t, t; 2)$. If $j = 2t$ and $s' = t + 1$, then $\ell(f) = t + 1$ and (1.3.12) holds by Lemma 1.33. Otherwise $2s' \leq j + 1$ and by Theorem 1.43 we conclude that $\ell(f) = s'$ and for every i with $s' \leq i$, $2i \leq j$ one has $\operatorname{rk} Cat_f(j - i, i; 2) = s'$, while for i smaller then s' clearly $\dim(A_f)_i = i + 1 \leq s'$ since $I_i = 0$. Therefore $s = s'$ and the Hilbert function $H(A_f)$ equals (1.3.12). This proves Part (i).

   Part (ii) follows from (i).

   Part (iii) follows from the inclusion $R\phi \subset I$ and the equalities

$$\operatorname{cod}_{R_v}(R_{v-s}\phi) = s = H(A_f)_u = \operatorname{cod}_{R_u}(I_u).$$

It remains to prove Part (iv). Since $H(A_f)_{j+2-s} = s - 1$ we have that

$$\dim_k I_{j+2-s} = \dim_k R_{j+2-2s}\phi + 1.$$

Let $\psi \in I_{j+2-s} - R_{j+2-2s}\phi$.

CLAIM. The homogeneous polynomials ϕ and ψ have no common zeros.

PROOF. Suppose on the contrary that the binary forms ϕ and ψ have a common linear factor $\lambda = ax + by$. Let $u = j + 2 - s$. Then $I_u \subset R_{u-1}\lambda$. Since $(I_u)^\perp = R_{j-u} \circ f$ we conclude that there is a nonzero form $\theta \in R_{j-u}$, such that $\theta \circ f$ is apolar to $R_{u-1}\lambda$. Equivalently R_{u-1} is apolar to $(\lambda\theta) \circ f$, i.e. $\lambda\theta \in I$. But $\deg(\lambda\theta) = j - u + 1 = s - 1$ which contradicts Lemma 1.33. q.e.d.

Now, we prove that for every v with $j + 2 - s \leq v \leq j + 1$ we have

$$I_v = R_{v-s}\phi + R_{v-(j+2-s)}\psi. \qquad (1.3.13)$$

Let us first prove that $R_{v-s}\phi \cap R_{v-(j+2-s)}\psi = 0$. Suppose $a\phi = b\psi$ is an element of the intersection. Then $\psi|a$ and $\phi|b$ by the claim proved above. This is impossible unless $a = 0$ since $\deg a = v - s \leq j + 1 - s < \deg\psi$. Now, we have $\dim_k R_{v-s}\phi = v - s + 1$, $\dim_k R_{v-(j+2-s)}\psi = v + s - j - 1$ which sum up to $2v - j$. On the other hand $\dim_k H(A_f)_v = j - v + 1$, so $\dim_k I_v = 2v - j$. This proves (1.3.13).

It remains to prove that for every $v \geq j + 1$ one has

$$R_v = R_{v-s}\phi + R_{v-(j+2-s)}\psi. \qquad (1.3.14)$$

Since the powers of linear forms ℓ^v span R_v it suffices to prove that every general ℓ^v belongs to the ideal (ϕ, ψ). Changing the variables of $\mathbb{P}^1$ we can assume that $\ell = x$ and x does not divide neither ϕ nor ψ. Let $g = \frac{\phi}{x^s}$, $h = \frac{\psi}{x^{j+2-s}}$. Then g, h are prime polynomials in one variable $t = \frac{y}{x}$, so there exist polynomials $a(t), b(t)$ of degrees $\deg a(t) \leq j + 1 - s$, $\deg b(t) \leq s - 1$ such that $ag + bh = 1$. Multiplying this equality by x^{j+1} we obtain $x^{j+1} = \tilde{a}\phi + \tilde{b}\psi$ for some homogeneous polynomials $\tilde{a}, \tilde{b}$. This proves that $x^v \in (\phi, \psi)$ and consequently (1.3.14) as shown above. Part (iv) is proved. $\square$

THEOREM 1.45. *Let*

$$F = Z_0 X_1^j + \cdots + \binom{j}{d} Z_d X_1^{j-d} X_2^d + \cdots + Z_j X_2^j$$

be a generic binary form of degree j whose coefficients $Z_0, \dots Z_j$ are indeterminates over k. Let s, v be integers which satisfy $2s \leq j$, $s \leq v \leq j - v$. Let $I_{s+1}(Cat_F(j - v, v; 2))$ be the ideal generated by the $(s + 1) \times (s + 1)$ minors of $Cat_F(j - v, v; 2)$. Let $I(\overline{P_s})$ be the ideal of the affine irreducible variety $\overline{P_s} \subset \mathbb{A}(\mathcal{R}_j)$. Then

   i. $I(\overline{P_s}) = I_{s+1}(Cat_F(j - v, v; 2))$.

   ii. $\overline{P_s}$ is Cohen-Macaulay, normal and its singular locus is $\overline{P_{s-1}}$.

   iii. The projectivization $\mathbb{P}(\overline{P_s}) \subset \mathbb{P}(\mathcal{R}_j)$ is a rational variety of dimension $2s - 1$ and degree $\binom{j-s+1}{s}$.

   iv. The birational morphism $\pi_2 : X \to \pi_2(X) = \overline{P_s}$ of (1.3.6) is a desingularization of $\overline{P_s}$, and is an isomorphism over $\overline{P_s} - \overline{P_{s-1}}$.

PROOF. Let us denote by $\mathbf{V}_s(j - v, v; 2)$ the catalecticant scheme in $\mathbb{A}(\mathcal{R}_j)$ whose ideal is $I_{s+1}(Cat_F(j - v, v; 2))$. As we showed in the proof of Theorem 1.43

$$I_{s+1}(Cat_F(j - s, s; 2)) \subset I_{s+1}(Cat_F(j - v, v; 2)), \qquad (1.3.15)$$

and the reduced scheme $V_s(j - v, v; 2)$ whose ideal is the radical of $I_{s+1}(Cat_F(j - v, v; 2))$ coincides with $\overline{P_s}$. So, in order to prove (i) it suffices to show that $I_{s+1}(Cat_F(j - s, s; 2))$ is a prime ideal. We have

$$\operatorname{cod}_{\mathbb{A}(\mathcal{R}_j)} \mathbf{V}_s(j - s, s; 2) = j + 1 - 2s.$$

This is the codimension of the general determinantal variety $D_s(j-s+1, s+1)$ of rank $\leq s$ matrices in the affine space $\mathbb{A}^{(j-s+1)(s+1)}$ of $(j - s + 1) \times (s + 1)$ matrices (see e.g. [**ACGH**, p.67]). Now, $D_s(j - s + 1, s + 1)$ is a Cohen-Macaulay variety according to the theorem of Eagon-Northcott [**EaN**] (see also [**BruV**] or [**ACGH**, p.79]), thus $\mathbf{V}_s(j - s, s; 2)$ is a Cohen-Macaulay scheme as well, since it is an intersection of $D_s(j - s + 1, s + 1)$ by the affine space of catalecticant matrices of the same codimension (see e.g. [**ACGH**, p.84]).

Let $\mathfrak{q} = I_{s+1}(Cat_F(j - s, s; 2))$, $\mathfrak{p} = I(\overline{P_s})$. By the Unmixedness Theorem [**Ma2**, Theorem 32] or [**Ei2**, Corollary 18.14] the primary decomposition of $\mathfrak{q}$ in the ring $k[Z_0, \dots, Z_j]$ has only minimal associated primes, so in our case there is a unique associated prime ideal, namely $\mathfrak{p}$. Thus $\mathfrak{q}$ is a $\mathfrak{p}$-primary ideal. We need a lemma.

LEMMA. The scheme $\mathbf{V}_s(j-s, s; 2)$ is smooth at every closed point $f \in \overline{P_s} - \overline{P_{s-1}}$.

PROOF. Let $I = \operatorname{Ann}(f) \subset R$ be the ideal of polynomials apolar to f. We use here Theorem 3.2 which tells that the tangent space

$$T_f \mathbf{V}_s(j - s, s; 2) = (I_s I_{j-s})^{\perp}.$$

Now, if ϕ is the form of minimum degree apolar to f, by Theorem 1.44 $I_s = k\phi$, $I_{j-s} = R_{j-2s}\phi$, hence $I_s I_{j-s} = R_{j-2s}\phi^2$ and $\dim_k(I_s I_{j-s})^\perp = (j+1) - (j - 2s + 1) = 2s = \dim \overline{P_s}$. q.e.d.

If an irreducible scheme is generically smooth it is generically reduced. So, if $A = k[Z_0, \ldots, Z_j]/\mathfrak{q}$ and $\overline{\mathfrak{p}} = \mathfrak{p}/\mathfrak{q}$, then the localization $A_{\overline{\mathfrak{p}}}$ has no nonzero nilpotents. By the correspondence of primary decomposition under localization [AM] we conclude that $\mathfrak{p} = \mathfrak{q}$. This proves Part (i) of Theorem 1.45.

By the above lemma the singular locus of $\overline{P_s}$ is contained in $\overline{P_{s-1}}$ which is of codimension 2 by Lemma 1.38. Therefore $\overline{P_s}$ is normal according to Serre's Criterion (see e.g. [Ei2] or [ACGH, pp.98–99]). Let $D_s \subset \mathbb{A}^{(j-v+1)(v+1)}$ be the general determinantal variety of $(j - v + 1) \times (v + 1)$ matrices of rank $\leq s$, its ideal being generated by the $(s + 1) \times (s + 1)$ minors (by the Second Fundamental Theorem of the theory of invariants for GL_s, [Weyl, DP]; in fact we need the corank 1 case, $v = s$ which is simpler, see e.g. [Ke1]). The singular locus of D_s is D_{s-1} and the tangent space at every $A \in D_{s-1}$ is the whole space of $(j - v + 1) \times (v + 1)$ matrices (see e.g. [Ha, pp.184–185]). Equivalently, every polynomial in the ideal of D_s vanishes to order ≥ 2 at every point of D_{s-1}. The ideal of $\overline{P_s}$ equals the restriction of the polynomials of the ideal of D_s to the space of catalecticant matrices, according to Part (i). We conclude that all polynomials in the ideal of $\overline{P_s}$ vanish to order ≥ 2 at every point of $\overline{P_{s-1}}$. Thus $\overline{P_s}$ is singular along $\overline{P_{s-1}}$. Part (ii) is proved.

Let us specialize $v = s$, $D_s = D_s(j - s + 1, s + 1)$. Then the intersection with the space of catalecticant matrices is proper, i.e. the codimension is preserved, so the degree of $\mathbb{P}(\overline{P_s})$ is equal to the degree of the projective determinantal variety $M_s(j - s + 1, s + 1) = \mathbb{P}(D_s)$ in $\mathbb{P}^{(j-s+1)(s+1)-1}$ which is $\binom{j-s+1}{s}$ (see [Ha, p.243] for a simple proof). Let us consider the map

$$\pi_2 : X \to \mathbb{P}(\overline{P_s}), \qquad \pi_2([\phi], [f]) = [f] \qquad (1.3.16)$$

defined in (1.3.6). This map has an inverse defined over $\overline{P_s} - \overline{P_{s-1}}$ by

$$\pi_2^{-1}([f]) = ([\operatorname{Ker} Cat_f(j - s, s; 2)], [f]).$$

Furthermore the map π_1 from (1.3.6) makes X a projectivization of an algebraic vector bundle of rank s over $\mathbb{P}^s = \mathbb{P}(R_s)$, thus X is rational and nonsingular. Therefore $\mathbb{P}(\overline{P_s})$ is a rational variety with desingularization X. Notice that this argument gives another proof of the nonsingularity of $\overline{P_s} - \overline{P_{s-1}}$. This completes the proof of Theorem 1.45. $\qquad\square$

REMARK 1.46. Let $2s \leq j$. Let $T_s = H(s, j, 2)$ ($=$ the right-hand side of (1.3.12)). By Theorem 1.44(ii) we have $\overline{P_s} - \overline{P_{s-1}} = Gor(T_s)$ and this is a nonsingular variety as shown above. We obtain a partition of $\overline{P_s} = \overline{Gor(T_s)}$ into smooth varieties $Gor(T_i), 1 \leq i \leq s$, satisfying

$$\overline{Gor(T_s)} \supset \overline{Gor(T_{s-1})} \supset \ldots \supset \overline{Gor(T_1)},$$

for which $Gor(T_i) = \overline{Gor(T_i)} - \overline{Gor(T_{i-1})}$ is nonsingular, $Sing(\overline{Gor(T_i)}) = \overline{Gor(T_{i-1})}$ and

$$\overline{Gor(T_i)} = \bigcup_{l \leq i} Gor(T_l)$$

for every $i, 1 \leq i \leq s$ ($Gor(T_0) = \emptyset$). The last equality means that the partition satisfies the frontier property, so by definition it is a stratification.

Recall that a disjoint union of topological spaces $X = \bigcup_{i \in I} X_i$ satisfies the *frontier property* if for every pair of indices $i, \ell \in I$ provided $X_\ell \cap \overline{X_i} \neq \emptyset$ then $X_\ell \subset \overline{X_i}$.

REMARK 1.47. The map (1.3.16) is a desingularization of $\mathbb{P}(\overline{P_s})$. The latter is the same as the s-secant variety $Sec_s(C)$ to a rational normal curve $C \subset \mathbb{P}^j$ (see Definition 1.12).

REMARK 1.48. As mentioned in the previous remark $\mathbb{P}(\overline{P_s}) = Sec_s(C)$. The aim of this remark is to give another proof of the equality

$$\deg Sec_s(C) = \binom{j - s + 1}{s} \qquad (1.3.17)$$

without using Theorem 1.45(i) (for $v = s$). This in its turn yields an alternative proof of the equality of the ideals

$$I(\overline{P_s}) = I_{s+1}(Cat_F(j - v, v; 2)) \qquad (1.3.18)$$

for every v with $s \leq v \leq j - v$.

PROOF. Indeed, assume for the moment that (1.3.17) holds. It suffices to prove (1.3.18) for $v = s$ (because of (1.3.15)), which we do by comparing the degrees of the projective schemes $\mathbb{P}(\overline{P_s})$ and $\mathbb{P}(\mathbf{V}_s(j - s, s; 2))$.[3] As in the proof of Theorem 1.45 let $\mathfrak{p}, \mathfrak{q}$ be the ideals on the left and right-hand side of (1.3.17). Both are homogeneous ideals in $S = k[Z_0, \ldots, Z_j]$ and we showed that $\mathfrak{q}$ is $\mathfrak{p}$-primary. Now, by the definition of degree (see [**Har2**, Ch.I §7]) one has

$$\deg \mathbb{P}(\mathbf{V}_s(j - s, s; 2)) = e \deg \mathbb{P}(\overline{P_s}) \qquad (1.3.19)$$

---

[3]See [**ACGH**, pp.97–99] for a similar argument.

where e is the length of the $S_\mathfrak{p}$ module $S/\mathfrak{q}$. The left-hand side of (1.3.19) equals $\binom{j-s+1}{s}$ as shown in the proof of Part (iii), so by (1.3.17) one obtains $e = 1$, hence $\mathfrak{p} = \mathfrak{q}$.

In proving (1.3.17) we first observe that

$$\deg Sec_s(C) \leq \deg \mathbb{P}(\mathbf{V}_s(j-s,s;2)) = \binom{j-s+1}{s}.$$

So, it suffices to verify that

$$\deg Sec_s(C) \geq \binom{j-s+1}{s}. \qquad (1.3.20)$$

We aim to prove this inequality by degeneration of C.[4] We consider the $2 \times j$ matrix depending on a parameter $t \in \mathbb{A}^1$

$$\Omega(t) = \begin{pmatrix} Z_0 & Z_1 & \cdots & Z_{j-2} & tZ_{j-1} \\ Z_1 & Z_2 & \cdots & Z_{j-1} & Z_j \end{pmatrix}$$

and the one-parameter family of curves in $\mathbb{P}^{j-1}$

$$C_t = \{[Z] \in \mathbb{P}^{j-1} \mid \text{rk}\,\Omega(t) \leq 1\}.$$

The curves C_t for $t \neq 0$ are all rational normal curves (see [**Ha**, pp.100–102]), and are projectively equivalent to each other. For $t = 0$ one obtains a union $C_0 = C' \cup \ell_j$ where C' is a rational normal curve in the hyperplane $Z_j = 0$ and ℓ_j is the line $Z_0 = \ldots = Z_{j-2} = 0$, $C' \cap \ell_j = [0,\ldots,0,1,0]$. Repeating the same with C' (observe that the new deformation does not move $[0,\ldots,0,1,0]$), one obtains successive degenerations

$$C \rightsquigarrow C' \cup \ell_j \rightsquigarrow \ldots \rightsquigarrow \ell_1 \cup \ldots \cup \ell_j \qquad (1.3.21)$$

where $\ell_1 \cup \ldots \cup \ell_j$ is an open chain of lines, each ℓ_i intersecting the span $\langle \ell_1, \ldots, \ell_{i-1} \rangle$ in a point of ℓ_{i-1} not belonging to $\langle \ell_1, \ldots, \ell_{i-2} \rangle$.

If we could prove that a sufficiently general $(j-2s)$-plane W meets $Sec_s(C' \cup L_j)$ in at least $\binom{j-s+1}{s}$ points, then the same would be true for each t in some neighborhood of 0 and since all C_t are projectively equivalent we would prove (1.3.20). Repeating the same argument with each degeneration of (1.3.21) we reduce the problem to the curve $X = \ell_1 \cup \ldots \cup \ell_j$. Now, the components of dimension $2s - 1$ of $Sec_s(X)$ are the projective subspaces equal to the spans $\langle \ell_{i_1}, \ldots, \ell_{i_s} \rangle$, where $i_1 < i_2 < \ldots < i_s$ and $i_\alpha + 1 < i_{\alpha+1}$ for every α. The number of these components is $\binom{j-s+1}{s}$, so a general $j - 2s$ plane W meets $Sec_s(X)$ in $\binom{j-s+1}{s}$ points. This proves (1.3.20) as discussed above. □

---

[4]This argument is contained in [**Ro**, p.221].

So far we have assumed $\mathrm{char}(k) = 0$ or $\mathrm{char}(k) > j$ when working with binary forms of degree j. We want to remove this restriction and from now on we return to our usual hypothesis that $\mathrm{char}(k)$ is arbitrary. Thus, as we discussed already in Section 1.1, we shall work with the divided power ring $\mathcal{D}$ in 2 variables instead of the ring $\mathcal{R} = k[X, Y]$ and with the contraction action $R_v \times \mathcal{D}_j \to \mathcal{D}_{j-v}$ instead of the differentiation action. An additive decomposition of a binary form as sum of powers of linear forms is replaced by an additive decomposition (AD) of a DP-form in $\mathcal{D}_j$

$$f = c_1 L_1^{[j]} + \cdots + c_s L_s^{[j]}. \qquad (1.3.22)$$

with $c_i \in k$, $i = 1, \ldots, s$. We call (1.3.22) a *normalized* AD if for every $\alpha \neq \beta$ the linear forms L_α, L_β are non-proportional to each other and all $c_i \neq 0$. Then the integer s is called the *length* of a normalized AD. We use the divided powers monomials as a basis of $\mathcal{D}_j$ and for calculating the catalecticant matrix (cf. Definition 1.2, and, for further detail, Appendix A).

Defining generalized additive decompositions analogously to (1.3.1) as sums $\sum_{i=1}^m G_i \cdot L^{[j-d_i+1]}$ is not correct, since a generalization of Lemma 1.31 would not be true without some restrictions on $\mathrm{char}(k)$ (unless all $d_i = 1$). This is because some DP-forms apolar to x^n might not be of the form $G \cdot Y^{[j-n+1]}$ because of the multiplication rule in $\mathcal{D}$: $Y^{[i]} \cdot Y^{[j-n+1]} = \binom{i+j-n+1}{i} Y^{[i+j-n+1]}$. The binomial coefficient might be 0 in k; in this case the element $Y^{[i+j-n+1]}$ apolar to x^n does not belong to $\mathcal{D}_{n-1} \cdot Y^{[j-n+1]}$ (cf. Lemma 2.2). Under the restriction $\mathrm{char}\, k > j$, however, all the previous statements, including the definition of GAD and Jordan's Lemma remain true. In the rest of the Section, we consider arbitrary characteristic.

First, we have as a particular case of the Apolarity Lemma 1.15, or by adapting the proof of Lemma 1.31, the following statement:

LEMMA 1.49. *Let $\phi = \prod_{i=1}^s (b_i x - a_i y)$ be a polynomial with simple roots. Let $L_i = a_i X + b_i Y$. Then a DP-form $f \in \mathcal{D}_j$ with $j \geq s$ has an additive decomposition (1.3.22) if and only if ϕ is apolar to f.*

We define the length $\ell(f)$ of a nonzero element $f \in \mathcal{D}_j$ to be the minimum degree of a nonzero polynomial in R apolar to f. Then the statements of Lemma 1.33 hold by definition. We do not have at our disposition Jordan's Lemma, so the following analog of Proposition 1.36 needs a different proof.

PROPOSITION 1.50. *Suppose $j = 2t$ or $2t + 1$.*

i. *Let $f \in \mathcal{D}_j$, $f \neq 0$ and let $2\ell(f) \leq j+1$. Then there is a unique (up to multiplication by elements of k^*) homogeneous form ϕ of degree $\ell(f)$ apolar to f.*

ii. *Let*

$$f = c_1 L_1^{[j]} + \cdots + c_s L_s^{[j]}$$

be a normalized AD with $s \leq t+1$. Then $\ell(f) = s$.

PROOF. (i). Let $\mathrm{Ann}(f) = I = I_d + I_{d+1} + \cdots$. By definition $\ell(f) = d$. We want to prove that $\dim I_d = 1$ provided $d \leq j-d+1$. This is equivalent to proving that the catalecticant matrix $Cat_f(j-d, d; 2)$ has a nonzero $d \times d$ minor. We saw in the proof of Theorem 1.43 that the $d \times d$ minors of $Cat_f(j-d+1, d-1; 2)$ are minors of $Cat_f(j-d, d; 2)$. Since $I_{d-1} = 0$ by assumption, all $d \times d$ minors of the former matrix are nonzero. This proves Part (i).

(ii). We use the Apolarity Lemma 1.15. If $L_i = a_i X + b_i Y$, then $\phi = \prod_{i=1}^s (b_i x - a_i y)$ is apolar to f, so $\ell(f) \leq s$. If $s' < s$ then $s \leq t+1$ implies $s \leq j - s' + 1$. A set P of s distinct points in $\mathbb{P}^1$ imposes independent conditions on $|\mathcal{O}_{\mathbb{P}^1}(j - s')|$, hence by Lemma 1.15(iv) we have $I_{s'} = (\mathcal{I}_P)_{s'} = 0$. This proves $\ell(f) = s$.   □

EXAMPLE 1.51 If $f = X^{[3]} + (X+Y)^{[3]} = 2X^{[3]} + X^{[2]}Y + XY^{[2]} + Y^{[3]}$, then $\mathrm{Ann}(f) = (y(x-y), x^3 - 2y^3)$, of order two, as shown in Proposition 1.50. In characteristic three, $\mathrm{Ann}(f) = (y(x-y), (x^3 + y^3))$. The form $(x+y)^3$ is apolar to f, but one easily checks that f cannot be written as $G \cdot (X-Y)^{[2]}$ for any linear form $G \in \mathcal{D}_1$.

The DP-form $g = X^{[4]} + Y^{[4]} + (X+Y)^{[4]}$ of length $\ell(g) = 3$ has apolar ideal $\mathrm{Ann}(g) = (xy(x-y), x^3 + y^3 - 3x^2y)$ and many additive decompositions of minimum length 3 (compare with g in Example 1.34).

We let $P_s \subset PS(s, j; 2)$ be the locus of elements in $\mathcal{D}_j$ which have an additive decomposition $f = L_1^{[j]} + \cdots + L_s^{[j]}$ with all pairs L_α, L_β being non-proportional. Replacing the expression

(*) f is a form with a GAD of length s

by

(**) f is a DP-form with $\ell(f) = s$

all proofs on pages 25–34 hold for arbitrary characteristic in the setting of the divided power ring. The statement of Theorem 1.40 should be replaced by

PROPOSITION 1.52. *Every general binary DP-form f of odd degree $2t - 1$ has a unique additive decomposition*

$$f = L_1^{[j]} + \cdots + L_t^{[j]}$$

The statements of Lemma 1.38, Theorems 1.43, 1.44, 1.45 and Remarks 1.46, 1.47 are valid for arbitrary characteristic with the following standard changes in the hypotheses: replacement of $(*)$ by $(**)$ as above; replacement of " binary form of degree j" by " binary DP-form of degree j". Let us formulate three theorems which either follow immediately or are equivalent to Theorems 1.43, 1.44 and 1.45.

THEOREM 1.53. *Let char(k) be arbitrary. Let $j = 2t$ or $2t + 1$.*

i. *Let $f \in \mathcal{D}_j$ and $s = \mathrm{rk}\, Cat_f(j - t, t; 2)$. Then $s = \ell(f)$.*

ii. *Let $2s \leq j + 1$. Then $V_s(j - t, t; 2)$ is an irreducible variety and every sufficiently general element $f \in V_s(j - t, t; 2)$ has a unique additive decomposition*

$$f = L_1^{[j]} + \cdots + L_s^{[j]}.$$

iii. *Let f and s be as in (i). For every v with $s \leq v \leq j - v + 1$ we have $\ell(f) = s = \mathrm{rk}\, Cat_f(j - v, v; 2)$.*

THEOREM 1.54. *Let k be an arbitrary field. Let f be a binary DP-form of degree j. Let A_f be the associated Gorenstein Artin algebra $A_f = k[x, y]/\mathrm{Ann}(f)$. Let $s = \max\{\dim_k(A_f)_i\}$. Then*

$$H(A_f) = (1, 2, \ldots, s - 1, \overset{s-1}{s}, s, \ldots, \overset{j-s+1}{s}, s - 1, \ldots, 2, 1).$$

The ideal $I = \mathrm{Ann}(f)$ is generated by two homogeneous polynomials $\phi \in I_s$ and $\psi \in I_{j+2-s}$.

PROOF. As mentioned above Theorem 1.44 is valid for arbitrary characteristic of the algebraically closed field $\overline{k}$ with the proper changes of hypothesis. For the case of nonclosed fields observe that for $f \in \mathcal{D}_j(k)$ the catalecticant matrices $Cat_f(j - d, d; 2)$ are with entries from k, and their ranks are the same considered over k or over $\overline{k}$. This shows the formula for $H(A_f)$. The kernels of the catalecticant homomorphisms $C_f(j - d, d)$ are also vector spaces defined over k, so one can chose $\phi, \psi \in k[x, y]$ and repeat the proof of Theorem 1.44(iv) for the ideal $I = \mathrm{Ann}(f) \subset k[x, y]$. $\square$

REMARK 1.55. In the light of the connection between Artinian Gorenstein quotients R/I and principal inverse systems f, through $I = \mathrm{Ann}(f)$ (see Lemma 2.14), the above theorem is the graded case of the frequently quoted statement in commutative algebra,

Every graded Gorenstein 0-dimensional quotient of $k[x, y]$, or Goren-stein quotient of the regular local ring $k\{x, y\}$ is a complete intersection (CI).

This fact can be deduced from the Hilbert-Burch Theorem [**Ei2**]. There is a generalization due to Serre, that "every codimension two Gorenstein ideal of a regular local ring is a CI" [**Ei2**, Corollary 21.20].

THEOREM 1.56. *Let* $\mathrm{char}(k)$ *be arbitrary. Let* $C = v_j(\mathbb{P}^1) \subset \mathbb{P}^j$ *be the Veronese rational normal curve. Let* $2s \leq j$. *Then the* s-*secant variety* $Sec_s(C)$ *is a projectively normal, arithmetically Cohen-Macaulay variety of dimension* $2s - 1$ *and degree* $\binom{j-s+1}{s}$. *Its singular locus equals* $Sec_{s-1}(C)$. *Let* $s \leq u \leq j - u$. *Consider the generic catalecticant matrix*

$$Cat_F(u, j - u; 2) = \begin{pmatrix} Z_0 & Z_1 & \dots & Z_{j-u} \\ Z_1 & Z_2 & \dots & Z_{j-u+1} \\ \vdots & \vdots & & \vdots \\ Z_u & Z_{u+1} & \dots & Z_j \end{pmatrix}.$$

The graded ideal of $Sec_s(C)$ *is equal to the ideal* $I_{s+1}(Cat_F(u, j - u; 2))$ *generated by the* $(s+1) \times (s+1)$ *minors of the matrix* $Cat_F(u, j - u; 2)$.

The following corollary was first proved by L. Gruson and C. Peskine [**GruP**, Lemma 2.3], and shown independently by J. Watanabe (in an early version of [**Wa4**]).

COROLLARY 1.57. *Let* $2s \leq j$. *Let* u, v *be integers which satisfy* $s \leq u \leq j - u$, $s \leq v \leq j - v$. *Then*

$$I_{s+1}(Cat_F(u, j - u; 2)) = I_{s+1}(Cat_F(v, j - v; 2)).$$

The next corollary is due to D. Eisenbud [**Ei1**] and also independently to J. Watanabe [**Wa4**] (Proposition 4.3 of [**Ei1**] is our Theorem 1.56; both proofs use L. Gruson and C. Peskine's result cited above).

COROLLARY 1.58. *Let* $s \leq u \leq j - u$. *Then the ideal* $I_{s+1}(Cat_F(u, j - u; 2))$ *is prime and perfect* [5]

REMARK 1.59. The minimal resolution of the graded ideal $I_{s+1}(Cat_F(u, j - u; 2))$ can be obtained from the equality

$$I_{s+1}(Cat_F(u, j - u; 2)) = I_{s+1}(Cat_F(s, j - s; 2)).$$

---

[5]i.e. *I* perfect is equivalent to R/I being a perfect R-module; when R is a polynomial ring over a field, this is equivalent to R/I being Cohen-Macaulay [**BruH**, Corollary 2.2.15].

The latter ideal is a maximal minors ideal and its minimal resolution is given by the Eagon-Northcott complex [EaN]. See Remark 3.15 where a more general case is considered.

REMARK 1.60. Some of the results of this section, in particular, Theorem 1.45 are generalized by V. Kanev to the case $t_1 = 2, r \geq 2, \mathrm{char}(k) = 0$ in [Ka]. A. Conca has recently determined the primary decomposition of the powers $I_{s+1}(Cat_F(u, j - u; 2))^t$, of the determinantal ideals of the Hankel matrix, and he has shown that the Rees ring and symbolic Rees rings associated to $I_s(u, v, 1, F)$ are Cohen-Macaulay normal domains [Con]. Conca adapts standard monomial theory, and determines Gröbner bases for the powers and symbolic powers of the determinantal ideals of the Hankel matrix.

We discuss some still open questions concerning the determinantal varieties $\mathbf{V}_s(j - v, v; 2)$ in Chapter 9.

1.4. Detailed summary and preparatory results

We now give a detailed summary of the results in this Memoir, organized under five goals corresponding to the main chapters. We also state preparatory results of previous authors that we need.

First Goal — Section 2.2. DETERMINE THE NUMBER OF MINIMUM-LENGTH ADDITIVE DECOMPOSITIONS OF A FORM. When the field k is algebraically closed, a translation after Terracini [Ter1, Ter2] of the theorem of J. Alexander and A. Hirschowitz (Theorem 1.66) on the symmetric square of vanishing ideals at points of $\mathbb{P}^{r-1}$ yields

THEOREM 1.61. DIMENSION OF $PS(s, j; r)$ (J. Alexander, A. Hirschowitz, and A. Terracini). *Let $j \geq 3$. Suppose* $\mathrm{char}(k) \nmid j$. *Then we have* $\dim PS(s, j; r) = \min(rs, \dim_k R_j)$, *except for the four triples* $(s, j, r) = (5, 4, 3), (9, 4, 4), (14, 4, 5),$ *and* $(7, 3, 5)$ *where the dimension is one less, equal to* $\dim_k R_j - 1$.

This theorem is deduced from the Alexander-Hirschowitz Theorem (Theorem 1.66) in [I7] in the case of ordinary powers when $\mathrm{char}(k) = 0$ or $\mathrm{char}(k) > j$, and in the more general case of divided powers when $\mathrm{char}(k) \nmid j$ in Section 2.1. As a corollary, using that $\dim_k R_j = \binom{j+r-1}{r-1}$ and comparing $\dim PS(s, j; r)$ with $\dim_k R_j$, one obtains a solution to Waring's problem for general forms.

COROLLARY 1.62. WARING'S PROBLEM FOR GENERAL FORMS. *Suppose $j \geq 3$, $\mathrm{char}(k) = 0$ or $\mathrm{char}(k) > j$. Then a sufficiently general* homogeneous form $f(X_1, \ldots, X_r)$ of degree j can be represented as a sum of s powers of linear forms

$$f = L_1^j + \cdots + L_s^j$$

where $s = \lceil \frac{1}{r} \binom{j+r-1}{r-1} \rceil$ except in the cases $(j,r) = (4,3), (4,4), (4,5)$ where one needs $s = 6, 10, 15$ respectively, and in the case $(j,r) = (3,5)$ where one needs $s = 8$.

The same statement is valid for representation of a general DP-form $f \in \mathcal{D}_j$ as a sum of divided powers

$$f = L_1^{[j]} + \cdots + L_s^{[j]}$$

provided $\mathrm{char}(k) \nmid j$.

We also give in Theorem 2.6 a concise proof of Theorem 1.61 in case (j, s) satisfies

$$s \leq \binom{t+r-1}{r-1} - r + 1 \quad \text{if} \quad j = 2t \qquad \text{and}$$

$$s \leq \binom{t+r-1}{r-1} \qquad \text{if} \quad j = 2t+1$$

which are the cases we mostly use. The original proof of J. Alexander and A. Hirschowitz comprises over 100 pages [AlH1, AlH2, AlH3], so a short proof in these special cases may be of some interest, despite their previous result [6]. Our proof works in arbitrary characteristic, and within the required upper bounds on s yields the dimension of $PS(s, j; r)$ also for $\mathrm{char}(k)|j$, a case not covered by Theorem 1.61. The proof of Theorem 2.6 uses the Apolarity Lemma 1.15 as well as the General Position Theorem for curves, extended by D. Laksov to arbitrary characteristic $p \geq 0$ [Lak].

REMARK 1.63. We do not know whether the dimension formula for $PS(s, j; r)$ in Theorem 1.61 is valid if $\mathrm{char}(k)|j$, except within the above limits on s when our Theorem 2.6 applies.

We show in Section 2.2 the following theorem for uniqueness of additive decomposition (Theorem 2.10).

THEOREM. Let $j \geq 3$. Let k be an arbitrary infinite field of characteristic 0 or $\mathrm{char}(k) > j$. When $r = 2$ suppose $2s \leq j + 1$. When $r \geq 3$ suppose $s \leq \binom{t+r-1}{r-1} - r$ if $j = 2t$ and $s \leq \binom{t+r-1}{r-1}$ if $j = 2t + 1$. Then there is a Zariski open subset in $\mathcal{R}_1(k) \times \cdots \times \mathcal{R}_1(k)$ such that for

---

[6]They give a shorter proof in [AlH4]; a brief proof based on the original proof of J. Alexander and A. Hirschowitz has been given by K. Chandler [Cha3]

*any s-tuple of linear forms $L_1, \ldots, L_s$ from this set the representation
of the degree-j form*

$$f = c_1 L_1^j + \cdots + c_s L_s^j, \quad all \quad c_i \neq 0$$

*is the unique representation as sum of powers of $\leq s$ linear forms.
Here uniqueness is understood up to the transformations $c_i L_i^j = (c_i c^{-j})(c L_i)^j$ and up to permutation of the L_i, $i = 1, \ldots, s$. The same
statement holds for arbitrary infinite fields, if one replaces ordinary
powers by divided powers.*

Second Goal — Chapter 3. DETERMINE THE TANGENT SPACES TO
THE SCHEME $\mathbf{Gor}(T)$ AND TO THE DETERMINANTAL SCHEMES
$\mathbf{V}_s(u, v; r)$ OF CATALECTICANT MATRICES.

We will first show, concerning the catalecticant schemes, a simple
but important statement with many applications.

THEOREM 3.2. *Suppose that the closed point f_0 of the scheme
$\mathbf{V}_s(u, v; r)$ is not in $\mathbf{V}_{s-1}(u, v; r)$, and that $I = \mathrm{Ann}(f_0)$ is the corresponding ideal of R. Then the tangent space $\mathcal{T}_{f_0} \mathbf{V}_s(u, v; r) \subset \mathcal{D}_j$ to
the scheme $\mathbf{V}_s(u, v; r)$ at the point f_0 satisfies*

$$\mathcal{T}_{f_0} \mathbf{V}_s(u, v; r) = \{f \in \mathcal{D}_j \mid (I_u \cdot I_v) \circ f = 0\}. \tag{1.4.1}$$

We illustrate the use of this theorem by several examples: 3.6,
3.7, 3.8. In particular we show that when r = 3, the decomposition
of $A(\mathcal{D}_6)$ into a disjoint union of $Gor(T)$ does not satisfy the frontier
property (cf. Remark 1.46 for the binary case).

We also use (1.4.1) to show the following theorem for $\mathbf{Gor}(T)$.
If W is a vector subspace of R_j, then by $W^\perp \subset \mathcal{D}_j$ we denote its
perpendicular space in the exact pairing $R_j \times \mathcal{D}_j \to k$ of Definition 1.1.

THEOREM 3.9. *Let f be a closed point of $\mathbf{Gor}(T) \subset \mathbb{A}(\mathcal{D}_j)$. Let
$I = \mathrm{Ann}(f) \subset R$. Then the tangent space $\mathcal{T}_f = \mathcal{T}_f(\mathbf{Gor}(T))$ is given
by*

$$\mathcal{T}_f = ((I^2)_j)^\perp \subset \mathcal{D}_j. \tag{1.4.2}$$

This is related to a well known fact concerning the tangent space
$\mathcal{T}_{z_A} \mathbf{H} = \mathrm{Hom}(I, A)$ to the punctual Hilbert scheme $\mathbf{H} = \mathbf{Hilb}^n(\mathbb{A}^r)$ at
a point z_A parameterizing the Gorenstein scheme $\mathrm{Spec}(A)$, $A = A_f = R/\mathrm{Ann}(f)$, concentrated at the origin. Then

$$\mathcal{T}_{z_A} \mathbf{H} \cong I/I^2. \tag{1.4.3}$$

Theorem 3.9 is a basic tool of the sequel. We also give several direct
applications of Theorem 3.9. The first is to extremal Gorenstein Artin

algebras $A = R/I$. Introduced by P. Schenzel in [**Sch1**], these are Gorenstein Artin algebras whose socle degree $j(A)$ and order $\nu = \nu(I) = \min\{d \mid I_d \neq 0\}$ of defining ideal satisfy $j = 2(\nu - 1)$ (see Section 2.3 for the definitions). Compressed Gorenstein algebras are those having a maximal Hilbert function $H(j, r)$, given r and the socle degree j (see [**EmI1**], [**Gre**], [**FL**], and [**I3**]). One consequence of Theorem 3.9 is a concise proof that when $j = 2t$, the compressed Gorenstein algebras have Hilbert function

$$H(j, r)_i = \min(\dim_k R_i, \dim_k R_{j-i}), \qquad (1.4.4)$$

so they are extremal. This is equivalent to showing that the square catalecticant matrix $Cat_f(t, t; r)$ in general has nonzero determinant (Proposition 3.12). (Thus, $H(4, 3) = (1, 3, 6, 3, 1)$ is compressed and extremal, while $H(5, 4) = (1, 3, 6, 6, 3, 1)$ is compressed only). Another consequence is a conceptual proof that the determinant of the square catalecticant is irreducible (Proposition 3.13). Both these results were known.

Using the same technique we obtain that for $u \leq v$ the corank one catalecticant schemes $\mathbf{V}_{r_u-1}(u, v; r)$ (where $r_u = \dim_k R_u$) are reduced, irreducible, Cohen-Macaulay and normal (Theorem 3.14). In the case of binary forms this fact is one of the main ingredients of the proof of Theorem 1.56.

When $f = L_1^{[j]} + \cdots + L_s^{[j]}$ is a general sum of s divided powers, we relate the tangent spaces $T_f(\mathbf{Gor}(T))$ to $(\mathcal{I}_Z^2)_j$ where $\mathcal{I}_Z$ is the ideal in R defining the set of points $Z \subset \mathbb{P}(\mathcal{D}_1) = \mathbb{P}^{r-1}$ corresponding to $L_1, \ldots, L_s$. The key Lemma 3.16 assembles most of the information connecting $I = \mathrm{Ann}(f)$, I^2 and $\mathcal{I}_Z$, $\mathcal{I}_Z^2$ needed in the proofs of Chapter 4. Many of our results for $r = 3$ could be generalized to arbitrary r, if we knew

CONJECTURE 3.25. *Suppose that* $t \geq 2, \dim_k R_{t-1} \leq s < \dim_k R_t$, *let* $s = \dim_k R_{t-1} + a$, *let* $s^\vee = \dim_k R_t - s$, *and suppose that* $\mathcal{I}_Z$ *is the ideal of a scheme* Z *of* s *general points of* $\mathbb{P}^{r-1}$. *Then*

$$H(R/\mathcal{I}_Z^2) = \left(1, r, \ldots, r_{2t-1}, r_{2t} - \binom{s^\vee + 1}{2}, \right.$$

$$\left. \dim_k R_{2t+1} - s^\vee(\dim_k R_{t+1} - s) + r\binom{s^\vee}{2}, rs, rs, \ldots \right).$$

We prove Conjecture 3.25 when $r = 3$ (Proposition 4.8), and when $s^\vee \leq r$ (Lemma 4.12, Theorem 4.19). In degrees $i \leq 2t - 1$ and $i \geq 2t + 2$ Conjecture 3.25 follows from the Alexander-Hirschowitz

Theorem 1.66 and Corollary 1.71. Computer calculations for $r = 4, 5, 6$ support Conjecture 3.25 (see Examples 3.19, 3.22).

Third Goal — Chapter 4. Determine the dimension of the variety $U_s(t, t; r)$ parameterizing degree $j = 2t$ forms f having s linearly independent t-th higher derivatives when $r = 3$. Determine the relation between $PS(s, j; r)$ and $Gor(T), T = H(s, j; r)$, and the dimension(s) of the component(s) of $V_s(u, v; r)$ containing $PS(s, j; r)$.

We first let $r = 3, j = 2t$ or $2t + 1$, and denote by $s^\vee$ the corank of the catalecticant matrix $Cat_f(t, j - t; r)$, $r = 3$ when s is the rank: $s^\vee = \dim_k R_t - s$. S. J. Diesel had shown,

THEOREM 1.64. Codimension of $Gor(T), T = H(s, j, 3)$ when $s^\vee$ is small (S.J.Diesel, Theorem 4.4 of [**Di**]). *When* $r = 3, j = 2t$, *and* $s^\vee \leq 2t/3 + 1$, *and* $T = H(s, j, 3)$, *then* $Gor(T)$ *is an irreducible component of* $V_s(t, t; 3)$ *having codimension in* $\mathbb{A}(\mathcal{D}_j)$ *given by*

$$\text{cod}(Gor(T)) = \binom{s^\vee + 1}{2}. \tag{1.4.5}$$

She conjectured the following generalization for $r = 3$.

CONJECTURE 1.65. Codimension of determinantal loci of catalecticants *(See* [**Di**]*). The determinantal locus satisfies*

i. $\text{cod}(V_s(t, t; 3)) = \binom{s^\vee + 1}{2}$, *for* $s \geq \dim_k R_{t-1}$;

ii. $\dim(V_s(t, t; 3)) = 3s$ *for* $s \leq \dim_k R_{t-1}$.

$$\tag{1.4.6}$$

This conjecture as stated is false as shown recently by Y. Cho and B. Jung: see below Example 4.27 (from [**I9**, Examples 7C,8]) and also Example 4.28 from [**ChoJ2**]). We prove (Theorems 4.1A, 4.1B),

THEOREM. *The locus* $V_s(t, t; 3)$ *contains an irreducible component* $Gor(T), T = H(s, j, 3)$ *having the dimension given by* (1.4.6). *When* $s \leq s_0 = \dim_k R_{t-1}$ *then* $\overline{PS(s, j; 3)} = \overline{Gor(T)}$. *When* $s > s_0, \overline{Gor(T)}$ *is the unique irreducible component of* $V_s(t, t; 3)$ *containing* $PS(s, j; 3)$.

We also prove some weaker statements about the catalecticant varieties $V_s(t, j - t; r)$, $j = 2t$ or $2t + 1$, for arbitrary r, when either $s \leq \dim_k R_{t-1}$ or $s^\vee \leq r$ (Theorems 4.10A, 4.13, 4.19). As mentioned above, we further show that when r is arbitrary, $j = 2t$ or $2t + 1$ and $s \leq \dim_k R_{t-1}$ the Zariski closure $\overline{PS(s, j; r)}$ is an irreducible component of the closure $\overline{Gor(T)}$, $T = H(s, j, r)$, with equality $\overline{PS(s, j; 3)} =$

$\overline{Gor(T)}$ when $r = 3$; and the scheme $\mathbf{Gor}(T)$ is smooth along an open dense subset of $PS(s, j; r)$ (Theorems 4.1A, 4.5A, 4.10A).

Suppose that $r = 3, j = 2t + 1, s_0 = \dim_k R_{t-1}, a = s - s_0, 0 \le a \le t + 1$, and $T = H(s, j, r)$. If f is a sum of s divided powers of general linear forms, we show,

THEOREM 4.5B.

$$3s + \lfloor a/2 \rfloor \le \dim Gor(T) \le \dim_k T_f = 3s_0 + \frac{a(a+5)}{2}$$

Since the 1996 manuscript [IK] of our work was circulated, J. O. Kleppe has shown the smoothness of the scheme $\mathbf{Gor}(T)$ for all Gorenstein sequences T when $r = 3$ (see [Kl2]) following our partial results and those of A. Geramita, M. Pucci, and Y. S. Shin. We report on these developments — including several new dimension formulas, for $Gor(T)$ when $r = 3$ — by J. O. Kleppe, A. Conca and G. Valla, and Y. Cho and B. Jung, and others [Kl2, KlM-R, CoV1, ChoJ1, ChoJ2] in Section 4.4. We have retained our partial results such as Theorem 4.5B when $r = 3$, because of the close connection to our development of some of these results to the cases $r \ge 3$ in Section 4.2, and because we emphasize beyond dimension an aspect of the component structure of $V_s(u, v; r)$: that $PS(s, j; r)$ belongs to a unique irreducible component. This requires a calculation of the tangent space to $f \in PS(s, j; r)$, which, when $r > 3$ — where there is no structure theorem for Gorenstein ideals — requires a study of points in $\mathbb{P}^{r-1}$, and of the mutliplication map $\text{Sym}^2(I_t) \longrightarrow (I^2)_j$, in the case $j = 2t$, and of an analogous multiplication map when $j = 2t + 1$ (see Lemma 4.12).

Tools that we use include the Buchsbaum-Eisenbud structure theorem for height three Gorenstein ideals [BE2], the result of S.J. Diesel [Di] that $Gor(T)$ is irreducible when $r = 3$ (see Theorem 1.72 below), and the minimal resolution for I^2 for I height three Gorenstein shown by A. Kustin and B. Ulrich, and by G. Boffi and R. Sanchez [KusU, BofS], the latter to show results such as Theorem 4.5B for $r = 3$. The more recent work on the smoothness and dimension formulas cited above use also J. O. Kleppe's homologically proven smoothness result - based on earlier work with R. M. Miró-Roig [KlM-R], and either the Kustin-Ulrich-Boffi-Sanchez minimal resolutions for I^2, or another for the exterior power due to K. Lebelt and J. Weyman (see [Kl2, We1]).

More generally, let $\mathcal{I}_P$ in R be the graded ideal defining the set of s general points P in $\mathbb{P}^{r-1}$, and let $\mathcal{I}_P^{(2)}$ denote the symbolic square.

We use frequently the following interpolation result of Alexander and Hirschowitz, usually when s satisfies the conditions of Theorem 2.6.[7] Under these restrictions we give in Theorem 2.6(iii) a simple proof of the Alexander-Hirschowitz Theorem when $\mathrm{char}(k) = 0$.

THEOREM 1.66 (J. Alexander and A. Hirschowitz). [8] POLYNOMI- ALS VANISHING TO ORDER 2 AT A GENERAL SET OF POINTS: HILBERT FUNCTION OF $R/\mathcal{I}_P^{(2)}$. *Suppose that k is an infinite field, that $P = \{p(1), ..., p(s)\}$ is a sufficiently general set of points in $\mathbb{P}^{r-1}$, and that the integer $j \geq 3$. Then $(\mathcal{I}_P^{(2)})_j$ — the j-th homogeneous component of the ideal $\mathcal{I}_P^{(2)}$, consisting of polynomials vanishing to order 2 at the points of the set P, has codimension $\min\left(rs, \binom{j+r-1}{r-1}\right)$ in R_j, except for the following four exceptional triples (s, j, r)*

$$(5, 4, 3), \quad (9, 4, 4), \quad (14, 4, 5), \quad (7, 3, 5)$$

for which the codimension is $\binom{j+r-1}{r-1} - 1$.

Our first proof of Theorem 4.1A depended on showing that if $r = 3$ and $s = s_0 (= \dim_k R_{t-1})$, then

$$(\mathcal{I}_P)^2 = \mathcal{I}_P^{(2)} \cap M^j, \tag{1.4.7}$$

where M is the maximal ideal of R. Our proof used a smoothing result of H. Kleppe for height three Gorenstein Artin ideals, and we used it to determine $I^2, I = \mathrm{Ann}(f)$, in terms of $\mathcal{I}_P$ and $\mathcal{I}_P^2$ if f is a general sum of s_0 powers of linear forms (Theorem 4.1C).

Subsequently to proving (1.4.7) for $r = 3$, we learned that A. Geramita and B. Harbourne had announced a more general asser- tion than (1.4.7): they showed that if $r = 3$ and $s = \dim_k R_{t-1}$, then $\mathcal{I}_P^a = \mathcal{I}_P^{(a)} \cap M^{at}$, where $\mathcal{I}_P^{(a)}$ is the ideal of homogeneous polynomials of R vanishing to orders a at P. We use here a further generaliza- tion by A. Geramita, A. Gimigliano, and Y. Pitteloud [GGP] and by K. Chandler [Cha1]. To state it we need some further notation, and some well known facts.

Recall that $M = (x_1, \dots, x_r) \subset R$. If $\mathcal{I}$ is a graded ideal of R, we let $\mathcal{I} : M$ denote the ideal $\mathcal{I} : M = \{f \in R \mid Mf \subset \mathcal{I}\}$. A graded ideal $\mathcal{I}$ of R is *saturated* if $\mathcal{I} : M = \mathcal{I}$. Let $\mathcal{I}$ be a graded ideal. Let

---

[7] The exception is Theorem 4.13 which require cases of Theorem 1.66 not shown in Theorem 2.6.

[8] See [Al] for degrees $j \geq 5$ and degrees 3 and 4 for $r \leq 5$, [AlH2] for degree 4 and $r > 5$, and [AlH3] for degree 3 when $r \geq 5$. See also [AlH4, Cha3].

$Z = \mathrm{Proj}(R/\mathcal{I})$. One has a standard homomorphism

$$\phi \; : \; R \longrightarrow \oplus_{n\geq 0} H^0(Z, \mathcal{O}_Z(n)). \tag{1.4.8}$$

Then $\mathcal{I}$ is saturated if and only if $\mathcal{I} = \mathrm{Ker}\,\phi$. In this case we write $\mathcal{I} = \mathcal{I}_Z$. If $\mathcal{I}$ is an arbitrary graded ideal its *saturation* $\mathrm{Sat}(\mathcal{I})$ is defined to be the kernel of ϕ [**Har2**, II Ex. 5.10]. Explicitly $\mathrm{Sat}(\mathcal{I}) = \oplus_{n\geq 1} \mathrm{Sat}(\mathcal{I})_n$ where

$$\mathrm{Sat}(\mathcal{I})_n \; = \; \left\{ \phi \in R_n \mid M^{N(\phi)}\phi \subset \mathcal{I} \text{ for some } N(\phi) \gg 0 \right\}.$$

As an exception for this section, we give proofs of the following Lemma and Theorem 1.69 below for the convenience of the reader and because of their importance for us. Recall that the *depth* of a local ring A_m is the common length of the maximal A_m-sequences in m [**BruH**, Def. 1.2.7].

LEMMA 1.67. *Let $\mathcal{I} \subset R = k[x_1,\ldots,x_r]$ be a graded ideal whose quotient $R/\mathcal{I}$ has Krull dimension one. Let $Z = \mathrm{Proj}\,R/\mathcal{I}$ be the corresponding 0-dimensional scheme.*

  i. *$R/\mathcal{I}$ is Cohen-Macaulay (CM) if and only if $\mathcal{I}$ is saturated, i.e. $\mathcal{I} = \mathcal{I}_Z$.*

  ii. *If $R/\mathcal{I}$ is CM then every general enough element $x \in (R/\mathcal{I})_1$ is a nonzerodivisor of the ring $R/\mathcal{I}$.*

PROOF. (i). $A = R/\mathcal{I}$ is CM iff for every maximal ideal $\mathfrak{m}$ of A the depth of the localization $A_\mathfrak{m}$ is one. In particular this holds for $\mathfrak{m}_0 = M/\mathcal{I}$. It is proved in [**Mum**, Lecture 13, Appendix] that $A_{\mathfrak{m}_0}$ has depth ≥ 1 if and only if

$$\phi \; : \; R/\mathcal{I} \longrightarrow \oplus_{n\geq 0}\Gamma(Z, \mathcal{O}_Z(n))$$

is injective. Thus A being CM implies $\mathcal{I} = \mathcal{I}_Z$. Conversely, if $\mathcal{I}$ is saturated, then

$$1 \leq depth\,(A_{\mathfrak{m}_0}) \leq \dim(A_{\mathfrak{m}_0}) = 1.$$

So, $depth\,(A_{\mathfrak{m}_0}) = 1$. Let z be an arbitrary closed point of $\mathrm{Spec}\,R/\mathcal{I} \subset \mathbb{A}^r$. Let $\mathfrak{m} = \mathfrak{m}_z$. Let $\bar{z}$ be the corresponding point of $Z = \mathrm{Proj}\,R/\mathcal{I}$. If $z = (z_1,\ldots,z_r)$ has a non-zero coordinate z_i, then $A_{\mathfrak{m}_z} = \mathcal{O}_{\bar{z},Z}[x_i, x_i^{-1}]$. The ring $\mathcal{O}_{\bar{z},Z}$ is Artinian, so CM, hence $\mathcal{O}_{\bar{z},Z}[x_i]$ is also CM [**Ma1**, Theorem 33], therefore the localization $\mathcal{O}_{\bar{z},Z}[x_i, x_i^{-1}]$ is CM as well.

(ii). Let $\mathfrak{p}_1,\ldots,\mathfrak{p}_m$ be the associated primes of (0) in A; the zero-divisors of A are $\cup\mathfrak{p}_i$. Then $depth\,A_{\mathfrak{m}_0} = 1$ implies $\mathfrak{m}_0 \neq \mathfrak{p}_i$ for $\forall i$. Since $\mathfrak{m}_0$ is generated by A_1 this is equivalent to $A_1 \subsetneq \mathfrak{p}_i$ for $\forall i$. This implies that each element of the Zariski open set $A_1 - \cup_{i=1}^m \mathfrak{p}_i$ is a nonzerodivisor of $A = R/\mathcal{I}$. $\square$

DEFINITION 1.68. REGULARITY AND SATURATION DEGREES. Let Z be a 0-dimensional subscheme of $\mathbb{P}^{r-1}$ and let $\mathcal{I} = \mathcal{I}_Z$ be as above. Let $\tilde{\mathcal{I}}$ be the ideal sheaf of $\mathcal{I}$. Then $\mathcal{I}$ is m-regular iff

$$H^1(\mathbb{P}^{r-1}, \tilde{\mathcal{I}}(t-1)) = 0 \quad \text{for} \quad t \geq m$$

Compared with the definition in [**Mum**, Lecture 14] one needs also $H^i(\mathbb{P}^{r-1}, \tilde{\mathcal{I}}(t-i)) = 0$ for $t \geq m$, $i \geq 2$. This holds since $\dim Z = 0$, as one verifies easily considering the long exact sequence of cohomology associated with (1.4.9) below. The *Castelnuovo–Mumford regularity* or *m-regularity* is the integer

$$\sigma(\mathcal{I}) = \min\{m \mid \mathcal{I} \text{ is } m - \text{regular}\}$$

(see [**Mum**, p. 99], [**BaM**, §3]). If L is a general element of R_1 then $m_L : (R/\mathcal{I})_i \longrightarrow (R/\mathcal{I})_{i+1}$, the multiplication by L, is an injective linear map by Lemma 1.67. We define

$$\tau(\mathcal{I}) = \min\{i \mid m_L \text{ is surjective to } (R/\mathcal{I})_{i+1}\}$$
$$= \min\{i \mid H(R/\mathcal{I})_i = s\}$$

where $s = \max_{i \in N}(\dim_k(R_i/\mathcal{I}_i))$.

If $\mathcal{I}$ is such an ideal of R, then $\mathcal{I}^{(a)}$ denotes the a-th symbolic power, which is also the saturation of $\mathcal{I}^a$. The *saturation degree* $n(\mathcal{I}^{(a)})$ of $\mathcal{I}^a$ is the smallest degree after which $\mathcal{I}^a$ equals $\mathcal{I}^{(a)}$:

$$n(\mathcal{I}^a) = \min\{\ell \mid i \geq \ell \Rightarrow (\mathcal{I}^a)_i = (\mathcal{I}^{(a)})_i\}.$$

The following theorem collects several results about graded ideals of 0-dimensional schemes. We borrowed its various parts from: A. Geramita and P. Maroscia [**GM**], F. Orecchia [**Or**], and [**I1**, Lemma 1.1]. The portion on Castelnuovo–Mumford m-regularity is a particularly simple case of [**Mum**, **EiG**].

THEOREM 1.69. PROPERTIES OF IDEALS OF ZERO-DIMENSIONAL SUBSCHEMES. *Let Z be a 0-dimensional subscheme of $\mathbb{P}^{r-1}$, let $\mathcal{I} = \mathcal{I}_Z$. Let $s = \deg(Z)$. Then the Hilbert function $H(R/\mathcal{I})_i$ is nondecreasing in i, and stabilizes at the value $s = H(R/\mathcal{I})_i$ for $i \geq \tau(\mathcal{I})$. The Castelnuovo–Mumford regularity $\sigma = \sigma(\mathcal{I})$ satisfies $\sigma = \tau(\mathcal{I}) + 1$. If $i \geq \sigma$ then $\mathcal{I}_i = R_{i-\sigma}\mathcal{I}_\sigma$. We have $\tau(\mathcal{I}) \leq s - 1$, and this bound is sharp.*

PROOF. By Lemma 1.67(ii) there is an element $x \in (R/\mathcal{I})_1$ which is a nonzerodivisor of $R/\mathcal{I}$. Considering the multiplication map $m_x : (R/\mathcal{I})_i \to (R/\mathcal{I})_{i+1}$ this shows that $H(R/\mathcal{I})$ is nondecreasing; since $s = \deg Z = \deg \text{Proj}(R/\mathcal{I})$ is the Hilbert polynomial of $R/\mathcal{I}$, the Hilbert function $H(R/\mathcal{I})$ stabilizes at s. Finally, from the standard exact sequence of sheaves

$$0 \longrightarrow \tilde{\mathcal{I}}(d) \longrightarrow \mathcal{O}_{\mathbb{P}^{r-1}}(d) \longrightarrow \mathcal{O}_Z(d) \longrightarrow 0 \qquad (1.4.9)$$

one obtains $\dim_k(R_d/\mathcal{I}_d) = s - H^1(\mathbb{P}^{r-1}, \tilde{\mathcal{I}}(d))$, and the regularity

$$\sigma(\mathcal{I}) \;=\; \min\{i \mid H(R/\mathcal{I})_i = s\} + 1.$$

That $\mathcal{I}_i = R_{i-\sigma}\mathcal{I}_\sigma$ when $i \geq \sigma$ is a particular case of a general fact proved in [**Mum**, Lecture 14] (see also [**GM**, p.530] and Proposition C.21 of Appendix C).

It remains to prove that $\tau(\mathcal{I}) \leq s - 1$ and that this bound is sharp. Let $x \in (R/\mathcal{I})_1$ be a nonzerodivisor, and consider the quotient $A = (R/\mathcal{I})/(x)$, an Artin algebra satisfying $\dim_k A = s$. Denote by $\mathfrak{m}$ its maximal ideal, the image of $M = (x_1, \ldots, x_r)$ in A. Then $\mathfrak{m}^s = 0$, or equivalently $M^s \subset \mathcal{I} + Rx$. Indeed, suppose, by way of contradiction that $\mathfrak{m}^s \neq 0$. Then we would have $\mathfrak{m}^i \supsetneq \mathfrak{m}^{i+1}$ for every i, $1 \leq i \leq s$, since otherwise, by induction, $\mathfrak{m}^i = \mathfrak{m}^N$ for every $N > i$, hence $\mathfrak{m}^i = 0$, since $\mathfrak{m}^N = 0$ for $N \gg 0$ ($\dim_k A < \infty$), thus contradicting the assumption $\mathfrak{m}^s \neq 0$. Now, we would have

$$\dim_k A = 1 + \left(\sum_{i=1}^{s} \dim_k(\mathfrak{m}^i/\mathfrak{m}^{i+1}) \right) + \dim_k \mathfrak{m}^{s+1} \geq s + 1,$$

a contradiction. Thus $H(A)_i = 0$ for $i \geq s$. Since $H(A) = \Delta H(R/\mathcal{I})$ (i.e. $H(A)_i = H(R/\mathcal{I})_i - H(R/\mathcal{I})_{i-1}$), this implies $H(R/I)_i = s$ for $i \geq s - 1$.

The graded ideal of a thick point Z_0 in a line with ideal $\mathcal{I}_{Z_0} = (x_1, \ldots, x_{r-2}, x_{r-1}^s)$ has a quotient with Hilbert function

$$H(R/\mathcal{I}_{Z_0}) \;=\; (1, 2, \ldots, s-1, \overset{s-1}{\widehat{s}}, s, \ldots)$$

with $\tau(\mathcal{I}_{Z_0}) = s - 1$. This example shows that the bound $\tau(\mathcal{I}) \leq s - 1$ is sharp. $\qquad \square$

REMARK. An algebraic definition of Castelnuovo–Mumford regularity, equivalent to that given above is the following ([**EiG**, Theorem 1.2] or [**Ei2**, p.505]). Suppose that the saturated ideal I of R has minimal resolution

$$0 \to E_r \to \cdots \to E_0 = R \to I \to 0$$

and let b_i be the maximum degree of E_i; then $\sigma(I) = \max\{b_i - i\}$.

The following result is shown by A. Geramita, A. Gimigliano, and Y. Pitteloud , [**GGP**] and by K. Chandler. That $\sigma(\mathcal{I}^a) \leq a\sigma(\mathcal{I})$ is false for ideals $\mathcal{I} = \mathcal{I}_Z$ when $\dim Z > 0$ — as shown first by Terai, then by B. Sturmfels, see [**Stu**, §1].

THEOREM 1.70. SATURATION DEGREE OF POWERS (A. Geramita, A. Gimigliano, Y. Pitteloud ((i) for $\mathcal{I}$ saturated is Corollary 1.4 of [**GGP**]); (i),(ii) K. Chandler [9] (Theorem 8 and Corollary 7 of [**Cha1**]).

i. *Let $\mathcal{I}$ be a graded ideal of R such that the Krull dimension $\dim(R/\mathcal{I}) = 1$. Let $\sigma = \sigma(\mathcal{I})$, and let L be an element of R_1 which is not a zero-divisor on $R/\mathcal{J}, \mathcal{J} = \mathrm{Sat}(\mathcal{I})$. The multiplication map $(R/\mathcal{I}^a)_t \xrightarrow{L} (R/\mathcal{I}^a)_{t+1}$ is an isomorphism for $t \geq a\sigma$ and a surjection for $t = a\sigma - 1$;. Furthermore, $n(\mathcal{I}^a) \leq a\sigma(\mathcal{I})$ and $\sigma(\mathcal{I}^a) \leq a\sigma(\mathcal{I})$.*

ii. *Assume that $\mathcal{I}$ is saturated, generated by degree e, and regular in degree $\sigma = e + \delta$. Then $\mathcal{I}^{(a)}$ is regular in degree $ae + \delta$.*

COROLLARY 1.71. SATURATION DEGREE OF $\mathcal{I}_P^2$. *If $s \leq \dim_k R_{t-1}$, if $P = \{p_1, \cdots, p_s\}$ are s general points of $\mathbb{P}^{r-1}$, and $\mathcal{I}_P$ is the graded ideal in R of polynomials which vanish at P, then $n(\mathcal{I}_P^2) \leq 2t$.*

PROOF. It is known (see Lemma 1.19 above, or [**GM**]) that for s general enough points $P \subset \mathbb{P}^{r-1}$, $\tau(\mathcal{I}_P) = \min\{i \mid s \leq \dim_k R_i\}$, so here $\tau(\mathcal{I}_P) \leq t-1$. Thus $\sigma(\mathcal{I}_P) = \tau(\mathcal{I}_P)+1 \leq t$, and by Theorem 1.70, $n(\mathcal{I}_P^2) \leq 2t$. □

These results are components in our proofs of Theorems 4.1A–4.1C, 4.5A, 1.61, 4.10A, 4.10B and 4.13.

In Section 4.3 we study those Gorenstein ideals whose initial generators form a complete intersection, following an approach of J. Watanabe [**Wa2**]. This allows us to extend Theorem 4.13 to corank r, obtaining

THEOREM 4.19. AN IRREDUCIBLE COMPONENT OF CERTAIN DE-TERMINANTAL LOCI, ARISING FROM A COMPLETE INTERSECTIONS AND A GENERIC FORM. *Suppose $r \geq 3$, $t \geq 2$, $0 \leq v \leq t - 1$. Let $r_t - r \leq s < r_t$. Put $s^\vee = a = r_t - s$. Let $j = 2t + v$; if $s^\vee = r$, suppose that $j \leq r(t - 1)$. Suppose $\mathrm{char}(k) = 0$ or $\mathrm{char}(k) > j$. Consider the variety C_T constructed as in Theorem 4.17 with $d_1 = \cdots = d_a = t$.*

---

[9]K. Chandler had first obtained a slightly weaker result, that $\mathcal{I}^a$ is $a\sigma(I) - 1$ regular; then Geramita, Gimigliano, and Pitteloud obtained the sharper bound for $\mathcal{I}$ saturated [**GGP**]; K. Chandler then improved her method in [**Cha1**] to obtain the result stated.

Then $\overline{C_T}$ is an irreducible component of $V_s(t, t+v, r)$, having dimension

$$\dim(C_T) = \dim_k R_j - s^\vee(\dim_k R_{t+v} - s) + (\dim_k R_v)(s^\vee)(s^\vee - 1)/2,$$

and the scheme $\mathbf{V}_s(t, t+v, r)$ is smooth along an open dense subset of $\overline{C_T}$. The projective variety $\mathbb{P}(\overline{C_T})$ is rational. If $s^\vee < r$, then $\overline{C_T}$ is the unique irreducible component of $V_s(t, t+v, r)$ which contains $PS(s, j; r)$.

We say little concerning the connection between $\overline{PS(s, j; 3)}$ and $\overline{Gor(T)}$ when both $r \geq 4$, and $s > \dim_k R_{t-1}$, but $s^\vee > r$; the gap here is in our knowledge of the Hilbert function $H(R/\mathcal{I}_P^2)$ in degrees $2\tau(\mathcal{I}_P)$ and $2\tau(\mathcal{I}_P) + 1$ (see Conjecture 3.25). When the degree $j \geq 2\sigma(\mathcal{I}_P)$ we have by Theorems 1.66, 1.69, and 1.70 that $H(R/\mathcal{I}_P^2)_j = rs$ in the cases we consider.

Fourth Goal — Chapters 5, 6. Using a "fibration" to a Hilbert scheme, describe the closure of $PS(s, j; 3)$. Show that for suitable T, the variety $Gor(T)$ has several irreducible components when $r \geq 5$.

S.J. Diesel showed

Theorem 1.72 (S. J. Diesel [Di]). Irreducibility of $Gor(T)$ when $r = 3$. *When $r = 3$, and T is a Gorenstein sequence, then $Gor(T)$ is irreducible. An open dense set of $Gor(T)$ consists of those Gorenstein ideals having the smallest possible number of generators, given T.*

In Section 5.1 we introduce the key concept of "annihilating scheme" of f - a zero-dimensional subscheme of $\mathbb{P}^{r-1}$ whose defining ideal is included in $\mathrm{Ann}(f)$ — and we show some of its properties (Lemma 5.3). We also give some basic examples — as of a form f not having a "tight" annihilating scheme, one of degree equal to the maximum value of H_f (Example 5.7).

In Section 5.2 we first resume some well-known material about flat families and limit ideals. We apply this to show that if f has a smoothable degree-s annihilating scheme — one that lies in a flat family, the general member of which is smooth — then $f \in \overline{PS(s, j; r)}$ (Lemma 5.17). We go on to define determinantal schemes, and the postulation Hilbert scheme $\mathbf{Hilb}^H(\mathbb{P}^n) \subset \mathbf{Hilb}^s(\mathbb{P}^n)$ parametrizing subschemes Z of $\mathbb{P}^n$ having Hilbert function $H(R/\mathcal{I}_Z) = H$, which we use later.

Section 5.3 contains several of the main results of the book. In Section 5.3.1 we show that if $r = 3$ and T contains a subsequence

(s, s, s), then $Gor(T)$ is fibred over a postulation stratum $Hilb^H(\mathbb{P}^2)$ of the punctual Hilbert scheme $\mathbf{Hilb}^s(\mathbb{P}^2)$, with fiber an open set in the affine space $\mathbb{A}^s$: here the form $f \in Gor(T)$ is mapped to the point of $Hilb^H(\mathbb{P}^2)$ parametrizing the annihilating scheme of f, whose defining ideal is generated by the lower degree generators of $I = Ann(f)$ (Theorem 5.31). We then show similar results for suitable subvarieties $Gor_{\mathrm{sch}}(T)$ of $Gor(T)$ when $T \supset (s, s)$ or $T \supset (s-a, s, s-a)$ (Theorems 5.39, 5.46). It follows from the smoothability of degree-s subschemes of $\mathbb{P}^2$ and our study of annihilating schemes (Theorem 5.17) that such $Gor(T)$ (when $T \supset (s, s, s)$, or $Gor_{\mathrm{sch}}(T)$ are completely inside the closure of $PS(s, j; 3)$ (Theorem 5.71). Also, these Gorenstein algebras $A_f = R/I$ satisfy the weak Lefschetz property (Corollary 5.49).

Since Gerd Gotzmann has determined the dimensions of the postulation strata $Hilb^H(\mathbb{P}^2)$ (see [**Got1**]) we can also give independently the dimension of such $Gor(T)$ when $r = 3$ (Corollary 5.50). Recall that when $r = 3$ there are now complete dimension results for $Gor(T)$, reported on in Section 4.4. The formulas we obtain for $T \supset (s, s, s)$ are apparently distinct, though, of course equivalent to other known formulas.

Although Section 5.3 is somewhat technical, we include examples and exposition to make it accessible to a reader with a basic commutative algebra background; some more specialized background material is included in Appendix B.

In Section 5.4, written to be accessible for a general reader, we revisit the main results of Section 5.3, with emphasis on the Problem 0.1 from the Introduction – finding conditions for a form to be represented as a sum $f = L_1^j + \cdots + L_s^j$: we answer this problem in most cases when $r = 3$.

In Section 5.5 we study the Betti strata of $Hilb^H(\mathbb{P}^2)$, using information about the Betti strata of $Gor(T)$, in particular a result announced by M. Boij. Our goal here is to illustrate the tight connection between the two, a consequence of the main theorems of Section 5.3.

In Section 5.6 we partially answer the question of identifying the forms f in the closure of $PS(s, j; 3)$: we determine the Hilbert functions H_f possible for elements in the closure (Proposition 5.66); and we give criteria that are effective in many cases for deciding whether $f \in \overline{PS(s, j; 3)}$ (Theorem 5.71). Nevertheless, given $f \in \mathcal{D}_j$, the problem of determining intrinsically whether f is in the closure of $PS(s, j; 3)$ is still open. Similarly when $r \geq 3$ no explicit description is known of the forms that are in the boundary $\overline{PS(s, j; r)} - PS(s, j; r)$ (i.e. an analog of GADs for binary forms). In Section 5.7 we connect the work on annihilating schemes with the problem of determining arithmetic Cohen-Macaulay subschemes Z of $\mathbb{P}^n$ containing a given arithmetically Gorenstein subscheme W as a "tight" codimension one subscheme - one whose degree is the maximum of the h-vector of W (Lemma 5.75). When codimension $W = 3$, we can determine in most cases for which arithmetically Gorenstein schemes W there exists a scheme Z on which W is a tight codimension one subscheme (Theorem 5.77).

In Chapter 6 we study the annihilating scheme more deeply. Recall that in Section 5.1 we showed that forms having a smoothable annihilating scheme of degree s are in the closure of $PS(s, j; r)$ (Lemma 5.17). Here we prove a weaker statement than the converse (Proposition 6.7). Nevertheless, this approach suffices to give examples of reducible $Gor(T)$. We let $T(j, r)$ denote the sequence

$$T(j, r) = (1, r, 2r - 1, 2r, \ldots, 2r, 2r - 1, r, 1).$$

of socle degree j. The first author had shown

THEOREM 1.73 (A. Iarrobino [I7, Prop. 10]). [10] FORMS NOT IN THE CLOSURE OF $PS(2r, j; r)$, $r = 7$. *There are degree-j forms f in r variables, $7 \leq r \leq 13$, $r \neq 8$, $j \geq 7$, such that f satisfies $H_f = T(j, r)$, and f is not in the closure of $PS(2r, j; r)$, although H_f is bounded by $2r$.*

It follows that if $j = 2t$ or $2t + 1$, then $V_{2r}(t, j - t; r)$ has several irreducible components. Using the special properties of these forms, we show (for the full statement see Section 6.3)

THEOREM 6.26. *If $r \geq 7$, $r \neq 8$, $j \geq 8$ and $T = T(j, r)$, then $Gor(T)$ has at least two irreducible components. One parameterizes forms $f = L_1^{[j]} + \cdots + L_s^{[j]}$ whose polar polyhedron $\mathbb{P}(\mathcal{D}_1)$ is a self-associated point set. Another parameterizes forms with nonsmoothable annihilating schemes.*

We give similar examples for $r = 5, 6$ (Corollaries 6.28, 6.29). These reducibility results contrast the irreducibility of $Gor(T)$ when $r = 3$ (Theorem 1.72). The proof of Theorem 6.26 requires three steps:

a. *Translation from annihilating schemes to forms.* If Z is either a "conic" Gorenstein scheme $Z \subset \mathbb{P}^{r-1}$ concentrated at a point p, or Z is a smooth subscheme of $\mathbb{P}^{r-1}$, then there is a form $f \in \mathcal{D}$, annihilated by $\mathcal{I}_Z$ from which we can recover Z (Lemma 6.1).

---

[10]The mysterious case $r = 8$, erroneously included in Prop. 10 of [I7], is not well understood — see Lemma 6.23

b. *Suitable annihilating schemes.* When $r \geq 5$, there are well known examples of nonsmoothable, conic, Gorenstein schemes Z concentrated at a point p of $\mathbb{P}^{r-1}$, that are defined by a graded ideal $\mathcal{I}_{Z,p}$ of $\mathcal{O}_p$ (See [**EmI1**], and Lemma 6.21 below). When $r \geq 7$, these schemes have local compressed Hilbert function $H = (1, r-1, r-1, 1) = H(3, r-1)$ such that $T(j, r) = \text{Sym}(H, j)$. On the other hand, a construction of A. Geramita, P. Maroscia, and L. Roberts [**GMR**] shows that there are smooth punctual schemes S with the same Hilbert function, $H(R/\mathcal{I}_S)) = H(R/\mathcal{I}_Z)$. When $7 \leq r$ by a result of A. Geramita and F. Orecchia, we can use for S the smooth scheme consisting of $2r$ general points lying on a rational normal curve in $\mathbb{P}^{r-1}$ (See [**GO**] or Example 6.9 below); but we find it useful to study more general self-associated points sets in $\mathbb{P}^{r-1}$: sets of $2r$ points imposing only $2r - 1$ conditions on quadrics (Example 6.10, Lemmas. 6.11, 6.13, 6.16).

c. *Comparison Theorem.* We show that a deformation of forms f in $Gor(T)$ implies a corresponding deformation of the minimum degree annihilating schemes. (Lemma 6.6, Theorem 6.26).

Recently, M. Boij has shown the reducibility of $Gor(T)$ for certain T when $r \geq 4$, thus covering the cases $r = 4, 8$ which had remained unsettled by our approach. Boij also uses the annihilating scheme method, but in general he permits annihilating schemes of arbitrary dimension [**Bo2**]. Y. Cho and the first author extend the approach above, to employ annihilating schemes that are locally Gorenstein [**ChoI1**]. In particular, Theorem 6.34 implies that when $r \geq 4$, many of the varieties $Gor(T), T = H(s, j, r)$, where s is the length of a compressed Gorenstein algebra in $r - 1$ variables, have several irreducible components (Corollary 6.36). We report briefly on these developments in Section 6.4.

Fifth Goal — Chapter 7. INVESTIGATE THE CONNECTIVITY AND COMPONENTS OF THE LOCI $V_s(u, v; r)$. We show that when $r = 3$, and T is a Gorenstein sequence bounded above by s, then (Theorem 7.3)

$$\overline{PS(s, j; 3)} \bigcap Gor(T) \neq \emptyset.$$

We then show for arbitrary r and k algebraically closed with $\text{char}(k) = 0$ or $\text{char}(k) > j$, that each component C of $\overline{Gor(T)} - \{0\}$ contains $PS(1, j; r)$ (Theorem 7.6). Clearly, if the intersection $\overline{Gor(T')} \cap Gor(T)$ is nonempty, then $T' \geq T$ termwise. We exhibit a simple obstruction to deforming from an element f in $Gor(T)$, to $\overline{Gor(T')}$: depending on the number of generators of the ideal $I = \text{Ann}(f)$ in degrees where T and T' differ (Theorem 7.9).

When $r = 3$, the degrees of generators and relations for general elements $f \in Gor(T)$ are known, because of the Buchsbaum-Eisenbud structure theorem and Diesel's Theorem (see [**BE2**] and [**Di**] as well as Theorem 5.25). We use this and the obstruction criterion to show that the locus $V_s(t, t; r)$, with $s = \dim_k R_{t-1}$ has at least $[t/4] + 1$ irreducible components (Theorem 7.15).

In Section 7.3 we apply some of these results concerning components, and others concerning the closure of $PS(s, j; r)$ to the s–secant varieties $Sec_s(v_j(\mathbb{P}^n))$ of the j-th Veronese embeddings of $v_j(\mathbb{P}^n)$ (Theorem 7.18). In particular, we give a numerical condition guaranteeing that a general element $\overline{f} \in Sec_s(v_j(\mathbb{P}^n))$, lies on a unique s-secant $s - 1$ secant plane (Theorem 7.18(B)). We also clarify a point concerning a result of G. Ellingsrud and S. A. Strömme. When $n = 2$, and s is small, they had given a formula for a product of the degree of the multisecant variety, and a number $p(s, j, n)$ giving the number of secant planes containing a general element $\overline{f}$ as above: we show that $p(s, j, n) = 1$ in their case, hence their formula gives the degree of the multisecant variety (Remark 7.20).

Other Goals EXPLORE THE RELATIONSHIP OF $Gor(T)$ TO THE PARAMETER SPACE FOR GRADED IDEALS; DETERMINE COMPONENTS OF $Gor(T)$ ARISING FROM HIGHER DIMENSIONAL FAMILIES OF "ANNIHILATOR SCHEMES. We explore some further goals in Chapter 8, in Chapter 9, and also in Appendix C, Section C.5. The first involves the relation between $Gor(T)$ and the parameter scheme $\mathbf{G}(T)$ for graded ideals (Chapter 8): when $r = 2$ the latter is a desingularization of $Gor(T)$ (Theorem 8.1); however, this statement is no longer true when $r \geq 3$ since $\mathbf{G}(T)$ is not smooth (Example 8.3). Many open problems are stated in Chapter 9, and applications are suggested to other areas, in particular to the study of germs of differentiable maps.

Appendix C was written by the first author and Steven L. Kleiman. It explains Macaulay's theorem and Gotzmann's theorems to provide a convenient reference and a solid base for the use of the Hilbert scheme made in the main text. Gotzmann's Persistence Theorem and his Hilbert scheme Theorem (Theorems C.17 and C.29) concern a graded ideal I that is generated by I_d and has the property that the growth in vector space dimension from $\dim_k I_d$ to $\dim_k I_{d+1}$ is the minimum possible permitted by Macaulay's theorem [**Mac3**]. The theorems assert that I continues to have minimum growth (persistence), and that I determines a point on the appropriate Hilbert scheme $\mathbf{Hilb}^P(\mathbb{P}^n)$ [**Got1**]. Gotzmann's Regularity Theorem gives an effective bound on the regularity degree of an arbitrary subscheme of $\mathbb{P}^n$ having Hilbert polynomial P. In addition, the appendix establishes Bayer's conjecture that, through a variation of Gotzmann's description, we can describe the Hilbert scheme by a simple set of determinantal equations on a single suitable Grassmanian (Proposition C.30).

Section C.5 applies these results to determine annihilating schemes for forms f such that H_f contains the subsequence (s, s) in degree $(i, i + 1)$ where $i \geq s$ (Proposition C.33). Then the section constructs examples of $Gor(T)$ having several irreducible components; one component parameterizes forms f having "annihilating schemes" of higher dimension than zero. The first such case has

$$T = (1, 4, 9, 15, 22, 15, 9, 4, 1)$$

(Example C.38). These results are related to work of M. Boij [**Bo2**], and also to applications of Gotzmann's theorems in a quite different direction by A. Bigatti, A. Geramita and J. Migliore [**BGM**].

CHAPTER 2

Sums of Powers of Linear Forms, and Gorenstein Algebras

2.1. Waring's problem for general forms

The aim of this section is to deduce Theorem 1.61 concerning the dimension of the family $PS(s, j; r)$ of power sums, from the Alexander-Hirschowitz Theorem 1.66, which solved the order two interpolation problem for a general set of s points of $\mathbf{P}^{r-1}$. We follow [I7] which considers the case of ordinary powers and $\operatorname{char}(k) = 0$, or $\operatorname{char}(k) > j$. Here we work out the more general case $\operatorname{char}(k) = 0$, or $\operatorname{char}(k) \nmid j$, which requires divided powers. We should mention that our proof does not work when $\operatorname{char}(k) \mid j$ and at present the authors do not know whether the dimension formula from Theorem 1.61 holds if $\operatorname{char}(k) \mid j$ except in the cases covered by Theorem 2.6. We also give a self-contained proof of Theorem 1.66 in most of the cases we will use later.

Recall from Definition 1.13 that given a set P of s points in $\mathbb{P}^{r-1}$ with representatives the r-tuples $p(i) \in k^r$, $i = 1, \ldots, s$ we denote by $L_P^{[j]}$ the span $\langle L_{p(1)}^{[j]}, \ldots, L_{p(s)}^{[j]} \rangle \subset \mathcal{D}_j$.

DEFINITION 2.1. IDEAL OF POLYNOMIALS VANISHING TO ORDER a AT P. If P is a set of points of $\mathbb{P}^{r-1}$, we denote by $\mathcal{I}_P^{(a)}$ the graded ideal in R of all polynomials vanishing to order at least a at each point of the set P. This is the a-th symbolic power of $\mathcal{I}_P = \mathcal{I}_P^{(1)}$. Thus,

$$\mathcal{I}_P^{(a)} = m_{p(1)}^a \cap \ldots \cap m_{p(s)}^a.$$

There is a close connection between the ideal $\mathcal{I}_P^{(a)}$ and the ideal generated by the vector spaces $L_P^{[j]}$. If V is a vector subspace of $\mathcal{D}_v$, we let $\mathcal{D}_u \cdot V$ be the vector space span of $\{g \cdot h \mid g \in \mathcal{D}_u, h \in V\}$ in $\mathcal{D}_{u+v}$. The following lemma is an improved version of J. Emsalem's and the first author's result [EmI2, Theorem 1], which required $\operatorname{char} k = 0$ or $\operatorname{char} k > j$.

LEMMA 2.2. INVERSE SYSTEM OF A SYMBOLIC POWER. *When* char$(k) = 0$ *or is larger than* j, *and* $\mathcal{R}$ *is the polynomial ring, the annihilator* $[I^{-1}]_j$ *in* $\mathcal{R}_j$ *of the degree-j piece of the ideal* $I = \mathcal{I}_P^{(a)}$ *of* R *satisfies*

$$[I^{-1}]_j = \mathcal{R}_{a-1} L_P^{j+1-a}. \tag{2.1.1}$$

Suppose char$(k) = 0$, *or if* char$(k) = p > 0$ *and* $a \geq 2$ *suppose the following* $a - 1$ *conditions are satisfied:* p *does not divide* $j + 2 - a$, $\binom{j+3-a}{2}, \ldots, \binom{j}{a-1}$. *Then for the annihilator in the divided power ring* $\mathcal{D}$ *one has*

$$[I^{-1}]_j = \mathcal{D}_{a-1} \cdot L_P^{[j+1-a]}. \tag{2.1.2}$$

In particular, if $a = 2$ *and* char$(k) = 0$, *or* char$(k) > 0$ *and* char$(k) \nmid j$ *one has*

$$[I^{-1}]_j = \mathcal{D}_1 \cdot L_P^{[j-1]}. \tag{2.1.3}$$

The conditions stated for char(k) *are necessary for* (2.1.2) *and* (2.1.3) *to hold.*

Notice that the case $a = 1$, when no restrictions on char(k) are required is the Apolarity Lemma 1.15 which is classical. Formula (2.1.1) was also shown by R. Ehrenborg and G.-C. Rota [**EhR**]; the case $a = 2$ is due to A. Terracini[**Ter1, Ter2**] (in the setting of Veronese varieties this is known as Terracini's Lemma)[1].

PROOF OF LEMMA 2.2. It suffices to prove (2.1.2). Then the case of ordinary powers would follow from Proposition A.12. With respect to the contraction pairing $\mathcal{R}_j \times \mathcal{D}_j \to k$ the annihilator in $\mathcal{D}_j$ of the intersection of spaces

$$[m_{p(1)}^a]_j \cap \ldots \cap [m_{p(s)}^a]_j.$$

is the sum of the annihilators of $[m_{p(d)}^a]_j$, $d = 1, \ldots, s$. Thus it suffices to prove (2.1.2) for $s = 1$. From Appendix A the contraction pairing, the multiplication in $\mathcal{D}$, and the operation $L \mapsto L^{[j]}$ of taking divided powers are all GL_r-equivariant, so it suffices to prove (2.1.2) for $s = 1$, $p(1) = (1, 0, \ldots, 0)$, $L_{p(1)} = X_1$. In this case $[m_{p(1)}^a]_j$ is the linear span of the forms of type $\psi x_2^{a_2} \cdots x_r^{a_r}$ with $a_2 + \cdots + a_r =$

---

[1]F. L. Zak, in his review of [**AlH3**], states that E. Lasker [**Las**] was the first to show that the Waring Problem for forms would be solved by proving what is now the Alexander-Hirschowitz Theorem 1.66. This requires knowledge of the case $a = 2$ of Lemma 2.2 or an equivalent. We learned of the Waring problem implication from R. Lazarsfeld.

a with arbitrary $\psi \in R_{j-a}$. Now, using the property $\langle \phi_1\phi_2, f \rangle = \langle \phi_1, \phi_2 \circ f \rangle$ of the contraction map (A.0.2), we see that an element $f \in \mathcal{D}_j$ annihilates $[m^a_{p(1)}]_j$ if and only if it is a linear combination of DP-monomials $X_1^{[b_1]} X_2^{[b_2]} \cdots X_r^{[b_r]}$ with $b_2 + \cdots b_r \leq a - 1$. Clearly, $\mathcal{D}_{a-1} \cdot X_1^{[j+1-a]}$ is contained in the linear span of these monomials. An easy calculation using the multiplication rule $X_1^{[u]} \cdot X_1^{[v]} = \frac{(u+v)!}{u!v!} X_1^{[u+v]}$ shows that, conversely, these monomials belong to $\mathcal{D}_{a-1} \cdot X_1^{[j+1-a]}$ if and only if the conditions of the Lemma are satisfied. $\square$

We now deduce Theorem 1.61, determining the dimension of $PS(s, j; r)$ when char $k = 0$ or char$(k) \nmid j$, from the Alexander-Hirschowitz vanishing Theorem 1.66. When char $k = 0$ this implication was made by A. Terracini [**Ter1**, **Ter2**] and apparently was known to E. Lasker [**Las**]. When char $k > j + 1$ it is shown in [**I7**, p. 1096].

PROOF OF THEOREM 1.61. Consider the map

$$\mu : \mathcal{D}_1 \times \cdots \times \mathcal{D}_1 \to \mathcal{D}_j, \quad \mu(L_1, \ldots, L_s) = L_1^{[j]} + \cdots + L_s^{[j]}.$$

For a fixed s-tuple $(L_1, \ldots, L_s)$ consider the corresponding point set $P \subset \mathbb{P}(\mathcal{D}_1) = \mathbb{P}^{r-1}$. We claim that the image of the derivative μ_* at $(L_1, \ldots, L_s)$ is the space

$$Im(\mu_*) = \mathcal{D}_1 \cdot L_P^{[j-1]} = \mathcal{D}_1 \cdot L_1^{[j-1]} + \cdots + \mathcal{D}_1 \cdot L_s^{[j-1]}. \quad (2.1.4)$$

It suffices to prove this for $s = 1$. Let $L = a_1 X_1 + \cdots + a_r X_r$. We have

$$L^{[j]} = \sum_{j_1 + \cdots j_r = j} a_1^{j_1} \cdots a_r^{j_r} X_1^{[j_1]} \cdots X_r^{[j_r]}.$$

The partial derivative $\frac{\partial}{\partial a_d} L^{[j]}$ equals

$$j_d \cdot \sum_{\substack{j_1 + \cdots j_r = j \\ j_d \geq 1}} a_1^{j_1} \cdots a_d^{j_d - 1} \cdots a_r^{j_r} X_1^{[j_1]} \cdots X_d^{[j_d]} \cdots X_r^{[j_r]}$$

$$= X_d \cdot X_d^{[j_d-1]} \sum a_1^{j_1} \cdots a_d^{j_d-1} \cdots a_r^{j_r} X_1^{[j_1]} \cdots \widehat{X_d^{[j_d]}} \cdots X_r^{[j_r]}$$

$$= X_d \cdot \sum_{i_1 + \cdots i_r = j-1} a_1^{i_1} \cdots a_r^{i_r} X_1^{[i_1]} \cdots X_r^{[i_r]}$$

$$= X_d \cdot L^{[j-1]}.$$

This shows (2.1.4).

Now, when char$(k) = 0$ or char$(k) \nmid j$, Lemma 2.2 shows that the annihilator of $\mathcal{D}_1 \cdot L_P^{[j-1]}$ in R_j with respect to the contraction pairing is equal to the degree-j piece of the ideal $\mathcal{I}_P^{(2)} = m_{p(1)}^2 \cap \ldots \cap m_{p(s)}^2 \subseteq R$. Thus, if P is a sufficiently general set of s points in $\mathbb{P}^{r-1}$, we have by the Alexander-Hirschowitz Theorem 1.66 that

$$\dim_k \mathcal{D}_1 \cdot L_P^{[j-1]} = \min(rs, \dim_k R_j) \qquad (2.1.5)$$

except for the four triples $(s, j, r) = (5, 4, 3), (9, 4, 4), (14, 4, 5)$, and $(7, 3, 5)$ where the dimension is $\dim_k R_j - 1$ and $rs \geq \dim_k R_j$. For a dominant map $\mu : X \to Y$ of two irreducible varieties one has for the tangent space at a sufficiently general point $x \in X$ that $\mu_* T_x X \subseteq T_{\mu(x)} Y$ and $\dim_k T_{\mu(x)} Y = \dim Y$; furthermore $\mu_* T_x X = T_{\mu(x)} Y$ if $\mathrm{char}(k) = 0$. This applied to $X = \mathcal{D}_1 \times \cdots \times \mathcal{D}_1$ and $Y = \overline{PS(s, j; r)}$, combined with the obvious inequality

$$\dim \overline{PS(s, j; r)} \leq \min(rs, \dim_k R_j)$$

shows the theorem if $\mathrm{char}(k) = 0$ and also in the case of positive characteristic, except in the four exceptional cases where one has to prove that $\overline{PS(s, j; r)}$ is not the whole of $\mathcal{D}_j$ (and consequently is a hypersurface in $\mathcal{D}_j$).

First consider the three exceptional cases with $j = \deg(f) = 4$. We claim that in each of these cases the catalecticant determinant $\det Cat_F(2, 2; r)$ is a non-zero polynomial which vanishes on $PS(s, j; r)$. The proof is identical to Clebsch's argument for the case $(s, j, r) = (5, 4, 3)$ sketched in the introduction. Namely, in each of the three cases $s = \dim |\mathcal{O}_{\mathbb{P}^{r-1}}(2)|$. So, if $L_1, \ldots, L_s$ is a general s-tuple in $\mathcal{D}_1$ with corresponding set $P \subset \mathbb{P}^{r-1}$, there exists a quadric defined by the equation $\phi(x) = 0$, $\phi \in R_2$ which contains P. Let $f = L_1^{[4]} + \cdots + L_s^{[4]}$. From Lemma 1.15(i) we have

$$\phi \circ \left(L_1^{[4]} + \cdots + L_s^{[4]} \right) = \phi(p_1) L_1^{[2]} + \cdots + \phi(p_s) L_s^{[2]} = 0.$$

which is equivalent to $\phi \in \mathrm{Ker}\, C_f(2, 2)$. Therefore the polynomial $\det Cat_F(2, 2; r)$ vanishes on $\overline{PS(s, 4; r)}$.

One proves that $\det Cat_F(2, 2; r)$ is not a zero polynomial by repeating the above argument with $s+1 = \dim_k H^0(\mathbb{P}^{r-1}, \mathcal{O}_{\mathbb{P}^{r-1}}(2))$ general linear forms $M_1, \ldots, M_{s+1}$. Let $g = M_1^{[4]} + \cdots + M_{s+1}^{[4]}$. Then from Lemma 1.15(iv), $\mathrm{Ann}(g)_2$ consists of quadratic forms vanishing on the general set of $s + 1$ points corresponding to $\{M_i\}$. Thus $\mathrm{Ann}(g)_2 = 0$, which means that $\det Cat_F(2, 2; r)$ does not vanish on g.

Now, consider the remaining case $(s, j, r) = (7, 3, 5)$. If $\mathrm{char}(k) = 0$ it was proved above that $\overline{PS(7, 3; 5)}$ is a hypersurface (this is a classical fact, see e.g. [Ri, Wak, EhR]). We prove by lifting to characteristic 0 that $PS(7,3;5)$ is contained in a hypersurface for arbitrary char$(k) = p > 0$ as well. The morphism μ, $\mu(L_1,\dots,L_s) = L_1^{[j]} + \cdots + L_s^{[j]}$ is defined over $\mathbb{Z}$. This means, one considers the affine schemes

$$X = \operatorname{Spec} \mathbb{Z}[a_{11},\dots,a_{1r},\dots,a_{s1},\dots,a_{sr}]$$
$$Y = \operatorname{Spec} \mathbb{Z}[b_{j_1,\dots,j_r}]_{j_1+\cdots+j_r=j}$$

and the morphism $\mu_{\mathbb{Z}} : X \to Y$ induced by the homomorphism

$$\Phi : \mathbb{Z}[b_{j_1,\dots,j_r}]_{j_1+\cdots+j_r=j} \longrightarrow \mathbb{Z}[a_{ij}] \quad \text{defined by}$$

$$\Phi(b_{j_1\cdots j_r}) = \sum_{i=1}^{s} a_{i_1}^{j_1} \cdots a_{i_s}^{j_s}.$$

For every field L we denote by X_L, Y_L, Φ_L, μ_L the corresponding extensions via the homomorphism $\mathbb{Z} \to L$ (or in scheme-theoretic language $X \times_{\operatorname{Spec}(\mathbb{Z})} \operatorname{Spec}(L)$, etc.). We claim there is a homogeneous polynomial of positive degree in the kernel of Φ. Indeed, if $\overline{\mathbb{Q}}$ is the algebraic closure of $\mathbb{Q}$, then $\overline{\mu_{\overline{\mathbb{Q}}}(X_{\overline{\mathbb{Q}}})} = \overline{PS(7,3;5)}$, $(k = \overline{\mathbb{Q}})$, which is a hypersurface in $R_5(\overline{\mathbb{Q}}) = Y_{\overline{\mathbb{Q}}}$. Taking the norm of the polynomial defining this hypersurface and multiplying by an appropriate rational number one obtains a non-zero homogeneous polynomial $G \in \mathbb{Z}[b_{j_1\cdots j_r}]$ with the property $\Phi(G) = 0$ whose G.C.D. of the coefficients is 1. Now, $G_p = C(mod\ p)$ is a non-zero polynomial in $\mathbb{Z}_p[b_{j_1\cdots j_r}]$ which belongs to the kernel of $\Phi_{\mathbb{Z}_p}$. Let k be an algebraically closed field of characteristic p, $\mathbb{Z}_p \subset k$. Then $\Phi_{\mathbb{Z}_p}(G_p) = \mu_{\mathbb{Z}_p}^*(G_p) = 0$ implies that G_p vanishes on $\mu_k(X_k) = \mu_k(\mathcal{D}_1 \times \cdots \times \mathcal{D}_1)$. This proves the theorem. $\square$

COROLLARY 2.3. *Let (s,j,r) be one of the exceptional triples $(5,4,3), (9,4,4), (14,4,5)$. Suppose* char$(k) \neq 2$. *Then* $\overline{PS(s,j;r)} = V_s(2,2;r)$, *and it is a normal hypersurface in* $\mathbb{P}(\mathcal{D}_j)$ *whose ideal is generated by the catalecticant determinant* $\det Cat_F(2,2;r)$.

PROOF. We proved above that $\overline{PS(s,j;r)}$ is a hypersurface which contains $\det Cat_F(2,2;r)$ in its ideal. It remains to observe that every catalecticant determinant $\det Cat_F(t,t;r)$ is an irreducible polynomial and the hypersurface it defines is normal (see e.g. Proposition 3.13). $\square$

REMARK 2.4. We do not know of any explicit equation for $\overline{PS(7,3;5)}$, nor its degree. There is an interesting argument in [RS] which explains why a general cubic form in 5 variables cannot be represented as a sum of 7 cubes of linear forms. Namely, 7 general points in $\mathbb{P}^4$ lie on a unique rational normal curve. Using this, K. Ranestad and F.-O. Schreyer deduce that the minimal resolution of the Artinian Gorenstein ring A_f for f general in $PS(7, 3; 5)$ differs from the minimal resolution of A_F for a general cubic form F.

REMARK 2.5. There are other interesting directions of research on Waring's problem that we do not touch in this book but we wish to mention. B. Reichstein studied in [**Rei**] the problem of representing a cubic form as sum of cubes of linear forms. This is a case when the catalecticant matrices do not give much information (cf. Section 1.2). Another very interesting direction is the study of the variety of sums of powers representing a general form. This is a compactification of the parametric variety for the representations of a given general form $f = L_1^j + \cdots + L_s^j$ when s is large. It is a natural contravariant of the form and was calculated in some cases (see the papers of B. Reichstein and Z. Reichstein [**ReiR**], S. Mukai [**Muk**], K. Ranestad and F.-O. Schreyer [**RS**], and of A. Iliev and K. Ranestad [**IlR**]).

2.2. Uniqueness of additive decompositions

Recall that $PS(s, j; r)$ denotes the constructible subset of $\mathbb{A}(\mathcal{D}_j)$, parameterizing those homogeneous elements $f \in \mathcal{D}_j$, representable as $f = L_1^{[j]} + \cdots + L_s^{[j]}$ for some linear forms $L_1, \ldots, L_s \in \mathcal{D}_1$.

THEOREM 2.6. UNIQUENESS OF THE DECOMPOSITION OF A FORM, AND DIMENSION OF $PS(s, j; r)$. *Suppose* $j = 2t$ *or* $2t - 1$, $j \geq 3$.

  i. *If the integer s satisfies*

$$s \leq \dim_k R_t - r + 1 = \binom{t + r - 1}{r - 1} - r + 1 \text{ when } j = 2t, \text{ or}$$

$$s \leq \dim_k R_{t-1} = \binom{t + r - 2}{r - 1} \text{ when } j = 2t - 1, \tag{2.2.1}$$

  then the dimension of $PS(s, j; r)$ satisfies

$$\dim PS(s, j; r) = rs. \tag{2.2.2}$$

  ii. *Furthermore, if $s \leq \binom{t+r-1}{r-1} - r$ then for any sufficiently general set of linear forms $L_1, \ldots, L_s$ in $\mathcal{D}$,*

$$f = L_1^{[j]} + \cdots + L_s^{[j]}$$

  is the unique representation of the DP-form $f \in \mathcal{D}_j$ as a sum of j-th divided powers of $\leq s$ linear forms, up to multiplication of L_i by j-th roots of unity and permutation of the L_i, $i = 1, \ldots, s$. The hypothesis (i) *implies the hypothesis of* (ii) *if j is odd and $r \geq 3$.*

iii. Let char$(k) = 0$ and let $\mathcal{R} = \mathcal{D}$ be the polynomial ring in r variables. Then (i) and (ii) hold for polynomials and ordinary powers. Let $L_1, \ldots, L_s$ be sufficiently general linear forms from $\mathcal{R}_1$ and let $P = \{p_1, \ldots, p_s\} \subset \mathbb{P}(\mathcal{D}_1) = \mathbb{P}^{r-1}$ be the corresponding points. Then under the assumptions of (i) one has

$$\dim_k \mathcal{R}_1 L_P^{j-1} \;=\; \dim_k(\mathcal{R}_1 L_1^{j-1} + \cdots + \mathcal{R}_1 L_s^{j-1}) \;=\; rs, \qquad (2.2.3)$$

or equivalently

$$\mathrm{cod}_{R_j}(\mathcal{I}_P^{(2)})_j = rs. \qquad (2.2.4)$$

PROOF. We let $j = m + n$, where $m = t = n$ if $j = 2t$ and $m = t$, $n = t - 1$ if $j = 2t - 1$. The hypothesis of (i) can be read as

$$s \leq \min(\dim_k R_m - r + 1, \dim_k R_n).$$

Indeed, if $j = 2t - 1$, then

$$\binom{t+r-1}{r-1} - \binom{t+r-2}{r-1} = \binom{t+r-2}{r-2}.$$

If $r \geq 3$, then the right-hand number $\geq t + r - 2 \geq r$ (this proves the last claim of (ii)). If $r = 2$, then both numbers in (i) coincide ($= t$).

PROOF OF (i). Let $L_1, \ldots, L_s \in \mathcal{D}_1$ and let

$$f = L_1^{[j]} + \cdots + L_s^{[j]}.$$

Let $\{\tilde{p}_1, \ldots, \tilde{p}_s\} \subset k^r$ be the coordinates of the linear forms L_i, $i = 1, \ldots, s$ in the basis $X_1, \ldots, X_r$ and let $\{p_1, \ldots, p_s\} \subset \mathbb{P}^{r-1}$, $p_i = [\tilde{p}_i]$ be the corresponding points in $\mathbb{P}(k^r)$. According to Lemma 1.15, if

$$p_1, \ldots, p_s \quad \text{impose independent conditions on} \quad |\mathcal{O}_{\mathbb{P}^{r-1}}(n)|, \qquad (2.2.5)$$

then one has that the linear system of degree-m forms $|\mathrm{Ann}(f)_m|$ is equal to $|(J_P)_m| = |\mathcal{O}_{\mathbb{P}^{r-1}}(m)(-\sum_{i=1}^{s} p_i)|$.

Our goal is to compare P with the base locus of $|\mathrm{Ann}(f)_m|$. We need a lemma.

LEMMA 2.7. BASE POINTS OF CERTAIN LINEAR SYSTEMS. There exists a Zariski open, dense subset $U \subset \mathbb{P}^{r-1} \times \cdots \mathbb{P}^{r-1}$ (s times) such that for $(p_1, \ldots, p_s) \in U$ we have

i. If $s \leq \dim_k R_m - r + 1$, then the linear system

$$|\mathcal{O}_{\mathbb{P}^{r-1}}(m)(-p_1 - \cdots - p_s)|$$

has finitely many base points.

ii. If $s \leq \dim_k R_m - r$, then the set of base points is:

$$Bs \, |\mathcal{O}_{\mathbb{P}^{r-1}}(m)(-p_1 - \cdots - p_s)| = \{p_1, \ldots, p_s\}. \qquad (2.2.6)$$

PROOF. Let $N = \dim_k R_m$, let $v_m : \mathbb{P}^{r-1} \to \mathbb{P}^{N-1}$ be the Veronese map and let $X = v_m(\mathbb{P}^{r-1})$. Let $s_0 = N - r + 1$. Let $W \subset \mathbb{P}^{r-1} \times \cdots \times \mathbb{P}^{r-1}$ (s_0 times) be a Zariski open dense subset which consists of points $(p_1, \ldots, p_{s_0})$ such that $\{v_m(p_i)\}$ are linearly independent. Every general enough subspace $E \subset \mathbb{P}^{r-1}$ of codimension $r-1$ intersects X in a finite number of points which span E. Thus the map into the Grassmannian,

$$e : \quad W \to Grass(s_0 - 1, \mathbb{P}^{N-1})$$
$$e(p_1, ..., p_{s_0}) = \langle v(p_1), \ldots, v(p_{s_0}) \rangle$$

is dominant. It suffices to prove (i) for $s = s_0$, which is immediate from the above, since

$$Bs \, |\mathcal{O}_{\mathbb{P}^{r-1}}(m)(-p_1 - \cdots - p_{s_0})| = v_m^{-1}(e(p_1, \ldots, p_{s_0}) \cap X).$$

To prove (ii) we notice that by Bertini's Theorem and the General Position Theorem ([**ACGH**] if char$(k) = 0$, Laksov [**Lak**] if char$(k) > 0$, see also [**Rat**]), if E is a sufficiently general subspace in $\mathbb{P}^N$ having codimension $r-1$, then every set of s_0 points in the intersection $E \cap X$ are linearly independent. This shows that if $s < s_0$ and $p_1, \ldots, p_s$ are sufficiently general, then

$$\langle v_m(p_1), \ldots, v_m(p_s) \rangle \cap X = \{v_m(p_1), \ldots, v_m(p_s)\},$$

thus proving (ii). This completes the proof of Lemma 2.7.   □

COMPLETION OF THE PROOF OF THEOREM 2.6. We continue the proof of (i). Now suppose by way of contradiction that $\dim PS(s, j; r) < rs$. Let $L_1, \ldots, L_s$ be an s-tuple with corresponding s-tuple of points $p_1, \ldots, p_s$ in $\mathbb{P}^{r-1}$ which satisfy the assumptions of (2.2.5) and Lemma 2.7. Then the form

$$f = L_1^{[j]} + \cdots + L_s^{[j]}$$

has at least a one-dimensional family of representations as sums of divided powers,

$$f = M_1(\tau)^{[j]} + \cdots + M_s(\tau)^{[j]},$$

where $L_i = M_i(\tau_0)$. Let $\{b_i(\tau)\}$ be the points in $\mathbb{P}^{r-1}$ corresponding to $\{M_i(\tau)\}$. For τ in a Zariski open neighborhood of τ_0 the points $b_1(\tau), \ldots, b_s(\tau)$ satisfy (2.2.5). Hence the points belong to the base locus of

$$|\mathrm{Ann}(f)_m| = |\mathcal{O}_{\mathbb{P}^{r-1}}(m)(-p_1 - \cdots - p_s)|.$$

Lemma 2.7 shows that for τ in a neighborhood of τ_0 we have $M_i(\tau) = c_i(\tau)L_i$. Now, if we impose the Zariski open condition on $L_1, \ldots, L_s$

$$p_1, \ldots, p_s \text{ impose independent conditions on } |\mathcal{O}_{\mathbb{P}^{r-1}}(j)| \qquad (2.2.7)$$

we have by Lemma 1.15 that $L_1^{[j]}, \ldots, L_s^{[j]}$ are linearly independent, so $c_i(\tau)^j = 1$ which is impossible. Notice that in fact condition (2.2.7) is implied by condition (2.2.5) according to Theorem 1.69.

PROOF OF (ii). That f cannot belong to $\overline{PS(s-1, j; r)}$ if $L_1, \ldots, L_s$ are sufficiently general follows from the dimension formula (2.2.2). Also by (i) if $L_1, \ldots, L_s$ are sufficiently general, then there are finitely many presentations of f as a sum of s j-th powers. Hence all s-tuples $M_1, \ldots, M_s$ such that

$$f = L_1^{[j]} + \cdots + L_s^{[j]} = M_1^{[j]} + \cdots + M_s^{[j]}$$

must satisfy (2.2.5), (2.2.7) and the conditions of Lemma 2.7. By Part (ii) of Lemma 2.7, after renumbering the M_u we have $M_i = c_i L_i$. We have $c_i^j = 1$ since $L_1^{[j]}, \ldots, L_s^{[j]}$ are linearly independent.

PROOF OF (iii). The vector space $\mathcal{R}_1 L_1^{j-1} + \cdots + \mathcal{R}_1 L_s^{j-1}$ is the image of the tangent space at $(L_1, \ldots, L_s)$ of the map

$$\mu : (\mathcal{R}_1)^s \to \mathcal{R}_j, \quad \mu(M_1, \ldots, M_s) = M_1^j + \cdots + M_s^j.$$

Thus part (iii) follows from (2.2.2) and Lemma 2.2 (case $a = 2$). This completes the proof of Theorem 2.6. $\qquad\square$

COROLLARY 2.8. LOWER BOUND FOR THE DIMENSION OF DETER-MINANTAL LOCI OF CATALECTICANTS. *If $j = 2t$ and s satisfies (2.2.1) then the dimension of $V_s(t, t; r)$ satisfies*

$$V_s(t, t; r) \geq rs. \qquad (2.2.8)$$

EXAMPLE 2.9. NONUNIQUENESS OF POLAR PENTAGONS FOR QUARTICS, $r = 3$. Let $j = 4$, $r = 3$. Here the inequality (i) of Theorem 2.6 is $s \leq 4$. This is the best possible, since Clebsch showed that every quartic which is a sum of five fourth powers of linear forms has at least a one-dimensional family of such representations (see Introduction and Section 2.1). So $\dim PS(5, 4; 3) < 15$.

The geometry of the polar pentagons in this case is very nice and was the subject of intensive study in the 19th century. We refer to the paper of I. Dolgachev and V. Kanev [DK] for a modern account, references and amplification. We now want to obtain the uniqueness of additive decompositions in the case of algebraically non-closed fields.

THEOREM 2.10. *Suppose $j \geq 3$. If $r = 2$ let $2s \leq j + 1$. If $r \geq 3$ let $s \leq \binom{t+r-1}{r-1} - r$ when $j = 2t$, and $s \leq \binom{t+r-2}{r-1}$ when $j = 2t - 1$. Suppose k is an arbitrary infinite field. Then there is a nonempty, Zariski open subset in $\mathcal{D}_1(k) \times \cdots \times \mathcal{D}_1(k)$, such that for any s-tuple $L_1, \ldots, L_s$ from this set one has*

i. *$f = L_1^{[j]} + \cdots + L_s^{[j]}$ is the unique representation of the degree-j element $f \in \mathcal{D}_j$ as a sum of divided powers of $\leq s$ linear forms up to multiplication by j-th roots of unity belonging to k and permutation of L_i, $i = 1, \ldots, s$.*

ii. *$f = c_1 L_1^{[j]} + \cdots + c_s L_s^{[j]}$ with all $c_i \neq 0$ is the unique representation of f as a linear combination of divided powers of $\leq s$ linear forms up to the transformations $c_i L_i^{[j]} = (c_i c^{-j})(c L_i)^{[j]}$ and permutation of L_i, $i = 1, \ldots, s$.*

iii. *If char$(k) = 0$, if s satisfies the conditions of (2.2.1) and $P \subset \mathbb{P}^{r-1}(k)$ is a sufficiently general set of s k-rational points, then the codimension of $(\mathcal{I}_P^{(2)})_j$ in $k[x_1, \ldots, x_r]_j$ is rs.[2]*

PROOF. Let $\overline{k}$ be the algebraic closure of k. Let $P = \{p_1, \ldots, p_s\} \subset \mathbb{P}(\mathcal{D}_1(k)) = \mathbb{P}^{r-1}(k)$ be the set of points corresponding to $L_1, \ldots, L_s$. Let us first consider the binary case $r = 2$. In this case we have by Proposition 1.50 that the points $p_1, \ldots, p_s$ are the roots of the unique (up to constant) polynomial of degree s apolar to f, so the uniqueness is clear. We may thus assume that $r \geq 3$ and that the hypothesis of Theorem 2.6(ii) holds. In the course of its proof we showed that the set P is uniquely determined by the DP-form $f \in \mathcal{D}_j$ provided the following genericity conditions were satisfied: conditions (2.2.5),(2.2.7) and condition (2.2.6).

Let us prove that failure of these conditions to hold defines a closed subvariety $W = W_1 \cup W_2 \cup W_3$, $W \subsetneq (\mathbb{P}^{r-1}(\overline{k}))$ defined over k. This is obvious for W_1 and W_2 corresponding to conditions (2.2.5) and (2.2.7). As for the third condition, consider the Veronese variety $X = v_m(\mathbb{P}^{r-1})$ which is clearly defined over k. Consider the locally closed subset $Y \subset X \times \cdots \times X \times X$ ($s+1$ times) consisting of $(s+1)$-tuples $q_1, \ldots, q_s, q_{s+1}$ of distinct points such that $q_1, \ldots, q_s$ are linearly independent and q_{s+1} belongs to the span $\langle q_1, \ldots, q_s \rangle$. Clearly Z is defined over k. Now, (2.2.6) is satisfied for the s-tuples in $X^s - \overline{\pi_s(Z)}$, where π_s is the projection onto the first s factors. The variety $\overline{\pi_s(Z)}$ is defined over k [**Bor**, Corollary 14.5 from Ch. AG], moreover it is a

---

[2]Part (iii) is a particular case of the Alexander-Hirschowitz Theorem 1.66. We include it here for its elementary proof.

proper subvariety of X_s by Lemma 2.7. For W_3 equal to the preimage
of $\overline{\pi_s(Z)}$ in $\mathbb{P}^{r-1} \times \cdots \times \mathbb{P}^{r-1}$ and $W = W_1 \cup W_2 \cup W_3$ as above, we
see that $W \subsetneqq (\mathbb{P}^{r-1})^s$.

Now k is an infinite field, so the k-rational points are dense in
$(\mathbb{P}^{r-1})^s - W$. This shows Parts (i) and (ii) since by the above arguments one reconstructs the finite set $P \subset \mathbb{P}^{r-1}(k)$ as the base locus of
the linear system $|\operatorname{Ann}(f)_m|$. Then the remaining statements follow
from the fact that $L_1^{[j]}, \ldots, L_s^{[j]}$ are linearly independent (Lemma 1.15
(iii) and Condition (2.2.7)). Part (iii) is clear from the fact that the
image by μ of the k-rational points is dense in $\overline{PS(s, j; r)}$. $\square$

2.3. The Gorenstein algebra of a homogeneous polynomial

Let us recall from Definition 1.11 that to each homogeneous ele-
ment $f \in \mathcal{D}_j$ we associate the graded ideal $\operatorname{Ann}(f) = \{\phi \in R \mid \phi \circ f = 0\}$ of polynomials in R apolar to f, and set $A_f = R/\operatorname{Ann}(f)$. The algebra A_f is an Artinian Gorenstein algebra since the type of A, which
is the dimension of the socle $Soc(A) = (0 : m)$, is one. We recall the
relevant definitions. The socle of a local Artin algebra A (with maxi-
mal ideal $\mathfrak{m}$) is the ideal $Soc(A) = (0 : \mathfrak{m}) = \{g \in A \mid g\mathfrak{m} = 0\}$. If A is
an Artin algebra, its socle degree $j(A)$ is the maximum integer j such
that $A_j \neq 0$.

DEFINITION 2.11. The Artin algebra $A = R/I$ is *Gorenstein*, if its
socle has dimension one as a k-vector space, or, equivalently, if A has a
unique minimal ideal (which must be $(0 : m)$). Then I is a Gorenstein
ideal of R.

We have, from §60ff of [**Mac2**],

LEMMA 2.12 (F.H.S.Macaulay). GORENSTEIN ALGEBRA OF A
FORM. *There is a bijective correspondence*

$$\{f \in \mathcal{D}_j \bmod (k^*\text{-multiple})\} \overset{\sigma}{\to} \{A = R/I\} \qquad (2.3.1)$$

*with A a graded Artinian Gorenstein quotient of socle degree j. The
correspondence is given by $\sigma(f) = A_f = R/\operatorname{Ann}(f)$. The inverse map
$\sigma^{-1}(A) = f \bmod k^*$ takes the quotient $A = R/I$ to the k^*-class of a
homogeneous generator f of the simple R-submodule $\mathcal{D}_f = I^\perp$ in $\mathcal{D}$.
Furthermore,*

$$R \circ f = \mathcal{D}_f \cong \operatorname{Hom}_R(A_f, k). \qquad (2.3.2)$$

*The Hilbert function $H_f = H(A_f)$ is symmetric about $\frac{j}{2}$ and satisfies
$H_f = H(\mathcal{D}_f)$.*

We give several references for the proof: [**Mac2**, Chapter IV], [**Em**], [**I3**, p. 344], [**I6**, Lemma 1.2], [**Ei2**, p.527] or Lemma 2.14 below.

Recall that M denotes the maximal ideal $(x_1, \dots, x_r)$ of $R = k[X]$.

EXAMPLE 2.13. ANNIHILATOR OF A PURE POWER. When $f = (L_a)^{[j]}$ then the ideal $\mathrm{Ann}(f)$ satisfies

$$\mathrm{Ann}(f) = (\{\phi \mid \phi(a) = 0\} + M^{j+1}),$$

and the Hilbert function $H(A_f) = H_f = (1, 1, ..., 1)$.

We next state then show an extension of F.H.S. Macaulay's Lemma 2.12 quoted above (See §70ff of [**Mac2**] and Lemma 1.2 of [**I6**])[3]. Recall that if $f \in \mathcal{D}_j$, the map $C_f(u, v) : R_v \to \mathcal{D}_u$, is the linear map that associates to each $\phi \in R_v$ the contraction $\phi \circ f$ in $\mathcal{D}_u$. We denote by $\mathcal{D}_u(f)$ the image $\mathcal{D}_u(f) = R_v \circ f$ of $C_f(u, v)$ in $\mathcal{D}_u$, and we let $H_f = (1, t_1, \dots, t_i = \dim_k \mathcal{D}_i(f)$. If V is a vector subspace of R_j we let $V : R_i$ or $R_{-i}V, 1 \leq i \leq j$ denote the subspace

$$V : R_i = \{\psi \in R_{j-i} \mid \phi\psi \in V \quad \text{for all} \quad \phi \in R_i\}.$$

We denote by $\overline{V}$ the *ancestor ideal* of V in R. It is a partial saturation of the ideal (V) that can be defined in several equivalent ways:

$$\overline{V} = (V) \oplus \sum_{1 \leq i < j} V : R_i \quad \text{or}$$

$$\overline{V} = \max \{I \mid I \cap M^j = (V)\}$$

$$= \bigcup (I \mid I \cap M^j \subset (V)). \qquad (2.3.3)$$

LEMMA 2.14 (F.H.S. Macaulay). GORENSTEIN IDEALS AND MACAULAY'S PRINCIPAL INVERSE SYSTEMS. *Suppose that the algebra quotient $A = R/J$ is graded, standard (generated in degree one), and has finite dimension as a vector space over k (Artinian). Then A is Gorenstein of socle degree j iff $J = \mathrm{Ann}(f)$ for some homogeneous degree-j element $f \in \mathcal{D}_j$, in which case $A = A_f$. If so, the sequence H_f of ranks of catalecticant matrices and the Hilbert function $H(A_f)$ satisfy*

$$H_f = H(A_f). \qquad (2.3.4)$$

the Hilbert sequence $H(A)$ is symmetric with respect to $\frac{j}{2}$ if A is Gorenstein.

---

[3]Macaulay did not have available some of our notation, so it is hard to say exactly whether Lemmas 2.15 or 2.17 go beyond his work.

Lemma 2.15. Artinian Gorenstein ideals are determined by their socle degree piece. *If $i \leq j$, and $J = \mathrm{Ann}(f)$, then J_i satisfies*

$$J_i = \mathrm{Ann}(\mathcal{D}_i(f)) \cap R_i = (\mathrm{Ann}(R_{j-i} \circ f)) \cap R_i,$$
$$= J_j : R_{j-i}. \qquad (2.3.5)$$

Corollary 2.16. Gorenstein ideal of a differential form. *Suppose* $\mathrm{char}(k) = 0$. *If $f \in \mathcal{R}_j$, let*

$$J_i = \{\psi \in R_i \mid f(\partial \cdot /\partial x_1, \ldots, \partial \cdot /\partial x_r) \circ (R_{j-i}\psi) = 0\}. \qquad (2.3.6)$$

Then the ideal $J = J_0 \oplus J_1 \oplus \ldots \oplus J_j \oplus M^{j+1}$ satisfies,

$$J = \mathrm{Ann}(f).$$

Before proving Lemma 2.14, Lemma 2.15, and Corollary 2.16 we need a result on forms apolar to f, certainly well known to Macaulay.

Lemma 2.17. The forms apolar to f are determined by those of degree j. *Suppose that $f \in \mathcal{D}_j$ is a degree-j element, we let $J = \mathrm{Ann}(f)$, and set $J_j = \mathrm{Ann}(f) \cup R_j$. Then the ideal $J = \overline{J_j} + M^{j+1}$ is the ancestor ideal of the vector space J_j, modulo M^{j+1}:*

$$\mathrm{Ann}(f) = \overline{\mathrm{Ann}(f) \cap R_j} + M^{j+1}. \qquad (2.3.7)$$

Proof of Lemma 2.17. If $\psi \in J_i$ and $i \leq j$ we have $(R_{j-i}\phi) \circ f = R_{j-i} \circ (\phi \circ f) = 0$; thus, $R_{j-i}\phi \subset J_j$, and we have $J \subset \overline{J_j} + M^{j+1}$. Suppose by way of contradiction that the nonzero homogeneous form $\psi \in R_i$ satisfies $\psi \in \overline{J_j} - J_i$, and that ψ has maximal degree $i \leq j$ among such forms. Evidently, $i < j$. By assumption $R_1\psi \subset (\overline{J_j})_{i+1} = J_{i+1}$, so $(R_1\psi) \circ f = 0$. This implies that $\psi \circ f$ is a nonzero element of $\mathcal{D}_{j-i}$ satisfying $R_1 \circ (\psi \circ f) = 0$, hence $R_i \circ (\psi \circ f) = R_{i-1} \circ (R_1 \circ (\psi \circ f)) = 0$ which is not possible, as the apolarity pairing $(\phi, g) \mapsto \phi \circ g$, $R_i \times \mathcal{D}_i \to k$ is exact. This shows that $J = \overline{J_j} + M^{j+1}$ and proves Lemma 2.17. $\qquad \square$

Proof of Lemma 2.14. Suppose first that $J = \mathrm{Ann}(f)$, where f is homogeneous of degree j. Then $J \supset M^{j+1}$, and J_j has codimension one in R_j, so A_j has dimension one and is in $Soc(A)$. Since $J = \overline{J_j} + M^{j+1}$, it follows that $Soc(A) \subset A_j$, hence $Soc(A) = A_j$, which has vector space dimension one as a k-vector space, implying $A = R/J$ is Gorenstein.

Now suppose conversely that $A = R/J$ is a graded Artinian Gorenstein quotient of R, of socle degree j. Since $A_j \subset Soc(A)$, a dimension one vector space over k, it follows that $A_j = Soc(A)$, and that J_j has codimension one as a vector subspace of R_j. Let $\langle f \rangle$ be the annihilator of J_j in the exact pairing $R_j \times \mathcal{D}_j \to k$. By 2.3.7, $\text{Ann}(f) = \overline{J_j} + M^{j+1}$. Any ideal I of R satisfies

$$i < j \Rightarrow R_{j-i} I_i \subset I_j \quad \text{and} \quad I_i \subset \overline{I_j}. \qquad (2.3.8)$$

Since $J \cap M^j = J_j \cap M^j + M^{j+1}$ here, we have $J \subset \overline{J_j} + M^{j+1}$. Since A has no socle in degrees less than j, it follows that $J \supset \overline{J_j}$, and we have $J = \text{Ann}(f)$. The homomorphism $C_f(j - i, i) : R_i \to R_i \circ f$ has kernel $\text{Ann}(f)_i = J_i$, and image $R_i \circ f = \mathcal{D}_{j-i}$. It follows that $h_{j-i}(f) = \text{rk}(C_f(j - i, i)) = \dim_k(R_i/J_i) = H(A_f)_i$, showing 2.3.4. The equality (1.1.5) shows that $H_f = H(A_f)$ is symmetric with respect to $\frac{j}{2}$. $\qquad \square$

Proof of Lemma 2.15. If $i \leq j$, and $\phi \in J_i$, then $\phi \circ (R_{j-i} \circ f) = (R_{j-i}\phi) \circ f = 0$, so $J_i \subset (\text{Ann}(R_{j-i} \circ f)) \cap R_i$. Conversely, if $\phi \in (\text{Ann}(R_{j-i} \circ f)) \cap R_i$, then $R_{j-i}\phi \subset \text{Ann}(f) \cap R_j$, and by Lemma 2.17 we have $\phi \in J_i$. This shows 2.3.5 and Lemma 2.15. $\qquad \square$

Proof of Corollary 2.16. If $\phi \in R_j$, then clearly $\phi \in J_j$ iff $\phi \in \text{Ann}(f)_j$, as $f(\partial \cdot /\partial x_1, \ldots, \partial \cdot /\partial x_r) \circ \phi = \phi(\partial \cdot /\partial X_1, \ldots, \partial \cdot /\partial X_r) \circ f\}$. The Corollary now follows from Lemma 2.15. $\qquad \square$

Remark. A talk by C. Peskine (who uses the differential form approach) inspired us to include Corollary 2.16.

We now restrict to $r = 3$. Using Lemma 1.19, and the Minimal Resolution Theorem for the ideals of s general-enough points of $\mathbb{P}^2$, we determine the minimal generator degrees for the ideal $I = \text{Ann}(f)$, where $f = L_1^j + \cdots + L_s^j$ is the sum of s j-th powers of general-enough linear forms. This illustrates the connection between zero-dimensional subschemes of $\mathbb{P}^n$, and Gorenstein Artin algebras that we develop later.

We let $I = \text{Ann}(f)$, and denote by $v(I)$ the number of minimal generators of I. We let $D(I) = (d_1, \ldots, d_{v(I)})$, $d_1 \leq d_2 \leq \cdots$, denote the generator degrees of I. When $r = 3$, the generator degrees D of a graded Gorenstein ideal determines the Hilbert function and the relation degrees for I. Recall that $r_t = \dim_k R_t$. If s satisfies $r_{t-1} \leq s \leq r_t$, and $j = 2t + 1$, then $H(s, j, r) = (1, \ldots, r_{t-1}, s, s, r_{t-1}, \ldots, 1)$; if $j = 2t$, then $H(s, j, r) = (1, \ldots, r_{t-1}, s, r_{t-1}, \ldots, 1)$. If $\text{char}(k) < j$, we need to replace the sum f by a sum $f = L_1^{[j]} + \cdots + L_s^{[j]}$ below, and use the contraction action of R on $\mathcal{D}$ in place of differentiation.

The following result is a special case of the Cohen-Macaulay type theorem proven for all $\mathbb{P}^n$ by N. V. Trung and G. Valla — that for s

general points of $\mathbb{P}^n$ the Cohen-Macaulay type (dimension of the socle of a minimal reduction of the coordinate ring of the points) has the minimum possible value.

THEOREM 2.18. [GO, TrV] MINIMAL RESOLUTION FOR IDEALS DEFINING s-GENERAL POINTS OF $\mathbb{P}^2$. *Let* $s = \nu(\nu + 1)/2 + \alpha$, *where* $\alpha \leq \nu+1$. *There is an open dense subset of the variety* $U^s \subset Hilb^s(\mathbb{P}^2)$ *parametrizing sets* S *of* s *distinct points of* $\mathbb{P}^2$, *such that the Cohen-Macaulay type of* $A_S = R/\mathcal{I}_Z$ *is* $\beta = \alpha + min\{\nu - 2\alpha, 0\}$. *Set* $\gamma = min\{2\alpha - \nu, 0\}$. *The minimal resolution of such* A_S *is*

$$0 \longrightarrow R(-\nu - 2)^\alpha \oplus R(-\nu - 1)^{\beta-\alpha}$$
$$\longrightarrow R(-\nu)^{\nu+1-\alpha} \oplus R(-\nu - 1)^\gamma \longrightarrow A_S \longrightarrow 0 \quad (2.3.9)$$

PROPOSITION 2.19. MINIMAL RESOLUTION FOR $\text{Ann}(f)$, f A GENERAL-ENOUGH POWER SUM, WHEN $r = 3$. *Assume that* $\text{char}(k) = 0$ *or* $\text{char}(k) = p > j$, *and that* $f = L_1^j + \cdots + L_s^j$ *is a general enough sum of* s *j-th powers of linear forms, and let* $I = \text{Ann}(f)$. *Let* J *be a generic Gorenstein ideal of Hilbert function* $T = H(s, j, 3)$.

A. *If* $r_{t-1} \leq s \leq r_t$ *and* $j = 2t + 1$, *then the minimal number of generators* $v(I)$ *satisfies,*

$$v(I) = t + 1 + |2(s - r_{t-1}) - t| .$$

The ideal $I = \text{Ann}(f)$ *has* $s - r_{t-1}$ *(if* $s - r_{t-1}$ *is even) or* $s - r_{t-1} - 1$ *(if* $s - r_{t-1}$ *is odd) extra generators in degree* $t + 2$, *compared to* J; *in all other degrees* $i \neq t + 2$, *I has the same numbers of generators as* J. *The ideal* I *has the minimal number of generators possible for a Gorenstein ideal of Hilbert function* $H(s, j, 3)$ *iff* $s = r_{t-1}$, *and* $v(I)$ *is the maximal possible given* $H(s, j, 3)$ *iff* $s = r_t - 1$ *or* $s = r_t$.

B. *If* $r_{t-1} \leq s \leq r_t$ *and* $j = 2t$, *then there are several cases:*

  i. *If* $r_{t+1} - 3(r_t - s) \geq s$, *then* $v(I) = v(J)$.
  ii. *If* $r_{t+1} - 3(r_t - s) = s - a$, $a > 0$, *and* a *is even then* $v(I) = v(J)+a$, *and* I *has* a *extra generators in degree* $t+1$. *If* a *is odd, then* $v(I) = v(J)+a+1$, *and* I *has* a *extra generators in degree* $t + 1$, *as well as 1 extra generator in degree* $t + 2$, *compared to* J.

C. *If* $s \leq r_{t-1}$, *and* $j = 2t$ *or* $j = 2t + 1$, *then* $v(I) = v(J)$.

PROOF. This is a straightforward exercise, which we leave for the reader. The minimal resolution for I results from

  a. The known minimal resolution theorem for the vanishing ideal I_P at s general points of $\mathbb{P}^2$ (Lemma 2.18).

b. The identity $(I_P)_i = I_i$ for $i \leq t$ if $j = 2t$, for $i \leq t+1$ if $j = 2t + 1$, from Lemma 1.19.

c. The symmetry of the minimal resolution for I.

The statements comparing $v(I)$ and $v(J)$, or referring to an ideal of Hilbert function $H(s, j, 3)$ with maximum number of generators require also

d. The known minimal resolution for J, and the known minimal resolution for a Gorenstein ideal of Hilbert function $H(s, j, 3)$ having the maximum number of generators (see [BE2], [Di], and the proof of Theorem 5.25(v.g) below).

$\square$

EXAMPLE 2.20. Let $(s, j, r) = (8, 7, 3)$, so

$$H(8, 7, 3) = (1, 3, 6, 8, 8, 6, 3, 1).$$

If $I = \mathrm{Ann}(f)$, where f is a general enough length 8 power sum, then $I = \mathrm{Ann}(f)$ has generator degrees $(3, 3, 4, 5, 5)$. The generic ideal J in $Gor(T)$, $T = H(8, 7, 3)$ is a complete intersection of generator degrees $(3, 3, 4)$. A Gorenstein ideal in $Gor(T)$ with maximum number of generators has generator degrees $(3, 3, 4, 4, 5, 5, 6)$.

PROBLEM 2.21. Let T satisfy $T_{\max} = s$, and suppose D is the sequence of generator degrees for an ideal $I = \mathrm{Ann}(f)$, $f \in Gor(T)$. Let $Gor_D(T) \subset Gor(T)$ be the algebraic set parameterizing degree-j forms f such that $\mathrm{Ann}(f)$ has generator degrees D. When $r = 3$, $Gor_D(T)$ is irreducible (S. J. Diesel, [Di], see Theorem 5.25). For what triples (s, j, D) is $\overline{Gor_D(T)} = \overline{PS(s, j; 3)} \cap \overline{Gor(T)}$? When $T = H(s, j, 3)$ and $(s, j) = (9, 6)$ a dimension calculation shows there is equality for $D = D_{\min}(T)$; but if $(s, j) = (14, 8)$, $\dim Gor(T) = 44$, and $\dim PS(14, 8; 3) = 42$.

We refer the reader to Section 5.6 for some partial results (Theorem 5.71).

CHAPTER 3

Tangent Spaces to Catalecticant Schemes

In Section 3.1 we determine the tangent spaces to the schemes $V_d(u, v; r)$ of r-variable $\dim_k R_u$ by $\dim_k R_v$ catalecticant matrices (Theorem 3.2), and we use it to elucidate an example of S. J. Diesel where there are several components to $V_4(2, 4; 3)$ (Examples 3.6, 3.7). In Section 3.2 we determine the tangent space to $\mathbf{Gor}(T)$ (Theorem 3.9) and we give several applications. We give a new proof of the existence of compressed Gorenstein Artin algebras of the specified maximal Hilbert function $H(j, r)$; and we give a natural geometric proof of the known irreducibility and normality of the determinant of the symmetric catalecticant matrix (Propositions 3.12 and 3.13). We next generalize this result to show that the corank one catalecticant variety, $V_s(u, v; r)$ with $s = r_u - 1, u \leq v$ is irreducible, and normal, with singular locus the corank two catalecticants; we also show that its projectivization is rational (Theorem 3.14).

We use Theorems 3.2 and 3.9 to show a key Lemma relating the Gorenstein ideal $I = \mathrm{Ann}(f)$, $f = L_1^{[j]} + \cdots + L_s^{[j]}$, and a vanishing ideal of points in $\mathbb{P}^{r-1}$ (Lemma 3.16). We use these results and the computer algebra program "Macaulay" [BaS1] to give examples illustrating a pattern in the dimensions of the tangent space to $\mathbf{Gor}(T)$ at f, for f a sum of powers of s general linear forms (Examples 3.18, 3.19, 3.22). The examples suggest conjectures generalizing our main results concerning these dimensions to arbitrary triples (s, j, r) (Conjectures 3.20, 3.23, 3.25).

3.1. The tangent space to the determinantal scheme $V_s(u, v; r)$ of the catalecticant matrix

Throughout the following u, v are nonnegative integers, $j = u + v$. We use the notation introduced in Definition 1.4. If f is a homogeneous element in $\mathcal{D}_j$, we let $Cat_f(u, v; r)$ denote the matrix of the homomorphism $C_f(u, v) : R_v \to \mathcal{D}_u$, $g \to g \circ f$ (contraction) with respect to the standard lexicographic monomial bases for R_v and $\mathcal{D}_u$ (see Appendix C.1). If $I = \mathrm{Ann}(f)$, we let $[I_v]$ denote the matrix with column vectors obtained from writing a basis of I_v in terms of the standard lexicographically ordered basis $\{x^V \; : \; |V| = v\}$ for R_v; thus, $[I_v]$ is a $\dim_k(R_v) \times \dim_k(I_v)$ matrix with entries in k. If the sequence T satisfies $T_u \leq s$, we denote by $\pi_{u,v,T}$ the projection $\pi_{u,v,T} : \mathbf{Gor}(T) \to \mathbf{V}_s(u, v; r)$.

We determine the tangent space to $\mathbf{V}_s(u, v; r)$ at a point $p \in \mathbb{A}(\mathcal{D}_j)$ in terms of the ideal $I = \mathrm{Ann}(f_p)$ (Theorem 3.2). We also show that the rank-s locus $U_s(u, v; r)$ is dense in $V_s(u, v; r)$ (Lemma 3.5). The following result is immediate from the definitions and (1.1.5) (see also §70 of [**Mac2**]).

Lemma 3.1. **Kernel and image of the catalecticant ho-momorphism.** *The kernel of $C_f(u, v)$ is I_v. The image of $C_f(u, v)$ is $R_v \circ f$; and ${}^tC_f(v, u) = C_f(u, v)$. We have*

$$(\mathrm{Im}\, C_f(u, v))^\perp \;=\; \mathrm{Ker}(C_f(v, u)) \cap R_u. \tag{3.1.1}$$

Theorem 3.2. **Tangent space to the determinantal loci.** *Suppose that the closed point f_0 of the scheme $\mathbf{V}_s(u, v; r)$ is not in $\mathbf{V}_{s-1}(u, v; r)$, and that $I = \mathrm{Ann}(f_0)$ is the corresponding ideal of R. We let $A = Cat_{f_0}(u, v, r)$. Then the tangent space $T_{f_0}(\mathbf{V}_s(u, v; r)) \subset \mathcal{D}_j$ to the scheme $\mathbf{V}_s(u, v; r)$ at the point f_0 equals any of the following three vector spaces of $\mathcal{D}_j$:*

   i. $\{f|$ *for* $B = Cat_f(u, v; r)$ *we have* $B \cdot \mathrm{Ker}(A) \subset \mathrm{Im}(A)\}$;

   ii. $\{f|$ *for* $B = Cat_f(u, v; r)$ *we have* ${}^t[I_u] \cdot B \cdot [I_v] = 0\}$;

   iii. $\{f \in \mathcal{D}_j | (I_u \cdot I_v) \circ f = 0\}$.

Proof. The first statement follows from the description of the tangent space to a determinantal variety [**ACGH**, §II.2], since the catalecticant map $\mathcal{D}_j \to \mathrm{Hom}(R_v, \mathcal{D}_u)$, $f \mapsto Cat_f(u, v; r)$ is linear and injective. The second is immediate from Lemma 3.1 and the first. The third is equivalent to the second since the linear span of the vectors $B \cdot [I_v]$ equals by definition $I_v \circ f$, so the span of the entries of the matrix ${}^t[I_u] \cdot B \cdot [I_v]$ is (see (A.0.2))

$$\langle I_u, I_v \circ f \rangle \;=\; \langle I_u I_v, f \rangle \;=\; (I_u I_v) \circ f.$$

Example 3.3. **Tangent space, $r = 2$.** We take $r = 2$, $u = v = 3$, and let $f_0 = X^{[6]} + Y^{[6]} + (X + Y)^{[6]}$ in $\mathcal{D} = k[X, Y]$; then $I_3 = \mathrm{Ann}(f_0)_3 = (x^2y - xy^2)$. If $f = aX^{[6]} + \cdots + gY^{[6]}$, then the tangent space T to $\mathbf{V}_3(3, 3; 2)$ at $Cat_{f_0}(3, 3; 2)$ is defined from the Hankel matrix by

$$\mathcal{T} = \left\{ B = \begin{pmatrix} a & b & c & d \\ b & c & d & e \\ c & d & e & f \\ d & e & f & g \end{pmatrix} \Bigm| B \begin{bmatrix} 0 \\ 1 \\ -1 \\ 0 \end{bmatrix} \subset \left\langle \begin{bmatrix} 1 \\ 0 \\ 0 \\ 0 \end{bmatrix}, \begin{bmatrix} 0 \\ 0 \\ 0 \\ 1 \end{bmatrix}, \begin{bmatrix} 0 \\ 1 \\ 1 \\ 0 \end{bmatrix} \right\rangle \right\}$$

$$= \left\{ B \Bigm| \begin{bmatrix} 0 & 1 & -1 & 0 \end{bmatrix} \begin{bmatrix} b-c \\ c-d \\ d-e \\ e-f \end{bmatrix} = 0 \right\}$$

$$= \{ B \mid 2d = c + e \}.$$

This is a space of codimension one, as we expect, since $\mathbf{V}_3(3, 3; 2)$ is defined by $\det(B) = 0$. The space is by Theorem 3.2(ii) also

$$\mathcal{T} = \{ f \mid (I_3)^2 \circ f = 0 \}$$
$$= \{ f \mid (xy(x - y))^2 \circ f = 0 \}$$
$$= \{ f \mid 2d = c + e \}.$$

REMARK 3.4. When $r = 2$ and $s \leq u \leq j - u$, L. Gruson and C. Peskine, and also J. Watanabe have shown that the scheme $\mathbf{V}_s(u, j - u; 2)$ is independent of u ([**GruP**, Lemma 2.3], and [**Wa4**]; see also Section 1.3). When $r > 2$, even the dimension of the algebraic sets $V_s(u, j - u; r)$ depends on u, by Theorems 4.1A, 4.5A (see also [**Di**]).

Recall that $U_s(u, v; r)$ is the locus of DP-forms $f \in \mathcal{D}_j$ whose catalecticant matrix $Cat_f(u, v; r)$ has rank s.

LEMMA 3.5. DENSITY OF RANK-s CATALECTICANTS. If $1 \leq s \leq \min(\dim_k R_u, \dim_k R_v)$, then $U_s(u, v; r)$ is dense in $V_s(u, v; r)$.

PROOF. We have

$$V_s(u, v; r) = \bigcup_{i=0}^{s} U_i(u, v; r),$$

so by induction on i, it suffices to prove that for $1 \leq i \leq s - 1$

$$U_i(u, v; r) \subset \overline{U_{i+1}(u, v; r)}. \qquad (3.1.2)$$

Let $f \in U_{i+1}(u, v; r)$. Since $\langle L^{[k]} \mid L \in \mathcal{D}_1 \rangle = \mathcal{D}_k$ for every $k \geq 1$ (see Corollary 1.16), there exists an open Zariski dense subset of $\mathbb{A}^r = \mathbb{A}(\mathcal{D}_1)$ which consists of forms L such that both

$$L^{[u]} \notin R_v \circ f, \qquad L^{[v]} \notin R_u \circ f. \qquad (3.1.3)$$

We claim that for every $c \in k$, $c \neq 0$, the DP-form

$$f_c = f + cL^{[j]}$$

belongs to $U_{i+1}(u, v; r)$, which proves (3.1.2). Indeed,

$$\dim_k(R_v \circ f_c) \leq \dim_k(R_v \circ f) + 1,$$

so it suffices to prove that $I(f_c)_v \subsetneq I(f)_v$. Let $\phi \in I(f_c)_v$. Then by Lemma 1.15(i),

$$\phi \circ f + c \cdot \phi(L)L^{[u]} = 0.$$

Thus by (3.1.3), $\phi \circ f = 0$, $\phi(L) = 0$, so

$$I(f_c)_v \subset I(f)_v \cap I(L^{[j]})_v = (R_u \circ f + \langle L^{[u]} \rangle)^\perp.$$

By (3.1.3) the right space is strictly contained in $(R_u \circ f)^\perp = I(f)_v$. This completes the proof. $\square$

We now give an example, due to S.J. Diesel [**Di**], of a determinantal variety of a non square catalecticant having two irreducible components, when $r = 3$. In Example 3.8 we study the behavior of the tangent space on each of the components. For coherence, we include the reasoning from [**Di**] giving the dimension of the components.

EXAMPLE 3.6. REDUCIBLE DETERMINANTAL LOCUS OF A CATA-LECTICANT MATRIX, $r = 3$. (S.J. Diesel [**Di**]). The locus $V_4(2, 4; 3)$ has two components of the same dimension, 12, corresponding to $T_1 = (1, 3, 4, 4, 4, 3, 1)$, and $T_2 = (1, 3, 4, 5, 4, 3, 1)$.

The first component, $\overline{C_1}$, $C_1 = \pi_{2,4}(Gor(T_1))$ is a complete intersection component, whose general point corresponds to a complete intersection ideal $\text{Ann}(f) = (g_1, g_2, g_3)$, where $(\{\deg g_i\}) = (2, 2, 5)$. (*Warning:* special points of $Gor(T_1)$ need not be complete intersection (CI) ideals). The dimension calculation for the variety $CI(T) \subset \mathbb{P}(\mathcal{D}_j)$ of graded CI ideals of Hilbert function T, of socle degree j, from Example 3 of [**I2**], is

$$\dim CI(T) = \sum_{a=\deg g_i} t_a. \tag{3.1.4}$$

Thus, $\dim \mathbb{P}Gor(T_1) = 2(4) + 1(3) = 11$, and $\dim Gor(T_1) = 12$.

The second component, $\overline{C_2}$, $C_2 = \pi_{2,4}(Gor(T_2))$, corresponds to dual forms f determining ideals $J = \text{Ann}(f)$ whose degree two part $J_2 = \langle g_1, g_2 \rangle$ always has a common factor – there is a linear relation between g_1 and g_2. Thus $J_2 = hV$ where $h \in R_1$ and V is a two-dimensional subspace of R_1. It follows that J_2 is parameterized by

$\mathbb{P}(R_1) \times Grass(2, R_1) = \mathbb{P}^2 \times \mathbb{P}^2$. Once J_2 is chosen, the dual form f determining $J = Ann(f)$ can be chosen arbitrarily from the subspace $\langle R_4 J_2 \rangle^{\perp}$ of $\mathcal{D}_6$. Since $\dim_k \langle R_i J_2 \rangle^{\perp} = 4 + i$, we have $\dim_k \langle R_4 J_2 \rangle^{\perp} = 8$. Thus, $\mathbb{P}Gor(T_2)$ is fibred by $\mathbb{P}^7 = \mathbb{P}(\langle R_4 J_2 \rangle^{\perp})$ over $\mathbb{P}^2 \times \mathbb{P}^2$, so also has dimension 11, and $\dim Gor(T_2) = 12$. We have

$$V_4(2, 4; 3) - V_3(2, 4; 3) = Gor(T_1) \cup Gor(T_2), \quad \text{where}$$

$$Gor(T_1) \cap \overline{Gor(T_2)}' \neq Gor(T_1), \tag{3.1.5}$$

since it is not possible to specialize from a family $\{J(t) \mid t \neq 0\}$ of ideals such that the vector space $J(t)_2$ has a common factor to a general point J of $Gor(T_1)$ where J_2 has no common factor. Here we denote by $\overline{Gor(T_2)}'$ the restriction of the closure to $V_4(2, 4; 3) - V_3(2, 4; 3)$. The intersection

$$\overline{Gor(T_1)} \cap Gor(T_2) = \emptyset, \tag{3.1.6}$$

since rank $Cat_{f_\alpha}(2, 2; 3)) \leq 4$ is a closed condition on the variety $\mathbb{P}(R_6)$ parameterizing the coefficients α of f_α. The two components of $V_4(2, 4; 3)$ satisfy, correspondingly,

$$C_1 \cap \overline{C_2} \neq C_2, \quad \text{and} \quad \overline{C_1} \cap C_2 = \emptyset. \tag{3.1.7}$$

S. J. Diesel also constructed examples of $V_s(i, j - i; 3)$ having components of different dimensions. Using a similar proof, one can show that $V_4(2, j - 2; 3), j \geq 6$ has two irreducible components,

$$\overline{C_1}, \quad C_1 = \pi_{2,j-2}(Gor(T_1)), \quad T_1 = (1, 3, 4, 4, \ldots, 4, 3, 1),$$

$$\overline{C_2}, \quad C_2 = \pi_{2,j-2}(Gor(T_2)), \quad T_2 = (1, 3, 4, 5, 6, \ldots, 3, 1).$$

They have dimensions $\dim(C_1) = 12$, and $dim(C_2) = j + 6$, respectively.

EXAMPLE 3.7. THE DECOMPOSITION OF $\mathbb{A}(\mathcal{D}_6)$ BY THE VARIETIES $Gor(T)$ DOES NOT SATISFY THE FRONTIER PROPERTY. The form $f = X^{[6]} + Y^{[2]} Z^{[4]}$ is in $Gor(T_1) \cap \overline{Gor(T_2)}$ (in the notation of Example 3.6), but by (3.1.5) the general element of $Gor(T_1)$ is not in $\overline{Gor(T_2)}$. This shows that when $r = 3$, the decomposition of the affine space

$$\mathbb{A}^{28} = \mathbb{A}(\mathcal{D}_6) = \bigcup_{T \in \mathcal{H}(6,3)} Gor(T)$$

into irreducible varieties does not satisfy the frontier property. The ideal $I = Ann(f)$ has five generators (see below), and the catalecticant matrix $Cat_f(2, 4; 3)$, is in $C_1 \cap \overline{C_2}$. We next consider the tangent spaces T_f to $\mathbf{Gor}(T_1)$ at f, and $T_f(\mathbf{V}_4(2, 4; 3))$ at $Cat_f(2, 4; 3)$ in Example 3.8.

The following Example uses Theorem 3.2 and also Theorem 3.9 below. We include it here to motivate and illustrate these results on the tangent spaces to the schemes $\mathbf{V}_s(u, v; r)$ and to $\mathbf{Gor}(T)$.

EXAMPLE 3.8. TANGENT SPACE TO THE REDUCIBLE $\mathbf{V}_4(2, 4; 3)$. We consider first $r = 3$, $T_1 = (1, 3, 4, 4, 4, 3, 1)$, and the map: $\pi_{2,4,T_1}$: $\mathbf{Gor}(T_1) \to \mathbf{V}_4(2, 4; 3)$ at the point $f = X^{[6]} + Y^{[2]}Z^{[4]}$. Consider the matrix $B_f = Cat_f(2, 4; 3)$ representing $\pi_{2,4}(f)$ in the scheme $\mathbf{V}_4(2, 4; 3)$. We have, by Example 3.6 and the discussion below,

$$\dim Gor(T_1)_f = \dim \mathbf{V}_4(2, 4; 3)_f = 12.$$

The ideal $I = \mathrm{Ann}(f)$ is

$$I = (xy, xz, y^3, z^5, x^6 - y^2 z^4). \tag{3.1.8}$$

Hence $I_2 = \langle xy, xz \rangle$, and $I_4 = R_2 \langle xy, xz \rangle + R_1 y^3$. The tangent space $T_B = T_f(\mathbf{V}_4(2, 4; 3))$ to the scheme $\mathbf{V}_4(2, 4; 3)$ at B_f is $T_B = \langle I_2 I_4 \rangle^\perp$ where

$$I_2 I_4 = x^2 \langle y, z \rangle^4 \oplus x^3 \langle y, z \rangle^3 \oplus x^4 \langle y, z \rangle^2 \oplus xy^3 \langle y, z \rangle^2. \tag{3.1.9}$$

Since $\dim_k \langle I_2 I_4 \rangle = 15$, we have

$$\dim_k T_B = \mathrm{cod}_k (I_2 I_4) = \dim_k R_6 - 15 = 13, \tag{3.1.10}$$

showing that the variety $\mathbf{V}_4(2, 4; 3)$ is singular at B.

By Theorem 3.9, the tangent space T_f to $\mathbf{Gor}(T_1)$ satisfies, $T_f = \langle (I^2)_6 \rangle^\perp$, where

$$(I^2)_6 = I_2 I_4 + I_3 I_3 = I_2 I_4 \oplus \langle y^6 \rangle, \tag{3.1.11}$$

of dimension 16. Thus $\dim_k T_f = 12$, so f is a nonsingular point of $\mathbf{Gor}(T_1)$. The extra information in $\mathbf{Gor}(T_1)$ over that in $\mathbf{V}_4(2, 4; 3)$ at f removes the singularity of $\mathbf{V}_4(2, 4; 3)$ at f. The linear map $(\pi_{2,4,T_1})_*$: $T_f \to T_B$ is just the inclusion of T_f in T_B as subspaces of $\mathcal{D}_j$. In contrast, if $f' = X^{[6]} + Y^{[3]}Z^{[3]})$ in $\mathbf{Gor}(T_2)$, $T_2 = (1, 3, 4, 5, 4, 3, 1)$, then the ideal $J = \mathrm{Ann}(f')$ is

$$J = (xy, xz, y^4, z^4, x^6 - y^3 z^3). \tag{3.1.12}$$

The product $J_2 J_4$ satisfies

$$J_2 J_4 = x^2 \langle y, z \rangle^4 \oplus x^3 \langle y, z \rangle^3 \oplus x^4 \langle y, z \rangle^2 \oplus x \langle y, z \rangle \langle y^4, z^4 \rangle = (J^2)_6,$$

of dimension 16, so the tangent space $T_{B'} = (J_2 J_4)^\perp$ to $\mathbf{V}_4(2, 4; 3)$ at $B' = Cat_{f'}(2, 4; 3)$ has dimension 12. Since the tangent space map

$(\pi_{2,4,T_2})_* : \mathcal{T}_{f'} \to \mathcal{T}_{B'} = ((J^2)_6)^{\perp}$ is an isomorphism, it follows that the map $\pi_{2,4,T_2} : \mathbf{Gor}(T_2) \to \mathbf{V}_4(2,4;3)$ is also an isomorphism at f'. A similar result is obtained by considering a general enough element f'' in $\mathbf{Gor}(T_1)$: we conclude that $\pi_{2,4,T_1} : \mathbf{Gor}(T_1) \to \mathbf{V}_4(2,4;3)$ is an isomorphism at f''.

3.2. The tangent space to the scheme $\mathbf{Gor}(T)$ parametrizing forms with fixed dimensions of the partials

We use Theorem 3.2 to find the tangent space to the scheme $\mathbf{Gor}(T)$ (see Definition 1.10), and we give several applications and examples. Recall that the order $d = \nu(I)$ of a graded ideal I is the minimum d such that $I_d \neq 0$.

THEOREM 3.9. *The tangent space $\mathcal{T}_{f_0}$ to the scheme $\mathbf{Gor}(T)$ at a point $f_0 \in \mathcal{D}_j$ defining an ideal $I = \mathrm{Ann}(f_0)$ of order d satisfies*

$$\mathcal{T}_{f_0} = \{f \in \mathcal{D}_j | I_i \cdot I_{j-i} \circ f = 0 \text{ for } d \leq i \leq j/2\}$$
$$= (I_d \cdot I_{j-d} + I_{d+1} \cdot I_{j-(d+1)} + \cdots + I_{[j/2]} \cdot I_{j-[j/2]})^{\perp} \cap \mathcal{D}_j$$
$$= ((I^2)_j)^{\perp} \cap \mathcal{D}_j. \tag{3.2.1}$$

The dimension of the tangent space satisfies

$$\dim_k \mathcal{T}_{f_0} = \dim_k(R_j/(I^2)_j) = H(R/I^2)_j. \tag{3.2.2}$$

PROOF. Immediate from Part iii of Theorem 3.2, as the tangent space to $\mathbf{Gor}(T)$ at f_0 is the intersection of the tangent spaces to $\mathbf{V}_{t_i}(i, j - i; r)$ at f_0, $i = 1, 2, ..., [j/2]$. The last equality of (3.2.1) follows from I being a graded ideal. □

REMARK 3.10. In the well-known isomorphism, $\mathrm{Hom}(I, A) \cong I/I^2$, for I graded Gorenstein, the degree zero portion $\mathrm{Hom}_0(I, A)$ under the usual k^*-action – corresponding to deformations of I with the same Hilbert function – gets mapped to the degree-j portion of $(I/I^2)_j$. Since I_j has codimension one in R_j, the dimension of $\mathcal{T}_f$ is one more than that of $\mathrm{Hom}_0(I, A)$, corresponding to $\mathcal{T}_f$ being the tangent space to $\mathbf{Gor}(T)$, the affine cone over $\mathbb{P}\mathbf{Gor}(T)$, the projective scheme parameterizing Gorenstein ideals of Hilbert function T. The punctual Hilbert scheme $\mathbf{Hilb}^n(\mathbb{A}^r)$ parameterizes Artin quotients of R having dimension n as a k-vector space. The module $\mathrm{Hom}(I, A)$ is the tangent space to $\mathbf{Hilb}^n(\mathbb{A}^r)$, $n = \dim_k R/I$, at the point parameterizing the ideal $I \subset R$. Thus it is natural that the degree zero portion $\mathrm{Hom}_0(I, A)$ is the tangent space to $\mathbb{P}\mathbf{Gor}(T)$, as we have shown. See also Remark 4.3.

DEFINITION 3.11. A Gorenstein Artin algebra $A = R/I$ of socle degree j is compressed if it has the maximum possible Hilbert function, given the socle degree, and the embedding dimension r.

Several authors have shown that this maximum is $H(A) = H(j,r)$ where $H(j,r)_i = \min(\dim_k R_i, \dim_k R_{j-i})$ ([**EmI1**], [**Gre**], [**FL**], [**I3**]). Our description of the tangent space to the scheme $\mathbf{Gor}(T)$ gives a new, concise proof that compressed Gorenstein Artin algebras have the expected Hilbert function.

PROPOSITION 3.12. HILBERT FUNCTION OF A COMPRESSED GO-RENSTEIN ALGEBRA. *For f a general homogeneous element of $\mathcal{D}_j$, the Hilbert function $H_f = H(R/\operatorname{Ann}(f))$ is $H(j,r)$.*

PROOF. If the Hilbert function $T = H_f$ is not $H(j,r)$, then some pair I_i, I_{j-i} are both nonzero, so $I_i \cdot I_{j-i} \neq 0$, and the tangent space T_f must have dimension no larger than $N = \dim_k \mathcal{D}_j - 1$ by (3.2.2). Thus, $Gor(T)$ cannot be dense in $\mathcal{D}_j$, and $Gor(T)$ has dimension less than N. Since $\mathbb{A}^N$ is not the union of a finite number of lower-dimensional varieties, it follows that $Gor(T)$, $T = H(j,r)$ is dense in $\mathbb{A}^N$. $\square$

When $u = v = t$, so $j = 2t$ we let D denote the determinant $D = \det(Cat_F(t,t;r))$: it is different from zero by Lemma 1.17. Thus, $\mathbf{D}(j,r) = \operatorname{Spec}(R/(D))$ is the catalecticant hypersurface $\mathbf{V}_s(t,t;r)$, $s = r_t - 1$ (recall that $r_i = \dim_k R_i$).

PROPOSITION 3.13. IRREDUCIBILITY OF THE CATALECTICANT DE-TERMINANT. *The determinant $D = \det(Cat_F(t,t;r))$ is irreducible. The variety $V_s(t,t;r)$, $s = r_t - 1$, is normal.*

PROOF. The scheme $\mathbf{Gor}(T)$, $T = H(s,j,r)$, $s = \dim_k R_t - 1$, is an open subscheme of the hypersurface $\mathbf{D}(j,r) \subset \mathbb{A}(\mathcal{D}_j)$. Let f be any closed point of $\mathbf{Gor}(T)$, and let $I = \operatorname{Ann}(f)$. The tangent space T_f to $\mathbf{Gor}(T)$ at f is just $\langle (I_t)^2 \rangle^\perp \cap \mathcal{D}_j$. Since $\dim_k I_t = 1$, the tangent space satisfies,

$$\dim T_f = \dim_k \mathcal{D}_j - 1.$$

This shows that f is a nonsingular point of the hypersurface $\mathbf{D}(j,r)$. Denoting by $D(j,r)$ the algebraic set $D(j,r)_{red}$, we have

$$D(j,r) = \bigcup_{T' \leq T} Gor(T).$$

Here $T' \leq T$ means that $t'_i \leq t_i$ for every i. We next show that if $T' \leq T$ and $T' \neq T$, then the codimension of the locus $Gor(T')$ in

$\mathbb{A}(\mathcal{D}_j)$ satisfies

$$\operatorname{cod} Gor(T') \geq 3.$$

Indeed, let T' be one such Hilbert function, let $p \in Gor(T')$, let $f = f_p$, and $I = \operatorname{Ann}(f)$. Then $\dim_k(I_t) \geq 2$ (if T' and T differ in degree $i < t$, then $I_t \supset R_{t-i} I_i$ and $\dim_k I_t \geq r$).

Let gv, hv be two linearly independent elements of I_t where g, h are relatively prime. Then

$$v^2 \langle g^2, gh, h^2 \rangle \subset (I^2) j$$

is a 3-dimensional subspace. Therefore the codimension in $\mathcal{D}_j$ of $T_f = (I^2)_j^{\perp}$ is at least 3 by Theorem 3.9.

We have shown that $\mathbf{D}(j, r)$ is a reduced hypersurface whose singular locus has codimension at least two in it. Thus, D is irreducible. The normality of $V_{r_t-1}(t, t; r)$ follows from Serre's Criterion. $\qquad\square$

The proposition we just proved has a generalization, a particular case of which we already encountered in Section 1.3 (Theorem 1.45).

THEOREM 3.14. *Let $u \leq v$, let $s = r_u - 1 = \dim_k R_u - 1 = \binom{u+r-1}{r-1} - 1$. Then the corank-one catalecticant scheme $\mathbf{V}_s(u, v; r)$ is irreducible of dimension $r_j - r_v + r_u - 1$. It is reduced, Cohen-Macaulay, normal and $\operatorname{Sing}(V_s(u, v; r)) = V_{r_u-2}(u, v; r)$. The variety $\mathbb{P}V_s(u, v; r) \subset \mathbb{P}(\mathcal{D}_j)$ is rational of degree $\binom{r_v}{r_u-1}$.*

PROOF. The proof is essentially the same as in the special case $r = 2$ which we considered in Section 1.3. For completeness we include the general case.

Let $s = r_u - 1$, let $\mathbf{V}_s = \mathbf{V}_s(u, v; r)$, $V_s = (\mathbf{V}_s(u, v; r))_{red}$. From the equality ${}^t Cat_F(u, v; r) = Cat_F(v, u; r)$ we have $\mathbf{V}_s = \mathbf{V}_s(v, u; r)$. The catalecticant variety $V_s = (\mathbf{V}_s(v, u; r))_{red}$ consists of elements $f \in \mathcal{D}_j$ for which there is a form $\phi \in R_u$ apolar to f. We consider the variety $X \subset \mathbb{P}(R_u) \times \mathcal{D}_j$

$$X = \{(\phi, f) \mid \phi \circ f = 0\}$$

with its two projections

$$X \xrightarrow{\ \pi_2\ } \mathcal{D}_j$$
$$\pi_1 \downarrow \qquad\qquad\qquad\qquad (3.2.3)$$
$$\mathbb{P}(R_u)$$

The image of π_2 is by definition V_s. The fiber of π_1 over a point $[\phi]$ is $(\phi R_{j-u})^{\perp} \subset \mathcal{D}_j$. This is a vector space of dimension $\dim_k R_j -$

$\dim_k R_{j-u} = r_j - r_v$. So, X is an irreducible variety of dimension

$$\dim X = r_j - r_v + r_u - 1,$$

therefore V_s is also irreducible. The map π_2 has inverse π_2^{-1} over the rank-s open dense subset $U_s(v, u; r)$ (Lemma 3.5) by Cramer's formulas. This yields the formula for $\dim V_s$ and proves that $\mathbb{P}(V_s)$, being birational to a projectivization of an algebraic vector bundle over $\mathbb{P}(R_u)$, is a rational variety.

Consider the general corank 1 determinantal locus $D_{r_u-1}(r_u, r_v) \subset D(r_u, r_v) = \mathbb{A}^{r_u r_v}$. It is well-known that its ideal is generated by the $r_u \times r_u$ minors of the generic $r_u \times r_v$ matrix, it is of codimension $r_v - r_u + 1$ and is Cohen-Macaulay. The ideal of $\mathbf{V}_s(u, v; r)$ consists by definition of the polynomials in the ideal of $D_{r_u-1}(r_u, r_v)$ restricted to the subspace of catalecticant matrices. The formula for $\dim V_s$ shows that the codimensions of the two schemes are equal, hence $\mathbf{V}_s$ is also Cohen-Macaulay (see e.g. [ACGH, p.84]).

Using [Ei2, Theorem 18.15] or the argument we used on page 33 we see that in order to prove $\mathbf{V}_s$ is reduced it suffices to verify that it is generically smooth. We prove that $\mathbf{V}_s$ is smooth along the rank-s open subscheme $\mathbf{U}_s = \mathbf{U}_{r_u-1}(u, v; r)$. Indeed, for a closed point $f \in \mathbf{U}_s$ we have for $I = \mathrm{Ann}(f)$, $\dim_k I_u = 1$, $I_u = \langle \phi \rangle$, $\dim_k I_v = r_v - r_u + 1$. By Theorem 3.2

$$\mathcal{T}_f \mathbf{V}_s = (I_u I_v)^{\perp} = (\phi I_v)^{\perp} \subset \mathcal{D}_j.$$

Since ϕI_v is a subspace of R_j of dimension $r_v - r_u + 1$ we obtain that the codimension of $\mathcal{T}_f \mathbf{V}_s$ in $\mathcal{D}_j$ is $r_v - r_u + 1 = \mathrm{cod}_{\mathcal{D}_j}(V_s)$. Thus f is a nonsingular point of $\mathbf{V}_s$.

We obtain that the ideal defining $\mathbf{V}_s$, generated by the $r_u \times r_u$ minors of $Cat_F(u, v; 2)$, is radical, and is the same as the ideal $I(V_s)$ of the algebraic set V_s. The same argument as on page 33, using [Ha, pp.184–185] shows that

$$\mathrm{Sing}(V_{r_u-1}(u, v; r)) = V_{r_u-2}(u, v; r). \qquad (3.2.4)$$

The irreducible variety V_s is a scheme-theoretic proper intersection of the determinantal variety $D_{r_u-1}(r_u, r_v)$ by a linear subspace, which implies that their projectivizations have the same degrees. Hence the formula from [Ha, p.243] yields that $\deg \mathbb{P}(V_s) = \binom{r_v}{r_u-1}$.

It remains to prove that V_s is normal. This would follow using (3.2.4) from Serre's Criterion [Ei2, Theorem 18.15], or [ACGH, pp.98–99], provided we knew that $V_{r_u-2}(u, v; r)$ has codimension ≥ 2 in V_s. Since $U_{s-1} = U_{r_u-2}(u, v; r)$ is open dense in $V_{r_u-2}(u, v; r)$ (Lemma 3.5) it suffices to give an upper bound for $\dim U_{s-1}$. Let

$f \in U_{s-1}$. Then $\dim_k I_u = 2$. If $I_u = \langle \phi, \psi \rangle$, let $\phi = \alpha\beta$, $\psi = \alpha\gamma$, where β and γ are prime polynomials of degree w, $1 \leq w \leq u$. the space of DP-forms in $\mathcal{D}_j$ apolar to $\langle \alpha\beta, \alpha\gamma \rangle$ equals $[\alpha(\beta R_v + \gamma R_v)]^{\perp}$. Since β, γ are prime we have $\dim(\beta R_v + \gamma R_v) = 2r_v - r_{v-w}$. Thus $\dim \langle \alpha\beta, \alpha\gamma \rangle^{\perp} = r_j - 2r_v + r_{v-w}$. Let $\mathbb{G}_w \subset \mathbb{G} = Grass(2, R_u)$ be the closed subvariety consisting of planes of the form $\alpha \langle \beta, \gamma \rangle$ with $\deg \alpha = u - w$, $\deg \beta = \deg \gamma = w$, $1 \leq w \leq u$. We have

$$\dim \mathbb{G}_w = \dim \mathbb{P}(R_{u-w}) + \dim Grass(2, R_w)$$
$$= r_{u-w} - 1 + 2(r_w - 2).$$

Considering for every w a diagram analogous to (3.2.3) with a bottom line $\mathbb{G}_w$ we conclude that the dimension of U_{s-1} (thus of $\dim V_{s-1}$) is bounded from above by $\max_{1 \leq w \leq u}(d_w)$, where

$$d_w \doteq r_j - 2r_v + r_{v-w} + r_{u-w} - 1 + 2(r_w - 2).$$

Let us subtract a d_w from $\dim V_s = r_j - r_v + r_u - 1$. We obtain

$$(r_u - r_w - r_{u-w}) + (r_v - r_{v-w} - r_w) + 4. \tag{3.2.5}$$

Considering the multiplication map $\mathbb{P}(R_w) \times \mathbb{P}(R_{u-w}) \to \mathbb{P}(R_u)$ one obtains the inequality $r_u - r_w - r_{u-w} \geq -1$ which becomes equality only if $w = 0$ or $w = u$ or $r = 2$. We conclude that the integer in (3.2.5) is ≥ 2. This proves that the codimension of $V_{r_u-2}(u, v; r)$ in $V_{r_u-1}(u, v; r)$ is ≥ 2. By Serre's Criterion this implies the normality of $V_{r_u-1}(u, v; r)$. $\qquad \Box$

REMARK 3.15. The minimal resolution of the graded ideal generated by the $r_u \times r_u$ minors $I_{r_u}(Cat_F(u, v; 2))$ is the Eagon-Northcott complex, where for calculating the differentials one uses the generic catalecticant matrix $Cat_F(u, v; 2)$ instead of the generic $r_u \times r_v$ matrix (see e.g. [La1, p.216, p.229] and [Ke1]). The condition required in [La1, p.216] is that the diagram (3.2.3) yields a resolution of the singularities of the scheme $\mathbf{V}_s$. This is the case since X is smooth and $\mathbf{V}_s = (\mathbf{V}_s)_{red} = V_s$.

Note that Theorem 3.14 does not extend to corank two, even when $r = 3$. The corank-two locus $V_4(2, j - 2; 3), j \geq 6$ consists of two irreducible components, of dimensions 12 and $j + 6$, respectively, by Example 3.6: when $j > 6$ the locus has components of different dimension, so is not Cohen-Macaulay. However, in three variables, the corank two locus of the symmetric catalecticant is easily seen to be irreducible, of codimension 3, as a consequence of Diesel's Theorem 1.66 and the known condition for T to be a Gorenstein sequence (see Section 4.4): a similar result to Theorem 3.14 could be shown in that case, using the structure theorem for determinantal ideals of a generic symmetric matrix [**Kut, JPW**]. See Chapter 9 for a further discussion.

The tangent space, and sets of points of $\mathbb{P}^{r-1}$. We now give a key lemma connecting the defining ideal of a set of general points in $\mathbb{P}^{r-1}$ and the tangent space $\mathcal{T}_f(\mathbf{Gor}(T))$ at a point parameterizing the power sum $f \in PS(s, j; r)$.

Recall that if $a > 0$, and $V \subset R_v$, then we define

$$R_a V = \langle gh \mid g \in R_a, h \in V \rangle.$$

If s is a positive integer we define two integers

$$d_{s,r} = \min\{i \mid \dim_k R_{i-1} \leq s < \dim_k R_i\}, \text{ and}$$

$$\tau_{s,r} = \min\{i \mid s \leq \dim_k R_i\}. \tag{3.2.6}$$

Then $d_{s,r} = \tau_{s,r}$ unless $s = \dim_k R_{\tau-1}$, when $d_{s,r} = \tau_{s,r} + 1$. If P is a general set of s points in $\mathbb{P}^{r-1}$ and $d = d_{s,r}$, $\tau = \tau_{s,r}$, then $H(R/\mathcal{I}_P)_i = H(s, j, r)_i$ for $i \leq j - \tau_{s,r}$, and $\mathcal{I}_P = (\mathcal{I}_P)_d + (\mathcal{I}_P)_{d+1} + \cdots$, with $(\mathcal{I}_P)_d \neq 0$. The regularity $\sigma(\mathcal{I}_P) = \tau + 1$, and $\mathcal{I}_P = ((\mathcal{I}_P)_d, (\mathcal{I}_P)_{\tau+1})$ (see Theorem 1.69).

LEMMA 3.16. SQUARE OF THE GORENSTEIN IDEAL DETERMINED BY A SUM OF POWERS, AND OF THE CORRESPONDING VANISHING IDEAL AT POINTS OF $\mathbb{P}^{r-1}$. *Suppose that* $T = H(s, j, r)$, $j = 2t$ *or* $2t + 1$, $s < \dim_k R_t$, *and that* $f = L_1^{[j]} + \cdots + L_s^{[j]}$, *where* $L_1, \ldots, L_s$ *are general enough elements of* $\mathcal{D}_1$, *and let* $I = \mathrm{Ann}(f)$. *Suppose that* $P = \{p_1, \ldots, p_s\}$ *are the corresponding points of* $\mathbb{P}^{r-1}$, *and that* $\mathcal{I} = \mathcal{I}_P$ *is their defining ideal.*

A. *When* $j = 2t$ *then*

$$(\mathcal{I}^2)_j = \mathcal{I}_t^2 = I_t^2 = (I^2)_j. \tag{3.2.7}$$

When $j = 2t + 1$, *then*

$$(\mathcal{I}^2)_j = \mathcal{I}_t \mathcal{I}_{t+1} = I_t I_{t+1} = (I^2)_j. \tag{3.2.8}$$

B. *Suppose that* $s \leq \dim_k R_{t-1}$. *We have, letting* $\sigma = \tau_{s,r} + 1$,

$$(I^2)_j = R_{j-2\sigma} I_\sigma^2 = R_{j-2\sigma}[(\mathcal{I}_P)_\sigma]^2 = (\mathcal{I}_P^2)_j = (\mathcal{I}_P^{(2)})_j. \tag{3.2.9}$$

The tangent space at f *to* $\mathbf{V}_s(t, j - t; r)$ *satisfies for* $s \leq \dim_k R_{t-1}$

$$\mathcal{T}_f(\mathbf{V}_s(t, j - t; r)) = \mathcal{T}_f(\mathbf{Gor}(T)) = (\mathcal{I}_P^{(2)})_j^{\perp}. \tag{3.2.10}$$

If also $\mathrm{char}(k) = 0$ *or* $\mathrm{char}(k) \nmid j$ *we have,*

$$\mathcal{T}_f(\mathbf{Gor}(T)) = \mathcal{D}_1 L_P^{[j-1]}, \tag{3.2.11}$$

namely the vector space span $\mathcal{D}_1 L_P^{[j-1]} = \langle \mathcal{D}_1 L_1^{[j-1]}, \ldots, \mathcal{D}_1 L_s^{[j-1]} \rangle$.

C. *If* $s \leq \dim_k R_{t-1}$, *if* char(k) *is arbitrary, and if* v *satisfies* $\sigma \leq v \leq t$, *then*

$$\mathcal{T}_f(\mathbf{V}_s(v, j-v; r)) \cong \mathcal{T}_f(\mathbf{Gor}(T))$$
$$\cong (R_{j-2\sigma} I_\sigma^2)^\perp = (R_{j-2\sigma}[(\mathcal{I}_P)_\sigma]^2)^\perp$$
$$= (R_{j-2\sigma}(\mathcal{I}_P^{(2)})_{2\sigma})^\perp = ((\mathcal{I}_P^{(2)})_j)^\perp. \qquad (3.2.12)$$

PROOF. By Lemma 1.17, $H(R/I) = H(s, j, r)$. Let d, τ, σ be $d_{s,r}$, $\tau_{s,r}$, and $\tau_{s,r} + 1$, respectively. By Lemma 1.19, the ideal $I = \mathrm{Ann}(f)$ satisfies $I_i = (\mathcal{I}_P)_i$ if $H(s, j, r)_i = s$, or if $i < d$, hence $I_i = (\mathcal{I}_P)_i$ for $i \leq j - \tau$. Since $H(R/I) = T$, I_i is 0 for $i < d$, and

$$(I^2)_j = \sum_{d \leq i \leq j/2} I_i I_{j-i}$$
$$= \sum_{d \leq i \leq j/2} (\mathcal{I}_P)_i (\mathcal{I}_P)_{j-i}$$
$$= (\mathcal{I}_P^2)_j.$$

If $d = t$, the equalities (3.2.7), (3.2.8) are obvious, so suppose $s \leq \dim_k R_{t-1}$. Since L is general, so is P. By Theorem 1.69 the ideal $\mathcal{I}_P$ is $\sigma = \tau + 1$ regular, and for any $i \geq \sigma$, $\mathcal{I}_i = R_{i-\sigma} \mathcal{I}_\sigma$. Thus we have for v satisfying $\sigma \leq v \leq t$

$$I_v I_{j-v} = (\mathcal{I}_P)_v (\mathcal{I}_P)_{j-v} = R_{v-\sigma} (\mathcal{I}_P)_\sigma R_{j-v-\sigma} (\mathcal{I}_P)_\sigma$$
$$= R_{j-2\sigma} (\mathcal{I}_P)_\sigma (\mathcal{I}_P)_\sigma = R_{t-\sigma} (\mathcal{I}_P)_\sigma R_{j-t-\sigma} (\mathcal{I}_P)_\sigma \qquad (3.2.13)$$
$$= (\mathcal{I}_P)_t (\mathcal{I}_P)_{j-t} = I_t I_{j-t}.$$

Thus,

$$(\mathcal{I}_P^2)_j = \sum_{d \leq i \leq t} (\mathcal{I}_P)_i (\mathcal{I}_P)_{j-i} = \sum_{d \leq i \leq t} (\mathcal{I}_P)_i R_{j-i-t} (\mathcal{I}_P)_t$$
$$= \sum_{d \leq i \leq t} R_{j-t-i} (\mathcal{I}_P)_i (\mathcal{I}_P)_t \qquad (3.2.14)$$
$$= \begin{cases} (\mathcal{I}_P)_t^2 = (I_t)^2 & \text{if } j = 2t, \\ (\mathcal{I}_P)_t (\mathcal{I}_P)_{t+1} = I_t I_{t+1} & \text{if } j = 2t+1. \end{cases}$$

By Corollary 1.71 of Theorem 1.70, $(\mathcal{I}_P^2)_i = (\mathcal{I}_P^{(2)})_i$ for $i \geq 2\sigma$, and in particular for $i = j$. This completes the proof of (3.2.7), (3.2.8), and (3.2.9).

The equality of tangent spaces $T_f(\mathbf{V}_s(t, j - t; r))$ and $T_f(\mathbf{Gor}(T))$ follows from (3.2.7), (3.2.8) and Theorems 3.2 and 3.9. The identification of the tangent space $T_f = (\mathcal{I}_P^{(2)})_j^{\perp}$ with $\mathcal{D}_1 L_P^{[j-1]}$ in (3.2.11) is the case $a = 2$ of Lemma 2.2. By Theorem 3.2, the tangent space $T_f(\mathbf{V}_s(v, j - v; r)) = (I_v I_{j-v})^{\perp}$, which by (3.2.13) is $T_f(\mathbf{Gor}(T))$. This completes the proof of (3.2.12) and of Lemma 3.16.   □

The following Examples 3.17–3.19 illustrate the use of the tangent space to bound the dimension of $\mathbf{Gor}(T)$. The values found by "Macaulay" calculation in Examples 3.17–3.18 for $r = 3$ are generalized by Theorem 4.1A,4.1B, and are further generalized by the results of J. O. Kleppe and others (see Section 4.4). Example 3.19 for $r = 4$ goes beyond Theorem 4.10A which covers only the case $s \leq \dim_k R_{t-1}$.

EXAMPLE 3.17. DIMENSION OF THE TANGENT SPACE TO $\mathbf{Gor}(T)$. We suppose that $t = 3$, $j = 6$, and consider the Hilbert functions $T = H(s, 6, 3)$. We give in Table 3.1 the values obtained by the "Macaulay" algebra program for the Hilbert function of $H(R/I^2)$, where $I = \mathrm{Ann}(f)$, $f = L_1^{[j]} + \cdots + L_s^{[j]}$, $L = L_1, \ldots, L_s$ are general enough linear forms, and where s varies from 4 to 9. In the table, the left column is s, the rank of $Cat_f(3, 3; 3)$, the second is the corank, the third is the Hilbert function $T = H_f = H(R/I)$, $I = \mathrm{Ann}(f)$, the fourth is $H' = H(R/I^2)$, and the last is the codimension of $\mathbf{Gor}(T)$ in $\mathbb{A}^N$.

s	corank	T t	$H' = H(R/I^2)$ $j = 6$	cod
4	6	1 3 4 4 4 3 1	1 3 6 10 12 12 **12** 10 6 4 3 1	16
5	5	1 3 5 5 5 3 1	1 3 6 10 14 16 **15** 13 7 5 2	13
6	4	1 3 6 6 6 3 1	1 3 6 10 15 21 **18 18** 6 6 0	10
7	3	1 3 6 7 6 3 1	1 3 6 10 15 21 **22** 18 9 3 0	6
8	2	1 3 6 8 6 3 1	1 3 6 10 15 21 **25** 21 9 1 0	3
9	1	1 3 6 9 6 3 1	1 3 6 10 15 21 **27** 27 6 0	1
10	0	1 3 6 10 6 3 1	1 3 6 10 15 21 **28** 36 0	0

TABLE 3.1. Dimension of tangent space to $\mathbf{Gor}(T)$, $r = 3$, $j = 6$.

By Theorem 3.9 we have for $(j, r) = (6, 3)$ as in Table 3.1,

$$\text{cod}(\mathbf{Gor}(T)) \geq \dim_k(I^2)_j = \dim_k(R_6) - H'_j = 28 - H'_6. \quad (3.2.15)$$

The dimension of the tangent space $T_f(V_s(3, 3; 3))$ is given in Table 3.1 by the $j = 6$ column of H', in boldface; the values illustrate Theorem 4.1A,4.1B below. The values in Table 3.1 of $H'_j, \ldots, H'_{j+3}$ for $s = 6$, are in boldface: they illustrate Theorem 4.1C below.

EXAMPLE 3.18. TANGENT SPACE TO **Gor**(T) WHEN $r = 3$, $j = 8$. We suppose that $r = 3$, $t = 4$, $j = 8$, so $T = H(s, 8, 3)$. Then the calculated values of $\dim_k(T_f)$ for $f = L_1^{[j]} + \cdots + L_s^{[j]}$, L_i general are

$j = 8 / \backslash s$	6	7	8	9	10	11	12	13	14	15
$\dim_k T_f$	18	21	24	27	30	35	39	42	44	45

For $s \geq 12$, these values are known by Diesel's Theorem 1.64 above; for $s \leq 10$, they follow from Theorem 4.1A; for $s \geq 10$ they are shown in the proof of Theorem 4.1B. Note the pattern in the first differences of $\dim_k T_f$ with respect to s.

Calculating the tables. We used the "Macaulay" script $<points$ to create the ideal I_Z, $Z = (p_1, \ldots, p_s)$, where $p_1, \ldots, p_s$ is chosen generally enough by requesting "Macaulay" to input a "random" matrix. We then squared the ideal, and found $\mathbf{Hilb}(R/I_Z^2)_j$.

EXAMPLE 3.19. TANGENT SPACE TO POWER SUM FORMS OF **Gor**(T), $r = 4$, $T = H(s, j, 4)$, $j = 2t$. In Table 3.2 we give the tangent space dimension for **Gor**(T), $T = H(s, j, 4)$, $j = 2t$ for all $s \geq \dim_k R_{t-1}$ when $t = 3$, $t = 4$, and $t = 5$. These are the cases not covered by Theorem 4.10A. The values of $\dim_k T_f$ for $s = 4, 5$, when $j = 4$, and $10 \leq s \leq 13$ when $j = 6$, and $20 \leq s \leq 23$ when $j = 8$ along with Lemma 3.16 and the method of Theorem 4.1A show that $\overline{PS(s, j; 4)}$ is a component of $\overline{Gor(T)}$ in those cases. The values of $\dim_k T_f$ for $5 \leq s$ when $j = 4$, for $13 \leq s$ when $j = 6$, and for $24 \leq s$ when $j = 8$, and an argument analogous to that of Theorem 4.1B below show that there is a single component of $\overline{Gor(T)}$ containing $PS(s, j; 4)$ in these cases: that component has codimension $\binom{s^{\vee}+1}{2}$, in $\mathbb{A}(\mathcal{D}_j)$. There is a pattern to the dimensions, which we formulate as a conjecture.

CONJECTURE 3.20. DIMENSION OF THE TANGENT SPACE T_f TO **Gor**(T), $T = H(s, j, r)$, j EVEN. *Suppose that $r \geq 3$, $j = 2t$, s is arbitrary, $T = H(s, j, r)$ and that f is a general element of*

$PS(s,j;r)$. Then the tangent space T_f to $\mathbf{Gor}(T)$ at f has dimension $d = \max(rs, \dim_k R_j - s^\vee(s^\vee + 1)/2)$. The locus $PS(s,j;r)$ is contained in a unique irreducible component of $\overline{Gor(T)}$. The dimension of this component is equal to d. If $d = rs$ this component coincides with $\overline{PS(s,j;r)}$. The same statement holds for $V_s(t,t;r)$.

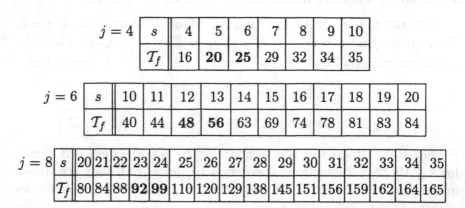

$j = 4$	s	4	5	6	7	8	9	10
	T_f	16	**20**	**25**	29	32	34	35

$j = 6$	s	10	11	12	13	14	15	16	17	18	19	20
	T_f	40	44	**48**	**56**	63	69	74	78	81	83	84

$j=8$	s	20	21	22	23	24	25	26	27	28	29	30	31	32	33	34	35
	T_f	80	84	88	**92**	**99**	110	120	129	138	145	151	156	159	162	164	165

TABLE 3.2. Calculated values of $\dim_k T_f$, the tangent space to $\mathbf{Gor}(T)$, $T = H(s,j,4)$. The transition values are in boldface. (See Example 3.19, Conjecture 3.20, and Remark 3.21.)

REMARK 3.21. We have checked the first statement in Conjecture 3.20 using the "Macaulay" algebra program also when $r = 5$, for $j = 4, 6$. Note that the method of proof of Theorem 4.1A, 4.1B below suggests that to verify the first statement of Conjecture 3.20 for a particular pair (j,r), $j = 2t$, and all s, it would be sufficient to verify that $\dim_k T_f$ satisfies the Conjecture for two adjacent "transition" values, namely $s = s_{j,r}$ and $s_{j,r} - 1$, where

$$s_{j,r} = \min\left\{s \mid rs + \binom{s^\vee + 1}{2} \le \dim_k \mathcal{D}_j\right\} \tag{3.2.16}$$

When $r = 3$, there is a single transition value $s_{j,3} = \dim_k R_{t-1}$ for which there is equality in (3.2.16). We prove Conjecture 3.20 for $r = 3$ and for $s^\vee < r$ in Theorems 4.1A, 4.1B, and 4.13 below. It is known for $r = 3$ ([**K12**], see also Section 4.4).

Example 3.22. Dimension of the tangent space T_f to $\mathbf{Gor}(T)$, $T = H(s, j, r)$, j ODD. Consider $r = 4$, $j = 2t + 1$, and $s \geq \dim_k R_{t-1}$. We tabulate for $j = 9$, $27 \leq s \leq 35$, the dimension of T_f, f a general form of $PS(s, j; 4)$. The values are consistent with Conjecture 3.23 below, which we have also checked for $r = 5$, $j = 5, 7$.

$r = 4$ $j = 9$	s	27	28	29	30	31	32	33	34	35
	T_f	108	112	118	130	144	160	178	198	220

Conjecture 3.23. Dimension of the tangent space T_f to $\mathbf{Gor}(T)$, $T = H(s, j, r)$, j ODD. Suppose $r \geq 3$, $j = 2t + 1$, and $\dim_k R_{t-1} \leq s \leq \dim_k R_t$, $T = H(s, j, r)$ and $s^\vee = \dim_k R_t - s$, suppose f is a general element of $PS(s, j; r)$, and let $T_f = T_f(\mathbf{Gor}(T))$. Then $d = \dim_k T_f$ satisfies

$$d = \max\left(rs, \dim_k R_j - s^\vee(\dim_k R_{t+1} - s) + r\binom{s^\vee}{2}\right) \qquad (3.2.17)$$

Furthermore the locus $PS(s, j; r)$ is contained in a unique irreducible component of $\overline{\mathbf{Gor}(T)}$. The dimension of this component is equal to d. If $d = rs$ this component coincides with $\overline{PS(s, j; r)}$. The same statement holds for $V_s(t, t + 1; r)$.

Remark 3.24. The part of Conjecture 3.23 concerning $\dim_k T_f$ is shown for $r = 3$ in Theorem 4.5B, and Conjecture 3.23 is shown for $s^\vee < r$ in Theorem 4.13. When $s^\vee = 1$, T_f has the codimension in $T(\mathcal{D}_j)$ of the general determinantal variety $D_{r_t-1}(r_t, r_{t+1})$ of corank ≥ 1 matrices in the space of all $r_t \times r_{t+1}$ matrices (see Theorem 3.14). However, when $s^\vee > 1$, the codimension is not that of a corank $s^\vee$ determinantal locus of a generic $r_t \times r_{t+1}$ matrix. When $s \geq \dim_k R_{t-1}$ and $T = H(s, j, r)$, $j = 2t + 1$ Theorem 2.6 shows that $rs \leq \dim \mathbf{Gor}(T)$; but Conjecture 3.23 predicts a higher dimension. The formula (3.2.17) has a simple explanation in terms of the multiplication map $R_1 \otimes I_t \otimes I_t \rightarrow (I_j)^2$ having the maximum possible rank — see the proof of Theorem 4.13.

By Lemmas 1.19 and 3.16, if $j = 2t$ or $2t + 1$, $s \leq \dim_k R_t$ and if f is general in $PS(s, j; r)$, we have $\dim_k T_f = \dim_k(R_j / (\mathcal{I}_P^2)_j)$, where P is a set of s general points of $\mathbb{P}^{r-1}$. Thus the formulas in Conjectures 3.20 and 3.23 can be rephrased entirely as a conjecture about the postulation of $\mathcal{I}_P^2$. We set $r_i = \dim_k R_i$.

CONJECTURE 3.25. HILBERT FUNCTION OF THE SQUARE OF A VANISHING IDEAL AT s GENERAL POINTS OF $\mathbb{P}^{r-1}$. *Suppose that* $\dim_k R_{t-1} \leq s < \dim_k R_t$, *let* $s^\vee = \dim_k R_t - s$, *and suppose that* $\mathcal{I}_Z$ *is the ideal of a scheme* Z *of* s *general points of* $\mathbb{P}^{r-1}$. *Then the Hilbert function* $H(R/\mathcal{I}_Z^2)$ *satisfies*

$$H(R/\mathcal{I}_Z^2) = \left(1, r, \ldots, r_{2t-1}, r_{2t} - \binom{s^\vee + 1}{2}\right),$$

$$\dim_k R_{2t+1} - s^\vee(\dim_k R_{t+1} - s) + r\binom{s^\vee}{2}, rs, rs, \ldots). \quad (3.2.18)$$

REMARK 3.26. We prove Conjecture 3.25 when $r = 3$ (Proposition 4.8), and when $s^\vee \leq r$ (Lemma 4.12, Theorem 4.19). In degrees $i \leq 2t - 1$ and $i \geq 2t + 2$ Conjecture 3.25 follows from Theorem 1.66 and Corollary 1.71. The middle degrees $i = 2t$ and $i = 2t + 1$ are the formulas of Conjectures 3.20 and 3.23. By Lemma 4.12 the left side of (3.2.18) is termwise greater or equal the right side, provided the multiplication map $m_{t,1} : (\mathcal{I}_Z)_t \otimes R_1 \to R_{t+1}$ is injective.

CHAPTER 4

The Locus $PS(s, j; r)$ of Sums of Powers, and Determinantal Loci of Catalecticant Matrices

Recall that $\overline{PS(s, j; r)}$ denotes the Zariski closure in $\mathbb{A}(\mathcal{D}_j)$ of the family of DP-forms that can be written as sums of divided j-th powers of s linear forms, elements of $\mathcal{D}_1$. The sequence $T = H(s, j, r)$ of (4.1.1) is the Hilbert function H_f of a general element f in the irreducible variety $PS(s, j; r)$ (Lemma 1.17).

In Section 4.1 we show that when $j = 2t$ the Zariski closure of $Gor(T)$, $T = H(s, j, 3)$, equals the closure of $PS(s, j; 3)$ if $s \leq \dim_k R_{t-1}$ (Theorem 4.1A); the closure of $Gor(T)$ is the unique irreducible component of $V_s(t, t; 3)$ containing $PS(s, j; 3)$ when $s > \dim_k R_{t-1}$ (Theorem 4.1B). We then study the locus $V_s(t, t + 1; 3)$ (Theorems 4.5A, 4.5B).

In Section 4.2, we study $V_s(t, t; r)$ and $V_s(t, t + 1; r)$ when $r \geq 3$. We show that when $s \leq \dim_k R_{t-1}$, the closure of $PS(s, j; r)$ is an irreducible component of $\overline{Gor(T)}$ (Theorem 4.10A). For arbitrary r when $s^{\vee} = \dim_k R_t - s$ satisfies $s^{\vee} < r$ we find the unique irreducible component of $\mathbf{Gor}(T)$ containing $PS(s, j; r)$ and calculate its dimension (Theorem 4.13).

For the proofs of these results, we need the Alexander-Hirschowitz Theorem (Theorem 1.66), usually in the cases covered by Theorem 2.6, and Lemma 3.16, which uses the Geramita-Gimigliano-Pitteloud or Chandler Theorem (Theorem 1.70 above). When $r = 3$ we use a variation by A. Kustin of a result of Kustin-Ulrich (Theorem B.3).

In Section 4.3 we consider Gorenstein ideals whose lowest degree generators are a complete intersection, and are able to extend some of the results of Section 4.2 to determinantal loci of catalecticant matrices of corank $s^{\vee} = r$ (Theorem 4.19).

Since these results were first promulgated in the 1996 manuscript [**IK**], J. O. Kleppe, improving earlier work of A. Geramita, M. Pucci, and Y. Shin [**GPS**], showed that when $r = 3$ each $\mathbf{Gor}(T)$ is smooth [**K12**]; J. O. Kleppe and as well several other authors A. Conca and G. Valla, and Y. Cho and B. Jung gave formulas for the dimension of

the tangent space T_f, to a point $f \in \mathbf{Gor}(T)$ ([**Kl2, CoV1, ChoJ1**]).
Together with our formula of Corollary 5.50 – for the special case
$T \supset (s, s, s)$ – there are now several formulas for $\dim(Gor(T))$, and
they are all distinct! Y. Cho and B. Jung showed as a consequence of
their formula that there are high-dimensional components of $V_s(t, t; 3)$,
for $t \geq 7$ and $s \approx r_t - t$, giving a negative answer to Diesel's conjecture
1.65 (see [**ChoJ2**] and Example 4.28 below). We report on these
developments in Section 4.4.

4.1. The case $r = 3$

Recall that $H(s, j, r)$ is the sequence

$$H(s, j, r)_i = \min(s, \dim_k R_i, \dim_k R_{j-i}), \; 0 \leq i \leq j. \quad (4.1.1)$$

We will assume in this chapter that $s \leq \dim_k(R_t), t = \lfloor j/2 \rfloor$. We will
suppose that $f \in Gor(T)$ where $T = H(s, j, 3)$, and $I = \mathrm{Ann}(f) \subset R$.
If also $f \in PS(s, j; 3)$ is general enough, so f has by Theorem 2.10 a
unique decomposition $f = L_1^{[j]} + \cdots + L_s^{[j]}$, we denote by P or P_L the set
$P = \{p(1), \ldots, p(s)\}$ of points in $\mathbb{P}^2 = \mathbb{P}(V^*), V = \langle X_1, X_2, X_3 \rangle$ that
correspond to the lines $L_1, \ldots, L_s$ (Definition 1.13). We let $\mathcal{I}(P_L)$ (or
$\mathcal{I}_P$) denote the vanishing ideal $\mathcal{I}_P = m_{p(1)} \cap \cdots \cap m_{p(s)}$ in R, and we
denote by $\mathcal{I}_P^{(a)}$ or $\mathcal{I}(P)^{(a)}$ the ideal $\mathcal{I}_P^{(a)} = m_{p(1)}^a \cap \cdots \cap m_{p(s)}^a$ of functions vanishing to order at least a at each point of P (Definition 2.1).
If $r = 3$ and t is fixed, we let $s_0 = \dim_k R_{t-1} = \binom{t+1}{2}$. Recall from
(3.2.6) that

$$\tau_{s,r} = \min\{i \mid s \leq \dim_k R_i\}.$$

THEOREM 4.1A. IDENTITY OF CLOSURES OF $Gor(T)$, $T = H(s, j, 3)$ AND OF $PS(s, j; 3)$, j EVEN, s NOT LARGE. *If* $r = 3$, $j = 2t$,
$t \geq 2$, $s \leq s_0 = \dim_k R_{t-1}$ *and* $T = H(s, j, 3)$ *we have*

$$\overline{PS(s, j; 3)} = \overline{Gor(T)} \quad (4.1.2)$$

and this is an irreducible component of $V_s(t, t; 3)$ *of dimension* $3s$.
The scheme, $\mathbf{Gor}(T)$ *is generically smooth. If also* $f = L_1^{[j]} + \cdots + L_s^{[j]}$
is a general enough DP-form in $PS(s, j; 3)$, *then we have for all* i
satisfying $\tau_{s,r} \leq i \leq j - \tau_{s,r}$, *the equality* $I_i = \mathcal{I}(P_L)_i$. *Furthermore, if*
$\sigma = \tau_{s,r} + 1$ *and* $\sigma \leq i \leq j - \sigma$ *then*

$$I_i = R_{i-\sigma} I_\sigma = R_{i-\sigma}(\mathcal{I}(P_L))_\sigma. \quad (4.1.3)$$

THEOREM 4.1B. RELATION BETWEEN CLOSURES OF $Gor(T), T = H(s, j, 3)$, AND OF $PS(s, j; 3)$, j EVEN, s LARGE. *If* $r = 3$, $j = 2t$,
$s \geq s_0 = \dim_k R_{t-1}$, *and* $s^\vee = \dim_k R_t - s$, *then every irreducible*

component of $V_s(t, t; 3)$ which contains $PS(s_0, j; 3)$ has codimension $\binom{s^\vee + 1}{2}$ in $\mathbb{A}(R_j)$. One of these components of $V_s(t, t; 3)$ equals $\overline{Gor(T)}$, $T = H(s, j, 3)$. This is the unique component of $V_s(t, t; 3)$ which contains $PS(s, j; 3)$. The scheme $\mathbf{Gor}(T)$ is generically smooth along $PS(s, j; 3)$.

THEOREM 4.1C. HILBERT FUNCTION OF $(Ann(f))^2$, f A SUM OF POWERS, WHEN s IS SPECIAL. *Suppose $r = 3, j = 2t + v, v \geq 0, s = s_0 = \dim_k R_{t-1}$, and $T = H(s_0, j, 3)$. Suppose $f = L_1^{[j]} + \cdots + L_s^{[j]}$ is a general enough DP-form in $PS(s_0, j; 3)$. Then $I = Ann(f)$ has Hilbert function $H(R/I) = T$, and satisfies*

$$I_t = \mathcal{I}(P_L)_t, \ldots, I_{t+v+1} = \mathcal{I}(P_L)_{t+v+1}. \tag{4.1.4}$$

Also, for $t \leq i \leq t + v + 1$, we have

$$I_i = R_{i-t}I_t. \tag{4.1.5}$$

The ideal I^2 satisfies,

$$M^{2t} \supset I^2 \supset M^{j+v+4} \tag{4.1.6}$$

$$(I^2)_{2t} = (\mathcal{I}(P_L)^{(2)})_{2t}, \ldots, (I^2)_{j+1} = (\mathcal{I}(P_L)^{(2)})_{j+1},$$
$$(I^2)_{j+2} = (\mathcal{I}(P_L))_{j+2}, \ldots, (I^2)_{j+v+3} = (\mathcal{I}(P_L))_{j+v+3}, \tag{4.1.7}$$

and has Hilbert function

$$H(R/I^2) = (1, 3, \ldots, \dim_k R_{2t-1}, \overset{length\ v+2}{3s, \ldots, 3s}, \overset{length\ v+2}{s, \ldots, s}). \tag{4.1.8}$$

where $h_i = 3s$ for $2t \leq i \leq j + 1, h_i = s$ for $j + 2 \leq i \leq j + v + 3$.

PROOF OF THEOREM 4.1A. According to Corollary 1.18 an open dense subset of $PS(s, j; r)$ is contained in $Gor(T)$. When $r = 3$, $Gor(T)$ is irreducible by Diesel's result, Theorem 1.72 above. By Theorem 2.6 the dimension of $PS(s, j; r)$ is rs. Thus, in order to show $\overline{PS(s, j; 3)} = \overline{Gor(T)}$ it suffices to show

$$\dim \overline{Gor(T)} \leq 3s \tag{4.1.9}$$

Let $\mathcal{T}_f = \mathcal{T}_f(Gor(T))$. By (3.2.10) in Lemma 3.16, $\dim_k \mathcal{T}_f = \dim_k(\mathcal{I}_P^{(2)})_j^\perp$. For a general enough element f of $PS(s, j; 3)$, by the Alexander-Hirschowitz result (Theorem 1.66), we obtain

$$\dim_k \mathcal{T}_f = 3s. \tag{4.1.10}$$

Note that the assumptions on (s, j) avoid the exceptional case $(s, j) = (5, 4)$ when $r = 3$ in Theorem 1.66. If $char(k) = 0$, we can obtain (4.1.10) using Part (iii) of Theorem 2.6.

In order to show that $\overline{PS(s,j;3)}$ is an irreducible component of $V_s(t,t;3)$, it suffices to show that the tangent space $\mathcal{T}_f(\mathbf{V}_s(t,t;3)) \cong \mathcal{T}_f(\mathbf{Gor}(T))$ at a general point $f \in PS(s,j;3)$. The identity of $\mathcal{T}_f = \mathcal{T}_f(\mathbf{Gor}(T))$ with $\mathcal{T}_f(\mathbf{V}_s(t,t;3))$, and also the remaining statements in Theorem 4.1A are shown in Lemma 3.16.   □

PROOF OF THEOREM 4.1B. Let f_0 be a general enough DP-form in $PS(s_0,j;3)$ and let $I = \mathrm{Ann}(f_0)$. We have shown in Part (A) that

$$\dim_k (I^2)_j = \dim_k \mathcal{D}_j - 3s_0$$

$$= \binom{2t+2}{2} - 3\binom{t+1}{2} = \binom{t+2}{2}$$

On the other hand, $\dim_k(I_t) = \dim_k R_t - s_0 = t + 1$. Thus,

$$\dim_k(I^2)_j = \binom{t+2}{2} = \dim_k \mathrm{Sym}^2(I_t) \qquad (4.1.11)$$

By (3.2.7) we have $(I^2)_j = (I_t)^2$, so the multiplication map

$$\mathrm{Sym}^2(I_t) \longrightarrow (I_t)^2 = (I^2)_j$$

is surjective, and therefore by (4.1.11) is an isomorphism. Suppose $s \geq s_0$, and let $f \in \mathcal{D}_j$. We let $I_f = \mathrm{Ann}(f)$, and $K_f = Ker(m_f)$, where $m_f : \mathrm{Sym}^2((I_f)_t) \to (I_f^2)_j$ is the multiplication map. Let $Y = \{f \in \mathcal{D}_j \mid K_f \neq 0\}$. We will show

CLAIM. Y is a closed algebraic subset of $\mathbb{A}^N = \mathbb{A}(\mathcal{D}_j)$.

PROOF OF CLAIM. Y is a constructible subset of $\mathbb{A}^N$. In order to show that Y is closed, it suffices to prove that for any smooth affine curve X with a morphism $\phi : X \to \mathbb{A}^N$, if $x_0 \in X$ satisfies $f(X - x_0) \subset Y$, then $\phi(x_0) \in Y$. Replacing X by a neighborhood of x_0, we may assume that for every $x \in X - \{x_0\}$ the spaces $(I_{f_x})_t$, $f_x = \phi(x)$, and $K_{\phi(x)}$ have constant dimensions a, b, respectively. The maps

$$\psi_1 : X - \{x_0\} \longrightarrow Grass(a, \mathcal{D}_t), \quad \psi_1(x) = (I_{f_x})_t,$$
$$\psi_2 : X - \{x_0\} \longrightarrow Grass(b, \mathrm{Sym}^2 \mathcal{D}_t), \quad \psi_2(x) = K_\phi(x)$$

extend to x_0, and letting $f = \phi(x_0)$ we have

$$\phi_1(x_0) \subset (I_f)_t,$$
$$\phi_2(x_0) \subset \mathrm{Sym}^2(\psi_1(x_0)) \subset \mathrm{Sym}^2((I_f)_t).$$

Clearly, $\psi_2(x_0) \subset K_f$, thus $f = \phi(x_0) \in Y$.   □

PROOF OF THEOREM 4.1B, CONTINUED. Now let $W \subset V_s(t, t; 3)$ be an irreducible component which contains $PS(s_0, j; 3)$. Since the closure $\overline{U_s(t, t; 3)}$ of the rank-s locus $U_s(t, t; 3)$ equals $V_s(t, t; 3)$ (Lemma 3.5), and W contains a sufficiently general point f of $PS(s_0, j; 3)$, which by (4.1.11) is not in Y, it follows that

$$W \not\subset V_{s-1}(t, t; 3) \cup Y.$$

Thus, by the claim, for f in a dense open subset of W we have both $\dim_k(I_f)_t = s^\vee$, and

$$m_f : \mathrm{Sym}^2((I_f)_t) \rightarrow (I_f)_t^2$$

is an isomorphism. This shows that the tangent space $T_f(V_s(t, t; 3))$ has codimension $\binom{s^\vee+1}{2}$. Since the locus Δ_s of symmetric $r_t \times r_t$ matrices of rank less or equal s, has the same codimension $\binom{s^\vee+1}{2}$ in the affine space of all symmetric $r_t \times r_t$ matrices (see Theorem 1.26), we conclude that the catalecticant matrix $C = Cat_f(t, t; 3)$ is a nonsingular point of Δ_s, and that the space of catalecticant matrices intersects Δ_s transversally at C. This proves that each irreducible component of $V_s(t, t; 3)$ which contains $PS(s_0, j; 3)$ has codimension $\binom{s^\vee+1}{2}$ in $\mathbb{A}(\mathcal{D}_j) = \mathbb{A}^N$.

To complete the proof of Theorem 4.1B we need to show that when $s \geq s_0$, there is a unique component of $V_s(t, t; 3)$ containing $PS(s, j; 3)$, namely $\overline{Gor(T)}, T = H(s, j, 3)$. First, $Gor(T)$ is an open subset of $V_s(t, t; 3)$, as it is the intersection of open sets defined by the nonvanishing of some maximal rank minor in each $Cat_F(u, j - u; 3), 2u < j$, with the rank-$s$ locus $U_s(t, t; 3)$ of $V_s(t, t; 3)$: we use here the unimodality of Gorenstein sequences T when $r = 3$ (see [BE2, St3] or Theorem 5.25, Part (i)). By Diesel's Theorem 1.72, $Gor(T)$ is irreducible, so $\overline{Gor(T)}$ is an irreducible component of $V_s(t, t; 3)$. By Lemma 1.17, an open dense subset U of $PS(s, j; 3)$ satisfies $U \subset Gor(T)$, and clearly $PS(s_0, j; 3) \subset \overline{PS(s, j; 3)}$. Thus $\overline{Gor(T)}$ is one of the components containing $PS(s_0, j; 3)$, and Y does not contain $PS(s, j; 3)$. Therefore, every general enough element of $PS(s, j; 3)$ is a nonsingular closed point of the scheme $\mathbf{V}_s(t, t; 3)$. This shows that $\overline{Gor(T)}, T = H(s, j, r)$ is the only irreducible component of $V_s(t, t; 3)$ which contains $PS(s, j; 3)$. This completes the proof of Theorem 4.1B. $\qquad\square$

PROOF OF THEOREM 4.1C. Let H' denote the sequence at the right of (4.1.8). We show first that $H(R/I^2) > H'$ termwise. We then show that the vector space dimension of R/I^2 is $4n, n = \dim_k R/I$, which is the sum of the sequence H'. To begin, Lemma 3.16, implies (4.1.4), (4.1.5), which in turn imply the first set of equalities in (4.1.7),

$(I^2)_i = (\mathcal{I}(P_L)^{(2)})_i$, $2t \le i \le j + 1$. For these i, $\dim_k R_i/(\mathcal{I}^{(2)})_i = 3s$ by the Alexander-Hirschowitz theorem, or Theorem 2.6. Also, if $j + 2 \le i \le 2(j + 1 - t) + 1 = j + v + 3$, we have

$$(I^2)_i \subset (\mathcal{I}_P)_i,$$

so for these i, $\dim_k(R_i/(\mathcal{I}^{(2)})_i) \ge s$. We have now shown that $H(R/I^2)$ is at least as large as the sequence in (4.1.8).[1] Since I is Gorenstein, the tangent space to the punctual Hilbert scheme $\mathbf{Hilb}^n \mathbb{P}^2$ at $A = R/I$ satisfies $\mathrm{Hom}(I, R/I) \cong I/I^2$. From a result of H. Kleppe [Kl.H] we have that height three Gorenstein Artin algebras are smoothable (see Definition 5.16). Thus, $\dim_k(R/I) = n$ implies that $\dim_k \mathrm{Hom}(I, R/I) = 3n$, and $\dim_k(R/I^2) = 4n$. It is easy to check that the length $|H'| = \sum h_i'$ is $4n$. This proves (4.1.8), so completes the proof of (4.1.7), and of Theorem 4.1C.   □

REMARK 4.2. Calculation with "Macaulay" indicates that Theorem 4.1C should extend to r variables but we were only able to extend Theorem 4.1A (see Theorem 4.10A, 4.10B).

In our proof of Theorem 4.1B, a key step is (4.1.11). In four variables, if $r = j = 4$, $t = 2$, $s_0 = \dim_k R_{t-1} = 4$, and $s^\vee = 6$, the variety $PS(4, 4, 4)$ has dimension 16, but codimension $35 - 16 = 19$ in $\mathbb{A}^{35} = \mathbb{A}(\mathcal{D}_4)$, smaller than $21 = \binom{s^\vee + 1}{2}$. This prevents extension of (4.1.11) in the proof to $r > 3$.

REMARK 4.3. TANGENT SPACES TO $\mathbb{P}\mathrm{Gor}(T)$. In the proof of Theorem 4.1C we have used two tangent spaces, corresponding to two natural embeddings of $\mathbb{P}\mathrm{Gor}(T)$. The first and primary one for this article is the tangent space $T_f = T_f\mathrm{Gor}(T))$ at a point z_p parameterizing $f = f_p$ on $\mathbf{Gor}(T)$. Here we consider $\mathbf{Gor}(T)$ as a subscheme of the affine space $\mathbb{A} = \mathbb{A}(\mathcal{D}_j)$, the space parametrizing degree-j DP-forms. By taking classes   mod k^*, we may consider the tangent space $T_{f,\mathbb{P}\mathrm{Gor}(T)} = T_f/k^*$ for the corresponding inclusion of $\mathbb{P}\mathrm{Gor}(T)$ in $\mathbb{P}(\mathcal{D}_j)$

---

[1]The Nice specialists J. Briançon et al showed directly a related result. If I_t defines a zero-dimensional scheme Z in $\mathbb{P}^2$ of degree s, the ideal $(I_t)^2$ defines a scheme of degree at least 3s. Their proof is as follows. We may suppose Z is concentrated at a single point p and $\mathcal{O}_p = k\{x, y\}$. If the quotient $A = \mathcal{O}_p/K$ of $k\{x, y\}$ has dimension s over k, then an A-homomorphism $\phi : A^{p-1} \to A^p$ such that all $(p - 1) \times (p - 1)$ minors are zero, must have a kernel with $\dim_k Ker(\phi) \ge \dim_k A$. This implies $Tor^1(A, A)$ has dimension at least $\dim_k A$, which is equivalent to $\dim_k K/K^2 \ge 2 \dim_k A$, implying $\dim_k \mathcal{O}_p/K^2 \ge 3 \dim_k A = 3s$. More generally for $K \subset \mathcal{O}_p = k\{x, y\}$, by deforming to a complete intersection, they show $\dim_k \mathcal{O}_p/K^n \ge \binom{n+1}{2} \dim_k \mathcal{O}_p/K$.

But the variety $\mathbb{P}Gor(T)$ can also be given a scheme structure as a subscheme of the scheme $\mathbf{G}(T) = \mathbf{GradAlg}(\mathbf{T})$ parametrizing all graded ideals of R defining a quotient algebra of Hilbert function T (see [**Kl1, Kl2**]). It is not clear if the two scheme structures are the same, but the tangent spaces for the two structures at corresonding closed points are the same (ibid., see also p. 117 below). Since the inclusion $R \to \mathcal{O}_p$ of R into the local ring $\mathcal{O}_p$ at the origin of affine space $\mathbb{A}^r \subset \mathbb{P}^r$, respects graded ideals, the variety G_T may be given an induced scheme structure as a subscheme of the Hilbert scheme $\mathbf{Hilb}^n(\mathbb{P}^r)$, $n = |T| = \Sigma t_i$. With this structure, the tangent space $T_A = \mathrm{Hom}_A(I, A)$ is the tangent space of the point $p_A \in \mathbf{Hilb}^n(\mathbb{P}^r)$ parameterizing $A = \mathcal{O}_p/I$. If A is Gorenstein Artin, the degree-0 part of T_A is T_f/k^*, $T_A \cong I/I^2$, and $\mathrm{Hom}_0(I, A) \cong I_j/(I^2)_j$. The closures of the algebraic set $\mathbb{P}Gor(T)$ are quite different in the two embeddings, to $\mathbb{P}(\mathcal{D}_j)$ and to $Hilb^n(\mathbb{P}^r)$ The closure of $\mathbb{P}Gor(T)$ in $Hilb^n(\mathbb{P}^r)$ consists of graded algebra quotients of $\mathcal{O}_p$ having Hilbert function T, but not necessarily Gorenstein. The closure $\overline{\mathbb{P}Gor(T)} \subset \mathbb{P}(\mathcal{D}_j)$ consists of Gorenstein algebras, of the same socle degree j, but having Hilbert functions $T' \geq T$ termwise. We relate the two closures in Chapter 8.

REMARK 4.4. Several authors have studied the minimal resolutions of the ideal I^2 when I is defined by Pfaffians of an alternating matrix: G. Boffi and R. Sánchez in [**BofS**], and A. Kustin and B. Ulrich for I^k, $k \geq 1$ in [**KusU**]. Their results have been used by A. Conca and G. Valla, and by Y. Cho and B. Jung to calculate the tangent space T_f for a general element f of $Gor(T)$ when $r = 3$ and T is an arbitrary Gorenstein sequence [**CoV1, ChoJ1**]. We recall a version of the Kustin-Ulrich result in Appendix B and use it in Theorem 4.5B below to study $Gor(T)$ with $T = H(s, j, 3)$ when j is odd.

When j is even, we could have calculated in this way $\dim(Gor(T))$, $T = H(s, j, 3)$ for any s, when $j = 2t$, verifying the values we have just obtained. The dimension of the tangent space at a general element f of $Gor(T)$ bounds the dimension above. We bound it below by the dimension $3s$ of $PS(s, j; 3)$ when $s \leq s_0$, and when $s \geq s_0$ by

$$\dim(V_s(t, t; r)) = r_j - \mathrm{cod}(V_s(t, t; r)) \geq r_j - \binom{s^\vee + 1}{2}.$$

The above inequality is a general result concerning the dimension of the corank s locus of a square symmetric matrix (see Theorem 1.26). Since our earlier manuscript [**IK**] was promulgated in 1996, A. Geramita, M. Pucci and Y. S. Shin, first improved the result by showing that when $T = H(s,j,3)$ and $T \supset (s,s,s)$, then **Gor**(T) is smooth ([GPS], see also Remark 7.19 below). Then J. O. Kleppe showed that when $r = 3$ and T arbitrary, **Gor**(T) is smooth ([K12], see also Section 4.4).

We now suppose that j is odd, and set $j = 2t+1$. The case $a = t$ of Theorem 4.5B was also shown by S. J. Diesel (announced in [Di]), by finding the dimension of the family of alternating matrices determining an ideal $I = \text{Ann}(f)$, $f \in$ **Gor**(T). See Section 4.4 Theorem 4.21 for a generalization of Theorem 4.6B to all T when $r = 3$.

THEOREM 4.5A. EQUALITY OF CLOSURES OF $Gor(T)$, $T = H(s,j,3)$, AND OF $PS(s,j;3)$, j ODD, s NOT TOO BIG. *If* $r = 3$, $j = 2t + 1$, $s \leq \dim_k R_{t-1} + 1$ *and* $T = H(s,j,3)$, *then* $\overline{Gor(T)} = \overline{PS(s,j;3)}$ *and is a component of* $V_s(t,t+1;3)$ *having dimension* $3s$.

THEOREM 4.5B. BOUNDS FOR THE DIMENSION OF $Gor(T)$, $T = H(s,j,3)$, j ODD, s LARGE. *If* $r = 3$, $j = 2t+1$, $t \geq 2$, $s = \dim_k R_{t-1} + a$, $0 \leq a \leq t + 1$ *and* $T = H(s,j,3)$, *and* f *is a general enough closed point of* **Gor**(T), *then if* $T_f = T_f(\textbf{Gor}(T))$ *we have, letting* $s_0 = \dim_k R_{t-1}$,

$$3s + \lfloor a/2 \rfloor \leq \dim(\textbf{Gor}(T)) \leq \dim_k T_f$$

$$= 3s_0 + \frac{a(a+5)}{2} = 3 \cdot \binom{t+1}{2} + \frac{a(a+5)}{2} \qquad (4.1.12)$$

When $a \in \{0,1,2,t-1,t,t+1\}$ *then* $\dim Gor(T) = \dim_k T_f$.

PROOF OF THEOREM 4.5A. We have $s \leq \dim_k R_{t-1}$. The first part is proven in the same way as Theorem 4.1A, taking $f = L_1^{[j]} + \cdots + L_s^{[j]}$ with $L_1, \ldots, L_s$ general enough linear forms, and using Lemma 3.16 and the knowledge of $H(R/\mathcal{I}_P^{(2)})$ to show that if $I = \text{Ann}(f)$, then $\dim_k T_f = 3s$. This, and the irreducibility of $Gor(T)$ show that $\overline{Gor(T)} = \overline{PS(s,j;3)}$. $\qquad \square$

PROOF OF THEOREM 4.5B. Corollary 5.50 in Section 5.3 shows the lower bound. For the upper bound we use the Kustin-Ulrich resolution for I^2, $I = \text{Ann}(f)$, f general $\in Gor(T)$ (see Theorem B.3, Appendix B), and denote by $r_i = \dim_k R_i$. Since

$$T = \ldots r_{t-2}, r_{t-1}, r_{t-1} + a, r_{t-1} + a, r_{t-1}, r_{t-2}, \ldots$$

the relevant portion of the third differences is

$$\Delta^3 T = \ldots 0, a - t - 1, t - 2a, 0, 2a - t, t + 1 - a, 0, \ldots \qquad (4.1.13)$$

showing there is a v-dimensional vector space $V = I_t$, $v = t + 1 - a$ of generators for I in degree t. When $t \leq 2a$ there is a $(2a - t)$ dimension space W of degree $t+1$ generators, and corresponding $2a-t$ relations in degree $t+3$, $t+1-a$ relations in degree $t+4$; in addition, if a is odd, there is an extra generator-relation pair in degree $t+2$, to make an odd number $a+2$ of generators in all (see [**BE2**] and [**Di**]). Thus, when $t \leq 2a$, the Kustin-Ulrich resolution for I^2 truncated to degree less or equal j is

$$\ldots 0 \to (R_1 \operatorname{Sym}^2 V \oplus R_1 V \otimes W)_{\leq j} \to I^2_{\leq j},$$

thus we have

$$\dim_k(I^2)_j = 3 \cdot \binom{t+2-a}{2} + (2a-t)(t+1-a) \tag{4.1.14}$$
$$= (t+1-a)(t+6+a)/2.$$

We conclude that $\dim_k T_f = r_j - \dim_k(I^2_j)$ satisfies (4.1.12) when $t \leq 2a$. When $t \geq 2a$, there is a $t-2a$ dimension space E of relations in degree $t+1$, in place of W, and the resolution for $I^2 \leq j$ satisfies,

$$\ldots 0 \to (R_1 V \otimes E)_{\leq j} \to (R_1 \operatorname{Sym}^2 V)_{\leq j} \to I^2_{\leq j},$$

leading again to (4.1.14) and (4.1.12) in this case.

When $a = 0$, the dimension result is given by Theorem 4.5A, and when $a = t+1$ it is trivial. The first difference with respect to a of the formula $F(t, a)$ for $\dim_k T_f$ in (4.1.12) is

$$F(t, a) - F(t, a-1) = a + 2. \tag{4.1.15}$$

Thus, when $s = r_t - 1$, the codimension of $Gor(T)$ in $\mathbb{A}(\mathcal{D}_j)$ is at least $F(t, t+1) - F(t, t) = t+3$, by (4.1.12). But $Cat(t, t+1; 3)$ is an $r_t \times r_{t+1}$ matrix, and any component of the corank one locus has codimension at most $1 \cdot (r_{t+1} - (r_t - 1)) = t + 3$ (See also Theorem 3.14 for a more precise statement). This shows that for $T = H(r_t - 1, 2t + 1, 3)$ the codimension of $Gor(T)$ in $\mathbb{A}(\mathcal{D}_j)$ is $t + 3$. The case $s = r_t - 2$ is treated in Theorem 4.13. If $a = 1$ we have $PS(s, j; 3) \subset Gor(T)$; $\dim PS(s, j; 3) = 3s$ according to Lemma 1.61 and $\dim_k T_f = 3s$ by (4.1.12). Thus $\overline{Gor(T)} = \overline{PS(s, j; 3)}$ also in this case. If $a = 2$, the lower bound in (4.1.12) for $\dim \mathbf{Gor}(T))$ equals the upper bound. This completes the proof of Theorem 4.5B. $\qquad\square$

REMARK. As we mentioned above, recently J. O. Kleppe proved that when $r = 3$ the schemes $\mathbf{Gor}(T)$ are smooth [**Kl2**] (see also Section 4.4). Thus under the hypothesis of Theorem 4.5B we have in fact the equality

$$\dim Gor(T) = 3\binom{t+1}{2} + \frac{a(a+5)}{2}.$$

$t=4,\quad s=$	10	11	12	13	14	15
$\dim_k T_f$	30	33	37	42	48	55

$t=5,\quad s=$	15	16	17	18	19	20	21
$\dim_k T_f$	45	48	52	57	63	70	78

TABLE 4.1. Values of $\dim_k T_f$, f a general element of $Gor(T)$, T= $H(s,2t+1,3)$. (See Theorem 4.5B).

EXAMPLE 4.6. DIMENSION OF TANGENT SPACE TO $\mathbf{Gor}(T)$, $T = H(s,j,3)$, $j = 9,11$. When $t = 4$ or 5, Theorem 4.5B implies that the dimensions of the tangent spaces for $\mathbf{Gor}(T)$ when $T = H(s,j,3)$, $j = 9,11$ are given by the Table 4.1. Note the pattern of the first differences (3,4, ...) of these dimensions, as given by (4.1.15).

REMARK 4.7. Under the assumptions of Theorem 4.5B let f be a closed point of $\mathbf{Gor}(T)$ and let $I = \mathrm{Ann}(f)$. As we see from the Kustin-Ulrich resolution of I^2 (see Appendix B) in order to calculate $(I^2)_{2t+1}$ we need the minimal resolution of I up to degree $t + 1$. Consider the following conditions of "genericity"

    i. For $a \geq t/2$ one has $t + 1 - a$ generators in degree t, $2a - t$ generators in degree $t + 1$, and no relations in degrees less or equal $t + 1$,

    ii. For $a \leq t/2$ one has $t+1-a$ generators in degree t, no generators in degree $t + 1$ and $t - 2a$ relations in degree $t + 1$,

It is easy to see that (i) and (ii) occur if and only if the

$$m : R_1 \otimes I_t \to I_{t+1}$$

satisfies the following condition,

m *is injective if* $a \geq t/2$, *and it is surjective if* $a \leq t/2$.   (4.1.16)

This condition is equivalent to

$$\dim_k(R_1 I_t) = \min(3 \cdot \dim_k I_t, \dim_k I_{t+1}) \qquad (4.1.17)$$

The proof of Theorem 4.5B shows that if for a given f the ideal $I = \mathrm{Ann}(f)$ satisfies the condition (4.1.17), then the dimension formula for the tangent space in (4.1.12) holds for $\dim_k T_f(\mathbf{Gor}(T))$.

We now apply the above results to determine $H(R/\mathcal{I}_Z^2)$ when Z is a set of general points in $\mathbb{P}^2$. Recall that if $H = (\ldots,h_i,\ldots)$, the difference function $\Delta H = (\ldots,\delta_i = h_i - h_{i-1},\ldots)$; also $r_i = \dim_k R_i$.

PROPOSITION 4.8. SQUARE OF THE VANISHING IDEAL OF s GENERAL POINTS OF $\mathbb{P}^2$. *Suppose that* $\dim_k R_{t-1} \leq s < \dim_k R_t$, *let* $s = \dim_k R_{t-1} + a$, *let* $s^\vee = \dim_k R_t - s = t + 1 - a$, *and suppose that* $\mathcal{I}_Z$ *is the ideal of a scheme* Z *of* s *general points of* $\mathbb{P}^2$. *Then the Hilbert function* $H(R/\mathcal{I}_Z^2)$ *satisfies*

$$H(R/\mathcal{I}_Z^2) = (1, 3, \ldots, r_{2t-1}, r_{2t} - \binom{s^\vee + 1}{2},$$

$$3 \cdot r_{t-1} + \frac{a(a+5)}{2}, 3s, 3s, \ldots)$$

$$\Delta H(R/\mathcal{I}_Z^2) = (1, 2, \ldots, \delta_{2t-1} = 2t, 2t + 1 - \binom{s^\vee + 1}{2},$$

$$-a(t - a - 1), -\binom{a}{2}, \delta_{2t+3} = 0, \ldots)$$

PROOF. Let $f = L_1^{[j]} + \cdots + L_s^{[j]}$, $j = 2t + 1$, with $L_1, \ldots, L_s$ corresponding to Z as in Definition 1.13, and set $I = \text{Ann}(f)$. Then by Lemma 1.19 we have $(\mathcal{I}_Z)_i = I_i$ for $i = t, t + 1$. Now a result of A. Geramita and P. Maroscia (Theorem 2.6 of [GM]) shows that

$$\dim_k R_1(\mathcal{I}_Z)_t = \min(3 \cdot \dim_k(\mathcal{I}_Z)_t, \dim_k(\mathcal{I}_Z)_{t+1}).$$

Thus, condition (4.1.17) of Remark 4.7 is satisfied. The proof of Theorem 4.5B shows that $\dim_k(I^2)_{2t} = \binom{s^\vee + 1}{2}$, and $H(R/I^2)_j = \dim_k T_f$ from (4.1.12), but these are identical with $\dim_k(\mathcal{I}_Z^2)_{2t}$ and $H(R/\mathcal{I}_Z^2)_j$, respectively. Corollary 1.71 of the Geramita-Gimigliano-Pitteloud's or Chandler's Theorem (Theorem 1.70) shows that $(\mathcal{I}_Z^2)_i = (\mathcal{I}_Z^{(2)})_i$ for $i \geq j + 1$. The Alexander-Hirschowitz result (Theorem 1.66) determines the codimension of $\mathcal{I}_Z^{(2)}$ in R_i, $i \geq j + 1$. This completes the proof. $\square$

In local singularity theory there has been some interest in the ideal $(J^{(2)}/J^2)$, as "primitive ideal" of a singularity. If J is the ideal at the vertex of a cone over the subscheme Z of $\mathbb{P}^2$, here is a relation between $(J^{(2)}/J^2)$ and the ideal $\mathcal{I}_Z^{(2)}/\mathcal{I}_Z^2$.

COROLLARY 4.9. HILBERT FUNCTION OF THE SYMBOLIC SQUARE MOD SQUARE OF THE VANISHING IDEAL AT s GENERAL POINTS OF $\mathbb{P}^2$. *Under the hypotheses of Proposition 4.8, the Hilbert function*

$H(\mathcal{I}_Z^{(2)}/\mathcal{I}_Z^2)$ satisfies $H(\mathcal{I}_Z^{(2)}/\mathcal{I}_Z^2)_i = r_i - \min(r_i, 3s)$ if $i \leq 2t - 1$, and

$$H(\mathcal{I}_Z^{(2)}/\mathcal{I}_Z^2) = (0, \ldots, r_i - \min(r_i, 3s), \ldots,$$

$$r_{2t} - \binom{s^\vee + 1}{2} - 3s, \binom{a}{2}, 0, \ldots),$$

except that when $(t, s, a, s^\vee) = (2, 5, 2, 1)$, $H(\mathcal{I}_Z^{(2)}/\mathcal{I}_Z^2)_{2t} = 0$. The order $\nu(\mathcal{I}_Z^{(2)}/\mathcal{I}_Z^2)$ is asymptotic to $t\sqrt{3}$.

Proof. The formula for $H(\mathcal{I}_Z^{(2)}/\mathcal{I}_Z^2)$ is immediate from the Alexander-Hirschowitz theorem (Theorem 1.66) and Proposition 4.8. The last statement is an elementary consequence of Theorem 1.66. $\square$

4.2. Sets of s points in $\mathbb{P}^{r-1}$ and Gorenstein ideals

We now generalize Theorems 4.1A, 4.1C, and 4.5A to $r \geq 3$. We let T denote the sequence $H(s, j, r)$ defined in (4.1.1). Recall the notation $\tau = \tau_{s,r} = \min\{i \mid s \leq \dim_k R_i\}$, $\sigma = \tau + 1$ and $d = d_{s,r}$ satisfying $\dim_k R_{d-1} \leq s < \dim_k R_d$ (see Lemma 3.16). Our relative lack of knowledge concerning $H(R/\mathcal{I}_Z^2)$, Z a set of s general points of $\mathbb{P}^{r-1}$, $r \geq 4$, leads to weaker statements.

Theorem 4.10A. Gorenstein algebras arising from a sum of powers form an irreducible component of $\overline{Gor(T)}$, $T = H(s, j, r)$, s not too large. Suppose $r \geq 3$, $j = 2t$ or $2t + 1$, $s \leq \dim_k R_{t-1} = \binom{t+r-2}{r-1}$, and let $T = H(s, j, r)$. Equivalently, let $T = H(s, j, r)$ and suppose T contains the subsequence (s, s, s). Then $\overline{PS(s, j; r)}$ is an irreducible component of $\overline{Gor(T)}$ and of $V_s(t, j - t; r)$. The schemes $\mathbf{Gor}(T)$ and $\mathbf{V}_s(t, j - t; r)$ are smooth along an open, dense subset of $\overline{PS(s, j; r)}$.

Theorem 4.10B. When $PS(s, j; r)$ is a component of the catalecticant locus. Suppose further that $\sigma = \tau_{s,r} + 1 \leq t$ and v satisfies, $\sigma \leq v \leq t$. Then $\overline{PS(s, j; r)}$ is a component of $V_s(v, j - v; r)$. Furthermore, if $f = L_1^{[j]} + \cdots + L_s^{[j]} \in PS(s, j; r)$ is general, and $I = \text{Ann}(f)$, then I has order $d_{s,r}$. If P_L is the set of points in $\mathbb{P}^{r-1}$ corresponding to $L_1, \ldots, L_s$, and $\mathcal{I}(P_L)$ the ideal in R vanishing at P_L then

$$I_i = \mathcal{I}(P_L)_i \quad \text{when} \quad i \leq j - \tau, \tag{4.2.1}$$

and

$$(I^2)_{2d} = (\mathcal{I}(P_L)^2)_{2d}, \ldots, (I^2)_{j+1} = (\mathcal{I}(P_L)^2)_{j+1},$$
$$(I^2)_{j+2} \subset (\mathcal{I}(P_L))_{j+2}, \ldots, (I^2)_{2j+3-2\sigma} \subset (\mathcal{I}(P_L))_{2j+3-2\sigma}. \tag{4.2.2}$$

When $2\sigma \leq i$, we have

$$(\mathcal{I}(P_L)^2)_i = (\mathcal{I}(P_L)^{(2)})_i. \tag{4.2.3}$$

If $\sigma \geq 2$, the Hilbert function $H(R/I^2)$ satisfies, taking $c = j+2-2\sigma$, $\alpha = \dim_k((\mathcal{I}(P_L)_{\sigma-1})^2)$ and $\beta = \dim_k(\mathcal{I}(P_L)_{\sigma-1}\mathcal{I}(P_L)_\sigma)$,

$$(R/I^2) \geq (1, r, \ldots, \dim_k R_{2\sigma-2} - \alpha,$$
$$\dim_k R_{2\sigma-1} - \beta, \overset{\text{length } c}{rs, \ldots, rs}, \overset{\text{length } c}{s, \ldots, s}). \tag{4.2.4}$$

with equality in degrees i less or equal $j+1$. When $s = \dim_k R_{\sigma-1}$ then $I_{\sigma-1} = 0$, and $\alpha = \beta = 0$.

PROOF. Identities (4.2.1) and (4.2.2) follow from Lemma 3.16. By Theorem 1.69, $\mathcal{I}(P_L)$ has Castelnuovo regularity $\sigma = \tau(\mathcal{I}(P_L)) + 1$. This and the Geramita-Gimigliano-Pitteloud and Chandler Theorem (Theorem 1.70 and Corollary 1.71) imply (4.2.3). Equality in (4.2.4) for degrees i, $2\sigma \leq i \leq j+1$ now follows from Theorem 1.69, and either Theorem 2.6 (char$(k) = 0$) or the Alexander-Hirschowitz result Theorem 1.66. The inequality in (4.2.4) for degrees i, $j+2 \leq i \leq 2j+3-2\sigma$, follows from (4.2.2), and the generation of $\mathcal{I}(P_L)$ by $(\mathcal{I}(P_L)_d, \mathcal{I}(P_L)_\sigma)$. Now using Lemma 3.16, the statements in Theorems 4.10A, 4.10B concerning $\overline{PS}(s, j; r)$ are proved in the same way as Theorem 4.1A □

EXAMPLE 4.11. COMPARISON BETWEEN $\mathcal{I}_P^2$ AND $\mathcal{I}_P^{(2)}$. We let $r = 3$, $j = 4$, and consider the set of points $P = \{a(1) = (1, 0, 0), a(2) = (0, 1, 0), a(3) = (0, 0, 1)\}$, defining the vanishing ideal $\mathcal{I}_P$ in R. Then

$$\mathcal{I}_P = (y, z) \cap (x, z) \cap (x, y) = (xy, xz, yz),$$
$$\mathcal{I}_P^{(2)} = (y^2, yz, z^2) \cap (x^2, xz, z^2) \cap (x^2, xy, y^2) = (xyz, x^2y^2, x^2z^2, y^2z^2),$$
$$\mathcal{I}_P^2 = (x^2yz, xy^2z, xyz^2, x^2y^2, x^2z^2, y^2z^2).$$

Thus, $\mathcal{I}_P^2 = M^4 \cap \mathcal{I}_P^{(2)}$, but of course $\mathcal{I}_P^2 \neq \mathcal{I}_P^{(2)}$. If $I = \text{Ann}(f)$, with $f = X^{[4]} + Y^{[4]} + Z^{[4]}$ in $\mathcal{D}$, then $I = (xy, xz, yz, x^4 - y^4, y^4 - z^4)$, $H(R/I) = (1, 3, 3, 3, 1)$, and $I_2 = (\mathcal{I}_P)_2$, $I_3 = (\mathcal{I}_P)_3$. The ideal

$$I^2 = (\mathcal{I}_P^2, x^5y, x^5z, xy^5, xz^5, y^5z, yz^5, M^8) = (\mathcal{I}_P^2, M^6 \cap \mathcal{I}_P, M^8),$$

and $H(R/I^2) = (1, 3, 6, 10, 9, 9, 3, 3, 0)$. The example extends readily to an arbitrary number of variables, illustrating Theorems 4.10A and 4.10B.

We next study $V_s(t, t; r)$ and $V_s(t, t+1; r)$ for arbitrary $r \geq 3$ when $s^\vee = \dim_k R_t - s$ is less than r. We denote $\dim_k R_u$ by r_u.

LEMMA 4.12. HILBERT FUNCTION OF THE SQUARE OF THE VANISHING IDEAL AT s POINTS OF $\mathbb{P}^{r-1}$, FOR SPECIAL VALUES OF s. *Let* $t \geq 2$. *Suppose that* $r_t - r < s \leq r_t$. *Let* Z *be a set of* s *general points in* $\mathbb{P}^{r-1}$ *and let* $\mathcal{I}_Z$ *be its ideal. Then the Hilbert function of* $\mathcal{I}_Z^2$ *satisfies*

$$H(R/\mathcal{I}_Z^2) = (1, r, \ldots, r_{2t-1}, r_{2t} - \binom{s^\vee + 1}{2},$$

$$r_{2t+1} - s^\vee(r_{t+1} - s) + r\binom{s^\vee}{2}, rs, rs, \ldots). \quad (4.2.5)$$

If s *satisfies only* $r_{t-1} < s \leq r_t$ *and* Z *is general, and if the multiplication map* $m_{t,1} : (\mathcal{I}_Z)_t \otimes R_1 \to R_{t+1}$ *is injective, then the left side of* (4.2.5) *is greater or equal the right side, with equality in degrees* $i \geq 2t + 2$.

PROOF. Put $\mathcal{I} = \mathcal{I}_Z$, and suppose $r_{t-1} < s \leq r_t$. For $i \geq 2t + 2$ we have by Corollary 1.71 that $(\mathcal{I}^2)_i = (\mathcal{I}^{(2)})_i$, so by Theorem 1.66 we conclude that $\dim_k(R/\mathcal{I}^2)_i = rs$ (there are no exceptions since $i \geq 6$). Since $r_{t-1} \leq s$ we have $\mathcal{I}_{t-1} = 0$, so

$$(\mathcal{I}^2)_{2t} = \mathcal{I}_t \mathcal{I}_t, \quad (\mathcal{I}^2)_{2t+1} = \mathcal{I}_t \cdot \mathcal{I}_{t+1}. \quad (4.2.6)$$

The case $s^\vee = 0$ of the Lemma is clear since then $\mathcal{I}_t = 0$.

Suppose $r > s^\vee \geq 1$. As in the proof of Lemma 2.7 we see that if $a_1, \ldots, a_s$ are general enough points in $\mathbb{P}^{r-1}$, then the images $v_t(a_1), \ldots, v_t(a_s)$ of the Veronese map span a subspace E which intersects transversally $v_t(\mathbb{P}^{r-1})$. Therefore, if $Z = \{a_1, \ldots, a_s\}$ then $\mathcal{I}_t = (\mathcal{I}_Z)_t$ has a basis $\{f_1, \ldots, f_m\}$, $m = s^\vee$, which forms a regular sequence of R. Now let us consider the Koszul resolution of the complete interesection ideal $(f_1, \ldots, f_m)$ in R (see [Ma2, p. 135]). Truncating to degrees $\leq 2t + 1$ we obtain that the multiplication map

$$m_{t,1} : \mathcal{I}_t \otimes R_1 \to R_{t+1}$$

is injective, and for the multiplication maps

$$m_{t,t} : \mathcal{I}_t \otimes R_t \to R_{2t}$$

$$m_{t,t+1} : \mathcal{I}_t \otimes R_{t+1} \to R_{2t+1}$$

we have

$$\mathrm{Ker}(m_{t,t})=\langle\{f_i \otimes f_j - f_j \otimes f_i \mid 1{\leq}i{<}j{\leq}m\}\rangle \tag{4.2.7}$$

$$\mathrm{Ker}(m_{t,t+1})=\langle\{f_i \otimes f_j x_k - f_j \otimes f_i x_k \mid 1{\leq}i{<}j{\leq}m, 1{\leq}k{\leq}r\}\rangle. \tag{4.2.8}$$

This shows that these kernels have dimension

$$\binom{s^\vee}{2}, \quad r\binom{s^\vee}{2}, \tag{4.2.9}$$

respectively. Furthermore, these kernels belong to $\mathcal{I}_t \otimes \mathcal{I}_t$, $\mathcal{I}_t \otimes \mathcal{I}_{t+1}$, respectively, implying

$$\dim_k \mathcal{I}_t \cdot \mathcal{I}_t = (s^\vee)^2 - \binom{s^\vee}{2} = \binom{s^\vee+1}{2},$$

$$\dim_k \mathcal{I}_t \cdot \mathcal{I}_{t+1} = s^\vee(r_{t+1} - s) - r\binom{s^\vee}{2}.$$

When s satisfies only $r_{t-1} < s \leq r_t$ then (4.2.6) remains valid, but the kernels of $m_{t,t}$ and $m_{t,t+1}$ could properly contain the subspaces of $\mathcal{I}_t \otimes R_t$ and $\mathcal{I}_t \otimes R_{t+1}$, respectively, listed in (4.2.7), (4.2.8). The subspace of (4.2.7) is isomorphic to $\Lambda^2(\mathcal{I}_t)$ and has dimension $s^\vee(s^\vee - 1)/2$. The subspace in (4.2.8) is easily seen to have dimension $r(s^\vee)(s^\vee - 1)/2$ if $m_{t,1}$ is injective. It follows that the left side of (4.1.17) is at least as great termwise as the right side, if $m_{t,1}$ is an injection. This completes the proof. $\qquad\square$

THEOREM 4.13. THE CATALECTICANT VARIETY $V_s(t, j - t; r)$ ALONG THE POWER SUM LOCUS $PS(s, j; r)$ WHEN THE CORANK $s^\vee < r$. *Suppose* $r \geq 3$, $t \geq 2$, *and that* $j = 2t$ *or* $2t + 1$. *Let* $s^\vee = \dim_k R_t - s$, *and suppose that* $0 < s^\vee < r$. *Then* $PS(s, j; r)$ *is contained in a unique irreducible component* X *of* $V_s(t, t; r)$ *if* $j = 2t$, *and in a unique irreducible component* Y *of* $V_s(t, t+1; r)$ *if* $j = 2t+1$. *Let* $f \in PS(s, j; r)$ *be sufficiently general and let* $T_f = T_f(\mathbf{V}_s(t, t; r))$ *or* $T_f' = T_f(\mathbf{V}_s(t, t+1; r))$. *Then*

i. $\dim X = \dim_k T_f = \dim_k R_{2t} - \binom{s^\vee+1}{2}$,
ii. $\dim Y = \dim_k T_f' = \dim_k R_{2t+1} - s^\vee(\dim_k R_{t+1} - s) + r\binom{s^\vee}{2}$.

The varieties X *and* Y *are irreducible components of* $\overline{Gor(T)}$, $T = H(s, j, r)$, *respectively, in the cases* $j = 2t$, $j = 2t + 1$. *The corresponding projective varieties* $\mathbb{P}(X)$ *and* $\mathbb{P}(Y)$ *are rational.*

PROOF. Let $f = L_1^{[j]} + \cdots + L_s^{[j]}$ where the set $Z = P_L \subset \mathbb{P}^{r-1}$ is sufficiently general. Let $I = \mathrm{Ann}(f)$. By Lemma 1.19 we have

$$I_t = (\mathcal{I}_Z)_t \quad \text{if} \quad j = 2t \quad \text{or} \quad 2t+1, \quad \text{and}$$

$$I_{t+1} = (\mathcal{I}_Z)_{t+1} \quad \text{if} \quad j = 2t+1.$$

Using Theorem 3.2 and Lemma 4.12 we obtain the formulas for $\dim_k T_f$ and $\dim_k T'_f$ in (i) and (ii). First, let us consider the case $j = 2t$. Every component of $V_s(t, t; r)$ has codimension no greater than $\binom{s^\vee+1}{2}$ in $\mathbb{A}(\mathcal{D}_j)$ (see Theorem 1.26). This shows that f is a nonsingular point of the scheme $\mathbf{V}_s(t, t; r)$. So there is a unique irreducible component X of $V_s(t, t; r)$ which contains $PS(s, j; r)$ and X is of codimension $\binom{s^\vee+1}{2}$ in $\mathbb{A}(\mathcal{D}_j)$.

Now let $j = 2t+1$. We want to show that $V_s = V_s(t, t+1; r)$ has an irreducible component which contains $PS(s, j; r)$ and is of dimension at least $\dim_k T'_f$. From (1.1.5) we know that $V_s = V_s(t+1, t; r)$. Let us consider the closed subset

$$\Gamma \subset Grass(s^\vee, R_t) \times \mathbb{A}(\mathcal{D}_j)$$

$$\Gamma = \{(\Delta, f) \mid \Delta \circ f = 0\}.$$

Let $\mathbb{G} = Grass(s^\vee, R_t)$ and let π_1, π_2 be the two projections of Γ onto the corresponding factors.

$$\Gamma \xrightarrow{\pi_2} \mathbb{A}(\mathcal{D}_j)$$
$$\pi_1 \downarrow$$
$$\mathbb{G}$$

Clearly $\pi_2(\Gamma) = V_s$.

CLAIM. The following properties hold:

a. There is a unique irreducible component Γ_0 of Γ such that $\pi_1(\Gamma_0) = \mathbb{G}$. The generic fibre of $\pi_1|_{\Gamma_0} : \Gamma_0 \to \mathbb{G}$ is a vector space of dimension

$$\dim_k R_j - s^\vee \cdot \dim_k R_{t+1} + r\binom{s^\vee}{2} \geq 1.$$

b. The map $\pi_2 : \Gamma_0 \to \mathbb{A}(\mathcal{D}_j)$ is birational onto its closed image $\pi_2(\Gamma_0)$. The variety $\pi_2(\Gamma_0)$ contains $PS(s, j; r)$, and $\dim \pi_2(\Gamma_0) = \dim_k T'_f$.

PROOF OF CLAIM. Let $\Delta \in Grass(s^\vee, R_t)$. The fiber $\pi_1^{-1}(\Delta)$ is isomorphic to the space $(\Delta \cdot R_{t+1})^\perp \subset \mathbb{A}(\mathcal{D}_j)$. Let d be the minimum dimension of $(\Delta \cdot R_{t+1})^\perp$ and let $U \subset Grass(s^\vee, R_t)$ be the set of Δ

such that $\dim_k((\Delta \cdot R_{t+1})^\perp) = d$. Then U is a Zariski open, dense subset in $Grass(s^\vee, R_t)$ and $\Gamma_0 = \pi_1^{-1}(U)$ is an irreducible variety. Now, let us prove that $d = \dim_k R_j - s^\vee \cdot \dim_k R_{t+1} + r\binom{s^\vee}{2}$ and that $d > 0$. Let f be a sufficiently general element of $PS(s, j; r)$ and let $\Delta_f = \text{Ann}(f)_t = (\mathcal{I}_Z)_t$. We showed in the proof of Lemma 4.12 that the multiplication map

$$m_{t,1}(\Delta_f) \; : \; \Delta_f \otimes R_1 \to R_{t+1}$$

is injective. This implies that the multiplication map

$$m_{t,1}(\Delta) \; : \; \Delta \otimes R_1 \to R_{t+1}$$

is injective for every sufficiently general $\Delta \in Grass(s^\vee, R_t)$. For such a Δ let $e_1, \ldots, e_m$, $m = s^\vee$ be a basis of Δ. Then the linear span

$$\langle e_i \otimes e_j x_k - e_j \otimes e_i x_k \mid 1 \le i < j \le m, \; 1 \le k \le r \rangle \qquad (4.2.10)$$

is of dimension $r\binom{s^\vee}{2}$ and is contained in the kernel of the multiplication map

$$m_{t,t+1}(\Delta) \; : \; \Delta \otimes R_{t+1} \to R_{2t+1}.$$

This proves that

$$d \ge \dim_k R_j - s^\vee \cdot \dim_k R_{t+1} + r\binom{s^\vee}{2}. \qquad (4.2.11)$$

Now, if we consider $\Delta = \Delta_f$ as above we have by Lemma 4.12 that $\text{Ker}(m_{t,t+1}(\Delta_f))$ is equal to the linear span (4.2.10). This proves by semicontinuity that we have equality in (4.2.11) and that $\Delta_f \in U$. Finally, f is a nonzero element of the vector space $(\Delta_f \cdot R_{t+1})^\perp$ of dimension d, thus $d \ge 1$. This proves Part (a) of the Claim. For every $g \in V_s$ the fiber $\pi_2^{-1}(g)$ is isomorphic to $Grass(s^\vee, \text{Ann}(g)_t)$. From the properness property of the projective variety $\mathbb{G}$ we have that $\pi_2(\Gamma_0)$ is closed in $\mathbb{A}(\mathcal{D}_j)$. Since for every sufficiently general $f \in PS(s, j; r)$ we have that $\Delta_f = \text{Ann}(f)_t$ is of dimension $s^\vee$ and it belongs to U, we conclude that $PS(s, j; r) \subset \pi_2(\Gamma_0)$. Furthermore there is a Zariski open dense subset of $\pi_2(\Gamma_0)$, which intersects $PS(s, j; r)$, over which π_2 has an inverse, given by Cramer's formulas for the solutions of a homogeneous system of linear equations whose coefficients are the matrix $Cat_F(t + 1, t; r))$. In particular,

$$\dim \pi_2(\Gamma_0) \; = \; s^\vee \cdot s + d \; = \; \dim_k T'_f.$$

This proves part (b) of the claim. $\qquad \square$

PROOF OF THEOREM 4.13, CONTINUED. Let Y be an irreducible component of V_s which contains $\pi_2(\Gamma_0)$. Then Y contains $PS(s,j;r)$ and $\dim Y \geq \dim_k T'_f$. An argument similar to that of the case $j = 2t$ shows that $\dim Y = \dim T'_f$, so $Y = \pi_2(\Gamma_0)$. One can make the same construction with $V_s(t,t;r)$ for $j = 2t$. This yields similarly $\pi_2(\Gamma_0) = X$ and an analog of Part (b) of the Claim above holds. By semicontinuity, for every sufficiently general $g \in X$, $g' \in Y$ one has $H(A_g) = H(s,2t,r)$, $H(A'_g) = H(s,2t+1,r)$, since by Lemma 1.17 this property holds for general enough elements in $PS(s,j;r)$. This shows that in both cases X, Y are subsets and therefore irreducible components of $\overline{Gor(T)}$. Finally, the map $\pi_1 : \pi_1^{-1}(U) \to U$ is an algebraic vector bundle, thus by Part (b) of the Claim $\mathbb{P}(X)$ (resp. $\mathbb{P}(Y)$) are birationally isomorphic to its projectivization, thus they are rational varieties. This completes the proof of Theorem 4.13. $\square$

REMARK. A portion of Theorem 4.13 is extended to the case $s^\vee = r$ in the next Section 4.3, Corollary 4.18, Theorem 4.19.

4.3. Gorenstein ideals whose lowest degree generators are a complete intersection

By taking a different viewpoint, begining with complete intersections, we are able to generalize some of the results of Section 4.2. Suppose that $A = R/I$, $I = (f_1, \ldots, f_r)$ is a general enough graded Artinian complete intersection $C = R/I$, and let $D = (d_1, \ldots, d_r)$, $d_1 \leq d_2 \leq \cdots \leq d_r$ be the generator degrees of I, and $w = \sum d_i - r$ be the socle degree of C. A result of R. Stanley determines the Hilbert function of the quotient S of C by a general enough homogeneous element f. Given an integer a, $0 < a \leq w$, if f is a general element of C_a, and k is any integer, then Stanley showed that multiplication by f

$$m_f : C_k \to C_{k+a}$$

has maximal rank – is injective or surjective (see [St2], also [Wa2, I3]). It follows that the almost complete intersection $S = R/(I,f)$ has the expected Hilbert function:

$$H(S)_i = \begin{cases} H(C)_i & \text{if } i < a, \\ H(C)_i - H(C)_{i-a} & \text{if } a \leq i < (w+a)/2, \\ 0 & \text{if } i \geq (w+a)/2. \end{cases}$$

By linking in I, we construct from S a new Gorenstein algebra $B = R/J$, $J = (I,f) : I$, whose socle degree is $j = w - a$. We show that its

Hilbert function $H(B)$ satisfies

$$H(B) = \text{Sym}(H(C), j) \; : \; \left(H(B)_i = \begin{cases} H(C)_i & \text{if } i \leq j/2 \\ H(C)_{j-i} & \text{if } i > j/2 \end{cases} \right).$$

the symmetrization of $H(C)$ about $j/2$ (Theorem 4.17). We also show that if $T = H(B)$ then the Gorenstein ideals J formed this way determine an irreducible component C_T of $\mathbb{P}Gor(T)$. These results, and as well Lemmas 4.14, 4.15, and Theorem 4.16 below are a corollary of Junzo Watanabe's work [**Wa2**].

We say that a Gorenstein Artin algebra A of socle degree j satisfies the Strong Lefschetz (SL) property if there is a linear element L such that for each k and each $i \leq j/2$, the multiplication map $m_{L^k} : A_i \to A_{i+k}$ has maximal rank.[2] We also say that (A, L) has the Strong Lefschetz property. R. Stanley had shown in particular that a complete intersection ideal generated by monomials satisfies the Strong Lefschetz property (see ibid. and the proof of Lemma 4.15 below). J. Watanabe showed,

THEOREM 4.14. STRONG LEFSCHETZ PROPERTY FOR CERTAIN QUOTIENTS. (J. Watanabe [**Wa2**, Proposition 3.4, Theorem 3.8]). *The pair (A, L) satisfies the SL property iff there exists a representation $\rho : sl_2 \to End_k(A)$, of the special linear Lie algebra sl_2 with $\rho \left(\begin{smallmatrix} 0 & 1 \\ 1 & 0 \end{smallmatrix} \right) = m_L$, such that the weight space decomposition coincides with the natural grading decomposition.*

If A is a SL algebra, then so is $A/(0 : F)$, where F is a general element of A_i.

We prove directly the following related result.

LEMMA 4.15. GORENSTEIN ALGEBRAS DETERMINED BY A COMPLETE INTERSECTION AND A FORM. *Suppose that r is arbitrary, that $\text{char}(k) = 0$ or that $\text{char}(k) > j$, and that $I = (f_1, \ldots, f_a)$, $a \leq r$, is a complete intersection generated by monomials: so $f_i = x_i^{d_i}$. Suppose t, u, v are nonnegative integers, $j = u + v$, and if $a = r$, that $j \leq \sum d_i - r$. Let F be a general enough element of $[I_j]^\perp \subset \mathcal{R}_j$. Consider the homomorphism*

$$C_F(v, u) : \mathcal{R}_u \to \mathcal{R}_v, \quad g \to g \circ F$$

The image of $C_F(v, u)$ lies in $[I_v]^\perp$, and $C_F(v, u)$ is zero on I_u. If $A_u = \mathcal{R}_u/I_u$, the homomorphism

$$C_F'(v, u) : A_u \to [I_v]^\perp$$

---

[2]This is termed the strong Stanley property (SSP) in [**Wa2**].

has maximal rank (is injective or surjective). If $v \leq j/2$, then $\dim A_u \geq \dim A_v = \dim[I_v]^\perp$, and $C'_F(v, u)$ is surjective. The Hilbert function $H(R/\operatorname{Ann}(F)) = \operatorname{Sym}(H(R/I), j)$.

PROOF. Note first that the assumption on j assures that $[I_j]^\perp \neq 0$. If $g \in R_u$ and $f \in I_v$, then $fg \in I_j$, and

$$0 = fg \circ F = f \circ (g \circ F),$$

implying that the image of $C_F(v, u)$ lies in $[I_v]^\perp$. Without loss of generality, we may use the contraction action of R_u on F, since a change of bases (multiplying by nonzero constants) in R_u and in $\mathcal{R}_v$ does not affect whether $C'_F(v, u)$ has maximal rank. Also, without loss of generality we may add monomial generators $f_{a+1}, \ldots, f_r$ of degree bigger than j to the ideal I, and thus assume that $A = R/I$ is zero-dimensional: this change does not affect the homomorphism $C'_F(v, u)$. We let w be the socle degree of the quotient $A = R/I$ (so $mA_w = 0$ but $A_w \neq 0$); then $w = \sum d_i - r$, since I is a complete intersection. Let $[I_w]^\perp = \langle H \rangle$, $H \in \mathcal{R}_w$. Then there are isomorphisms:

$$A_{w-i} \cong [I_i]^\perp : g \to g \circ H.$$

Thus, $F = f \circ H$ for some $f \in A_{w-j}$.

We claim that under the identification $A_{w-v} \cong [I^v]^\perp$, the homomorphism $C'_F(v, u)$ is just the multiplication homomorphism

$$m_f : A_u \to A_{w-v} : g \to gf.$$

This follows directly from the definition: if $G \in R_u$ satisfies $G \bmod I_u = g$, and $f' \in R_{w-j}$ satisfies $f' \bmod I_{w-j} = f$, then

$$C'_F(v, u)(g) = G \circ F = G \circ (f' \circ H) = Gf' \circ H = gf \circ H,$$

so, under the above identification we have $C'_F(v, u) = m_f$.

R. Stanley showed that m_f has maximal rank on each A_u as follows. He regards A as the cohomology ring of a product of projective spaces

$$A \cong H^*(X), \quad X = \mathbb{P}^{d_1 - 1} \times \cdots \times \mathbb{P}^{d_r - 1},$$

where we shift the grading so that each x_i has degree 2 in $H^*(X)$. By the Hard Lefschetz Theorem, if we take $f = L^{w-j}$, where L is a general enough linear element of A, then m_f has maximal rank. This completes the Stanley proof.

The Hilbert function $H(A)$ is monotone nondecreasing if $a < r$, and unimodal with center $w/2$ if $a = r$: this implies $\dim A_u \geq \dim A_v$

if $v \leq j/2$. That $H(R/\operatorname{Ann}(F)) = \operatorname{Sym}(H(R/I), j)$ follows from this, and the maximal rank property of $C'_F(v, u)$.

$\square$

The following Theorem generalizes the construction of compressed Gorenstein algebras. Such an algebra has maximum Hilbert function given the socle degree and r (see Definition 3.11 and Proposition 3.12). We now determine the maximum Hilbert function of a Gorenstein algebra of given socle degree, whose ideal contains a complete intersection of given generator degrees $d_1, \ldots, d_a$. Here, "maximum" is understood to be termwise maximum; when $I = 0$ the algebras B defined below are compressed.

THEOREM 4.16. MAXIMUM HILBERT FUNCTION OF A GOREN-STEIN ALGEBRA WHOSE IDEAL CONTAINS A COMPLETE INTERSEC-TION. *Under the numerical hypotheses of Lemma 4.15, suppose that $I = (f_1, \ldots, f_a)$ is a complete intersection having generator degrees $(d_1, \ldots, d_a)$, and such that $(f_1, \ldots, f_a)$ are either monomial $f_i = x_i^{d_i}$, $i = 1, \ldots, a$ or general enough. If F is a general enough element of $[I_j]^{\perp}$ then the Gorenstein algebra $B = R/J$, $J = \operatorname{Ann}(F)$ satisfies*
   a. *$J \supset I$, and $J_v = I_v$ for $v \leq j/2$.*
   b. *The Hilbert function $H_B = \operatorname{Sym}(H_A, j)$.*
The sequence $\operatorname{Sym}(H_A, j)$ is the maximum Hilbert function possible for an Artinian graded Gorenstein algebra of socle degree j containing a CI ideal of generator degrees $d_1, \ldots, d_a$.

PROOF. The statements (a) and (b) follow from deforming the monomial ideal and form F of Lemma 4.15, and the openness of the condition that each $C'_F(v, u)$ have maximal rank. That $\operatorname{Sym}(H_A, j)$ is the maximal possible Hilbert function as claimed follows from the symmetry of H_B and $\operatorname{Ann}(F) \supset I$, and its occurence in the monomial case (Lemma 4.15). $\square$

REMARK. It is not known if the Strong Lefschetz property holds for all graded Artinian complete intersections. If so, $(f_1, \ldots, f_a)$ could in Theorem 4.16 be taken to be an arbitrary CI, and the proof of Lemma 4.15 would extend. If not, the class of algebras considered in Theorem 4.16 could be larger than the class satisfying a SL propery of Theorem 4.14.

When $a = r$, $A_i \cong I_{w-i}^{\perp}$; letting $j = w - i$, this gives the connection between $A/(0 : F)$ of Watanabe's Theorem 4.14 and $R = Ann(F)$ of Theorem 4.16.

THEOREM 4.17. AN IRREDUCIBLE COMPONENT OF CERTAIN $\overline{Gor(T)}$, ARISING FROM COMPLETE INTERSECTIONS AND A GENERIC FORM. *Suppose that the sequence* $T = (1, r, \ldots, 1)$ *is symmetric about* $j/2$, *and satisfies*

$$T = \mathrm{Sym}(H, j),$$

where $H = (1, \ldots, h_i, \ldots)$ *is the Hilbert function of a graded complete intersection algebra* $A = R/I$. *Suppose further that* $D = (d_1, \ldots, d_a)$, $a \leq r$ *is the degree sequence of the generators of* I, *that* $j \geq 2 \cdot \max\{d_i\}$, *and that if* $a = r$, *then* $j \leq \sum d_i - r$. *Define the sequence* $E = (e_1, \ldots, e_s)$ *by* $e_k = \#\{i \mid d_i = k\}$, *Then* $\overline{Gor(T)}$ *contains an irreducible component* C_T, *an open dense subset of which parametrizes polynomials* f *with ideals* $J = \mathrm{Ann}(f)$ *such that* $J_{\leq j/2}$ *is a CI of the type* $(d_1, \ldots, d_a)$. *Furthermore,*

$$\dim C_T = \left(\sum_{i \leq j/2} e_i h_i \right) + h_j. \qquad (4.3.1)$$

PROOF. Let U_T parametrize the Gorenstein algebras $R/\mathrm{Ann}(F)$, $F \in [I_j]^\perp$ such that $H(R/\mathrm{Ann}(F)) = \mathrm{Sym}(H(A), j)$. By the assumption on j, $I = (I_{\leq j/2})$; if $F \in U_T$ then $J = \mathrm{Ann}(F)$ satisfies $I_{\leq j/2} = J_{\leq j/2}$, hence I can be recovered from J. The variety U_T is fibred over $CI(H)$ by the choice of $F \in \mathcal{A}_j = I_j^\perp$, so the fibre has dimension h_j. Since $\dim CI(H) = \sum_{i \leq j/2} e_i h_i$, we see that $\dim U_T$ is equal to the right-hand side of (4.3.1). Suppose that $B = \mathrm{Ann}(F)$ is a general enough element of U_T and $\{f_t \mid t \in X\}$ is a subfamily of $Gor(T)$, with $f_0 = F$. The condition that the earliest generators of $J(t) = \mathrm{Ann}(f_t)$ is a CI is an open condition on X; that each homomorphism $C'_F(v, u) : A_u \to [I_v]^\perp, v \leq j/2, u = j - v$ has maximal rank also defines an open condition on X, given that the ranks of A_u and A_v are fixed (as $J(t)_{\leq j/2}$ is a CI). It follows that there is an open neighborhood X_0 of $0 \in X$ with $X_0 \subset U_T$. Thus, $C_T = \overline{U_T}$ is an irreducible component of $Gor(T)$. $\qquad \square$

COROLLARY 4.18. AN IRREDUCIBLE COMPONENT OF CERTAIN CATALECTICANT LOCI. *In the notation of Theorem 4.13, suppose* $r \geq 3, t \geq 2$, $j = 2t$ *or* $2t + 1$, *and* $s^\vee = r$. *Suppose further that* $(j, r) \neq (4, 3), (5, 3), (7, 3)$ *or* $(5, 4)$. *Then* $V_s(t, j - t; r)$ *has an irreducible component of dimension given by the formulas in (i) or (ii), respectively of Theorem 4.13.*

PROOF. Let $T = \text{Sym}(H, j)$ with H the Hilbert function of an Artinian graded CI generated in degree t. Theorem 4.17 gives a component C_T of $Gor(T) \subset V_s(t, j - t; r)$, $s = \dim_k R_t - r$, having dimension $\dim(C_T) = (rs + h_j)$, where

$$h_j = \begin{cases} r_j - r(r_t) + \binom{r}{2} & \text{when } j = 2t \\ r_j - r(r_{t+1}) + r\binom{r}{2} & \text{when } j = 2t + 1. \end{cases}$$

The equality of $(rs + h_j)$ with the dimension formulas (i), (ii) of Theorem 4.13, is easy to check. The omitted cases are to assure that $j \leq \sum d_i - r = r(t - 1)$. Since $j = 2t$ or $2t + 1$, only two steps of the Koszul resolution of the CI are needed to give h_j.   □

The following Theorem generalizes Theorem 4.13 and Corollary 4.18.

THEOREM 4.19. AN IRREDUCIBLE COMPONENT OF CERTAIN DETERMINANTAL LOCI, ARISING FROM COMPLETE INTERSECTIONS AND A GENERIC FORM. *Suppose* $r \geq 3$, $t \geq 2$, $0 \leq v \leq t - 1$. *Let* $r_t - r \leq s < r_t$. *Put* $s^\vee = a = r_t - s$. *Let* $j = 2t + v$; *if* $s^\vee = r$, *suppose that* $j \leq r(t - 1)$. *Suppose* $\text{char}(k) = 0$ *or* $\text{char}(k) > j$. *Consider the variety* C_T *constructed as in Theorem 4.17 with* $d_1 = \cdots = d_a = t$. *Then* $\overline{C_T}$ *is an irreducible component of* $V_s(t, t + v, r)$, *having dimension*

$$\dim(C_T) = r_j - s^\vee(r_{t+v} - s) + \frac{r_v s^\vee(s^\vee - 1)}{2}, \qquad (4.3.2)$$

and the scheme $\mathbf{V}_s(t, t + v, r)$ *is smooth along an open dense subset of* $\overline{C_T}$. *The projective variety* $\mathbb{P}(\overline{C_T})$ *is rational. If* $s^\vee < r$, *then* $\overline{C_T}$ *is the unique irreducible component of* $V_s(t, t + v, r)$ *which contains* $PS(s, j; r)$.

PROOF. Let $f_1, \ldots, f_a$ be general enough elements of R_t, $I = (f_1, \ldots, f_a)$, $I_t = \langle f_1, \ldots, f_a \rangle = \Delta$ be the corresponding general enough subspace of R_t. Using the Koszul resolution of I and the assumption $v \leq t - 1$ we conclude that the multiplication maps

$$m : I_t \otimes R_v \rightarrow R_{t+v}$$

$$1 \otimes m : \Lambda^2(I_t) \otimes R_v \rightarrow I_t \otimes R_{t+v}$$

are injective, and the following sequence is exact:

$$0 \rightarrow (1 \otimes m)\Lambda^2(I_t) \otimes R_v \rightarrow I_t \otimes R_{t+v} \rightarrow I_t \cdot R_{t+v} \rightarrow 0. \qquad (4.3.3)$$

Therefore,

$$\dim(I_t \cdot R_{t+v})^\perp = r_j - a \cdot r_{t+v} + r_v \cdot \binom{a}{2}. \qquad (4.3.4)$$

This integer is positive, since from our assumptions it follows that $(I_t \cdot R_{t+v})^\perp = (I_j)^\perp \neq 0$.

Consider the closed subset $\Gamma \subset Grass(a, R_t) \times \mathbb{A}(\mathcal{D}_j)$,

$$\Gamma = \{(\Delta, f) \mid \Delta \circ f = 0\},$$

and let us denote by p_1, p_2 the projections on the first and second factor. For a general $\Delta \in Grass(a, R_t)$ the fibre $p_1^{-1}(\Delta)$ equals $(\Delta \cdot R_{t+v})^\perp$, so it has the dimension given by (4.3.4). This shows that there exists a unique irreducible component $\Gamma_0 \subset \Gamma$, so that the projection $p_1 : \Gamma_0 \to Grass(a, R_t)$ is surjective. Furthermore,

$$\dim(\Gamma_0) = a(r_t - a) + r_j - a r_{t+v} + r_v \binom{a}{2}$$
$$= r_j - s^\vee(r_{t+v} - s) + r_v \binom{s^\vee}{2} \qquad (4.3.5)$$

Let f be a sufficiently general element of $(I_j)^\perp = (\Delta \cdot R_{t+v})^\perp$ and let $J = \mathrm{Ann}(f)$. Then by Theorem 4.16 we have $J_t = \Delta$ since $2t \leq j = 2t + v$. This implies that the image of

$$p_2 : \Gamma_0 \to \mathbf{V}_s(t, t + v; r)$$

has the same dimension as Γ_0, equal to (4.3.5), and by Cramer's formulas (cf. page 107) that $p_2 : \Gamma_0 \to p_2(\Gamma_0)$ is a birational isomorphism.

Let f be a sufficiently general element of $p_2(\Gamma_0)$, set $J = \mathrm{Ann}(f)$, $\Delta = J_t$, and $I = (\Delta)$, the ideal generated by Δ. By Theorem 4.16, $I \subset J$ and $I_u = J_u$ for $2u \leq j = 2t + v$. If $A = R/I$, $H = H(A)$ is the Hilbert function of A, and $T = \mathrm{Sym}(H, j)$, we conclude that $p_2(\Gamma_0) = \overline{C_T}$. This proves that

$$\dim \overline{C_T} = r_j - s^\vee(r_{t+v} - s) + r_v \binom{s^\vee}{2}.$$

Furthermore the variety Γ_0 is birational to an algebraic vector bundle over an open subset of $Grass(a, R_t)$, so $\mathbb{P}(\overline{C_T})$ is birational to its projectivization, implying that $\mathbb{P}(\overline{C_T})$ is a rational variety.

Now let us prove that the tangent spaces at f to the schemes $\mathbf{Gor}(T)$ and $\mathbf{V}_s(t, t+v; r)$ coincide and have dimension equal to $\dim \overline{C_T}$.

Indeed, by Theorem 3.9,

$$(T_f \mathbf{Gor}(T))^\perp = (J^2)_j = \sum_{\substack{u+v=j \\ u \leq v}} J_u J_v. \qquad (4.3.6)$$

Since $2u \leq j$ in the above formula we have $J_u = J_t \cdot R_{u-t}$, so

$$(J^2)_j = J_t J_{j-t} = I_t J_{t+v}. \qquad (4.3.7)$$

This proves by Theorems 3.2 and 3.9 that $T_f(\mathbf{Gor}(T)) = T_f(\mathbf{V}_s(t, t+v; r))$. Now the kernel of the multiplication map

$$m_t : I_t \otimes J_{t+v} \rightarrow I_t J_{t+v} \qquad (4.3.8)$$

is contained in the kernel of $I_t \otimes R_{t+v} \rightarrow R_{2t+v}$. The latter is $(1 \otimes m) \cdot \Lambda^2(I_t) \otimes R_v$ by the exact sequence (4.3.3), and it is clearly contained in $I_t \cdot J_{t+v}$. Therefore, the kernel of m_t equals $(1 \otimes m) \cdot \Lambda^2(I_t) \otimes R_v$. We have

$$\dim J_{t+v} = r_{t+v} - \dim(R/J)_{j-t}$$
$$= r_{t+v} - \dim(R/J)_t = r_{t+v} - s.$$

Summing up, we obtain,

$$\dim T_f(\mathbf{Gor}(T)) = \dim T_f(\mathbf{V}_s(t, t+v; r))$$
$$= r_j - \dim(J^2)_j \qquad (4.3.9)$$
$$= r_j - s^\vee(r_{t+v} - s) + r_v \binom{s^\vee}{2} = \dim(\overline{C_T}).$$

This proves that $p_2(\Gamma_0) = \overline{C_T}$ is an irreducible component of both $\overline{Gor(T)}$ and $V_s(t, t+v; r)$, since the corresponding schemes $\mathbf{Gor}(T)$, respectively $\mathbf{V}_s(t, t+v; r)$, are nonsingular along a Zariski-open subset of $\overline{C_T}$.

Finally, assume $s^\vee < r$, set $s = r_t - s^\vee$, suppose that $f = L_1^{[j]} + \cdots + L_s^{[j]}$ is general enough, and let $J = \text{Ann}(f)$. For every i with $i \geq t$, $j - i \geq t$, equivalently $t \leq i \leq t + v$, we have by Lemma 1.19 that $J_i = (\mathcal{I}_P)_i$ where $P = \{p_1, \ldots, p_s\}$ is the set of points in $\mathbb{P}^{r-1}$ which correspond to $L_1, \ldots, L_s$. Since $\#Z = s > r_t - r$, we conclude as in the proof of Lemma 2.7 (see also Lemma 4.12) that for general P the space $\Delta = J_t \in Grass(a, R_t)$ is general, so $f \in (\Delta \cdot R_{t+v})^\perp$ is an element of $p_2(\Gamma_0) = \overline{C_T}$ and by the above arguments using the Koszul resolution, f is a nonsingular point of $\mathbf{V}_s(t, t+v; r)$. This shows that $PS(s, j; r) \subset \overline{C_T}$ and $\overline{C_T}$ is the unique irreducible component of $V_s(t, t+v; r)$ which contains $PS(s, j; r)$. This proves the Theorem. $\square$

EXAMPLE 4.20. IRREDUCIBLE COMPONENT OF A DETERMINAN-
TAL LOCUS, $r = 4$. If we take $r = 4$, $d_1 = \cdots = d_4 = 4$, $t = 4$, $s = 31$,
$j = 10$ in Theorem 4.17, then $s^\vee = 4$ and the complete intersection
Hilbert function $H(R/I)$, $I = (f_1, \ldots, f_4)$ is

$$H = (1, 4, 10, 20, 31, 40, 44, 40, 31, 20, 10, 4, 1).$$

The Hilbert function $T = \mathrm{Sym}(H, 10)$ of the Gorenstein algebras we
consider satisfies

$$T = (1, 4, 10, 20, 31, 40, 31, 20, 10, 4, 1).$$

By Theorem 4.17, $\dim(\overline{C_T}) = 31(4) + 10 = 134$, and $\overline{C_T}$ is an irreducible component of $Gor(T)$. If $f \in \overline{C_T}$ and $J = \mathrm{Ann}(f)$, we have

$$\mathcal{T}_f(\mathbf{Gor}(T)) \cong [(J^2)_j]^\perp = [I_4 \cdot J_6]^\perp.$$

By (4.3.9) in the proof of Theorem 4.19, we have for f general
enough in $\overline{C_T}$ that $\mathcal{T}_f(\mathbf{Gor}(T)) = \mathcal{T}_f(\mathbf{V}_s(4, 6; r))$, and

$$\dim \mathcal{T}_f(\mathbf{Gor}(T)) = r_{10} - s^\vee(r_6 - s) + r_2 \binom{s^\vee}{2}$$
$$= 286 - 4(53) + 10(6) = 134,$$

which is the dimension of $\overline{C_T}$. Furthermore $\overline{C_T}$ is an irreducible component of $\mathbf{V}_s(4, 6; r)$ as well as of $\mathbf{Gor}(T)$.

Note that here the vector space $\mathcal{T}_f(\mathbf{Gor}(T)) \cong [I_4 \cdot J_6]^\perp$ is not the
same as $[I_4 \cdot I_6]^\perp = [(I^2)_j]^\perp$, which has dimension 186; the explanation
is that $\dim J_6 = r_6 - 31 = 53$, while $\dim I_6 = r_6 - 44 = 40$, so J_6
properly contains I_6.

Now take $s' = 40$, and consider $Gor(T) \subset V_{40}(5, 5; 4)$; then $t' = 5$
and $s'^\vee = r_5 - 40 = 16$. By the formula in Remark 4.4, any irre-
ducible component W of $V_{40}(5, 5; 4)$ has codimension in $\mathbb{A}(\mathcal{D}_j)$ at most
$\binom{s'^\vee + 1}{2} = 136$. Since $\dim_k(\mathcal{D}_j) = 286$, we must have $\dim(W) \geq 150$.
Thus $\overline{C_T}$ has codimension at least $(150 - 134) = 16$ in any component
of $V_{40}(5, 5; 4)$ containing it!

4.4. The smoothness and dimension of the scheme $\mathbf{Gor}(T)$ when $r = 3$: a survey

We now report on results greatly extending those in Section 4.1.
Our intention is to state them in conveniently accessible form, so the
interested reader might consult the original sources. For a few, we
summarize some of the ideas entering into the proofs, but we do not
give the proofs. We assume throughout this section, that $r = 3$, unless
we explicitly state otherwise.

The main advance depends on J. O. Kleppe's result [**K12**, Corollary 1.5]. Although the paper has a blanket assumption of char(k) ≠ 2, the restriction does not appear to be needed in his proof.

THEOREM 4.21. *Suppose k is an algebraically closed field. Let T be the Hilbert function of a graded Artininan Gorenstein quotient of $R = k[x_1, x_2, x_3]$. Then* $\mathbb{P}\mathbf{Gor}(T)$ *is a smooth irreducible scheme.*

His result is a zero-dimensional version of similar results shown for height three arithmetically Gorenstein subschemes of $\mathbb{P}^n$ by J. O. Kleppe and R. Miró-Roig [**KlM-R**]. A precursor was the result of A. Geramita, M. Pucci, and Y. S. Shin, showing the smoothness of **Gor**(T) in many of the cases where we had shown generic smoothness: in particular, for $T = H(s, j, 3) \supset (s, s, s)$ [**GPS**]. This partial result depended on knowing the dimension of the variety $Gor(T)$, and checking that this was the same as the dimension of the tangent space. Kleppe's proof was independent of dimension results. The key steps in Kleppe's proof, based on algebraic deformation theory are

1. A comparison of the scheme $\mathbb{P}\mathbf{Gor}(T)$ with the scheme $\mathbb{P}\mathbf{Gor}_{GradAlg}(T)$, a subscheme of the scheme **GradAlg**(**T**). The latter represents the functor $GRADALG_T$ which associates to a scheme $Spec(S)$ the subschemes of $\mathbb{A}^3 \times S$ flat over S, and defined by graded ideals – see also Chapter 8); the subscheme $\mathbb{P}\mathbf{Gor}_{GradAlg}(T)$ represents the subfunctor which associates to a scheme $Spec(S)$ only the Gorenstein ideals. The representability of $GRADALG_T$ is shown in [**K12**, Theorem 1.1]. One needs also that the corresponding inclusion of closed points $PGor_{GradAlg}(T)$ in $GradAlg(T)$ is open, (ibid. Prop. 1.3).[3] Kleppe then shows that if $f \in R_j$, the scheme **GradAlg**(**T**) has the same tangent space, $Hom_R(I, A)_0 \cong (I/I^2)_j$ at the closed point $I = \mathrm{Ann}(f)$, as does $\mathbb{P}\mathbf{Gor}(T)$ at f.

2. For a graded Gorenstein Artin quotient of $R \to A \cong R/I$ of R we have

$$\dim Hom_R(I, A)_0 - \dim(H^2(R, A, A)_0) \leq$$
$$\dim \mathbb{P}\mathbf{Gor}_{GradAlg}(T) \leq \dim Hom_R(I, A)_0, \qquad (4.4.1)$$

where $H^2(R, A, A)_0$ is the obstruction space to deformations of X, and $A = R/\mathcal{I}(X)$ ([**K12**, Theorem 1.4]).

---

[3]Kleppe has also sketched an argument that there is a topological isomorphism of the two schemes $\mathbb{P}\mathbf{Gor}(T)$ and $\mathbb{P}\mathbf{Gor}_{GradAlg}(T)$, and that, therefore their dimensions at corresponding points are the same (e-mail note to authors of Sept 11, 1998).

3. A proof that $H^2(R, A, A)_0 = 0$. This is a result of Huneke
[**Hun1, Hun2**], where he shows more generally that if R is a
regular local ring, and A is a Gorenstein quotient of dimension
zero, then $H^2(R, A, A)_0 = 0$ if A is in the linkage class of a
complete intersection - which is true for all Gorenstein ideals in
codimension three. Kleppe gives a self contained proof of the
liaison invariance of the "obstruction part" of $H^2(R, A, A)_0$ in
the graded case needed here [**Kl2**, Proposition 1.7, and Remark
1.8(ii)]. [4]

REMARK 4.22. Kleppe also shows that equation (4.4.1) holds also
for Artinian Gorenstein quotients of a polynomial ring in an arbitrary
number $r \geq 3$ of variables [ibid, Theorem 1.4]; and he shows in this
generality that the obstruction part of $H^2(R, A, A)_0$, defined as

$$\text{obs}(A) = \dim(Hom_R(I, A)_0) - \dim_{R \to A}(GradAlg(H)),$$

is invariant under "algebraic liaison" – liaison via a complete intersection [**Kl2**, Proposition 1.7, Remark 1.8.ii.]. Thus, the knowledge of
both the tangent space dimensions for A and the linked A', as well as
the local dimension of $GradAlg(H)$ or, equivalently, of $\mathbb{P}\mathbf{Gor}(H)$ at
the point f defining the algebra $A = R/\text{Ann}(f)$, together determine
the dimension of $\mathbb{P}\mathbf{Gor}(H')$, $H' = H_{f'}$, at the linked point f'.

M. Boij has in progress a second, independent proof of Kleppe's
smoothness theorem, as a consequence of his study of the finer Betti
number strata of $Gor(T)$, [**Bo5**]. Boij's approach is geometric. See
below for a statement of his announced result on the dimension of the
Betti number strata. Although M. Boij's result concerning smoothness
is incomplete at the time of our writing, his approach promises to be
a useful complement to the algebraic methods used by J. O. Kleppe
and R. Miró-Roig.

Kleppe's smoothness theorem shows that any formula for the tan-
gent space dimension for $\mathbf{Gor}(T)$ for $r = 3$ implies the same formula for
the dimension of $Gor(T)$. About four such formulas are in print, and
they are all different! (See [**Kl2, CoV1, ChoJ1**] and also Corollary
5.50 below). That is, there is not a ready translation from one formula
to the other. Each may have some advantages in doing a practical calculation. We give first Kleppe's and A. Conca and G. Valla's formulas:
the latter is written in terms of the Hilbert function T. Y. Cho and

---

[4]Recently, J. O. Kleppe, J. Migliore, and R. Miró-Roig have generalized the
vanishing of $H^2(R, A, A)$ provided A is "licci" (in the linkage class of a complete intersection), to the case A generically Gorenstein in higher dimensions [**KMMNP**].

B. Jung's formula is in terms of the "alignment character" of T, a sequence of integers that can be constructed simply enough from T (see [**GPS, ChoJ1, ChoJ2**]. The Cho-Jung formula is particularly useful when the alignment character is small, which happens when the order $\nu(T)$ is close to $j/2$, and they have used it to provide counterexamples to Diesel's Conjecture 1.65 (see below).

Kleppe's dimension formula is based on two results. First, we have that I is syzygetic ([**SiV**]); this implies the exactness of the following sequence if char $k \neq 2$

$$0 \to \Lambda^2 I \to I \oplus I \to I^2 \to 0.$$

Second, J. Weyman has provided an exact sequence giving a minimal resolution of $\Lambda^2 I$, in terms of the Buchsbaum-Eisenbud minimal resolution of I,

$$0 \to R(-j-3) \xrightarrow{\lambda^T} \sum_{1 \leq i \leq v} R(-e_i) \xrightarrow{\Psi}$$

$$\sum_{1 \leq i \leq v} R(-d_i) \xrightarrow{\lambda} I \to 0 \quad (4.4.2)$$

where $v = 2k+1$, $d_1 \leq d_2 \leq \cdots \leq d_v$, $e_i = j+3-d_i$, Ψ is an alternating matrix, and λ is given by the diagonal Pfaffians of Ψ (see [**BE2**] or Appendix B Theorem B.2). Let $F_0, F_1, F_2 = R(-j-3)$ denote the free R-modules appearing in the exact sequence (4.4.2). Then Weyman's resolution of $\Lambda^2 I$ is (from [**KlM-R**, Remark 2.5]),

$$0 \to D_1 F_1 \otimes F_2 \to (F_0 \otimes F_2) \oplus D_2 F_1 \to$$

$$F_0 \otimes D_1 F_1 \to \Lambda^2 F_0 \to \Lambda^2 I \to 0. \quad (4.4.3)$$

This leads to the Miró-Roig, Kleppe Dimension Theorem ([**KlM-R**, Theorem 2.6], [**Kl2**, Theorem 2.3]). Let $\mu_i = \mu_i(T) = \dim_k(I_i) = r_i - t_i$. Recall that $r_i = \dim_k(R_i) = \binom{i+2}{2}$. They use the convention that $\binom{b+n}{n} = 0$ for $b < 0$, and define

$$\delta(q) = \delta(I, q) = \sum_{1 \leq i < j \leq v} \left(\binom{e_j - d_i + q + 2}{2} - \binom{d_i - e_j + q + 2}{2} \right)$$

$$= \sum_{1 \leq i < j \leq v} \left(r_{e_j - d_i + a} - r_{d_i - e_j + a} \right). \quad (4.4.4)$$

THEOREM 4.23. (J.O. Kleppe and R. Miró-Roig) *Assume* char $k \neq$
2. *The dimension of* $\mathbb{P}Gor(T)$ *satisfies*

$$\dim(\mathbb{P}Gor(T)) = \delta(0) - \sum_{1 \leq i \leq v} \mu_{d_i}$$

$$= \sum_{1 \leq i < j \leq v} (r_{e_j - d_i} - r_{d_i - e_j}) \; - \; \sum_{i=1}^{v} \mu_{d_i}. \quad (4.4.5)$$

Using duality, J. O. Kleppe obtains the Corollary

COROLLARY 4.24. *If* char $k \neq 2$, *the dimension of* $\mathbb{P}Gor(T)$ *satisfies*

$$\dim(\mathbb{P}Gor(T)) = r_j - 1 - \sum_{i=1}^{v} \mu_{e_i - 3} + \delta(-3). \quad (4.4.6)$$

It is not hard to see that these formulas depend only on the Hilbert
function, and not on the minimal resolution stratum of $f \in Gor(T)$.
Further, Kleppe showed [K12, Proposition 2.5]

PROPOSITION 4.25. *If I is a Gorenstein height three ideal of*
$R = k[x_1, x_2, x_3]$, *the Hilbert function of R/I^2 depends only on the*
Hilbert function of R/I.

This showed a conjecture of Geramita et al in [GPS], proven
there in some special cases. However, there was no explicit formula
for $\dim(Gor(T))$ or for $H(R/I^2)$ in terms of the Hilbert function
$T = H(R/I)$. A. Conca and G. Valla supplied just such a formula
in [CoV1]. Given T we let $P(T) = (p_0, p_1, \dots)$ denote the third difference sequence $p_i = t_i - 3t_{i-1} + 3t_{i-2} - t_{i-3}$. They showed ($T = H$
in their notation)

THEOREM 4.26. (A. Conca and G. Valla) *The dimension of*
$\mathbb{P}Gor(T)$ *satisfies*

$$\dim(\mathbb{P}Gor(T)) = \begin{cases} (h_t + 3h_{t-1} - \sum_{i=0}^{j} h_i p_i)/2 & \text{if } j = 2t; \\ (3h_t + h_{t-1} - \sum_{i=0}^{j} h_i p_i)/2 & \text{if } j = 2t + 1. \end{cases}$$
$$(4.4.7)$$

They also gave formulas for the Hilbert function $H(R/I^2)$: the
above equation is just the integer $H(R/I^2)_j - 1$, by Theorems 4.21
and 3.9.

We restate Conca-Valla's formula in a form we find convenient for
applications. The three formulas below follow imediately from (4.4.7),
using the fact $\Delta^3(T)$ is skew-symmetric about $(j+3)/2$ and an explicit

calculation of the third difference in degrees $\tau + 1, \tau + 2, \ldots, j + 2 - \tau$ (cf. (5.3.4), (5.3.20), (5.3.32)).

Case 1. If $T = (1, 3, \ldots, s - a - b, s - a, \overset{\tau}{s}, s, s, \ldots, 3, 1)$, $a \geq 1$, $j \geq 2\tau + 2$, then

$$\dim \mathbb{P}Gor(T) =$$

$$= 2s - \frac{1}{2} \left(\sum_{i \leq \tau} (T_i - T_{i-3}) \Delta^3(T)_i + a^2 + (a + b)(b - 2a) \right). \quad (4.4.8)$$

Case 2. If $T = (1, 3, \ldots, s - a - b, s - a, \overset{t}{s}, s, s - a, \ldots, 3, 1)$, $a \geq 1$, $j = 2t + 1$, $\tau = t$, then

$$\dim \mathbb{P}Gor(T) =$$

$$= 2s - \frac{1}{2} \left(\sum_{i \leq t} (T_i - T_{i-3}) \Delta^3(T)_i + a + (a + b)(b - 2a) \right). \quad (4.4.9)$$

Case 3. If $T = (1, 3, \ldots, s - a - b, s - a, \overset{t}{s}, s - a, \ldots, 3, 1)$, $a \geq 1$, $j = 2t$, $\tau = t$, then

$$\dim \mathbb{P}Gor(T) =$$

$$= 2s - \frac{1}{2} \left(\sum_{i \leq t} (T_i - T_{i-3}) \Delta^3(T)_i + 3a + b(b - 3a) \right). \quad (4.4.10)$$

Using the dimension formula (4.4.7), we give an example showing that there are components of $V_s(t, t; 3)$ having dimension larger then expected: this is [**I9**, Example 7C]

EXAMPLE 4.27. If $j = 2t$, $t = s^\vee + k$, $s = r_t - s^\vee$, and $T = H(s, j, 3)$, $T' = (1, 3, \ldots r_{t-2}, r_{t-1} - 1, s, r_{t-1} - 1, \ldots, 1)$, then we have

$$\dim(\mathbb{P}Gor(T')) - \dim(\mathbb{P}Gor(T)) = s^\vee - 7 - 2k. \quad (4.4.11)$$

This can be shown readily, using the Conca-Valla dimension formula. Taking $k = 0$ and $s^\vee$ at least 8, we see from Equation (4.4.11) that the catalecticant variety $V_s(t, t; 3)$ has irreducible components of dimension bigger than the "expected" dimension for the corank $s^\vee$ locus of a generic symmetric matrix, here $N - s^\vee(s^\vee + 1)/2$. This answers negatively a conjecture of Diesel [**Di**, Conjecture 4.3] (= Conjecture 1.65).

The Conca-Valla formula is particularly convenient for determining the difference in dimensions of $\mathbb{P}Gor(T)$ and $\mathbb{P}Gor(T')$ when T and T'

are close. It can also be used to recover the following result of Y. Cho and B. Jung generalizing Example 4.27 (see [I9]).

EXAMPLE 4.28. [ChoJ2, Theorem 3.1]. Let $j = 2t$, fix $a, s^\vee$ satisfying $2a+1 \le s^\vee \le a+t$. Let $s = r_t - s^\vee$, $T(a, s^\vee) = (1, 3, \ldots r_{t-2}, r_{t-1} - a, s, r_{t-1} - a, r_{t-2}, \ldots 3, 1)$ and $T = T(0, s^\vee) = H(s, j, 3)$. Then

$$\dim(\mathbb{P}Gor(T(a, s^\vee))) - \dim(\mathbb{P}Gor(T(0, s^\vee))) =$$
$$a(3s^\vee - 2t - 5a/2 - 9/2). \qquad (4.4.12)$$

REMARK 4.29. Here, when $s^\vee \le t$, $\mathbb{P}Gor(T(0, s^\vee))$ has the expected codimension $s^\vee(s^\vee + 1)/2$ in $\mathbf{P}^N$, $N = r_j - 1$. It follows from the Cho-Jung formula 4.4.12 that if $2/3 \le k \le 1$, $s^\vee = kt$, and $a \approx ct$, $c = (3k - 2)/5$, then

$$\dim(\mathbb{P}Gor(T(a, s^\vee))) - \dim(\mathbb{P}Gor(T(0, s^\vee)) \ge 5c^2t^2/2 - 9ct/2.$$

When $k = 1$, $s^\vee = t$, $s = r_{t-1} + 1$, this excess dimension is asymptotic to $r_j/20$. Thus, their calculation shows the existence of components of $V_s(t, t; 3)$ of very large dimension. The problem of determining the dimension of $V_s(t, t; 3)$, i.e. the largest of the dimensions of the irreducible components of $V_s(t, t; 3)$, for general s, t is open, despite the presence of several distinct dimension formulas for $Gor(T)$.

We complete our presentation of dimension formulas by giving that of Y. Cho and B. Jung [ChoJ1]. For this we need the "alignment character" associated to a height three Gorenstein sequence T.

DEFINITION 4.30. ALIGNMENT CHARACTER AND PARTITION. Suppose T is a height three Gorenstein sequence of socle degree $j = 2t$ or $j = 2t + 1$, and order $m = \nu(T)$, and consider the first difference sequence $\Delta(T)_i = T_i - T_{i-1}$. Set $S(T) = \Delta(T)_{\le j/2}$, augmented by zeroes in higher degrees, so

$$S(T) = (1, 2, \ldots, \nu(T), h_\nu, \ldots, h_t, 0, \ldots), \quad \text{with}$$
$$h_{\nu(T)-1} = \nu \ge h_\nu \ge \ldots h_{\lfloor j/2 \rfloor}. \qquad (4.4.13)$$

Note that $S(T)$ is the Hilbert function of a graded ideal in $R' = k[x, y]$, which is the meaning of $S(T)$ being an O-sequence (see Appendix C and Theorem 5.25). The alignment character $ALCHAR(T) = (b_1, \ldots, b_m)$ is defined as the lengths of horizontal bars in the bar graph of the function $S(i) = S(T)_i$, listed in order of increasing size (see for example [GPS, ChoJ1]). Note that $n = \sum_i b_i$ is the vector space dimension of any Artinian algebra quotient of R' having Hilbert function $S(T)$.

We give an equivalent definition of the alignment character. Let M_S be the lex-segment monomial ideal in R' of Hilbert function $H(R'/M) = S$ (see Appendix C, Definition C.1): its degree-i piece is

$$(M_S)_i = \langle x^i, x^{i-1}y, \ldots, x^{S(i)}y^{i-S(i)} \rangle. \tag{4.4.14}$$

Write $m = \nu(T)$, then there is a sequence of integers, $b_1 \leq b_2 \leq \ldots \leq b_m$ such that

$$M_S = \{x^m, x^{m-1}y^{b_1}, \ldots, x^0 y^{b_m}\}. \tag{4.4.15}$$

The alignment character of T, and a partition $P(T)$ are defined from $M_S, S = S(T)$ by

$$ALCHAR(T) = (b_1, \ldots b_m)$$
$$P(T) = (\alpha_1, \ldots, \alpha_m): \quad \alpha_i = b_i - i. \tag{4.4.16}$$

The partition $P(T)$ can also be defined directly from S: it is the dual partition to $S(m), \ldots, S(t)$ ([**I9**, §3], see Lemma B.15).

EXAMPLE 4.31. If $T = (1, 3, 5, 5, 3, 1)$, then $S(T) = (1, 2, 2, 0)$, and $M_S = (x^2, xy^2, y^3)$, so $ALCHAR(T) = (2, 3)$ and $P(T) = (2-1, 3-2) = (1, 1)$. Here $m = t = 2$, and $P(T)$ is the dual partition to $S(m) = 2$.

REMARK 4.32. The alignment character was used by A. Geramita, and others, and T. Harima [**GPS, GHS1, Hari1, Hari2**] in discussing sets of $n = \sum_i b_i$ distinct points in $\mathbb{P}^2$ associated to height three Gorenstein sequences, arranged on m lines. As an invariant of the 2-variable Hilbert function S, the sequence $(b_1, \ldots, b_m)$ was already used by J. Briançon and others who studied the Hilbert scheme in two variables [**Bri, I1**], where it is the lengths of the rows in the standard cobasis for an ideal of Hilbert function S.

The monomial ideal M_S has maximum number of generators and relations among ideals in R' of Hilbert function S. A. Geramita, M. Pucci, and Y. S. Shin show that a set Z of points in $\mathbb{P}^2$ arranged according to the alignment configuration, also has maximum Betti numbers possible for those zero-dimensional schemes in $\mathbb{P}^2$ whose Hilbert function H has first difference S [**GPS**]. The connection to Gorenstein ideals is that when the ideal sheaf $\mathcal{I}_Z$ of the punctual scheme Z is "glued" with its dualizing module, then the resulting Gorenstein Artin algebra has Hilbert function T, and has the maximum Betti numbers possible for elements of $Gor(T)$ [**GPS, Hari1, Hari2**]. We use this connection later in the proof of Theorem 5.25(v.g).

A. Geramita, T. Harima, and Y. S. Shin have recently extended the concept of alignment character of S to n-*type vectors* associated to a O-sequence in higher dimensions, and to a related geometrical study of certain sets of distinct points – k-configurations – in $\mathbb{P}^n$. Again, they show extremality of the Betti numbers for zero-dimensional schemes in $\mathbb{P}^n$ of a given Hilbert function S, and they study the connection to extremality questions for the Betti numbers of elements of $Gor(T)$ [**GHS1, GHS2**]. The latter is a much harder problem, still unsolved when $r \geq 4$.

The partition $P(T)$ derived above from the alignment character is the same that appears in Diesel's article [**Di**, p. 382], as the first half of the partition $Q(T)$, derived from the generator degrees $D_{max}(T) = (d_1, d_2, \ldots, d_{m+1}, \ldots d_{2m+1}), m = \nu(T)$ of a Gorenstein ideal having maximum Betti numbers possible for Hilbert function T. Diesel defines

$$Q(T) = (\alpha_1, \ldots, \alpha_{2m}), \alpha_i = d_{i+1} - m, \qquad (4.4.17)$$

where $\alpha_i = d_{i+1} - m$. When $i \leq m$ then, from (4.4.16) $d_{i+1} - m = (m - i + b_i) - m = b_i - m$, so is a part of the partition $P(T)$. Some parts of $P(T)$ may be zero; the largest possible size of a part, when $m = \nu(T)$ and $j = j(T) = 2t$ or $2t + 1$ is fixed, is $t + 1 - \nu(T)$. Diesel uses the partition $P(T)$ in counting Gorenstein sequences [**Di**, §3.4] (see also Appendix B, Theorem B.17). A further discussion of these connections can be found in [**I9**, Section III].

We now can state the dimension formula of Y. Cho and B. Jung [**ChoJ1**]. There is a variation in their preprint [**ChoJ2**], that is separated into cases for j even and j odd, and appears easier to use. We give their original version, except that we replace $b_i - i$ found throughout the formula by α_i (following [**ChoJ2**]). We adopt the notation $[n] = r_n = \binom{n+2}{2}$, with the convention that $[n] = 0$ if $n < 0$. Recall as a consequence of Theorem 3.9 that the codimension of $Gor(T)$ in $\mathbb{A}(\mathcal{D}_j) = \mathbb{A}^N, N = r_j$, satisfies $\operatorname{cod} Gor(T) = \dim_k (I^2)_j$, where $j = j(T)$ is the socle degree of Artinian Gorentein algebras $A = R/I$ of Hilbert function T.

THEOREM 4.33. [**ChoJ1**]. *If $I \subset R$ defines a height three Gorenstein Artin algebra of Hilbert function T, then the codimension of*

$Gor(T)$ in $\mathbb{A}(\mathcal{D}_j)$ satisfies

$$\text{cod}\, Gor(T) = \dim_k(I^2)_j =$$

$$[j - 2m] + \sum_{i=1}^{m} \left([j - 2m - \alpha_i] + [\alpha_i - 2] + [j - 2m - 2\alpha_i]\right)$$

$$- \sum_{i=1}^{m} \left([\alpha_i - 3] - [j - 2m - \alpha_i - 1]\right)$$

$$+ \sum_{1 \leq i < s \leq m} \left([j - 2m - \alpha_i - \alpha_s] + [\alpha_s - \alpha_i - 2]\right)$$

$$- \sum_{1 \leq i < s \leq m} \left(2 \cdot [\alpha_s - \alpha_i - 3] + \right.$$

$$\left. [j - 2m - \alpha_i - \alpha_s - 2] + [\alpha_s - \alpha_i - 4]\right)$$

$$- \sum_{1 \leq i,s \leq m} [j - 2m - \alpha_i - \alpha_s - 1]. \tag{4.4.18}$$

IDEA OF THE PROOF. First, $\dim_k(I^2)_j$ depends only on T: this may be shown directly from the Kustin-Ulrich formula, or follows from A. Conca and G. Valla's results or from the smoothness theorem [**K12, GPS, CoV1**]. The formula (4.4.18) then follows from the minimal free resolution for a Gorenstein ideal $I \mid H(R/I) = T$ having maximum Betti numbers (see also the proof of Theorem 5.25(vg)), the Kustin-Ulrich result Theorem B.3, and smoothness of $\mathbf{Gor}(T)$.   □

We now state the result announced by M. Boij concerning the dimension of the Betti number strata [**Bo5**]. He uses the integers $\beta_i = \#\{$ of generators of degree $i\} = \#\{d_u = i\}$, and he denotes by $Gor(\beta, T)$ the Betti number stratum parametrizing all Gorenstein ideals of Hilbert function T having β_i generators of degree i. S. J. Diesel had shown that these strata are irreducible ([**Di**], see also Theorem 5.25).

THEOREM 4.34. (M. Boij) *The codimension of* $Gor(\beta, T)$ *in* $Gor(T)$ *satisfies*

$$\text{cod}(Gor(\beta, T)) = \begin{cases} \frac{1}{2} \sum_{i=1}^{j+2} \beta_i \beta_{j+3-i} & \text{if } j \text{ is even} \\ \frac{1}{2} \sum_{i=1}^{j+2} \beta_i \beta_{j+3-i} - 1/2\beta_{(j+3)/2} & \text{if } j \text{ is odd,} \end{cases} \tag{4.4.19}$$

where j *is the top degree of* T.

When the Artinian Gorenstein algebra quotient of a local ring is not homogeneous, the Hilbert function is in general not symmetric, as shown by the CI example $A = k\{x, y\}/(xy, x^2 - y^3)$ for which $H(A) = (1, 2, 1, 1)$. In two variables the family $ZGor(T)$ of (nonhomogeneous) CI ideals of any given Hilbert function T is well enough understood so that there are dimension formulas, and some notions of structure of the parameter spaces [**Bri, Che, Gra, I1, IY, KW, Ya1, Ya1**]. However when the number of variables r satisfies $r \geq 3$, not even the set of possible Hilbert functions for nonhomogeneous Gorenstein Artin quotients of R is known, despite the presence when $r = 3$ of the Buchsbaum-Eisenbud structure theorem.

Each such Gorenstein ideal $I = \text{Ann}(f), f \in \mathcal{D}_{\leq j}$, has associated to it naturally a unique graded Gorenstein ideal $\text{Ann}(f_j)$, and as well a sequence of "reflexive modules", quotients of a descending sequence of ideals in the associated graded algebra of $A = R/\text{Ann}(f)$ [**I6**]. Thus one would expect that the strong recent results on the graded $Gor(T)$, would lead to progress in the nongraded case. J. O. Kleppe has shown just such a result when $r = 3$ and T is symmetric [**K12**, Theorem 3.5]. It is known in general that the associated graded algebra $A^* = Gr_m(A)$ of an Gorenstein Artin algebra A with symmetric Hilbert function is also Gorenstein (see [**Wa2**, Proposition 9], [**I6**, Proposition 1.7]).

To state Kleppe's result, we denote by $ZGor(T)$ the parameter space for possibly nonhomogeneous Gorenstein quotients of the (completed) local ring $R_p = k\{x_1, \ldots, x_3\}$ having the symmetric Hilbert function T; it may be regarded as a subspace of the algebraic variety $Alg_{R_p}(T)$ parametrizing algebra quotients of R_p having Hilbert function T, or as a subvariety of the Hilbert scheme $\mathbf{Hilb}^{\ell}(\mathbb{P}^2), \ell = |T|$. We use the notation of Equation (4.4.4) and Theorem 4.23, but take $\mu_i' = \dim_k I_i'$, where I' is the ideal in $R_p[x_4]$ having the same graded Betti numbers as the ideal I^* defining A^*.

THEOREM 4.35. (J. O. Kleppe). *The local dimension of $ZGor(H)$ above any point A^* of $Gor(T)$ is independent of A^* and satisfies*

$$\dim ZGor(H) = 3\ell + \sum_{1 \leq i \leq v} \mu_{d_i - 1}' - \delta_4(-1)$$

$$= \binom{j+3}{3} - \ell - \sum_{1 \leq i \leq v} \mu_{e_i - 3}' + \delta_4(-3). \qquad (4.4.20)$$

This result was known in the special cases of A a complete intersection, or A compressed [**I2, EmI1, FL, I3**]. Kleppe notes that the last two terms in the second part of (4.4.20) are zero in the compressed case, so $\delta_4(-3) - \sum_{1 \leq i \leq v} \mu_{e_i - 3}'$ forms a correction term to the formula of [**EmI1, FL**] (see equation (6.4.1) below), when extending it to A that are noncompressed.

Kleppe also shows that the formulas for $\dim(Gor(T))$ determine formulas for the dimension of the parameter space of height three Gorenstein ideals of $\mathbb{P}^3$ having given Hilbert function H [**Kl2**, Proposition 3.1].

Part II

Catalecticant Varieties and the
Punctual Hilbert Scheme

CHAPTER 5

Forms and Zero-Dimensional Schemes I: Basic Results, and the Case $r = 3$

One of the main points in the theory of binary forms (see Section 1.3) is associating to a form f an apolar form ϕ of minimum degree, or equivalently an annihilating codimension one subscheme $\mathfrak{d} = (\phi)$ of minimum degree. Then f has a generalized additive decomposition which depends on the points and the multiplicities of $\mathfrak{d}$. Furthermore the degree of $\mathfrak{d}$, which we denoted by $\ell(f)$, equals the maximum rank of the catalecticant matrices $Cat_f(u, j-u; 2)$, $1 \leq u \leq j-1$, which we denote by $\ell\mathrm{diff}(f)$ (Theorem 1.44); and $\ell(f) = \min\{s \mid f \in \overline{PS(s, j, 2)}\}$. In more variables $r \geq 3$, each of these three notions of length generalizes, and they are distinct.

In the previous chapters an analogous apolarity relation between sums of divided powers $f = L_1^{[j]} + \cdots + L_s^{[j]}$ and the corresponding annihilating set – the polar polyhedron – $Z = \{p_1, \ldots, p_s\} \subset \mathbb{P}(\mathcal{D}_1) = \mathbb{P}^{r-1}$, where p_i correspond to L_i, was studied and used many times. The meaning of annihilating is that every polynomial ϕ vanishing on Z annihilates f: $\phi \circ f = 0$. In this and the next chapter we go further and consider the apolarity relation between DP-forms in $\mathcal{D}_j$ when $r \geq 3$, and zero dimensional subschemes in $\mathbb{P}^{r-1}$. The basic concept is that of annihilating scheme of a DP-form. As in the case of sets (= smooth annihilating schemes) a zero dimensional scheme Z is annihilating for f if every polynomial ϕ from the graded ideal of Z annihilates f. Non-smooth annihilating schemes are the analog for $r \geq 3$ of the apolar polynomials of minimum degree with multiple roots in the case of binary forms. The binary case also shows that considering non-smooth annihilating schemes is unavoidable when trying to represent a given form of any number of variables ≥ 2 as a power sum of linear forms with minimum number of summands.

Our main results in this chapter are contained in Section 5.3. We consider an arbitrary DP-form f of degree j in 3 variables, and denote by $s = \ell\mathrm{diff}(f)$ the maximum rank of the associated catalecticant matrices

$s = \max\{\mathrm{rk}\, Cat_f(i, j - i; 3) \mid 1 \leq i \leq j - 1\}$; and our goal is to find intrinsic conditions, necessary and sufficient for f to have an annihilating scheme of the minimum possible degree, namely s. The three main theorems 5.31, 5.39 and 5.46 (see also the summarizing Theorem 5.55) give such conditions in terms of the ranks and other invariants of the catalecticant matrices. Furthermore under the hypothesis of these theorems it turns out that the annihilating scheme Z_f of degree s is unique, and its ideal is generated by the forms of lower degrees, apolar to the given DP-form f, so Z_f can be recovered explicitly from f. For a "general enough" f the annihilating scheme is smooth and one obtains a (divided) power sum representation of f with s summands, explicitly recoverable from f. This answers partially Problem 0.1 posed in the Introduction when $r = 3$. The intrinsic conditions found (see Theorem 5.55) may be considered as an analog for 3 variables of the classical rank conditions on the Hankel matrices in the binary case.

While Section 5.3 uses the language and deep results from commutative algebra, and is one of the most technically involved parts of the book, we extract and summarize in Section 5.4 the relevant results about the problem of power sum representations, and formulate them using only catalecticant matrices. We also give in this section an algorithm for calculating such a representation of minimum length s of a given form.

The theorems for existence of annihilating schemes, proved in Section 5.3 are based on the theory of height 3 Gorenstein ideals, most notably the Buchsbaum-Eisenbud Structure Theorem B.2, which yields the minimal resolution of such ideals in terms of alternating matrices, and also on results of S. Diesel, and T. Harima, which describe completely the possible sequences of generators degrees of these ideals when the Hilbert function H is fixed. We in addition give an alternate derivation of some of our preparatory Lemmas on generator-relation degrees, using a result of A. Conca and G. Valla that determines the degrees of a maximal generator sets compatible with H in terms of the second differences of H.

The canonical way the annihilating schemes are recovered from the DP-forms – generating their ideals by apolar forms – permits us to construct a dominant morphism $p : Gor_{\mathrm{sch}}(T) \rightarrow Hilb^H \mathbb{P}^2$. Here $Gor_{\mathrm{sch}}(T)$ is the subvariety of $Gor(T)$ distinguished by the intrinsic conditions we discussed above, and $Hilb^H \mathbb{P}^2$ is the locally closed subset of the punctual Hilbert scheme $\mathbf{Hilb}^s \mathbb{P}^2$ consisting of points corresponding to schemes with fixed Hilbert function H. Here $H =$

$(T_{\leq j/2}, s, s, \ldots)$. This morphism is particularly useful when T is a Gorenstein sequence containing (s, s, s). Then $Gor_{sch}(T) = Gor(T)$ and one obtains a dominant morphism $p : Gor(T) \to Hilb^H \mathbb{P}^2$ fibered by open sets in $\mathbb{A}^s$.

This morphism may be used in two ways. One is obtaining information about $Gor(T)$ from known facts about $Hilb^H \mathbb{P}^2$ – such an application is a formula for $\dim Gor(T)$ when $T \supset (s, s, s)$ derived from Gotzmann's formula for $\dim Hilb^H \mathbb{P}^2$ [Got1] (see Corollary 5.50). One can also use the morphism in the opposite way to derive information about $Hilb^H \mathbb{P}^2$ from known facts about $Gor(T)$. In this way all recently obtained formulas for $\dim Gor(T)$, surveyed in Section 4.4, yield formulas for $\dim Hilb^H \mathbb{P}^2$. One such application — of the Conca–Valla formula — turned out to be particularly useful. In Section 5.5 we study in this way the Betti strata of $Hilb^H \mathbb{P}^2$ — the locally closed subsets with fixed sequences of generators degrees.

A natural question is: What is the image of $p : Gor(T) \to Hilb^H \mathbb{P}^2$? Example 5.10 below shows that p is not surjective; for $T = (1, 3, 4, 3, 1)$, the map p is not defined at $f = X_1 X_3^{[3]} + X_2^{[3]} X_3$, since $((Ann\, f)_{\leq 2}) = (x_1^2, x_1 x_2)$ — a line and a point. On the other hand, the smooth schemes are in the image, since they are the polar polyhedrons of divided power sums. Y. Cho and the first author showed in [ChoI1] that locally Gorenstein schemes are included in the image (see also Lemma 6.1 for a particular class of non-smooth schemes in the image and the discussion in Section 6.4 of this question for arbitrary $r \geq 3$). Our work leaves open the following, when p is defined.

PROBLEM. Is it true that the image of $p : Gor(T) \to Hilb^H \mathbb{P}^2$ consists of the locally Gorenstein zero dimensional schemes?

Here is a brief description of the remaining sections. In Section 5.1 we introduce the concept of an annihilating scheme Z for a DP-form f (Definition 5.1); we then give some of its properties (Lemma 5.3) and connect this notion with polar polyhedrons in Corollary 5.4. Several examples illustrate the need for some further hypotheses to obtain the uniqueness of Z, given f (Examples 5.7, 5.8). We then in Section 5.2 discuss limit schemes and limit ideals, and give an essential Lemma concerning the limit of smooth schemes (Lemma 5.12). We show here that if a form f has a smoothable degree-s annihilating scheme, then it is in the closure of $PS(s, j; r)$ (Lemma 5.17); and we give a class of forms annihilated by complete intersections, satisfying $s = \max\{Cat_f(i, j-i; 3)\}$, but that are not in the closure of $PS(s, j; r)$ (Proposition 5.18). In this section we also define the "postulation"

Hilbert schemes $\mathbf{Hilb}^H \mathbb{P}^{r-1}$ and summarize in Theorem 5.21 some of their properties.

Sections 5.3 – 5.5 were discussed above. We mention also an example suggested by P. Rao, that a map from $Gor(T)$ to $Hilb^s(\mathbb{P}^2)$ need not extend to the closure of $Gor(T)$ (Remark 5.73).

It is natural to expect (in analogy with the binary forms case) that the study of the annihilating schemes would clarify what is the explicit form of the elements in the boundary $\overline{PS(s,j;r)} - PS(s,j;r)$, i.e. the correct analog of generalized additive decompositions in the binary case. This remains so far unclear to us and is an interesting open problem.

In Section 5.6 we first collect and compare the various notions of length of a form f used in the book. Recall that these notions are the same for binary forms, but are different for $r \geq 3$. We then report on the boundary $\overline{PS(s,j;3)} - PS(s,j;3)$. In particular, we partially answer Problem 0.2 of the Preface, and Goal 4 of the Detailed Summary, by determining the possible sequences of ranks $(\mathrm{rk}\, Cat_f(i, j - i; 3) \mid 0 \leq i \leq j)$ (= possible dimensions of the spaces of partial derivatives when $\mathrm{char}(k) = 0$ or $> j$) that occur for DP-forms f on the boundary; and we give some intrinsic criteria for deciding whether $f \in \overline{PS(s,j;3)}$. These criteria are not complete, but are effective in many cases. The question of whether a DP-form f belongs to $\overline{PS(s,j;3)}$ and whether f has an annihilating scheme of degree $\leq s$ are closely connected. The second property implies the first (see the proof of Theorem 5.71). Whether the converse holds is a particular case of the more general Problem 6.8. Nevertheless, our question of identifying the closure by intrinsic conditions on f remains only partially answered.

In Section 5.7 we show that the key arguments from Section 5.3 can be extended to an arbitrary number of variables $r \geq 3$. In this way the apolarity relation $\mathcal{I}_Z \subset \mathrm{Ann}(f)$ when $r = 3$ may be extended when $r \geq 4$ to a relation $\mathcal{I}_Z \subset \mathcal{I}_W$ between a codimension 2 arithmetically Cohen-Macaulay (ACM) subscheme of $\mathbb{P}^{r-1}$ and an arithmetically Gorenstein subscheme $W \subset Z$ of codimension 1 (Lemma 5.75). In particular, we determine in most cases when a codimension 3 Gorenstein scheme is contained as a "tight codimension one subscheme" on a codimension 2   ACM subscheme of $\mathbb{P}^n$ (Theorem 5.77).

We suppose throughout Chapters 5 and 6 that Z is a degree-s zero dimensional subscheme of $\mathbb{P}^{r-1}$ (thus, $\ell(Z) = \deg(Z) = s$), and that $\mathcal{I}_Z \subset R$ is its saturated homogeneous defining ideal, so $Z = \mathrm{Proj}(R/\mathcal{I}_Z)$. Note that $R/\mathcal{I}_Z$ is Cohen-Macaulay of dimension one.

We denote by $\tau(Z)$ the integer $\tau(Z) = \tau(\mathcal{I}_Z) = \min\{i \mid H(R/\mathcal{I}_Z)_i = s = \deg(Z)\}$, and by $\sigma(Z)$ the Castelnuouvo regularity of Z (Definition 1.68). Then the Hilbert function $H_Z = H(R/\mathcal{I}_Z) = (1, t_1, \ldots, t_\tau = s, s, \ldots)$ is a nondecreasing function, and the regularity degree satisfies $\sigma(Z) = \tau(Z) + 1$ (Theorem 1.69, or [GM]). We let $\mathrm{Sym}(H_Z, j)$ denote the symmetrization of the initial portion of H_Z about $j/2$. If $\tau = \tau(Z)$ and $j \geq 2\tau$, we have

$$\mathrm{Sym}(H_Z, j) = (1, t_1, \ldots, t_\tau = s, \ldots, t_1, 1). \tag{5.0.1}$$

5.1. Annihilating scheme in $\mathbb{P}^{r-1}$ of a form

We establish some basic facts concerning the relation between an annihilating scheme Z and a DP-form f. The annihilating 0-dimensional scheme Z for f was introduced in [I7, § 3]. When Z is smooth, it is the set of points in classical apolarity relation to f (Definition 1.13 and the Apolarity Lemma 1.15). Geometers sometimes consider the scheme $W_i(f)$ defined by the ideal $(\mathrm{Ann}(f)_i)$, especially when it is zero dimensional. Our concept of an annihilating scheme Z for f is related, although not the same.

DEFINITION 5.1. ANNIHILATING SCHEME. If $f \in \mathcal{D}_j$ and Z is a zero-dimensional subscheme of $\mathbb{P}^{r-1}$ with saturated defining ideal $\mathcal{I}_Z$, then Z is an annihilating scheme for f if $\mathcal{I}_Z \subset \mathrm{Ann}(f)$ (an equivalent condition is $f \in (\mathcal{I}_Z)_j{}^\perp$ according to Lemma 2.15). Recall that the "differential length" of f is $\ell\mathrm{diff}_k(f) = \max_i\{(H_f)_i\}$. We define the "scheme length" of f,

$$\ell\mathrm{sch}(f) = \min\{\deg(Z) \mid \dim(Z) = 0, \text{ and } \mathcal{I}_Z \subset \mathrm{Ann}(f)\}. \tag{5.1.1}$$

We say that Z is a "tight" annihilating scheme for f if Z is an annihilating scheme satisfying $\deg(Z) = \ell\mathrm{diff}_k(f)$.

EXAMPLE 5.2. Let $r = 2, \mathrm{char}\, k = 0, f \in \mathcal{R}_j$. In Section 1.3 we defined a number $\ell(f)$ which is exactly $\ell\mathrm{sch}(f)$ (see page 36 and Lemma 1.31). Furthermore it was proved that for $j = 2t$ or $2t + 1$ one has $\ell\mathrm{sch}(f) \leq t + 1$ and $\ell\mathrm{sch}(f) = \ell\mathrm{diff}(f)$ (Theorem 1.44). So, when $r = 2$ every annihilating scheme of minimum degree is tight. If $2\,\ell\mathrm{sch}(f) \leq j + 1$, then there is a unique tight annihilating scheme (Propositions 1.36 and 1.50). The validity of these statements for DP-forms and arbitrary characteristic is discussed on page 38.

We recall that if Z is a 0-dimensional subscheme of $\mathbb{P}^{r-1}$ of degree s, the number $\tau(Z)$ is the smallest degree i such that Z imposes s independent conditions on degree-i forms (cf. Theorem 1.69).

THEOREM 5.3. PROPERTIES OF ANNIHILATING SCHEMES.

A. *If Z is a subscheme of the 0-dimensional scheme Y ($\mathcal{I}_Z \supset \mathcal{I}_Y$) and Z is an annihilating scheme for f, then Y is an annihilating scheme for f.*

B. *If Z is smooth, $Z = \{p(1), \ldots, p(s)\}$, then Z is an annihilating scheme for f iff f has an additive decomposition*

$$f = c_1 L_{p(1)}^{[j]} + \cdots + c_s L_{p(s)}^{[j]} \tag{5.1.2}$$

where $c_i \in k$, and $L_{p(i)}$ is the linear form in $\mathcal{D}$ corresponding to $p(i)$ (Definition 1.13). Suppose f has polar polyhedron Z, i.e. all $c_i \neq 0$. Suppose $\deg(f) \geq 2\tau(Z)$. Then Z is a tight anninilating scheme of f.

C. *Suppose only for this part, $\mathrm{char}(k) = 0$ or $\mathrm{char}(k) > j$. Let $f \in k[X_1, \ldots, X_r]$ be a form of degree j. If Z is a union of fat points, (so $\mathcal{I}_Z = m_{p(1)}^{a_1} \cap \cdots \cap m_{p(t)}^{a_t}$), then Z is an annihilating scheme for f iff there exist homogeneous polynomials $B_1, \ldots, B_t$ in $\mathcal{D}$ of degrees $a_1 - 1, \ldots, a_t - 1$ such that, with $L_i = L_{p(i)}$,*

$$f = B_1 L_1^{j-(a_1-1)} + \cdots + B_t L_t^{j-(a_t-1)}. \tag{5.1.3}$$

D. *We have $\ell\mathrm{sch}(f) \geq \ell\mathrm{diff}_k(f)$. If equality is achieved by a degree-s annihilating scheme Z, then $(H_f)_i = s \Rightarrow \mathrm{Ann}(f)_i = (\mathcal{I}_Z)_i$.*

E. *Let $s = \ell\mathrm{diff}_k(f)$ and let $\tau = \min\{i \mid (H_f)_i = s\}$. Suppose Z is a tight annihilating scheme for f (so $\deg(Z) = \ell\mathrm{sch}(f) = \ell\mathrm{diff}_k(f)$). Then*

   i. *$\tau(Z) = \tau$ and $j \geq 2\tau(Z)$*

   ii. *$(\mathcal{I}_Z)_i = \mathrm{Ann}(f)_i$ for each $i \leq j - \tau$, $H_f = \mathrm{Sym}(H_Z, j)$, and the sequence H_f satisfies $(H_f)_{i-1} \leq (H_f)_i$ if $i \leq j - \tau$.*

   iii. *Z is the unique annihilating scheme for f of degree s if $j \geq 2\tau + 1$, or if $j = 2\tau$ and one of the following two equivalent conditions is satisfied*

     a. *The graded ideal $\mathcal{J} = R \cdot (\mathcal{I}_Z)_\tau \subset \mathcal{I}_Z$ has the property that $\mathcal{J}_n = (\mathcal{I}_Z)_n$ for $n \gg 0$.*

     b. *$(\mathcal{I}_Z)_\tau$ generates the sheaf $\tilde{\mathcal{I}}_Z(\tau)$.*

   Conditions (a) and (b) hold if $(\mathcal{I}_Z)_{\leq \tau}$ generates $\mathcal{I}_Z$.

F. *Let $s = \ell\mathrm{diff}_k(f)$ and let τ be as in (E). Suppose that the linear system $|\mathrm{Ann}(f)_\tau| \subset |\mathcal{O}_{\mathbb{P}^{r-1}}(\tau)|$ has a zero-dimensional base locus. Then there can be at most a finite number of tight annihilating schemes of f.*

G. *Any DP-form $f \in \mathcal{D}_j$ has a smooth annihilating scheme of degree no greater than $\dim_k R_j$.*

PROOF. (A) is immediate from the definition.

PROOF OF (B). That Z is annihilating is immediate from the formula in Lemma 1.15(i). Conversely, if Z is annihilating, then $f \in (\mathcal{I}_Z)_j{}^\perp$, so it has an additive decomposition (5.1.2) by Lemma 1.15(ii).

Every DP-form of the type (5.1.2) has $\ell\mathrm{diff}(f) \leq s$, since $(H_f)_i = \mathrm{rk}\, Cat_f(i, j-i) \leq s$ from the same formula of Lemma 1.15(i). Now let $\tau = \tau(Z)$ and suppose $j \geq 2\tau$. Then in degrees i, $\tau \leq i \leq j - \tau$ the set Z imposes s independent conditions on degree-i forms, since $(H_Z)_i = s$ for $i \geq \tau$ (Theorem 1.69). Hence if f has polar polyhedron Z, we conclude from the Apolarity Lemma 1.15(iv) that $\mathrm{Ann}(f)_i = (\mathcal{I}_Z)_i$ for $\tau \leq i \leq j - \tau$ and $(H_f)_i = (H_f)_{j-i} = (H_Z)_i = s$ when $i \in [\tau, j - \tau]$. This proves $\max_i\{(H_f)_i\} = s$, i.e. Z is tight for f.

PROOF OF (C). This statement restates the generalization of Lemma 2.2 to distinct vanishing degrees (Theorem 1 of [EmI2]).

PROOF OF (D). If Z is an annihilating scheme for f then $\mathcal{I}_Z \subset \mathrm{Ann}(f)$. By Theorem 1.69, $H(R/\mathcal{I}_Z)$ is nondecreasing in i and eventually stabilizes at the value s. So $H(R/\mathrm{Ann}(f))_i \leq H(R/\mathcal{I}_Z)_i \leq s$, and equality implies $\mathrm{Ann}(f)_i = (\mathcal{I}_Z)_i$.

PROOF OF (E). Let ν be the maximum number, $\nu \leq j$, such that $(H_f)_\nu = s$. Since $(H_f)_i$ is symmetric with respect to $j/2$, we have $j/2 \leq \nu \leq j$. By hypothesis $(\mathcal{I}_Z)_\nu \subset \mathrm{Ann}(f)_\nu$, and since Z is of degree s,

$$(H_f)_\nu \leq (H_Z)_\nu \leq s.$$

Thus we have equalities and $(\mathcal{I}_Z)_\nu = \mathrm{Ann}(f)_\nu$. The ideal $\mathcal{I}_Z$ is saturated, hence from $R_{\nu-i}\mathrm{Ann}(f)_i \subset \mathrm{Ann}(f)_\nu = (\mathcal{I}_Z)_\nu$ and $\mathcal{I}_Z \subset \mathrm{Ann}(f)$ we conclude that $\mathrm{Ann}(f)_i = (\mathcal{I}_Z)_i$ for $i \leq \nu$. We have $\nu \geq j/2$ and $(H_f)_i$ is symmetric with respect to $j/2$, so

$$H_f = \mathrm{Sym}(H_Z, j).$$

The sequence $(H_Z)_i$ is nondecreasing and stabilizes at s for $i \geq \tau(Z)$ (see Theorem 1.69). This proves the last claim of (ii). By the symmetry of H_f the integer $j - \nu$ is the minimum i for which $(H_f)_i = s$, so $j - \nu = \tau$, $\tau \leq j/2$ and $\tau = \tau(Z)$. This proves (i) and from the above, the equality

$$(\mathcal{I}_Z)_i = \mathrm{Ann}(f)_i \quad \text{for } i \leq j - \tau.$$

Proof of (iii). Let us first assume that $j \geq 2\tau + 1$. Then we claim that $\mathrm{Ann}(f)_{\leq \nu}$ generates $\mathcal{I}_Z$. Indeed, $(\mathcal{I}_Z)_{\leq \nu} = \mathrm{Ann}(f)_{\leq \nu}$ and since $j \geq 2\tau + 1$ we have $\nu = j - \tau \geq \tau + 1$, so we can use Theorem 1.69. Now, if Z' is an arbitrary degree-s annihilating scheme of f, then (i) and (ii) applied to Z' yield $\mathrm{Ann}(f)_{\leq \nu} = (\mathcal{I}_{Z'})_{\leq \nu}$, so $\mathcal{I}_Z \subset \mathcal{I}_{Z'}$ and therefore

$Z = Z'$ since these are 0-dimensional schemes of equal degree s. The same argument gives uniqueness if $j = 2t$ and $(\mathcal{I}_Z)_{\leq \tau}$ generates $\mathcal{I}_Z$ (the last condition of (iii)).

When $j = 2\tau$ let us prove uniqueness under the weaker conditions (a) or (b). The equivalence of (a) and (b) is standard and follows from the equality

$$\Gamma_*(\tilde{\mathcal{I}}_Z) := \oplus_{n \in \mathbb{Z}} \Gamma(\mathbb{P}^{r-1}, \tilde{\mathcal{I}}_Z(n)) = \mathcal{I}_Z.$$

(see [**Har2**, II §5]). If Z' is an arbitrary tight annihilating scheme of f, then $(\mathcal{I}_{Z'})_{\leq \tau} = \mathrm{Ann}(f)_{\leq \tau}$ by (ii). Condition (a) implies that the saturation of the ideal $R \cdot \mathrm{Ann}(f)_\tau$ equals $\mathcal{I}_Z$. Since $\mathcal{I}_{Z'}$ is saturated we have $\mathcal{I}_Z \subset \mathcal{I}_{Z'}$, implying $Z = Z'$ because of the equality $\deg Z = s = \deg Z'$.

PROOF OF (F). Let $I_\tau = \mathrm{Ann}(f)_\tau$. By hypothesis the graded ideal RI_τ determines a zero-dimensional scheme $Y = \mathrm{Proj}(R/RI_\tau)$. Suppose Z is a degree-s annihilating scheme of f. Then by (D) we have $(\mathcal{I}_Z)_\tau = I_\tau$. Hence $RI_\tau \subset \mathcal{I}_Z$ and Z is a closed subscheme of Y. There are at most a finite number of such schemes since $\dim Y = 0$.

PROOF OF (G). By Corollary 1.16 any DP-form has an additive decomposition of length $s \leq \dim_k R_j$. This and Part (B) prove (G). $\qquad \square$

COROLLARY 5.4. *Let*

$$f = c_1 L_1^{[j]} + \cdots + c_s L_s^{[j]}, \quad \text{all} \quad c_i \neq 0.$$

Let $Z = \{L_1, \ldots, L_s\} \subset \mathbb{P}(\mathcal{D}_1)$ *be the polar polyhedron and let* $\tau = \tau(Z)$. *Suppose* $j \geq 2\tau$. *Then* $\ell\mathrm{diff}(f) := \max_i\{\mathrm{rk}\, Cat_f(i, j-i; r)\} = s$ *and*

$$T = H_f := (\mathrm{rk}\, Cat_f(i, j-i; r))_{i=0,\ldots,j} = \mathrm{Sym}(H_Z, j).$$

One has $(\mathcal{I}_Z)_{\leq j-\tau} = \mathrm{Ann}(f)_{\leq j-\tau}$. *If* $j \geq 2\tau + 1$ *then* $\mathcal{I}_Z$ *is generated by* $\mathrm{Ann}(f)_{\leq \tau+1}$

PROOF. Immediate from Lemma 5.3, Parts (B) and (E), and Theorem 1.69. $\qquad \square$

EXAMPLE 5.5. Suppose $j = 2t$ or $2t + 1$ and let $s \leq \dim_k R_t := r_t$. Let $f = L_1^{[j]} + \cdots + L_s^{[j]}$, where $L_1, \ldots, L_s$ are general, let $Z = \{p_1, \ldots, p_s\}$ be the corresponding set of s general points in $\mathbb{P}^{r-1}$. Then, according to Lemma 1.19, $H_Z = (\min(s, r_i) \mid i = 0, 1, \ldots)$, and $\tau = \tau(Z) = \min\{i \mid s \leq r_i\}$, so $\tau \leq t$. Clearly $\ell\mathrm{diff}(f) = s$, $H_f = H(s, j, r) = \mathrm{Sym}(H_Z, j)$ and, from the above Corollary, if $j \geq 2\tau + 1$, then the graded ideal $\mathcal{I}_Z$ is generated by the apolar forms $\mathrm{Ann}(f)_{\leq \tau+1}$.

REMARK 5.6. Parts (E) and (F) of Lemma 5.3 complement Theorem 2.6, concerning uniqueness and finite number of additive decompositions of a form, by providing conditions for uniqueness/finite number of possibly non-smooth annihilating schemes. Even when the polar polyhedron Z is a general set of s points, Lemma 5.3 provides additional information about the forms generating the ideal $\mathcal{I}_Z$, as in the preceeding example. The case $j \geq 2\tau + 1$ is clear, and one obtains a generalization of the uniqueness statement of Theorem 2.6 in this case.

When $j = 2t$ and $\tau = t$ we leave it to the interested reader to check, using the argument with General Position Lemma from Lemma 2.7, that for a general $f \in PS(s, j; r)$, under the assumption of Theorem 2.6(i), the condition of Lemma 5.3(F) holds (this is Lemma 2.7(i)); also, provided char$(k) = 0$, under the assumptions of Theorem 2.6(ii), the condition of Lemma 5.3 E iii (b) holds. One uses the Bertini theorem to conclude that the base locus of $|\operatorname{Ann}(f)_\tau|$ equals Z not only set-theoretically — this fact was used to obtain uniqueness in Theorem 2.6 — but also scheme theoretically, which yields the condition of Lemma 5.3 E iii(b).

The condition that f has an annihilating scheme of degree s is more restrictive if s is smaller. We are interested in whether f has a tight annihilating scheme Z, because then we have a chance to show Z is unique (Lemma 5.3 E). Example 5.8 shows that a DP-form f need not even have a tight annihilating scheme. Example 5.7 below shows that even a tight annihilating scheme Z for f may not be unique if $(H_f)_{\sigma(Z)} < \ell(Z) = \ell\operatorname{diff}(f)$. Thus, Lemma 5.3(E) is sharp.

EXAMPLE 5.7. NONUNIQUENESS OF TIGHT ANNIHILATING SCHEME. Suppose $\mathcal{D} = \mathcal{R} = k[X, Y, Z]$ and take $f = X^{[2]} + Y^{[2]} + Z^{[2]} = \frac{1}{2}[(X+Y)^{[2]} + (X-Y)^{[2]}] + Z^{[2]}$. Thus $Z_1 = \{(1, 0, 0), (0, 1, 0), (0, 0, 1)\}$ and $Z_2 = \{(1, 1, 0), (1, -1, 0), (0, 0, 1)\}$ are annihilating schemes for f. They are tight, since $H_f = (1, 3, 1)$. But $\sigma(Z_i) = 2 = 2\tau(Z_i)$, and $\mathcal{I}_{Z_i}$ is not generated by $(\mathcal{I}_{Z_i})_{\leq \tau}$. Likewise, Example 2.9 yields a one-dimensional family of minimum degree annihilating schemes for a Clebsch quartic, a degree 4 form f in $\mathcal{R}$ having a length-5 additive decomposition. These examples show the need for the additional hypothesis in Lemma 5.3 E (iii) to obtain the uniqueness of Z.

EXAMPLE 5.8. FORM WITHOUT A TIGHT ANNIHILATING SCHEME. Let $r = 3$, $\mathcal{D} = \mathcal{R} = k[X, Y, Z]$, $f = X^{[3]}Y^{[3]}Z^{[3]}$. Then $I = \operatorname{Ann}(f) = (x^4, y^4, z^4)$, and $H_f = (1, 3, 6, 10, 12, 12, 10, 6, 3, 1)$, so $\ell\operatorname{diff}_k(f) = 12$. If a degree-12, 0-dimensional scheme Z satisfied $\mathcal{I}_Z \subset \operatorname{Ann}(f)$, then by

Lemma 5.3(D) we would have $(\mathcal{I}_Z)_4 = I_4 = \langle x^4, y^4, z^4 \rangle$, which generates an ideal whose radical is (x, y, z), a contradiction. The minimum degree of an annihilating scheme is unclear: the degree 16 scheme consisting of the points of intersection of the four lines $x^4 + ay^4 = 0$ with the four lines $x^4 + bz^4 = 0$, is certainly an annihilating scheme for f. [1]

Suppose $\mathrm{char}(k) = 0$ or $\mathrm{char}(k) > j$. The variety $CI(H)$ parametrizing complete intersections $I_g = (g_1, g_2, g_3)$, $H = H(R/I_g)$ of generator degrees $(4, 4, 4)$ can be considered to be an open dense subset of $Grass(3, R_4)$ via the map; $J \to J_4$; so $\dim(CI(H)) = 3(12) = 36$. The corresponding family C_T, $T = \mathrm{Sym}(H, 9)$ of forms $f \in \mathcal{R}_9$ annihilated by CI ideals in $CI(H)$ and having Hilbert function $H_f = T$ is an open dense subset of $Gor(T)$ (by Theorems 1.72 and 4.17). It has dimension 37 (add one since I_g determines f up to k^*-multiple). The same argument as above for $X^{[3]}Y^{[3]}Z^{[3]}$ shows that no $f \in C_T$ has a tight annihilating scheme. The smallest degree of an annihilating scheme for f is unclear to us.

Another related question is whether or not a form $f \in C_T$ belongs to $\overline{PS(12, 9; 3)}$. Here $\overline{PS(12, 9; 3)}$ has dimension 36, so a general $f \in C_T$ does not belong to this closure. As shown in Proposition 5.18 below $C_T \cap \overline{PS(12, 9; 3)}$ is empty, and this agrees with the non-existence of tight annihilating schemes for the elements of C_T.

Example 5.9. Forms constructed from complete intersections with no tight annihilating scheme. Let again $\mathrm{char}(k) = 0$ or $\mathrm{char}(k) > j$. More generally suppose, as in the hypothesis of Theorem 4.17, that we have $r \geq 3$, that the ideal $J = (g_1, \ldots, g_r)$ is a a complete intersection (i.e. $g_1, \ldots, g_r$ is a regular sequence) having generator degrees $d_1, \ldots, d_r$, and such that $g_1, \ldots, g_r$ are either of the form $g_i = x_i^{d_i}$, $i = 1, \ldots, r$, or are general enough. Let $j = \sum_{i=1}^r d_i - r$ and assume furthermore that $\max_i \{d_i\} \leq j/2$. Consider a nonzero element $f \in (J_j)^\perp \subset \mathcal{R}_j$ and recall from Theorem 4.17 that every element from the open dense subset C_T of an irreducible component of $Gor(T)$ is obtained in this way, where $T = \mathrm{Sym}(H, j)$ and $H = H(R/J)$. We claim that $\ell\mathrm{diff}_k(f) < \ell\mathrm{sch}(f)$, which is equivalent to the non-existence of a tight annihilating scheme of f (Lemma 5.3(D)). Indeed, suppose by way of contradiction that Z is a tight annihilating scheme. Then by Lemma 5.3(E) one has $(\mathcal{I}_Z)_i = \mathrm{Ann}(f)_i$ for $i \leq j/2$. Now,

---

[1] We are grateful to B. Jung for pointing out a mistake in the attempted proof of an earlier claim that 13 of these points would comprise an annihilating scheme: a calculation indicates that they do not.

Theorem 4.16 says that $\text{Ann}(f)_i = J_i$ for $i \leq j/2$, thus the condition $\max_i\{d_i\} \leq j/2$ implies that all generators of J belong to $\mathcal{I}_Z$, hence $J \subset \mathcal{I}_Z$. This is absurd, since the ring $R/\mathcal{I}_Z$, having Krull dimension one, cannot be a quotient of the Artinian ring R/J.

A different class of forms f for which $\ell\text{diff}_k(f) < \ell\text{sch}(f)$ is given in Corollary 7.16, when $r = 3$.

REMARK. Our main method of creating components of $Gor(T)$ – carried out in Chapter 6 – relies on being able to give a fibration from a family of DP-forms, to a family of unique tight annihilating schemes. The above examples show that a general f in some irreducible component of $Gor(T)$ might not have a tight annihilating scheme. It is this phenomenon that requires us later in Chapter 6 to show $\dim_k T_f Gor(T) = \dim(C)$ at a generic point f of at least one putative component C, in order to show that $Gor(T)$ has several irreducible components. Otherwise, both putative components could lie in some larger component of $Gor(T)$ whose general point g has no tight annihilating scheme.

Our last example of this section shows that if $Z \subset \mathbb{P}^{r-1}$ is not locally Gorenstein, then there may be no function f for which Z is a tight annihilating scheme, such that one can recover Z from f.

EXAMPLE 5.10. SCHEME Z IN $\mathbb{P}^2$ SUCH THAT A GENERAL FORM f IN $\mathcal{D}_j$ ANNIHILATED BY $\mathcal{I}_Z$ DOES NOT SATISFY $\text{Ann}(f)_t = (\mathcal{I}_Z)_t$. Suppose $t = r = 3$, $s = \dim_k R_2 = 6$ and let $Z \subset \mathbb{P}^2$ be the degree 6 scheme $Z = \text{Spec}(\mathcal{O}_p/m_p^3)$ defining m_p^3 at the point $p : (1,0,0) \in \mathbb{P}^2$. Then $(\mathcal{I}_Z)_3 = (y^3, y^2z, yz^2, z^3)$, of codimension 6 in R_3, so $\mathcal{I}_Z$ is 3-regular and satisfies $H(R/\mathcal{I}_Z) = (1, 3, 6, 6, \ldots)$. A general form f in $(\mathcal{I}_Z)_6^{\perp} \cap \mathcal{D}_6$ can be written $f = X^{[6]} + X^{[5]}L + X^{[4]}Q$ where L, Q belong to $\mathcal{D}'$, the divided power ring in variables Y, Z, and L, Q are homogeneous of degrees 1 and 2, respectively. Let $I = \text{Ann}(f)$, and let the vector space $W = \text{Ann}(Q) \cap R_2 \cap k[y, z]$; then $\dim_k(W) = 2$, and $W \subset I_2 = I_3 : R_1$ by Lemma 2.17. However $(\mathcal{I}_Z)_2 = (0)$ and satisfies $(\mathcal{I}_Z)_2 = (\mathcal{I}_Z)_3 : R_1$ since $\mathcal{I}_Z$ is saturated. It follows that I_3 cannot define m_p^3.

Similarly, one can see that there is *no* Gorenstein algebra $A = R/I$, of socle degree 6, with $I_3 = (\mathcal{I}_Z)_3$ defining m_p^3. This impossibility follows also from Theorem 4.1A and Theorem B.3 of Appendix B. Suppose otherwise, then $I = \text{Ann}(f)$ for some f, $I_3 = (\mathcal{I}_Z)_3$ and $H(R/I) = (1, 3, 6, 6, 6, 3, 1) = H(6, 6, 3)$. Since $(\mathcal{I}_Z)_3 : R_1 = 0$, then $(I_3)^2$ would define m_p^6, of colength 21, and $\dim_k(R_6/(I_3)^2)$ would be 21, so $\dim_k T_f = 21$, greater than the dimension of $\mathbf{Gor}(H(6, 6, 3))$.

But I has 4 generators in degree 3, and 3 relations in degree 4 (as does $\mathcal{I}_Z$), so has the minimum possible Betti numbers for $I = \text{Ann}(f)$, $f \in Gor(T)$. By Theorem B.3, I^2 has the same minimal resolution (Betti numbers) as that of $\text{Ann}(F)^2$ for a general element F of $Gor(T)$; but by Theorem 4.1A, $\dim_k(\mathcal{T}_F) = 3(6) = 18$, a contradiction.

REMARK. We will study the question of which zero-dimensional subschemes $Z \subset \mathbb{P}^n$ can be recovered from a general element $f \in \mathcal{D}_j$ annihilated by $\mathcal{I}_Z$ in Sections 6.1 and 6.4. It is known that when j is large enough, Z locally Gorenstein is sufficient [ChoI1], see also Lemma 6.1. It is not known if Z being a tight annihilating scheme for f implies Z locally Gorenstein.

5.2. Flat families of zero-dimensional schemes and limit ideals

In this section we give for the sake of the reader's convenience some fairly standard material about flat families and limit ideals which we use in this and the next chapters. We also give some applications to the closure of $PS(s, j; r)$, a topic we also discuss later. We start with the simpler case of one-parameter families and at the end of the section consider the general case.

Let C be an irreducible, nonsingular curve with a distinguished point o. Let $\mathcal{Z} \subset C \times \mathbb{P}^{r-1}$ be a subscheme such that the first projection π satisfies

$$\pi : \mathcal{Z} \to C \text{ is proper, flat of relative dimension } 0, \qquad (5.2.1)$$

i.e. $\mathcal{Z}$ is a flat family of 0-dimensional subschemes in $\mathbb{P}^{r-1}$. By the universal property of Hilbert schemes there is a unique morphism

$$\nu : C \longrightarrow \textbf{Hilb}^s\mathbb{P}^{r-1}$$

such that $\pi : \mathcal{Z} \to C$ is the pull-back of the universal family. Here $s = \deg(\mathcal{Z}_x)$ for each $x \in C$ and equals the (constant) Hilbert polynomial of the flat family $\mathcal{Z} \to C$. Notice that Condition (5.2.1) is equivalent to the following one.

The scheme $\mathcal{Z}$ is closed in $C \times \mathbb{P}^{r-1}$, the projection π is quasifinite (i.e. with finite fibers), and $\mathcal{Z}$ does not have associated closed points (i.e. points $x \in \mathcal{Z}$ such that the maximal ideal $\mathfrak{m}_{x,\mathcal{Z}}$ in $\mathcal{O}_{x,\mathcal{Z}}$ consists of zero divisors).

Indeed, closedness of $\mathcal{Z} \subset C \times \mathbb{P}^{r-1}$ is equivalent to properness of $\pi : \mathcal{Z} \to C$. Properness together with the quasifiniteness of π is equivalent to finiteness of π [Har2, III Ex. 11.2]. Now, since C is a smooth curve, π is flat if and only if each associated scheme point of $\mathcal{Z}$

maps dominantly to C [**Har2**, III Proposition 9.7], which is equivalent to the nonexistence of associated closed points of $\mathcal{Z}$, since π is a finite morphism.

Let $i \geq 0$. Then $\pi_* \mathcal{O}_{\mathcal{Z}}(i)$ is a coherent, flat sheaf over C, so it is a locally free sheaf of rank s [**Har2**, III §9]. Consider the usual restriction homomorphism

$$\rho_i \; : \; R_i \otimes_k \mathcal{O}_C \longrightarrow \pi_* \mathcal{O}_{\mathcal{Z}}(i).$$

Let $\mathcal{E}_i = \operatorname{Ker} \rho_i$, $\mathcal{F}_i = \operatorname{Im} \rho_i$. It is a standard fact that follows from the structure theorem of finitely generated modules over discrete valuation rings, that since C is a smooth curve the sheaves $\mathcal{E}_i$ and $\mathcal{F}_i$ are locally free of ranks $r_i - h_i$ and h_i respectively ($r_i = \dim_k R_i$), and moreover there is a Zariski open, dense subset $Y_i \subset C$, such that $\mathcal{E}_i \,|_{Y_i}$ and $\mathcal{F}_i \,|_{Y_i}$ are locally direct summands. The fact that π is finite (in particular affine) implies that for every $x \in C$ the right vertical homomorphism of the following commutative diagram is an isomorphism.

$$
\begin{array}{ccc}
(R_i \otimes_k \mathcal{O}_C) \otimes_{\mathcal{O}_C} k(x) & \xrightarrow{\;\rho_i(x)\;} & \pi_* \mathcal{O}_{\mathcal{Z}}(i) \otimes_{\mathcal{O}_C} k(x) \\
\cong \downarrow & & \downarrow \cong \\
R_i & \xrightarrow{\;res\;} & H^0(\mathcal{Z}_x, \mathcal{O}_{\mathcal{Z}_x}(i))
\end{array}
\qquad (5.2.2)
$$

Thus $\operatorname{Ker} \rho_i(x) = (\mathcal{I}_{\mathcal{Z}_x})_i$, the i-th graded component of the ideal of $\mathcal{Z}_x$. Furthermore when $y \in Y_i$, then $\operatorname{Ker} \rho_i(x) = \mathcal{E}_i \otimes_{\mathcal{O}_C} k(x) :=$ $\mathcal{E}_i(x)$, $\operatorname{Im} \rho_i(x) = \mathcal{F}_i(x)$, so $\dim(\mathcal{I}_{\mathcal{Z}_y})_i = r_i - h_i$.

DEFINITION 5.11. Under the assumption (5.2.1) we define a *limit ideal* $\mathcal{I}_o$ at $o \in C$ and write (abusing notation) $\mathcal{I}_o = \lim_{y \to o} \mathcal{I}_{\mathcal{Z}_y}$ by setting $\mathcal{I}_o = \oplus_{i \geq 1}(\mathcal{I}_o)_i$ where $(\mathcal{I}_o)_i$ is defined in one of the following two equivalent ways:

- $(\mathcal{I}_o)_i \; = \; \mathcal{E}_i(o)$
- $(\mathcal{I}_o)_i$ is the extension to $o \in C$ of the morphism
  $Y_i \to Grass(r_i - h_i, R_i)$ defined by $y \mapsto (\mathcal{I}_{\mathcal{Z}_y})_i$

LEMMA 5.12. *In the notation of Definition 5.11 the following properties hold.*

i. $\mathcal{I}_o = \lim_{y \to o} \mathcal{I}_{\mathcal{Z}_y}$ *is contained in the graded ideal* $\mathcal{I}_{\mathcal{Z}_o}$.

ii. *There is an open dense subset* $Y \subset C$ *and a sequence* $H = (h_0 = 1, h_1, h_2, \dots)$ *such that* H *is the Hilbert function of the graded algebra* $R/\mathcal{I}_{\mathcal{Z}_y}$ *for every* $y \in Y$. *The sequence* H *is nondecreasing and stabilizes at* s. *One has* $H(R/\mathcal{I}_o) = H$.

iii. $h_i = s$ *when* $i \geq s - 1$ *and* $(\mathcal{I}_o)_i = (\mathcal{I}_{\mathcal{Z}_o})_i$ *for every* $i \geq s - 1$.

iv. $\mathcal{I}_{\mathcal{Z}_o}$ *is equal to the saturation of* $\mathcal{I}_o = \lim_{y \to o} \mathcal{I}_{\mathcal{Z}_y}$.

PROOF. (i). This is immediate from (5.2.2) since $(\mathcal{I}_o)_i = \mathcal{E}_i(o) \subset$ Ker $\rho_i(o) = (\mathcal{I}_{Z_o})_i$.

(ii) and (iii). According to Theorem 1.69 the map $R_i \xrightarrow{\text{res}}$ $H_0(\mathcal{Z}_x, \mathcal{O}_{\mathcal{Z}_x}(i))$ is epimorphic when $i \geq s - 1$ for every $x \in C$. From (5.2.2) using Nakayama's lemma we conclude that the homomorphism ρ_i is surjective when $i \geq s - 1$. This shows that $h_i = s$ when $i \geq s - 1$, and we may choose $Y_i = C$ when $i \geq s - 1$. Now, $Y = \cap_{i=0}^{s-2} Y_i$ satisfies the property of (ii). That $H(R/\mathcal{I}_o) = H$ is clear. The equality $H = H(R/\mathcal{I}_{Z_y})$ shows H is nondecreasing according to Theorem 1.69. When $i \geq s - 1$ both $(\mathcal{I}_o)_i$ and $(\mathcal{I}_{Z_o})_i$ are of codimension s and $(\mathcal{I}_o)_i \subset (\mathcal{I}_{Z_o})_i$ by (i), hence they are equal.

(iv). This is a consequence of (iii).   □

REMARK 5.13. We note an application of this lemma. A flat family of 0-dimensional schemes $\pi' : \mathcal{Z}' \to C - \{o\}$ yields a morphism $\nu' :$ $C - \{o\} \to \mathbf{Hilb}^s \mathbb{P}^{r-1}$ which can be extended to $\nu : C \to \mathbf{Hilb}^s \mathbb{P}^{r-1}$, since $\mathbf{Hilb}^s \mathbb{P}^{r-1}$ is projective, thus obtaining an extended flat family $\pi : \mathcal{Z} \to C$. In order to calculate the limit scheme $\mathcal{Z}_o$ one takes the limit ideal $\mathcal{I}_o$. Then by the above lemma $\mathcal{I}_{Z_o} = \mathrm{Sat}(\mathcal{I}_o)$, the two ideals may differ only in degrees $\leq s - 2$, and $\oplus_{i \geq s-1}(\mathcal{I}_{Z_o})_i$ is generated by $(\mathcal{I}_o)_{s-1}$ and $(\mathcal{I}_o)_s$ by Theorem 1.69, since $\tau(\mathcal{I}_{Z_o}) \leq s - 1$. This makes the computation of $\mathcal{I}_{Z_o}$ effective.

EXAMPLE 5.14. Take a family $\mathcal{Z} \to C$ such that the fibers $\mathcal{Z}_x$ for $x \neq o$ are 3 non-collinear points in $\mathbb{P}^2$, while $\mathcal{Z}_o$ are 3 points on a line. Then $s = 3$, $H(R/\mathcal{I}_o) = (1, 3, 3, \ldots)$, while $H(R/\mathcal{I}_{Z_o}) = (1, 2, 3, 3, \ldots)$, so $\mathcal{I}_o \subsetneqq \mathcal{I}_{Z_o}$.

EXAMPLE 5.15. CALCULATION OF A LIMIT SCHEME. Let $r = 3$, $R = k[x, y, z]$, $\mathrm{char}(k) \neq 2, 3$, and consider the family of forms $\{F_t \in \mathcal{R}_4 \mid t \in \mathbb{A}^1\}$,

$$F_t = X^4 + (X + t^3 Y + tZ)^4 + (X + 8t^3 Y + 2tZ)^4,$$

whose polar polyhedron for $t \neq 0$ is the annihilating degree-3 scheme

$$\mathcal{Z}_t = \{(1 : 0 : 0), (1 : t^3 : t), (1 : 8t^3 : 2t)\}.$$

We aim to calculate the limit scheme $\lim_{t \to 0} \mathcal{Z}_t = \mathcal{Z}_0$ of degree $s = 3$. Clearly $(\mathcal{Z}_0)_{red} = (1 : 0 : 0)$. We want to apply the procedure in Remark 5.13. For $t \neq 0$ we have

$$\mathcal{I}_{\mathcal{Z}_t} = (y, z) \cap (t^2 z - y, tx - z) \cap (4t^2 z - y, 2tx - z).$$

If $\mathcal{I}_0$ is the limit ideal, then $(\mathcal{I}_0)_2$ and $(\mathcal{I}_0)_3$ generate $\mathcal{I}_{Z_0}$ in degrees $\geq s - 1 = 2$. Let us calculate these two limit spaces. From $H(R/\mathcal{I}_0) =$

$(1, 3, 3, 3, \dots)$ we have $\dim_k(\mathcal{I}_0)_2 = 3$, $\dim_k(\mathcal{I}_0)_3 = 7$. Products of two linear forms yield the following two elements in $(\mathcal{I}_0)_2$:

$$\lim_{t \to 0}(t^2 z - y)(4t^2 z - y) = y^2, \quad \lim_{t \to 0}(t^2 z - y)(2tx - z) = yz.$$

One obtains a third element $\notin \langle y^2, yz \rangle$ by

$$\lim_{t \to 0} \frac{1}{t} \left[(t^2 z - y)(2tx - z) - (4t^2 z - y)(tx - z) \right] = -xy.$$

Thus $(\mathcal{I}_0)_2 = \langle xy, y^2, yz \rangle$. We have $R_1(\mathcal{I}_0)_2 = \langle x^2 y, xy^2, xyz, y^3, y^2 z, yz^2 \rangle$ and to obtain the missing generator we consider

$$\lim_{t \to 0} z(tx - z)(2tx - z) = z^3.$$

Thus $\oplus_{i \geq 2}(\mathcal{I}_{Z_0})_i = \mathcal{I}_0 = (xy, y^2, yz, z^3)$. Clearly

$$\mathcal{I}_{Z_0} = \mathrm{Sat}(\mathcal{I}_0) = (y, z^3).$$

A second way to find Z_0 in this case is to notice that $y - z^3 \in \mathcal{I}_{Z_t}, t \neq 0$, hence $y - z^3 \in \mathcal{I}_{Z_0}$; since Z is of degree three, concentrated at $p = (1 : 0 : 0)$ we have that $(y, z)^3 \subset \mathcal{I}_{Z_0}$, so $(y, z^3) \subset \mathcal{I}_{Z_0}$ implying equality since their colengths are the same as ideals of $\mathcal{O}_p$.

We have the inclusion $\mathcal{I}_0 \subset \mathrm{Ann}(F_0)$ (holding in general for limit ideals of annihilating schemes). In some cases it could simplify the calculation of $\mathrm{Sat}(\mathcal{I}_0)$. Here however $F_0 = 3X^4$, $(\mathrm{Ann}(F_0))_{\leq 3} = (y, z)_{\leq 3}$, so the only information we obtain is $\mathcal{I}_{Z_0} \subset (y, z)$, which is clear anyway, since $(Z_0)_{red} = (1 : 0 : 0)$.

DEFINITION 5.16. We say that a 0-dimensional subscheme Z of $\mathbb{P}^{r-1}$ of degree s is *smoothable* if in the set-up of (5.2.1) we have $Z = \mathcal{Z}_o$ and the fibers $\mathcal{Z}_y$ over a dense open subset $Y \subset C$ are smooth, i.e. consist of s distinct points. We will frequently encounter this situation below and will informally write $Z(t)$ for the fiber $\mathcal{Z}_t$ over $t \in C$ and $\{Z(t) \mid t \in C\}$ for the flat family $\pi : \mathcal{Z} \to C$.

We now show that in some cases, we can identify points in the closure of $PS(s, j; r)$. We will refine the following result later (Proposition 6.7) in Section 6.1.

LEMMA 5.17. CLOSURE OF $PS(s, j; r)$. *Let $f \in \mathcal{D}_j$. If f has a smoothable annihilating scheme Z of degree s, then $f \in \overline{PS(s, j; r)}$.*

PROOF. If $s \geq r_j$, the conclusion follows from Lemma 5.3(G). Suppose $s < r_j$. Let $\{Z(t) \mid t \in C\}$ be an arbitrary smoothing of Z such that $Z(o) = Z$. We remark — although it is not needed for the proof — that since by Lemma 1.17 a general enough set of s points impose independent conditions on degree-j forms, we may choose the smoothing family so that this condition holds for a general element $Z(t)$ of the family. Let $V(t) = (\mathcal{I}_{Z(t)})_j$ for $t \neq o$. Replacing C by a Zariski open neighborhood of o we can assume that $\dim_k V(t) = d$ is constant, since the locus on C where the dimension is larger consists of a finite set of points ($d = r_j - s$ for $t \neq o$ if the family were chosen as above). Consider the limit subspace $V(o) \in Grass(d, R_j)$. By the Apolarity Lemma 1.15 we have for the perpendicular subspaces in $\mathcal{D}_j$ that for $t \neq o$,

$$V(t)^\perp \subset \overline{PS(s, j; r)}.$$

Therefore the limit space $V(o)^\perp$ belongs to $\overline{PS(s, j; r)}$. Since $V(o) \subset (\mathcal{I}_Z)_j$ according to Lemma 5.12(i) we conclude that $f \in (\mathcal{I}_Z)_j{}^\perp \subset V(o)^\perp$ belongs to $\overline{PS(s, j; r)}$.   □

REMARK. The argument used in the above lemma may be applied not only to $PS(s, j; r)$, but to other families of DP-forms. Namely, suppose we have an irreducible subset $U \subset Hilb^s \mathbb{P}^{r-1}$, and an irreducible subset $C_U \subset \mathcal{R}_j$ whose open, dense subset is traced out by the perpendicular spaces $(\mathcal{I}_Z)_j{}^\perp$ to the punctual schemes Z from a certain open subset of U. Then the same argument as in the above lemma shows that if $Z_0 \in \overline{U}$ and if $f \in (\mathcal{I}_{Z_0})_j{}^\perp$, then f belongs to the closure $\overline{C_U} \subset \mathbb{P}(\mathcal{D}_j)$.

The following Proposition is a complement to Theorem 4.13 and Corollary 4.18. When $(j, r) \neq (4, 3), (5, 3), (7, 3)$ or $(5, 4)$, then C_T below is an irreducible component of $V_s(t, j - t; r)$.

PROPOSITION 5.18. *Suppose* $\mathrm{char}(k) = 0$ *or* $\mathrm{char}(k) > j$. *Let* $j = 2t$ *or* $2t + 1$, $t \geq 2$, $s = r_t - r$. *Let* $T = H(s, j, r)$. *Let* C_T *be the irreducible subset of* $V_s(t, j - t; r)$ *consisting of forms* $f \in \mathcal{R}_j$, *apolar to CI ideals generated by* r *forms of degree* t, *and having Hilbert function* $H_f = H(s, j, r)$. *Then*

$$C_T \cap \overline{PS(s, j; r)} = \emptyset. \tag{5.2.3}$$

PROOF. Suppose, by way of contradiction, that for some $F \in C_T$ there is a curve $W \subset Gor(T)$ parametrizing forms $\{f_w \mid w \in W\}$ such that $f_w \in PS(s, j; r)$ when $w \neq w_0$ and $f_{w_0} = F$. We have the freedom to choose the curve, so that $W - \{w_0\}$ belongs to a given open subset of $PS(s, j; r)$. Since the hypotheses imply $s \geq r_{t-1}$ we may suppose by Lemma 1.17 and Theorem 2.6 that when $w \neq w_0$ we have $H_{f_w} = T = H(s, j, r)$, the polar polyhedron Z_w of f_w is unique, and the Hilbert function of $R/\mathcal{I}_{Z_w}$ equals (since $s \geq r_{t-1}$)

$$H = (1, r, \ldots, r_{t-1}, s, s, \ldots). \tag{5.2.4}$$

Replacing if necessary W by its finite covering (which may be needed if $\operatorname{char} k = p$), this yields a morphism $\mu : W - \{w_0\} \to \mathbf{Hilb}^s \mathbb{P}^{r-1}$. Replacing W by a local desingularization at w_0, $n : (C, o) \to (W, w_0)$, and using that $\mathbf{Hilb}^s \mathbb{P}^{r-1}$ is a projective scheme we extend $\nu = \mu \circ n$ to a morphism $\nu : C \to \mathbf{Hilb}^s \mathbb{P}^{r-1}$. We obtain a flat family of 0-dimensional degree-s schemes $\pi : \mathcal{Z} \to C$ and a family of forms $\{f_x \mid x \in C\}$, here $f_x := f_{n(x)}$, such that $\mathcal{Z}_x$ is a smooth annihilating scheme of f_x when $x \neq o$.

Let $\mathcal{I}_o$ be the limit ideal, $\mathcal{I}_o = \lim_{x \to o} \mathcal{I}_{Z_x}$. By Lemma 5.12 we have $\mathcal{I}_o \subset \mathcal{I}_{Z_o}$ and $H(R/\mathcal{I}_o) = H$ from (5.2.4). Now, in degree t we have $(\mathcal{I}_{Z_x})_t = \operatorname{Ann}(f_x)_t$ when $x \neq o$ and $\lim_{x \to o} \operatorname{Ann}(f_x)_t = \operatorname{Ann}(f_o)_t$ since $f_o \in Gor(T)$. Consequently $(\mathcal{I}_o)_t = \operatorname{Ann}(f_o)_t = \langle g_1, \ldots, g_r \rangle$, where $g_1, \ldots, g_r$ is the regular sequence in R_t generating the ideal $J \subset \operatorname{Ann}(F)$, $F = f_o$ by construction. We conclude from $(\mathcal{I}_o)_t \subset (\mathcal{I}_{Z_o})_t$ that $J \subset \mathcal{I}_{Z_o}$. This contradicts that $R/\mathcal{I}_{Z_o}$ has Krull dimension one, while R/J is Artinian. The contradiction completes the proof of (5.2.3). $\qquad\square$

REMARK. When $r = 3$ and $T = H(s, j, 3)$ Diesel's Theorem 1.72 shows that $\overline{PS(s, j; 3)}$ belongs to the border $\overline{C}_T - C_T$. We do not know whether this is the case when $r \geq 4$.

We conclude this section by a generalization of the set-up of (5.2.1), replacing C by an arbitrary Noetherian sheme [2] and defining the "postulation" Hilbert subscheme $\mathbf{Hilb}^H \mathbb{P}^n$ of $\mathbf{Hilb}^s \mathbb{P}^n$. Let us first recall the definition of determinantal scheme.

DEFINITION 5.19. DETERMINANTAL SCHEME. Let $\mathbf{X}$ be a scheme, $\mathcal{E}$, $\mathcal{F}$ locally free sheaves on $\mathbf{X}$ of ranks e, f respectively, and let $\rho : \mathcal{E} \to \mathcal{F}$ be a homomorphism. Let h be a positive integer, $h \leq f$. The exterior power $\wedge^{h+1} \rho : \wedge^{h+1} \mathcal{E} \to \wedge^{h+1} \mathcal{F}$ induces a homomorphism $\rho^\natural : \wedge^{h+1} \mathcal{E} \otimes \wedge^{h+1} \mathcal{F}^* \to \mathcal{O}_\mathbf{X}$ whose image is an ideal sheaf, generated locally by the $(h+1) \times (h+1)$ minors of ρ. We denote by $\mathbf{D}_h(\rho)$ the closed subscheme of $\mathbf{X}$, defined by this ideal sheaf; it is called the rank $\leq h$ determinantal subscheme associated with ρ.

Let us replace in (5.2.1) the smooth curve C by an arbitrary Noetherian scheme $\mathbf{X}$, i.e. we consider a subscheme $\mathcal{Z} \subset \mathbf{X} \times \mathbb{P}^{r-1}$ such

---

[2]As customary in the book we assume that the schemes are defined over an algebraically closed field k, however some of the definitions, e.g. that of determinantal scheme do not need this restriction

that the first projection $\pi : \mathcal{Z} \to \mathbf{X}$ is proper, flat of relative dimension 0, and each fibre has degree s. This is equivalent to having a morphism $\nu : \mathbf{X} \to \mathbf{Hilb}^s \mathbb{P}^{r-1}$, by the universal property of Hilbert schemes. The properness and quasifiniteness of π implies that π is finite [**Har2**, III Ex. 11.2]. By flatness, for every $i \in \mathbb{Z}$ the sheaves $\pi_* \mathcal{O}_{\mathcal{Z}}(i)$ are locally free. For each scheme point $x \in \mathbf{X}$ one has a canonical isomorphism

$$\pi_* \mathcal{O}_{\mathcal{Z}}(i) \otimes_{\mathcal{O}_{\mathcal{Z}}} k(x) \xrightarrow{\sim} H^0(\mathcal{Z}_x, \mathcal{O}_{\mathcal{Z}_x}(i)). \qquad (5.2.5)$$

This follows either from π being finite, thus affine, or one can refer to the general theorems of base change [**Har2**, III Theorem 12.11] and use $H^1(\mathcal{Z}_x, \mathcal{O}_{\mathcal{Z}_x}(i)) = 0$. In particular $\pi_* \mathcal{O}_{\mathcal{Z}}(i)$ is of rank s. Let $i \geq 0$ and let ρ_i be the canonical restriction homomorphism $(R = k[x_1, \ldots, x_r])$

$$\rho_i : R_i \otimes_k \mathcal{O}_{\mathbf{X}} \longrightarrow \pi_* \mathcal{O}_{\mathcal{Z}}(i)$$

Then from (5.2.5) we have a commutative diagram

$$(5.2.6)$$

$$\begin{array}{ccc}
(R_i \otimes_k \mathcal{O}_{\mathbf{X}}) \otimes_{\mathcal{O}_{\mathbf{X}}} k(x) & \xrightarrow{\rho_i(x)} & \pi_* \mathcal{O}_{\mathcal{Z}}(i) \otimes_{\mathcal{O}_{\mathbf{X}}} k(x) \\
\cong \downarrow & & \downarrow \cong \\
R_i \otimes_k k(x) & \xrightarrow{res} & H^0(\mathcal{Z}_x, \mathcal{O}_{\mathcal{Z}_x}(i))
\end{array}$$

Let Ω be the finite set of nondecreasing sequences of integers $H = (1, h_1, h_2, \ldots)$ such that $h_i = s$ for $i \geq s - 1$. These properties are those of Hilbert functions of 0-dimensional degree-s schemes $H(R/\mathcal{I}_Z)$, according to Theorem 1.69. For each $i \geq 1$ consider the determinantal subscheme $\mathbf{D}_{h_i}(\rho_i)$. Obviously $\mathbf{D}_{h_i}(\rho_i) = \mathbf{X}$ when $i \geq s - 1$. From (5.2.6) we see that the closed points of $\mathbf{D}_{h_i}(\rho_i)$ correspond to the 0-dimensional schemes $\mathcal{Z}_x$ in the family which impose (or postulate) $\leq h_i$ independent conditions·on the forms of degree i. We define

$$\mathbf{X}^{\leq H} = \bigcap_{i=1}^{s-2} \mathbf{D}_{h_i}(\rho_i).$$

Introducing in Ω a termwise partial order $H' \leq H$ we define

$$\mathbf{X}^H = \mathbf{X}^{\leq H} - \bigcup_{H' < H} \mathbf{X}^{\leq H'}.$$

Then $\mathbf{X}^H$ is a locally closed subscheme whose closed points correspond to those $\mathcal{Z}_x$ from the family with $H(R/\mathcal{I}_{\mathcal{Z}_x}) = H$.

EXAMPLE. Let $\mathbf{X} = C$ as in (5.2.1). Let $\eta \in C$ be the generic (scheme) point and $H = H(R/\mathcal{I}_{\mathcal{Z}_\eta})$. Then C^H is a locally closed subset which contains the generic point η, so C^H is a complement to a finite number of closed points. This is the open set $Y \subset C$ from Lemma 5.12.

DEFINITION 5.20. "POSTULATION" HILBERT SCHEME. Let $\mathbf{X} = \mathbf{Hilb}^s \mathbb{P}^n$, $n = r - 1$, and let $\pi : \mathcal{Z} \to \mathbf{Hilb}^s \mathbb{P}^n$ be the universal family. Let $H = (1, h_1, h_2, \dots)$ be a nondecreasing sequence of integers, stabilizing at s for $i \geq s - 1$. Then the scheme $\mathbf{X}^H$ introduced above is denoted by $\mathbf{Hilb}^H \mathbb{P}^n$. The associated reduced subscheme is denoted by $Hilb^H \mathbb{P}^n$. So, the closed points of $\mathbf{Hilb}^H \mathbb{P}^n$, i.e. the variety $Hilb^H \mathbb{P}^n$, parametrize the 0-dimensional subschemes of $\mathbb{P}^n$ with Hilbert function $H(R/\mathcal{I}_Z) = H$, and the Hilbert scheme is a disjoint union

$$\mathbf{Hilb}^s \mathbb{P}^n = \bigcup_{H \in \Omega} \mathbf{Hilb}^H \mathbb{P}^n.$$

For further discussion see [M-DP, pp. 117–120], [KW, Kl1, Kl2]. Before stating the next theorem, which collects several known facts about $Hilb^H \mathbb{P}^n$, we introduce some more notation. Let $Sm^s \mathbb{P}^n$ be the open, smooth subscheme of $\mathbf{Hilb}^s \mathbb{P}^n$ whose closed points parametrize sets of s points in $\mathbb{P}^n$. Let $Sm^H \mathbb{P}^n$ be the variety equal to $Sm^s \mathbb{P}^n \cap Hilb^H \mathbb{P}^n$.

THEOREM 5.21. PROPERTIES OF $Hilb^H \mathbb{P}^n$.

A. (P. Maroscia) $Hilb^H \mathbb{P}^n$ is non-empty if and only if the first difference ΔH is an O-sequence (see Appendix C). In this case $Sm^H \mathbb{P}^n$ is also non-empty.

B. $Hilb^H \mathbb{P}^2$ is non-empty if and only if for $\Delta H = (\delta_i)$ one has

$$\Delta H = (1, 2, \dots, \nu, \delta_\nu, \delta_{\nu+1}, \dots, \delta_\tau, 0, 0, \dots)$$

where $\nu = \delta_{\nu-1} \geq \delta_\nu \geq \delta_{\nu+1} \geq \cdots \geq \delta_\tau > 0$ and $\delta_i = 0$ for $i \geq \tau + 1$

C. (G. Gotzmann) $Hilb^H \mathbb{P}^2$ is irreducible and smooth.

D. $Sm^H \mathbb{P}^2$ is open, dense in $Hilb^H \mathbb{P}^2$.

PROOF. (A). P. Maroscia proved in [Mar] that there is a set Y of s points in $\mathbb{P}^n$ such that $H = H(R/\mathcal{I}_Y)$ if and only if ΔH is an O-sequence. It only remains to notice that if Z is an arbitrary 0-dimensional subscheme of $\mathbb{P}^n$ then according to Lemma 1.67 every general enough element $\bar{x} \in (R/\mathcal{I}_Z)_1$ is not a zero divisor in $R/\mathcal{I}_Z$, hence ΔH is the Hilbert function of the Artin algebra $R/(\mathcal{I}_Z + R\bar{x})$, thus is an O-sequence.

(B). From the proof of (A) we know that if $H = H(R/\mathcal{I}_Z)$, then $\Delta H = H(A)$, where A is the quotient of $R/\mathcal{I}_Z$ by $(\bar{x})$, where $x \in R_1$.

Choosing new variables we may assume that $x = x_3$. Thus A is a graded Artinian quotient of $k[x_1, x_2]$ and we may apply Corollary C.6.

(C) is proved in [**Got3**].

(D) follows from (A) and (C).   □

COROLLARY 5.22. *Every 0-dimensional degree-s subscheme $Z \subset \mathbb{P}^2$ with Hilbert function $H = H(R/\mathcal{I}_Z)$ is a limit (i.e. $Z = Z_o$ in (5.2.1)) of smooth degree-s subschemes with Hilbert function H.*

REMARK. The analogous statement is no longer true for $r = 3$, as $Hilb^s(\mathbb{P}^3)$ has several irreducible components for s large (see [**I4, I5**]). However, since Gorenstein schemes concentrated at a single point of $\mathbb{P}^3$ are smoothable, the analogous statement concerning $\mathbf{Hilb}^H(\mathbb{P}^3)$ could be true if restricted to the Gorenstein subschemes of $\mathbb{P}^3$. J. O. Kleppe remarks that smoothness is shown in his joint work with R. M. Miró-Roig for $\mathbf{Hilb}_{AG}^H(\mathbb{P}^3) \subset \mathbf{Hilb}^s(\mathbb{P}^3)$ parametrizing arithmetically Gorenstein subschemes Z of dimension zero, and having given degree s, and postulation H (See [**KlM-R**], [**Kl2**, Remark 1.6]).

5.3. Existence theorems for annihilating schemes when $r = 3$

It is well known that a Gorenstein height two ideal is a complete intersection – this is Serre's Theorem, see Remark 1.55 and [**Ei2**, Corollary 21.20]. One consequence, shown in Theorem 1.44, is that when $r = 2$ and the Gorenstein sequence $T \supset (s, s)$, if $f \in Gor(T)$, then the ideal $I = \text{Ann}(f) = (\phi, \psi)$ (a complete intersection) where $\deg(\phi) = s$. Thus, $T_f \supset (s, s)$ implies that f has a unique degree-s annihilating scheme (a codimension one subscheme in this case) whose defining ideal is generated by ϕ.

In this section we generalize this result to $r = 3$ and $\mathbb{P}^2$. We will show that when $r = 3$, and T is a Gorenstein sequence containing a subsequence (s, s, s), then there is a dominant morphism $p : Gor(T) \to Hilb^H(\mathbb{P}^2)$, whose fibers are dense open sets in an affine space $\mathbb{A}^s$ (Theorem 5.31). If T only contains a subsequence (s, s) then the dominant morphism p is defined on a proper subset $Gor_{\text{sch}}(T)$ of $Gor(T)$ (Theorem 5.39); a weaker result is obtained for other T (Theorem 5.46). It follows that one can identify $\overline{PS(s, j; 3)} \cap Gor(T)$ for most Gorenstein sequences T (Theorem 5.71). A result of Gerd Gotzmann — determining the dimension of $Hilb^H(\mathbb{P}^2)$ — allows us to determine the dimension of $Gor(T)$ if T contains (s, s, s), or to bound the dimension otherwise (Theorem 5.51 and Corollary 5.50).

Results of the authors ($r \geq 5$), M. Boij ($r \geq 4$), and of Y. Cho and the first author show that Theorem 5.31 is sharp in the sense that, if $r \geq 4$ there are Gorenstein sequences T having constant sub-sequences $(s, \ldots, s)$ of an arbitrary fixed length, such that $Gor(T)$ contains a component parametrizing forms f having no degree-s annihilating scheme (see Chapter 6; in particular when $r = 4$ M. Boij's Theorem 6.42 and Example 6.43).

5.3.1. The generator and relation strata of the variety $Gor(T)$ parametrizing Gorenstein algebras.

We assume that $r = 3$ throughout the section 5.3, except for the following basic Lemma. We fix a Gorenstein sequence T for r variables, and let $D = (d_1, \ldots, d_v)$, $d_1 \leq \ldots d_v$ denote a sequence of positive integers that occurs as the degrees of a minimal generating set for some Gorenstein ideal of Hilbert function T. We denote by $Gor_D(T)$ the subset of $Gor(T)$ parametrizing $f \in Gor(T)$ whose apolar ideal $I = \mathrm{Ann}(f)$ has a minimal generating set of the fixed degrees D. We let $Gor_{\geq D}(T)$ denote the subset of $Gor(T)$ parametrizing f such that $\mathrm{Ann}(f)$ has minimal generating degrees including the sequence D. We let $Gor_v(T) = \cup_{|D|=v} Gor_D(T)$, and $Gor_{\geq v}(T) = \cup_{v' \geq v} Gor_{v'}(T)$.

LEMMA 5.23. *The subset $Gor_D(T)$ is locally closed in $Gor(T)$: in particular it is an open subset of $Gor_{\geq D}(T)$, which is a closed subvariety of $Gor(T)$. Furthermore, $Gor_v(T)$ is a an open subset of $Gor_{\geq v}(T)$, which is a closed subvariety.*

PROOF. The number n_i of degree-i generators of $\mathrm{Ann}(f)$ satisfies $n_i = \dim_k(I_i/R_1 \cdot I_{i-1})$, so, fixing a constant α_i, the condition $n_i \geq \alpha_i$ is closed — as for each i, $\dim_k I_i = r_i - t_i$ is fixed for $I = \mathrm{Ann}(f), f \in Gor(T)$. The intersection of such conditions for $0 \leq i \leq j + 1$ is closed, implying that $Gor_{\geq D}(T)$ is a closed subvariety of $Gor(T)$. Now $Gor_D(T)$ is the complement of the finite union $\cup_{D' > D} Gor_{\geq D'}(T)$ in $Gor(T)$, so it is locally closed, as claimed. Since $Gor_{\geq v}(T)$ is the finite union $\cup_{|D'| \geq v} Gor_{\geq D'}(T)$, the second statement follows from the first. □

Henceforth, we let $r = 3$. We need several well-known results, for the sequel, which we collect in Theorem 5.25 below. We let $\underline{s}$ denote the sequence $(s, s, \ldots)$. Recall that $v(I)$ denotes the minimum number of generators of I. Let $T = (1, 3, \ldots, 3, 1)$ be a Gorenstein sequence and let $\mu(T)$ be the minimum possible number of generators for an ideal of Hilbert function T. We denote by $\Delta^k(T)$ the k-th difference function, satisfying

$$\Delta^k(T)_i = \Delta^{k-1}(T)_i - \Delta^{k-1}T_{i-1}.$$

We let
$$\Delta^k(T)_i^- = \max(0, -\Delta^k(T)_i),$$
$$\Delta^k(T)_i^+ = \max(0, \Delta^k(T)_i).$$

We let $\nu(T)$ be the order of ideals of Hilbert function $H(R/I) = T$. For the notion of order-sequence (O-sequence), see Appendix C. A Gorenstein sequence T of socle degree j is said to be *unimodal* if $1 \le i \le j/2$ implies $t_{i-1} \le t_i$: equivalently, if T has a single locally maximum value, that may be achieved for several sequential values of i. We denote by $\mathbb{P}Gor(T)$ the projectivization of $Gor(T)$. By Lemma 2.12 we may identify $\mathbb{P}Gor(T)$ with the algebraic set of graded Gorenstein ideals I having Hilbert function T:

$$\mathbb{P}Gor(T) = \{I = \operatorname{Ann}(f) \mid f \in Gor(T)\}.$$

Given a height three Gorenstein sequence T, let $s = \max\{T_i\}$, and let $H = (T_{\le j/2}, \underline{s})$ (the first part of T augmented in high degrees by the constant sequence $\underline{s}$). We need some further notation for operations on sequences.

DEFINITION 5.24. We consider nonincreasing (resp. nondecreasing) finite sequences of integers $C = (c_1 \ge c_2 \ge c_3 \ge \cdots)$, resp. $D = (d_1 \le d_2 \le d_3 \le \cdots)$. A partition is a nonincreasing sequence of positive integers. The concatenation of two nonincreasing (resp. nondecreasing) sequences is defined in the same way as for partitions [**Macd**, §I.1], e.g. $(1, 0, 0, -1) \cup (2, 0, -1) = (2, 1, 0, 0, 0, -1)$. Similarly to partitions one has $C \supset C'$ if C' is obtained from C by discarding several elements, and the same definition is used for nondecreasing sequences $D \supset D'$. We define the operations $\vee$ and $\wedge$ on the set of sequences D, by $D_1 \vee D_2$ is the union, and $D_1 \wedge D_2$ is the intersection: thus $(1, 0, 0, -1) \vee (2, 0, -1) = (2, 1, 0, 0, -1)$ and $(1, 0, 0, -1) \wedge (2, 0, -1) = (0, -1)$. A set of sequences is a lattice if it is closed under the operations of $\vee$ and $\wedge$. If n is an integer, we denote by $[n]$ the lattice that is the simply ordered set of integers $\{0, 1, \ldots, n\}$.

When $r = 3$, the subvariety $Gor_D(T)$ is an irreducible variety — a result of Diesel [**Di**]. The argument in [**Di**, Theorem 1.1] is: fixing the degrees $D = (d_1, \ldots, d_v)$ one also fixes the degrees of the homogeneous polynomials, entries of the the matrix Ψ, according to Theorem B.2. This yields that $Gor_D(T)$ is an image of an irreducible variety, so it is a constructible set with irreducible closure in $Gor(T)$ [3]. But $Gor_D(T)$ is locally closed in $Gor(T)$ by Lemma 5.23, so is a variety.

The next theorem uses the notation and results of the Buchsbaum-Eisenbud Structure Theorem B.2, for height three Gorenstein ideals. We consider a DP-form in 3 variables, $f \in Gor(T) \subset \mathcal{D}_j$ with $T_1 = 3$, and its ideal of apolar forms $I = \mathrm{Ann}(f)$; we consider a minimal set of homogeneous generators $\phi_1, \ldots, \phi_v$ of I of degrees $d_i = \deg \phi_i$, such that the sequence $D = (d_1, \ldots, d_v)$ is nondecreasing. Comparing with Theorem B.2 we have $v = g$, and up to sign one can take for ϕ_i the i-th submaximal Pfaffians of the alternating matrix Ψ (i.e. $\phi_i = \lambda_i = (-1)^i \mathrm{Pf}(\Psi_i)$). Furthermore, $2 \le d_i \le j+1$, since the condition $T_1 = 3$ implies that $d_i, e_i \ge 2$.

THEOREM 5.25. GENERATOR-RELATION STRATA FOR IDEALS IN $\mathbb{P}Gor(T)$. (See [BE2] and [Di])

i. The sequence $T = (1, 3, \ldots, t_i, \ldots, 3, t_j = 1)$ is a Gorenstein sequence (i.e. $Gor(T)$ is nonempty) if and only if T is symmetric about $j/2$, and

$$\Delta T = (1, 2, \ldots, h_i, \ldots, -1) \quad \text{satisfies for} \quad \nu = \nu(T)$$
$$h_i = i+1 \quad \text{when} \quad i \le \nu - 1 \quad \text{and}$$
$$\nu(T) = h_{\nu-1} \ge h_\nu \ge \cdots \ge h_{[j/2]} \ge 0,$$

(i.e. $(\Delta(T)_{\le j/2}, 0, 0, \ldots)$ is an O-sequence, see Corollary C.6). Equivalently, T is Gorenstein iff $T = \mathrm{Sym}(H, j)$, where $H = H_Z$ is the Hilbert function of a 0-dimensional subscheme Z of $\mathbb{P}^2$ which satisfies $\tau(Z) \le j/2$. Also, T is unimodal, and we have $\Delta^3(T)_i = -\Delta^3(T)_{j+3-i}$ (since T is symmetric). Furthermore, let $I \in \mathbb{P}Gor(T)$ and let us consider the minimal resolution of I from Buchsbaum-Eisenbud's Theorem B.2. Then for every i, $1 \le i \le j+2$ one has

$$\Delta^3(T)_i = \#\{ \text{ relations of degree } i\} \ - \ \#\{ \text{ generators of degree } i\}.$$

ii. The minimum number of generators $\mu(T)$ for ideals in $\mathbb{P}Gor(T)$ satisfies

$$\mu(T) = \begin{cases} \sum_{i \le j+2} \Delta^3(T)_i^- & \text{if the sum is odd} \\ 1 + \sum_{i \le j+2} \Delta^3(T)_i^- & \text{otherwise.} \end{cases} \tag{5.3.1}$$

---

[3] Note that there are examples in [Sha] of such images that are not algebraic varieties

       The second case can occur only if j is odd. Also, $\Delta^3(T)_i^- = 0$ for i satisfying $i \leq \nu(T) - 1$ or $j + 4 - \nu(T) \leq i \leq j + 2$.

iii. *Unless j is odd and the sum $\sum_{i \leq j+2} \Delta^3(T)_i^-$ is even, there is an open dense subset $U_{\min}(T) \subset Gor(T)$ such that if $f \in U_{\min}(T)$ then the minimal resolution for $I = \mathrm{Ann}(f)$ has $\Delta^3(T)_i^-$ generators and $\Delta^3(T)_i^+$ relations in degree i, and the degrees of the generators and relations belong to the interval $2 \leq i \leq j + 1$.*

iv. *If j is odd, and the sum $\sum_{i \leq j+2} \Delta^3(T)_i^-$ is even, then $I = \mathrm{Ann}(f)$ for $f \in U_{\min}(T)$ has an extra generator/relation pair in degree $(j+3)/2$, in addition to those specified in (iii).*

v. *Let $D = (d_i)$ and $E = (e_i)$, $i = 1, \ldots, v$ denote the generator degrees (nondecreasing) and relation degrees (nonincreasing) of an ideal $I = \mathrm{Ann}(f)$, $f \in Gor(T)$, and let C be the nonincreasing sequence $(c_i = e_i - d_i)$. Then we have*

      a. *The integers c_i are all even or all odd; v is odd.*

      b. *$e_i + d_i = j + 3$; and $c_i = j + 3 - 2d_i = 2e_i - j - 3$.*

      c. *$c_1 > 0$ and $c_i + c_{v+2-i} > 0$ for $i = 2, \ldots, (v+1)/2$.*

      d. *Varying $f \in Gor(T)$ each adjacent sequence $C' \supset C$ differs from C by the addition of a pair of integers $(n, -n)$, $n \geq 0$.*

      e. *The pair (C, j) determines the pair (D, E).*

      f. *(S.J. Diesel [**Di**, §3]) Denote by $Gor^C(T)$ the irreducible stratum [4] of $Gor(T)$ having fixed generator-relation degrees (D, E), determined by $C = E - D$. The sequences $C = (c_i)$ that are possible for ideals $I = \mathrm{Ann}(f)$, $f \in Gor(T)$ satisfy the property that $\overline{Gor^C(T)} \supset Gor^{C'}(T)$ if and only if $C' \supset C$. There is a unique minimal sequence $C_{\min}(T)$ corresponding to $D_{\min}(T)$ from (iii) and (iv) and a unique maximal (or saturated) sequence $C_{\max}(T)$ which satisfies*

$$v_{\max} = 2\nu(T) + 1 \quad \textit{and for} \quad v = v_{\max}$$

$$c_i + c_{v+2-i} = 2 \quad \textit{for each} \quad i = 2, \ldots, (v+1)/2. \quad (5.3.2)$$

One has $C_{\max}(T) = C_{\min}(T) \cup B^+ \cup B^-$ where B^+ is a nonincreasing sequence of integers ≥ 0 and $B^- = -B^+$. Let $\mathcal{L}(T)$ be the set of all sequences C possible for ideals $I \in \mathbb{P}Gor(T)$. There is a bijective correspondence between the nonincreasing subsequences $B' \subset B^+$ (including $\emptyset$) on one hand and $\mathcal{L}(T)$ on the other, given by $B' \mapsto C_{\min}(T) \cup B' \cup (-B')$. Let $B^+ =$

---

[4] This is the same as $Gor_D(T)$ considered above; the sequences C and D determine each other by (v.b)

$(n_1, \ldots, n_1, n_2, \ldots, n_2, \ldots)$ where $n_1 > n_2 > \ldots$. Let $\kappa_i(B)$ be the length of the block $(n_i, \ldots, n_i)$. Then the number of possible sequences for $Gor(T)$ is $\prod_i(\kappa_i + 1)$. Furthermore, $\mathcal{L}(T)$ is a lattice under the operations $\vee$ and $\wedge$, with structure $\mathcal{L}(T) = \prod_i([\kappa_i])$, the lattice product of strings $[\kappa_i]$ (see Defn. 5.24).

g. (A. Conca and G. Valla [CoV1]) Let m_i be the number of times i occurs in the sequence $D_{\max}$ (equal to the number of generators of degree i for an ideal with maximal sequence $C = C_{\max}$). Then

$$
m_i = \begin{cases} \max(0, -\Delta^2(T)_i , & \text{if } \nu(T) < i \le j + 2 - \nu(T); \\ 1 - \Delta^2(T)_i , & \text{if } i = \nu(T); \\ 0 , & \text{otherwise} - \text{where } \Delta^2(T) = 1. \end{cases}
$$

$$(5.3.3)$$

NOTE ON PROOF. Most of the Theorem, in particular (i)-(iv), (v.a)-(v.c),(v.e) follows from the graded Buchsbaum-Eisenbud structure theorem [BE2, BE3] (see also Appendix B, Theorem B.2). The condition (i) is stated also in [St3]. The equivalence of the two conditions on T in (i) was discussed in Theorem 5.21(B). The generator-relation sequences (D, E) that are possible are stated in [HeTV, Di], and studied further by T. Harima [Hari3].

We give some idea of Diesel's proofs [Di]. The frontier property for the closure of Betti-strata (v.d,f) is proved in [Di, §2]. To show it, one deforms the alternating matrix Ψ of (B.2.3), for an ideal in a lower stratum (having more generators), by allowing units to appear in suitable degree-0 entries of the deformed matrix. This procedure yields that if we had a positive number of generators in both degrees $(i, j + 3 - i)$ for some i, then discarding a pair $(i, j + 3 - i)$ from D would place us in a higher stratum. Using this one shows that the minimal degree sequence is unique and equals $D_{\min}$ given in (iii) and (iv) [The identity and uniqueness of the minimal degree sequence is due to [Wa1, BE2]]. The frontier property implies now that $Gor(T) = \overline{Gor_{D_{\min}}(T)}$, and one obtains Diesel's irreducibility Theorem 1.72 (cfr. [Di, Theorem 2.7]) from the irreducibility of $Gor_D(T)$ discussed above. A sequence $\tilde{C}$ is saturated if no addition $\tilde{C} \cup (n, -n)$ leads to a possible sequence for $Gor(T)$. In [Di, Theorem 3.2] it is shown that every saturated sequence of $\tilde{v}$ elements satisfies $\tilde{c}_1 > 0$ and $\tilde{c}_i + \tilde{c}_{\tilde{v}+2-i} = 2$ for each $i = 2, \ldots, (\tilde{v} + 1)/2$, moreover each sequence C has a unique saturation $\tilde{C} \supset C$. Applying this to $C_{\min}$ one obtains $C_{\max}$ and proves that it is in fact the unique maximal sequence.

We give a proof of the Conca-Valla maximal generator Lemma (Theorem 5.25(v.g)) in Appendix B, Lemma B.13.   □

EXAMPLE 5.26. Let $T = (1, 3, 4, 4, 4, 3, 1)$. Then $\nu(T) = 2$ and

$\Delta^3(T) = (1, 0, -2, 0, 1, -1, 0, 2, 0, -1)$, so

$D_{\min}(T) = (2, 2, 5)$, $E_{\min}(T) = (7, 7, 4)$, and $C_{\min}(T) = (5, 5, -1)$,

corresponding to complete intersections. Since $v_{\max} = 2\nu(T) + 1 = 5$, the only other element of $\mathcal{L}(T)$ is the saturated sequence $C_{\max}(T) = (5, 5, 3, -1, -3)$, from which we obtain $D_{\max}(T) = (2, 2, 3, 5, 6)$, $E_{\max}(T) = (7, 7, 6, 4, 3)$. Also, Theorem 5.25(v.g), implies that the enumerator sequence for $D_{\max}(T)$ can be read off from the sequence

$$\Delta^2(T) = (1, 1, -1, -1, 0, -1, -1, 1, 1),$$

as $((1 + 1)_2, 1_3, 1_5, 1_6)$.

EXAMPLE 5.27. If $T = (1, 3, 6, 8, 8, 6, 3, 1)$ then

$$D_{\min}(T) = (3, 3, 4), \quad E_{\min}(T) = (7, 7, 6), \quad C_{\min}(T) = (4, 4, 2).$$

The saturation has length $v = 7$ and to satisfy (5.3.2) it must be

$$C_{\max}(T) = (4, 4, 2, 2, 0, 0, -2) \quad \text{so}$$
$$D_{\max}(T) = (3, 3, 4, 4, 5, 5, 6) \quad \text{and}$$
$$E_{\max}(T) = (7, 7, 6, 6, 5, 5, 4).$$

One has $C_{\max}(T) = C_{\min}(T) \cup B^+ \cup (-B^+)$ where $B^+ = (2, 0)$. Thus $\mathcal{L}(T)$ has four elements, the middle ones being $C' = (4, 4, 2, 0, 0)$ and $C'' = (4, 4, 2, 2, -2)$, respectively, corresponding to $D' = (3, 3, 4, 5, 5)$, $E' = (7, 7, 6, 5, 5)$, and to $D'' = (3, 3, 4, 4, 6)$, $E'' = (7, 7, 6, 6, 4)$, respectively

5.3.2. The morphism from $Gor(T)$: the case $T \supset (s, s, s)$.

Assume now that $H = (1, 3, h_2, \ldots, h_{\tau-1}, \underline{s})$, is a nondecreasing sequence that occurs as the Hilbert function of a 0-dimensional subscheme Z in $\mathbb{P}^2$, let $\tau = \tau(Z)$ (so $h_{\tau-1} < s$), let $j \geq 2\tau + 2$, and let $T = \mathrm{Sym}(H, j)$. By Theorem 5.25(i) $Gor(T)$ is nonempty. We have

$$T = (1, 3, h_2, \ldots, h_{\tau-1} = s - a, s, \ldots, s, s, \overset{j+1-\tau}{s - a}, \ldots, 3, 1),$$
$$\Delta^2 T = (\ldots, \overset{\tau+1}{-a}, 0, \ldots, 0, \overset{j+1-\tau}{-a}, \ldots), \tag{5.3.4}$$
$$\Delta^3 T = (1, 0, \ldots, \overset{\tau+2}{a}, 0, \ldots, 0, \overset{j+1-\tau}{-a}, \ldots, \overset{j+2}{0}, -1).$$

REMARK 5.28. Under the assumption $j \geq 2\tau + 2$, the sequence T of (5.3.4) has a constant subsequence $(s, s, \ldots, s)$ of length $j + 1 - 2\tau$, at least three. The second difference $\Delta^2 T$ has a subsequence of zeroes of length $j - 1 - 2\tau$ (at least one) beginning in degree $\tau + 2$, and ending in degree $j - \tau$. If $j - 2 - 2\tau$ is positive, then $\Delta^3 T$ has a subsequence of zeroes that occur in degrees $\tau + 3$ to $j - \tau$.

We now prove a key lemma upon which many of the results in this chapter are based. We first introduce some more notation. If I is a graded Gorenstein ideal of R having Hilbert function T, then I has a minimal set of generators $(\phi_1, \ldots, \phi_v)$ where $v = v(I) = 2u + 1$, for some integer u; the generators are ordered so that the generator degrees $d_i = \deg(\phi_i)$ satisfy $d_1 \leq \cdots \leq d_v$. The relation degrees satisfy $e_i = j + 3 - d_i$. We define $J_I = (\phi_1, \ldots, \phi_{u+1})$. Recall that if $I \subset m_R$, we denote by $(I_{\leq n})$ the ideal generated by $I_1 + \cdots + I_n$.

LEMMA 5.29. GENERATOR AND RELATION DEGREES IF T CONTAINS (s, s, s). *Suppose that $T \supset (s, s, s)$, and let a, τ be defined by (5.3.4); suppose further that $I = \mathrm{Ann}(f)$, $f \in Gor(T)$. Then the ideal I has a relations in degree $\tau + 2$, a generators in degree $j + 1 - \tau$, and no further generators and relations in degrees i, $\tau + 2 \leq i \leq j + 1 - \tau$. Furthermore, we have*

$$\nu(T) = d_1 \leq d_2 \leq \cdots \leq d_{u+1} \leq \tau + 1;$$
$$j + 2 - \tau \leq e_{u+1} \leq \cdots \leq e_2 \leq e_1; \quad and \tag{5.3.5}$$

$$e_v \leq \cdots \leq e_{u+2} = \tau + 2 < j + 1 - \tau = d_{u+2} \leq \cdots \leq d_v. \tag{5.3.6}$$

The ideal $J_I = (\phi_1, \ldots, \phi_{u+1}) = (I_{\leq(\tau+1)})$.

PROOF. We give two proofs. The first one is an expanded exposition of our original proof from [**IK**], based on Diesel's Theorem 5.25(v.f). The second one uses additionally the nice result of Conca–Valla, Theorem 5.25(v.g), which allows to shorten the proof of the lemma: but the second proof is

a. more recent, and

b. our proof of the Conca-Valla result (Lemma B.13) depends on the study of point schemes in $\mathbb{P}^2$, so its use may be seen as less elementary.

The first part of the two proofs, the case $D_{\min}(T)$, is common to both.

FIRST PROOF. It suffices to prove only the statements for the generator degrees, since the equality $e_i = j + 3 - d_i$ implies the statements for the relation degrees.

Suppose first that I has the mininum possible number of generators $v(I) = \mu(T)$. By minimality of $v(I)$ and Theorem 5.25 (ii),(v.a – v.f) the corresponding sequences of generator, relation, and C degrees for I are $D_{\min}(T), E_{\min}(T), C_{\min}(T)$ which can be calculated from $D_{\min}(T)$. Furthermore no generator degree is equal to a relation degree in $D_{\min}(T)$, except possibly in degree $i = (j+3)/2$, and then only if j is odd and the sum $\sum_{i \leq j+2} \Delta^3(T)_i^-$ is even. When $1 \leq i \leq j+2$ and $i \neq (j+3)/2$, then $\Delta^3(T)_i^-$ (or, respectively $\Delta^3(T)_i^+$) is the length of the segment $i, i, \dots, i$ in the generator degree sequence $D_{\min}$ (respectively relation degree sequence $E_{\min}(T)$). From (5.3.4) $\Delta^3(T)_{j+1-\tau} = -a$, so I has a generators in degree $j+1-\tau$. Let w denote the total number of relations in degrees $\leq \tau + 2$. From (5.3.4) we have $\Delta^2(T)_{\tau+2} = 0$, furthermore

$$\Delta^2(T)_{\tau+2} = \sum_{i \leq \tau+2} \Delta^3(T)_i = 1 + \sum_{1 < i \leq \tau+2} \Delta^3(T)_i^+ - \sum_{1 < i \leq \tau+2} \Delta^3(T)_i^-.$$

Since $w = \sum_{1 < i \leq \tau+2} \Delta^3(T)_i^+$ it follows that $\sum_{1 < i \leq t+2} \Delta^3(T)_i^- = w+1$ is the number of generators of I in degrees $\leq \tau + 2$.

By the antisymmetry of $\Delta^3(T)$ with respect to $(j+3)/2$, the number of generators in degrees $i \geq j + 3 - (\tau + 2) = j + 1 - \tau$ equals w and their degrees correspond via $\ell \mapsto j + 3 - \ell$ to the degrees of the w relations in degrees $\leq \tau + 2$. This shows $\sum_{j+1-\tau \leq i \leq j+2} \Delta^3(T)_i^- = w$. Using $\Delta^3(T)_i = 0$ for $\tau + 3 \leq i \leq j - \tau$ from (5.3.4) we conclude that $\sum_{i \leq j+2} \Delta^3(T)_i^- = 2w + 1$ is odd. Applying Theorem 5.25(ii), there is no additional generator-relation pair in degree $(j+3)/2$, hence $\mu(T) = 2w+1$ (so $w = u$ in the case considered) and there are no generators and relations in degrees i, $\tau + 3 \leq i \leq j - \tau$. Since $\Delta^3(T)_{\tau+2} = a$ from (5.3.4), I has no generators in degree $\tau + 2$. This shows the lemma in the case of I having the minimum number of generators $v(I) = \mu(T)$.

It is at this point that the second proof diverges from the first one, which we continue. The equalities $c_i = j + 3 - 2d_i$ and $d_i = j + 1 - \tau$ for $w + 2 \leq i \leq w + a + 1$ yield $(\nu = \nu(T))$

$$C_{\min} = (j + 3 - 2\nu, \dots, c_{w+1}, c_{w+2} = \cdots$$

$$\cdots = c_{w+a+1} = 2\tau + 1 - j, \dots, c_{\mu(T)}). \quad (5.3.7)$$

with $c_{w+1} \geq j + 1 - 2\tau > 0$ and c_{w+2} negative.

Let us consider now an arbitrary $I = \text{Ann}(f)$ with $f \in Gor(T)$. As in Theorem 5.25(v) let $D = (d_1, d_2, \dots, d_v)$, $d_1 \leq d_2 \leq \cdots \leq d_v$, be the sequence of generator degrees, $E = (e_1, e_2, \dots, e_v)$, $e_1 \geq e_2 \geq$

$\cdots \geq e_v$ be the sequence of relation degrees, and $C = (e_1 - d_1, \ldots, e_v - d_v)$. According to Diesel's Theorem 5.25(v.f) the sequence C is obtained from $C_{\min}$ by consecutive adding of pairs $(\alpha_m, -\alpha_m)$, $\alpha_m \geq 0$. We claim that none of these α_m is in the interval $[0, j - 2\tau - 1]$. Suppose on the contrary that there is an $\alpha = \alpha_m$ with $0 \leq \alpha \leq j - 2\tau - 1$. Then, by Theorem 5.25(v.f) there is a Gorenstein ideal $I = \text{Ann}(f')$, $f' \in Gor(T)$ such that the associated sequence C' equals $C_{\min} \cup (\alpha, -\alpha)$. Indeed, if $C = C_{\min} \cup B' \cup (-B')$ then $(\alpha) \subset B' \subset B^+$, so $C_{\min} \cup (\alpha, -\alpha)$ also belongs to $\mathcal{L}(T)$.

The assumption $0 \leq \alpha \leq j - 2\tau - 1$ implies that the sequence C' of length $v' = \mu(T) + 2 = 2w + 3$ equals

$$C' = (j + 3 - 2v, \ldots, c_{w+1}, \alpha, -\alpha, c'_{w+4} = 2\tau + 1 - j, \ldots$$
$$\ldots, c'_{w+a+3} = 2\tau + 1 - j, \ldots, c'_{2w+3} = c_{\mu(T)}).$$

However, this sequence fails the criterion of Theorem 5.25(v.c) as

$$c'_{w+2} + c'_{v'+2-(w+2)} = c'_{w+2} - c'_{w+3} = \alpha - \alpha = 0.$$

This contradiction proves our claim that every $\alpha_m \geq j - 2\tau$.

Adding a pair $(\alpha, -\alpha)$ results in adding a pair of generators/relations of degrees $\frac{1}{2}(j + 3 - \alpha)$, $\frac{1}{2}(j + 3 + \alpha)$ symmetric about $(j + 3)/2$. The inequalities $\alpha_m \geq j - 2\tau$ yield that $D = (d_1, \ldots, d_v)$ is obtained from $D_{\min}$ by adding pairs of generator degrees $(d, j + 3 - d)$ with $d \leq \tau + 1$, $j + 3 - d \geq j + 2 - \tau$. This and the result for $D_{\min}$ show the statements of the lemma for D and consequently for E. That $J_I = (\phi_1, \ldots, \phi_{u+1}) = I_{\leq(\tau+1)}$ follows from $d_{u+1} \leq \tau + 1$.

SECOND PROOF. The first part of the proof – that the lemma holds for $D_{\min}$, $E_{\min}$ – is the same. Conca-Valla's result, Theorem 5.25(v.g) simplifies the second part. Namely, the calculation of $\Delta^2(T)$ and $\Delta^3(T)$ in (5.3.4) shows that there are the same number of elements of the sequences $D_{\min}$ and $D_{\max}$ equal to i, for each integer i, $\tau + 3 \leq i \leq j + 1 - \tau$. This implies that in the process of obtaining $D_{\max}$ from $D_{\min}$ by adding pairs of generator degrees, the interval $[\tau + 2, j + 1 - \tau]$ is not affected, since each of the added pairs is symmetric about $(j + 3)/2$. This proves the lemma, since by Diesel's Theorem 5.25(v.f) one has for every $D \in \mathcal{L}(T)$, that $D \subset D_{\max}$. $\square$

EXAMPLE 5.30. GENERATOR AND RELATION DEGREES. If $T = (1, 3, 6, 8, 8, 8, 6, 3, 1)$, then $\Delta^3(T) = (1, 0, 0, -2, -1, 2, -2, 1, 2, 0, 0, -1)$ and $j = 8$, so if $v(I) = \mu(T)$ then the generator degrees $D = (3, 3, 4, 6, 6)$, and the relation degrees $E = (8, 8, 7, 5, 5)$, so $C = (5, 5, 3, -1, -1)$. We have $v_{\max}(T) = 2v(T) + 1 = 7$. By Theorem 5.25, the only other $C' \in \mathcal{L}(T)$ is $C' = (5, 5, 3, 3, -1, -1, -3)$, corresponding to $D' = (3, 3, 4, 4, 6, 6, 7)$ and $E' = j + 3 - D' = (8, 8, 7, 7, 5, 5, 4)$. Both (D, E) and (D', E') satisfy (5.3.5) and (5.3.6).

THEOREM 5.31. MORPHISM FROM $Gor(T)$ TO $Hilb^H(\mathbb{P}^2)$ WHEN $T \supset (s, s, s)$. *Assume that H as above is a nondecreasing sequence that occurs as the Hilbert function of some degree-s 0-dimensional scheme Z_T in $\mathbb{P}^2$, let $\tau = \tau(Z_T)$, let $j \geq 2\tau + 2$, and let $T = \mathrm{Sym}(H, j)$. [Equivalently, assume that $T = (1, 3, \dots)$ is a Gorenstein sequence containing the subsequence (s, s, s), and let $H = (T_{\leq j/2}, \underline{s})$]. If $f \in Gor(T)$, $I = \mathrm{Ann}(f)$ and $I = (\phi_1, \dots, \phi_{2u+1})$, where the degrees $(d_i = \deg \phi_i)$ form a nondecreasing sequence, let $J_I = (\phi_1, \dots, \phi_{u+1}) = (I_{\leq(\tau+1)})$ (according to Lemma 5.29) and let $p(f) = Z_I = \mathrm{Proj}(R/J_I)$. Then p is a well-defined morphism of irreducible varieties $p : Gor(T) \to Hilb^H(\mathbb{P}^2)$, whose image is open, dense and whose fibers are open dense subsets of the affine space $\mathbb{A}^s$. We have*

    i. *If $v(I) = 2u + 1$, then $v(J_I) = u + 1$.*

    ii. *$Z = Z_I$ is the unique tight annihilating scheme of f and J_I is its graded ideal.*

    iii. *(M. Boij [**Bo1**]). The quotient ideal $I/\mathcal{I}_Z$ defines the dualizing sheaf $\mathcal{A}_Z$ of Z.*

    iv. *The image of p includes the open dense subset $Sm^H(\mathbb{P}^2)$.*

    v. *Suppose that I has minimal resolution (by the Buchsbaum-Eisenbud structure theorem — see [**BE2**] and Appendix B Theorem B.2)*

$$0 \to R(-j-3) \xrightarrow{\lambda^T} \sum_{1 \leq i \leq v} R(-e_i) \xrightarrow{\Psi} \sum_{1 \leq i \leq v} R(-d_i) \xrightarrow{\lambda} I \to 0$$

$$(5.3.8)$$

where $v = 2u + 1$, $d_1 \leq d_2 \leq \cdots \leq d_v$, $e_i = j + 3 - d_i$, Ψ is an alternating matrix, λ is a row vector equal to $(\phi_i = (-1)^i Pf(\Psi_i))$, where $Pf(\Psi_i)$ are the submaximal Pfaffians of Ψ (cfr. Appendix B). Then J_I has minimal resolution

$$0 \to \sum_{e_i \leq \tau+2} R(-e_i) \xrightarrow{\Psi'} \sum_{d_i \leq \tau+1} R(-d_i) \xrightarrow{\lambda'} J_I \to 0, \qquad (5.3.9)$$

involving $d_1, \dots d_{u+1}, e_{u+2}, \dots, e_{2u+1}$, and the restriction maps Ψ', λ'.

PROOF. The key step in the proof is to show that J_I is the saturated graded ideal of a 0-dimensional scheme $Z = Z_I$ which has Hilbert function H, thus proving the map p is set-theoretically well-defined (we will show later p is a morphism). Once we show this, then Z is clearly a tight annihilating scheme for f, and is unique by Lemma 5.3(E iii), since $\tau(Z) = \tau$ and $j \geq 2\tau + 1$. This shows (ii). M. Boij showed in [**Bo1**] that if $j \geq 2\tau$, $f \in (\mathcal{I}_Z)_j^{\perp}$, and $H_f = \mathrm{Sym}(H_Z, j)$ then $I = \mathrm{Ann}(f)$ satisfies $I/\mathcal{I}_Z \cong \mathcal{A}_Z$, the dualizing module to Z: this is (iii). Corollary 5.4 implies that the image of p includes $Sm^H(\mathbb{P}^2)$. That $Hilb^H(\mathbb{P}^2)$ is irreducible and $Sm^H(\mathbb{P}^2)$ is open, dense is from Theorem 5.21. This shows (iv). It remains to prove the crux, that J_I is the graded ideal of a degree-s scheme Z, and that J_I has the resolution (5.3.9); and also that p is a morphism whose fibres are open dense subsets of $\mathbb{A}^s$.

According to Lemma 5.29 each of the u smallest relation degrees, namely $e_v, \ldots, e_{u+2}$ are less then each of the u highest generator degrees, namely $d_v, \ldots, d_{u+2}$ of I. This implies that the alternating matrix Ψ has the form.

$$\Psi = \begin{matrix} d_1 \\ \vdots \\ d_v \end{matrix} \begin{pmatrix} \overset{e_1 \cdots e_{u+1};}{A} & \overset{e_{u+2} \cdots e_v}{B} \\ C & D = 0 \end{pmatrix} \tag{5.3.10}$$

where the matrices A, B, D have types $(u+1) \times (u+1)$, $(u+1) \times u$, $u \times u$ and $C = -B^t$. We take $\Psi' = B$. We need to show that indeed (5.3.9) gives the minimal resolution of J_I, and that J_I is saturated, and has height 2. Let $\phi_1, \ldots, \phi_v$ be the generators of I, $\phi_i = (-1)^i \mathrm{Pf}(\Psi_i)$.

CLAIM.

a. If $1 \leq k \leq u + 1$, the generator ϕ_k is up to sign the $u \times u$ minor of B, $\phi_k = (-1)^{k + \frac{u(u-1)}{2}} \det(B_k)$, where B_k is obtained by removing the k row from B.

b. If $u + 2 \leq k \leq v$, the generator ϕ_k satisfies

$$\phi_k = \sum_{1 \leq i < j \leq u+1} (-1)^{i+j+k+\frac{u(u-3)}{2}} a_{ij} \cdot \det(B_{ij,k}), \tag{5.3.11}$$

where $B_{ij,k}$ is the matrix obtained by removing the i, j rows and k-column from B.

c. $GCD(\phi_1, \ldots, \phi_{u+1}) = 1$.

PROOF OF (a), (b). They follow from Ψ being alternating, and the assumption on T creating the 0-block D in Ψ. If $1 \leq k \leq u + 1$ then Ψ_k has a 0-block of size $u \times u$ and we may apply (B.1.10). When $u + 2 \leq k \leq v$ the alternating $2u \times 2u$ matrix Ψ_k has a 0 block of size $(u - 1) \times (u - 1)$ and we may apply the Pfaffian expansion formula (B.1.11).

PROOF OF (c). Let $g = GCD(\phi_1, \dots, \phi_{u+1})$, and suppose $\deg g > 0$. Let $A^1, \dots A^{u+1}, B^{u+2}, \dots, B^v$ denote the columns of A and B. From the relation $\Psi \lambda^T = 0$ coming from (5.3.8) being a complex, we have

$$\sum_{i=1}^{u+1} \phi_i A^i + \sum_{i=u+2}^{v} \phi_i B^i = 0, \qquad (5.3.12)$$

For every integer k, $u+2 \le k \le v$ and each pair (i, j), $1 \le i < j \le u+1$, we consider the $(u-1) \times (u-1)$ submatrix $B_{ij,k}$ of B — obtained by removing the k-column and i, j rows of B. Deleting the coordinates with numbers i, j from (5.3.12) we obtain

$$B_{ij,k} \cdot (\phi_{u+2}, \dots, \hat{\phi}_k, \dots, \phi_v)^t \equiv 0 \mod (g, \phi_k), \qquad (5.3.13)$$

so the algebraic Cramer's rule implies

$$\det(B_{ij,k}) \cdot \phi_s \equiv 0 \mod (g, \phi_k), \qquad (5.3.14)$$

for every s with $u + 2 \le s \le v$. There are 2 alternatives.

1. Either there exists a triple i, j, k, such that $\det B_{ij,k} \notin (g, \phi_k)$ or

2. $\det B_{ij,k} \in (g, \phi_k)$ for each i, j, k.

Suppose that the first alternative occurs. Consider the quotient ring $\overline{R} = R/(g, \phi_k)$. By (5.3.14), and the definition of $g = GCD(\phi_1, \dots, \phi_{u+1})$, we have that I consists of zero-divisors of the class of $\det B_{ij,k}$ in $\overline{R}$, hence there is a prime containing I that is associated to (g, ϕ_k). Since $rad(I) = M = (x_1, x_2, x_3)$ this prime ideal equals M. We want to prove this is impossible. If $g | \phi_k$, then the associated prime ideals of the principal ideal $(g, \phi_k) = (g)$ are of height 1 by Krull's theorem, so M of height 3 cannot be an associated prime. Otherwise, since $\deg(g) \le \deg(\phi_k)$, one has $g = ac$, $\phi_k = bc$ with a, b relatively prime of positive degree. In this case the graded R-module $\overline{R} = R/(ac, bc)$ has projective dimension $projdim_R(\overline{R}) = 2$, since

$$0 \to R \xrightarrow{(b, -a)} R \oplus R \xrightarrow{(ac, bc)} R \to \overline{R} \to 0$$

is a minimal resolution of $\overline{R}$. The Auslander-Buchsbaum formula (see [Ei2, Theorem 19.9 and Ex. 19.8])[5]

$$depth(M, \overline{R}) + projdim_R(\overline{R}) = \dim R \qquad (5.3.15)$$

---

[5]Recall that if N is an R-module, and I an ideal of R, then $depth(I, N) =$ the common length of maximal N-sequences in I.

yields $depth\,(M, \overline{R}) = 1$. This implies that there is a homogeneous element of positive degree $\overline{e} \in \overline{R}$ which is not a zero divisor of $\overline{R}$. Therefore M is not an associated prime of (ac, bc). [6]

It remains to consider the second alternative. Here we first prove that if $\det B_{ij,k} \in (g, \phi_k)$ for some triple, then $g|\det B_{ij,k}$. Indeed, formula (5.3.11) is nothing else, but a convenient grouping of summands in the formula for the Pfaffian $\mathrm{Pf}(\Psi_k)$ (see (B.1.1)), so we have

$$\deg(a_{ij}) + \deg(\det(B_{ij,k})) = \deg \phi_k.$$

By (5.3.5) the degrees of the entries of A are positive, hence we have $\deg(\det(B_{ij,k})) < \deg(\phi_k)$, so $\det(B_{ij,k}) \in (g, \phi_k)$ only if $g|\det(B_{ij,k})$ as claimed. As a consequence, formula (5.3.11) yields $g|\phi_k$ for every $k = u + 2, \ldots, v$ thus $I \subset (g)$, which contradicts the assumption that the height of I is 3.

We see that either of the two alternatives leads to a contradiction with the assumption that $\deg(g) > 0$. This proves (c) and the Claim.

Let us return to the proof that J_I is the saturated ideal of a 0-dimensional scheme of degree s. That $GCD(\phi_1, \ldots, \phi_{u+1}) = 1$ implies $height(J_I) \geq 2$. Since J_I is an ideal in the Cohen-Macaulay ring $R = k[x_1, x_2, x_3]$ we have $grade(J_I) = height(J_I) \geq 2$. We want to apply the Hilbert-Burch Theorem (see [Ei2, Theorem 20.15] or [BruH, Theorem 1.4.17]). Using Claim (a) we see that (5.3.9) is the complex associated to the $(u + 1) \times u$ matrix B by Laplace expansion and $J_I = I_u(\Psi')$ — the ideal generated by the $u \times u$ minors of B. That $grade(J_I) \geq 2$ implies by the Hilbert-Burch Theorem that the complex (5.3.9) is exact. So $projdim_R(R/J_I) = 2$. Now the inequalities [7]

$$2 \leq grade(J_I) \leq projdim_R(R/J_I) = 2$$

(see [BruH, Proposition 1.4.16]) imply that there is equality throughout, so J_I is a perfect ideal of height 2. Therefore $\mathrm{Proj}(R/J_I) = Z$ is a 0-dimensional scheme. By [BruH, Theorem 2.1.5], or [Ei2, Exercise

---

[6]The analogue of this statement is false for three generator quotients of $k[x_1, \ldots, x_r], r \geq 4$: there, the primary decomposition of $\overline{0}$ may have m-primary components, by [Bu2]; and the minimal resolution of $\overline{R}$ may have, after the first two steps, arbitrary Betti numbers possible for any quotient by W. Bruns' Theorem [Bru]. We thank H. Schenck and J. Watanabe for their comments, and contribution to the proof here.

[7]Recall that by definition $grade(J) = \min\{i \mid \mathrm{Ext}_R^i(R/J, R) \neq 0\}$ = the common length of all maximal R-regular sequences in J (according to Rees' Theorem [BruH, Theorem 1.2.5]). This number is also denoted by $grade(M)$, where M is the R-module R/J.

19.9], R/J_I is Cohen-Macaulay, so Lemma 1.67 implies that J_I is saturated. Now, according to Lemma 5.29 the generators and relations of I and J_I are the same in degrees $\leq \tau + 2$. This yields that the Hilbert function H_Z satisfies

$$(H_Z)_{\leq \tau + 1} = (1, 3, T_2, \ldots, T_{\tau-1}, s, s).$$

Using Theorem 1.69 we obtain $\tau(Z) = \tau$ and

$$H_Z = (1, \ldots, T_{\tau-1}, \underline{s}).$$

Thus $\deg Z = s$ and $H_Z = H$, which completes the proof of all statements about J_I in the theorem.

Let us prove that p is a morphism. For each i, $1 \leq i \leq \tau + 1$ consider the vector bundle $\mathbb{J}_i \to Gor(T)$ contained in the trivial bundle $\mathbb{R}_i = Gor(T) \times R_i$ and consisting of pairs

$$\mathbb{J}_i = \{(f, \phi) \mid \phi \in \mathrm{Ann}(f)_i\}.$$

For every $f \in Gor(T)$ if $I = \mathrm{Ann}(f)$ and Z_I is as above we have $\mathbb{J}_{\leq \tau+1}(f) = (\mathcal{I}_{Z_I})_{\leq \tau+1}$. Furthermore, $\tau(Z_I) = \tau$, $\sigma(Z_I) = \tau + 1$, so $(\mathcal{I}_{Z_I})_{\tau+1}$ generates $(\mathcal{I}_{Z_I})_{\geq \tau+1}$ according to Theorem 1.69. This implies that setting $\mathbb{J}_{i+1} = \mathbb{R}_1 \cdot \mathbb{J}_i \subset \mathbb{R}_{i+1}$ we define by induction a vector subbundle of corank s in $\mathbb{R}_i$ for every $i \geq \tau + 2$.

Let $X = Gor(T)$ and $\mathbb{P}_X^{r-1} := X \times \mathbb{P}^{r-1}$. We have that $\mathbb{P}_X^{r-1} = \mathrm{Proj}(\oplus_{i \geq 0} \mathcal{O}_X(\mathbb{R}_i))$. If $\mathcal{J}_i = \mathcal{O}_X(\mathbb{J}_i)$, then $\mathcal{J} = \oplus_{i \geq 1} \mathcal{J}_i$ is a graded $\mathcal{O}_X$-sheaf of ideals and each graded component $\mathcal{J}_i$ is locally free and a locally direct summand of corank s of $\mathcal{O}_X(\mathbb{R}_i)$. This implies that $\mathrm{Proj}(\oplus_{i \geq 0} \mathcal{O}_X(\mathbb{R}_i)/\mathcal{J})$ is a closed subscheme of $X \times \mathbb{P}^{r-1}$, which is flat over X and its Hilbert function is constant $= s$. By universality of the Hilbert scheme we obtain a corresponding morphism

$$\pi : X \longrightarrow \mathbf{Hilb}^s \mathbb{P}^2.$$

By construction for each closed point $f \in X = Gor(T)$, the corresponding 0-dimensional scheme $\pi(f)$ is Z_I, $I = \mathrm{Ann}(f)$ as above, with Hilbert function H. This implies that the image of the variety $Gor(T) = X$ is contained in the locally closed subset $Hilb^H \mathbb{P}^2 \subset (\mathbf{Hilb}^s \mathbb{P}^2)_{red}$.[8] The morphism that we obtain from $Gor(T) \to Hilb^H \mathbb{P}^2$ is nothing else but p.

We now will show that the fibre $p^{-1}(Z)$ above a point $Z \in Hilb^H(\mathbb{P}^2)$ in the image of p, is open dense in $\mathbb{A}^s$. We suppose that Z is a degree-s 0-dimensional subscheme of $\mathbb{P}^2$ with Hilbert function $H(R/\mathcal{I}_Z) = H$,

---

[8]In fact $\mathbf{Hilb}^s \mathbb{P}^2$ is smooth by Fogarty's result [**Fog**], but we do not use this fact here.

that is in the image of p, and we denote by $\mathcal{I}_Z$ its saturated defining ideal. Then Z is a tight annihilating scheme of some $f_0 \in Gor(T)$, with $f_0 \in W = (\mathcal{I}_Z)_j^{\perp}$, and W an s-dimensional subspace of $\mathcal{D}_j$, whose elements are parametrized by an affine space $\mathbb{A}^s$. Consider the complement $U = \mathcal{D}_j - V_{s-1}(\tau, j - \tau; 3)$ of the rank $\leq s - 1$ determinantal locus. Then we claim that $p^{-1}(Z) = W \cap U$. Indeed, Z is an annihilating scheme for every $f \in (\mathcal{I}_Z)_j^{\perp} = W$, so $H_f \leq H_Z$ termwise, implying $\ell\mathrm{diff}_k(f) \leq s$. The further condition $f \in U$ is equivalent to $(H_f)_\tau \geq s$, implying $\ell\mathrm{diff}_k(f) = s$, so Z is a tight annihilating scheme of f. Lemma 5.3(D,E) then shows that $H_f = \mathrm{Sym}(H_Z, j)$ since $j \geq 2\tau(Z) = 2\tau$. Consequently, $f \in W \cap U$ implies $f \in Gor(T)$, and $p(f) = Z$. Conversely, if $f \in Gor(T)$ satisfies $p(f) = Z$, then, evidently, Z is an annihilating scheme of f, and $(H_f)_\tau = s$, implying $f \in W \cap U$. That $f_0 \in W \cap U$, implies that $p^{-1}(Z)$ is open, dense in $W \cong \mathbb{A}^s$. This argument describes the image $p(Gor(T)) \subset Hilb^H(\mathbb{P}^2)$ as the open subset consisting of those Z for which $(\mathcal{I}_Z)_j^{\perp}$ is not contained in $V_{s-1}(\tau, j - \tau; 3)$. This completes the proof of the theorem. $\square$

The last part of the proof of the theorem shows,

COROLLARY 5.32. *Under the assumptions of Theorem 5.31 on H, T, the image $p(Gor(T)) \subset Hilb^H \mathbb{P}^2$ is the open subset consisting of those Z for which $(\mathcal{I}_Z)_j^{\perp}$ is not contained in $V_{s-1}(\tau, j - \tau; 3)$.*

EXAMPLE 5.33. It follows from the corollary and Example 5.10 that if $T = (1, 3, 6, 6, \ldots 6, 3, 1)$, and $Z = \mathrm{Spec}(\mathcal{O}_p/m_p^3)$ defines an order two neighborhood at a point $p \in \mathbb{P}^2$, then Z is parametrized by a point of $Hilb^H(\mathbb{P}^2)$, but is not in the image of the morphism p. From the corollary and Theorem 6.31 of Y. Cho and the first author it follows that if Z is locally Gorenstein, of Hilbert function H, then the point parametrizing Z is in the image of p.

REMARK. What is needed for the proof of the Claim in Theorem 5.31 is that if $v = v(I) = 2u + 1$, then the $v \times v$ matrix Ψ has a $u \times u$ block of zeroes; this is the maximal possible size for a zero block, if $\phi_1, \ldots, \phi_v$ are to be non-zero generators. A sufficient condition for Ψ to have such a nonzero block is

$$e_{u+2} \leq d_{u+2}, \text{ or, equivalently, since } e_{u+2} = j + 3 - d_{u+2}$$
$$d_{u+2} \geq (j + 3)/2. \tag{5.3.16}$$

However, the degree condition is not a necessary condition, as we shall see in Theorem 5.46 below.

A generalization of a portion of Theorem 5.31 is shown by A. Ragusa and G. Zappalà: they show that if $W \subset \mathbb{P}^n$ is arithmetically Gorenstein of codimension three, and satisfies $c = \Delta^{n-1} H_W(t) = \Delta^{n-1} H_W(t+1)$ (so in $\mathbb{P}^2$ they consider $I_W = \mathrm{Ann}(f)$, $f \in Gor(T)$, $T \supset (s, s+c, s+2c)$ in degrees $(t-1, t, t+1)$, respectively), then each hypersurface of degree $d \leq t+1$ through W contains a fixed hypersurface h of degree c, and furthermore, $((I_W)_{\leq t+1}) : h$ defines an aCM codimension 2 subscheme of $\mathbb{P}^n$ [**RaZ**]. See also Section 5.7 below.

The following dimension formula is a consequence of Theorem 5.31 and the Conca-Valla dimension formula, Theorem 4.26. A dimension formula for $Hilb^H(\mathbb{P}^2)$ was first shown by G. Gotzmann [**Got3**]. The statement concerning minimal number of generators is well known, as we discuss further in Section 5.5.

COROLLARY 5.34. *Let*

$$H = (1, 3, \ldots, h_i, \ldots, s-a-b, s-a, \overset{\tau}{s}, s, \ldots)$$

be a sequence equal to the Hilbert function of a certain finite set of points in $\mathbb{P}^2$. Here $a \geq 1$ and we assume $h_{-i} = 0$ for $i > 0$. Let Z be a general set of s points belonging to $Hilb^H \mathbb{P}^2$. Then the minimal resolution of the ideal $\mathcal{I}_Z$ has $\Delta^3(H)_i^- = \max(0, -\Delta^3(H)_i)$ generators and $\Delta^3(H)_i^+ = \max(0, \Delta^3(H)_i)$ relations in each degree $i \geq 1$. The dimension of $Hilb^H \mathbb{P}^2$ is given by the following formula [9]

$$\dim Hilb^H \mathbb{P}^2 =$$

$$s + 1 - \frac{1}{2}\left(\sum_{i \leq \tau}(h_i - h_{i-3})\Delta^3(H)_i + a^2 + (a+b)(b-2a)\right). \quad (5.3.17)$$

PROOF. Take any $j \geq 2\tau + 2$ and consider $T = \mathrm{Sym}(H, j)$. Then according to Theorem 5.31 a general $Z \in Hilb^H \mathbb{P}^2$ is obtained as $Z = Z_I$ for a general $f \in Gor(T)$, $I = \mathrm{Ann}(f)$. So $\mathcal{I}_Z$ has a minimal resolution as in Theorem 5.31(v) and its generators and relations equal those of I in degrees $\leq \tau + 2$. Since $(H_Z)_{\leq j-\tau} = T_{\leq j-\tau}$ (by Lemma 5.3(E)) and $j - \tau \geq \tau + 2$ we have $\Delta^3(H)_i = \Delta^3(T)_i$ for $i \leq \tau + 2$. Now our statement about generators and relations follows from Theorem 5.25((iii) and (iv)). The formula (5.3.17) is immediate

---

[9] Formulas for $\dim Hilb^H \mathbb{P}^2$ are due to Gotzmann [**Got3**],[**Got5**, Appendix 3]. We discuss the latter on page 180. The point in giving the next alternative one is its expression using $\Delta^3(H)$, which is a sequence often having many zeros. Also it demonstrates the possibility of applications of Gorenstein Artin algebras to 0-dimensional schemes.

from Conca-Valla's formula (4.4.8) for $\dim \mathbb{P}Gor(T)$, and the relation $\dim \mathbb{P}Gor(T) = s - 1 + \dim Hilb^H \mathbb{P}^2$ coming from Theorem 5.31. $\quad \square$

EXAMPLE 5.35. Consider punctual schemes Z consisting of 9 points of $\mathbb{P}^2$ imposing 8 conditions on cubics; more precisely we suppose Z has the Hilbert function $H = (1, 3, 6, 8, 9, 9, \ldots)$. One has $\Delta^3(H)_0 = 1$, $\Delta^3(H)_3 = -2$, $\Delta^3(H)_6 = 1$ and $\Delta^3(H)_i = 0$ for $i \neq 0, 3, 6$. The defining ideal $\mathcal{I}_Z$ of a general enough scheme Z parametrized by a point of $Hilb^H \mathbb{P}^2$ has minimal resolution

$$0 \longrightarrow R(-6) \longrightarrow R(-3) \oplus R(-3) \longrightarrow \mathcal{I}_Z \longrightarrow 0,$$

which is in accordance with the fact that Z is the intersection of 2 cubics; so Z, being of codimension two, is a complete intersection, and the above resolution is the Koszul resolution. The formula (5.3.17) yields

$$\dim Hilb^H \mathbb{P}^2 = 9 + 1 - \frac{1}{2}(1 \cdot 1 + (8 - 1) \cdot (-2) + 1 + 3 \cdot 0) = 16$$

which is exactly $\dim Grass(2, R_3)$, as expected. Note that if $j \geq 10$, $T = \mathrm{Sym}(H, j)$, then the image of $p : Gor(T) \to Hilb^H(\mathbb{P}^2)$ includes the complete intersections Z — even "degenerate intersections" concentrated at a single point, such as $Z = \mathrm{Proj}(R/(x^3, y^3))$.

5.3.3. Morphism: the case $T \supset (s - a, s, s, s - a)$.

Suppose now that the Gorenstein sequence T has the form

$$T = (1, 3, \ldots, s - a - b, s - a, \overset{t}{s}, s, s - a, \ldots, 3, 1), \qquad (5.3.18)$$

$a \geq 1$, with $T_{t-1} = s - a$, $T_t = s$. Here $j = 2t + 1$, $T = \mathrm{Sym}(H, j)$ with

$$H = (1, 3, \ldots, s - a - b, s - a, \overset{t}{s}, s, \ldots). \qquad (5.3.19)$$

We have

$$\begin{aligned}
\Delta^2(T) &= (1, \ldots, a - b, -a, \overset{t+2}{-a}, a - b, \ldots), \\
\Delta^3(T) &= (1, \ldots, -2a + b, \overset{t+2}{0}, 2a - b, \ldots).
\end{aligned} \qquad (5.3.20)$$

We let $v_i(f) = v_i(\mathrm{Ann}(f)) = \#$ generators of $I = \mathrm{Ann}(f)$ in degree i.

Contrary to the case of Gorenstein sequences containing (s, s, s) we will see that a general $f \in Gor(T)$ has no annihilating scheme of degree s. Our goal is to characterize intrinsically those f which have such a scheme. As a first step we calculate the generator degrees for the ideal $I = \mathrm{Ann}(f)$, for a divided power sum f with polar polyhedron a general $Z \in Hilb^H \mathbb{P}^2$.

LEMMA 5.36. *Let H be the sequence (5.3.19) and let Z be a general enough set of s distinct points in $\mathbb{P}^2$ with $H_Z = H$ (i.e. general element in $Sm^H \mathbb{P}^2$). Let $f = \sum_{u=1}^{s} c_u L_u^{[j]}$ have the polar polyhedron $Z \subset \mathbb{P}(\mathcal{D}_1) = \mathbb{P}^2$, i.e. all $c_u \neq 0$. Let $I = \mathrm{Ann}(f)$. Then*

  i. $v_{t+2}(f) = a$ *and*
  ii. $v_i(f) = \Delta^3(T)_i^-$ *for all other i, $0 < i \leq j+2$.*

PROOF. With H as in (5.3.19) we have

$$\begin{aligned}
\Delta^2(H) &= (1, \ldots, a-b, \overset{t+1}{-a}, 0, 0, \ldots) \\
\Delta^3(H) &= (1, \ldots, -2a+b, \overset{t+2}{a}, 0, 0, \ldots)
\end{aligned} \tag{5.3.21}$$

with $\Delta^3(H)_{t+2} = a$ and $\Delta^3(H)_i = \Delta^3(T)_i$ if $i \leq t+1$. By Corollary 5.34 this yields a relations in degree $t+2$ and $\Delta^3(T)_i^-$ generators or $\Delta^3(T)_i^+$ relations in degrees $i \leq t+1$. Now we use Boij's Theorem B.12, setting $X = Z,$. If $i \leq t+1$, then $j+3-i \geq t+3$ and $\mathcal{I}_Z$ has no relations in degree $j+3-i$, i.e. $w_{j+3-i}(\mathcal{I}_Z) = 0$. So we obtain $v_i(f) = v_i(\mathcal{I}_Z) = \Delta^3(T)_i^-$. If $i = t+2$, then $j+3-i = t+2$, so $v_{t+2}(\mathcal{I}_Z) = 0$, $w_{t+2}(\mathcal{I}_Z) = a$. Thus Boij's formula (B.3.4) yields $v_{t+2}(f) = a$. Similarly if $i \geq t+3$ then $v_i(f) = w_{j+3-i}(\mathcal{I}_Z) = \Delta^3(T)_{j+3-i}^+ = \Delta^3(T)_i^-$ from the antisymmetry of $\Delta^3(T)$ about $(j+3)/2$.   □

If T satisfies (5.3.18) we let

$$\begin{aligned}
Gor_{sch}(T) &= \{f \in Gor(T) \mid v_{t+2}(f) \geq a\}, \quad \text{equivalently} \\
&= \{f \in Gor(T) \mid \mathrm{cod}_{I_{t+2}} R_1 I_{t+1} \geq a\}
\end{aligned} \tag{5.3.22}$$

and we consider the subset

$$\begin{aligned}
U_a(T) = \{f \in Gor(T) \mid {}&v_{t+2}(f) = a, \text{ and} \\
&v_i(f) = \Delta^3(T)_i^- \text{ for all other } i, 0 < i \leq j+2\}
\end{aligned} \tag{5.3.23}$$

Clearly $Gor_{sch}(T)$ is closed in $Gor(T)$. According to Lemma 5.36 the subset $U_a(T)$ is nonempty. It is defined by fixing a degree sequence D_a as in (5.3.23), so in fact $U_a(T) = Gor_{D_a}(T)$, thus it is irreducible and locally closed in $Gor(T)$ by Lemma 5.23. Furthermore, Diesel's frontier result (see Theorem 5.25 v.f) implies that $Gor_{sch}(T)$ is the closure of $U_a(T)$ in $Gor(T)$.

The following lemma is the analog of Lemma 5.29 for the case $T \supset (\ldots, s-a, s, s, s-a, \ldots)$. Recall that $\mu(T)$ is the minimum possible number of generators for a Gorenstein ideal of Hilbert function T.

LEMMA 5.37. GENERATORS OF $\mathrm{Ann}(f)$, $f \in Gor_{\mathrm{sch}}(T)$. *Suppose that T is a Gorenstein sequence of the form (5.3.18). If $f_0 \in U_a(T)$ then the number of generators of $I_0 = \mathrm{Ann}(f_0)$ satisfies*

$$v(I_0) = \begin{cases} \mu(T) + a & \text{if } a \text{ is even} \\ \mu(T) + a - 1 & \text{if } a \text{ is odd.} \end{cases} \tag{5.3.24}$$

If $f \in Gor_{\mathrm{sch}}(T)$ then $I = \mathrm{Ann}(f)$ has exactly a generators in degree $t + 2$; letting $v(I) = 2u + 1$, the sequence $(d_1, \ldots, d_{2u+1})$ of generator degrees of I satisfies

$$\nu(T) = d_1 \leq d_2 \leq \cdots \leq d_{u+1} \leq t + 1;$$
$$t + 2 = d_{u+2} = \cdots = d_{u+a+1} \leq \cdots \leq d_v. \tag{5.3.25}$$

One has $J_I := (\phi_1, \ldots, \phi_{u+1}) = (I_{\leq(t+1)})$. The variety $Gor_{\mathrm{sch}}(T)$ is irreducible, and is the closure of its open subvariety $U_a(T)$.

PROOF. Suppose $g \in U_{\min}(T) \subset Gor(T)$ is a DP-form such that $\mathrm{Ann}(g)$ has the minimum possible number of generators $v(\mathrm{Ann}(g)) = \mu(T)$, so its generator degree sequence is $D_{\min}$ from Theorem 5.25(iii, iv). Let D_a be the generator degree sequence of $\mathrm{Ann}(f_0)$, $f_0 \in U_a(T)$ as defined in (5.3.23). The sequence D_a is obtained from the sequence $D_{\min}$ by adding pairs $(t + 2 - \frac{n}{2}, t + 2 + \frac{n}{2})$ with $n \in \mathbb{Z}$, $n \geq 0$ (Theorem 5.25(v.f)). By the definition of $U_a(T)$ we see that these pairs can only be $(t + 2, t + 2)$ with $n = 0$. We have $\Delta^3(T)_{t+2} = 0$, so $D_{\min}$ has in degree $t + 2$ either zero or one generator depending on whether $\sum_{0 < i \leq j+2} \Delta^3(T)_i^-$ is odd or even (Theorem 5.25(ii)). Respectively, $a = 2m + \epsilon$, $\epsilon = 0$ or 1 where m is the number of pairs added. This yields $\epsilon = 0$ if a is even and then $v(I) = \mu(T) + a$, or $\epsilon = 1$ if a is odd and then $v(I) = \mu(T) + a - 1$. This proves (5.3.24). Notice that we have obtained the following relation between the sequence $C_{\min}$ and the sequence C_a associated with $f_0 \in U_a(T)$:

$$C_a = \begin{cases} C_{\min} \cup (0,0)^{\frac{a}{2}} & \text{if } a \text{ is even} \\ C_{\min} \cup (0,0)^{\frac{a-1}{2}} & \text{if } a \text{ is odd.} \end{cases} \tag{5.3.26}$$

The proof that the generator degree sequence $(d_1, \ldots, d_v)$ of any $f \in Gor(T)$ satisfies (5.3.25) is similar to the proof of Lemma 5.29. We first verify this property for $f_0 \in U_a(T)$. We have

$$\Delta^2(T)_{t+2} = -a, \quad \Delta^3(T)_{t+2} = 0,$$

so from $\Delta^3(T)_i^+ = \Delta^3(T)_{j+3-i}^-$ we obtain

$$w := \sum_{1 \leq i \leq t+1} \Delta^3(T)_i^- = 1 + a + \sum_{t+3 \leq i \leq j+2} \Delta^3(T)_i^-. \tag{5.3.27}$$

From the definition of the generator sequence D_a, we see that the left-hand side of (5.3.27) is the number of generators of degree $\leq t+1$. The number a on the right-hand side counts the generators of degree $t+2$, and the other sum is the number of generators of degree $\geq t+3$. So the total number is $v(f_0) = 2w + 1$ and we obtain (5.3.25) for D_a.

We now need to show that if $f \in Gor_{sch}(T)$, there can be no more than a generators for $\text{Ann}(f)$ in degree $t+2$. We show this in two ways. First, if there are more then a generators in degree $t+2$, by using Theorem 5.25 Part (v) we may choose $C \in \mathcal{L}(T)$ with $C = C_a \cup (0,0)$, where $C_a = C(\text{Ann}(f_0))$, $f_0 \in U_a(T)$. An inspection shows that

$$C_a = (c_1, \ldots, c_{w+1}, c_{w+2} = 0, \ldots, c_{w+a+1} = 0, \ldots, c_{2w+1}), \quad (5.3.28)$$

whence the addition of a $(0,0)$ pair violates Theorem 5.25 (v.c). This contradiction shows every $f \in Gor_{sch}(T)$ has $v_{t+2} = a$.

Alternatively, it is a direct result of the Conca-Valla criterion of Theorem 5.25(v.g), and the Diesel-Harima result that $D \subset D_{max}(T)$, that there can be no more than a generators of degree $t+2$ for an ideal $I \in \mathbb{P}Gor(T)$.

We now complete the proof of (5.3.25). If $f \in Gor_{sch}(T)$ is arbitrary, and C is the associated sequence $(e_i - d_i)$, then $C = C_a \cup B$. The fact that $v_{t+2}(f) = a$ implies that $B = (B', -B')$, where B' does not contain zeroes. Adding consecutively to C_a pairs $(n, -n)$, $n > 0$ results in adding pairs of generator degrees $(t+2-\frac{n}{2}, t+2+\frac{n}{2})$, so the condition (5.3.25) remains valid, thus proving it for all $f \in \overline{Gor}_{sch}(T)$.

Finally from Theorem 5.25(v.f) we see that $Gor_{sch}(T) = \overline{Gor_{D_a}(T)}$ (which is $\overline{U_a(T)}$), hence $Gor_{sch}(T)$ is irreducible. $\square$

LEMMA 5.38. *Let $Z \subset \mathbb{P}^2$ be a zero dimensional subscheme with Hilbert function*

$$H_Z = H = (1, 3, \ldots, s-a, \overset{\tau}{s}, s, \ldots)$$

with $a \geq 1$, $\tau(Z) = \tau$. Consider a minimal resolution of the graded ideal $\mathcal{I}_Z$, which according to the Hilbert-Burch theorem may be given by a $(u+1) \times u$ matrix M, and its maximal minors S.

$$0 \to \sum_{i=1}^{u} R(-e_i) \xrightarrow{M} \sum_{i=1}^{u+1} R(-d_i) \xrightarrow{S} \mathcal{I}_Z \to 0, \quad (5.3.29)$$

Then

i. $\Delta^3(H)_i = \#$ *of relations* $- \#$ *generators in degree i, for each $i \geq 1$.*

ii. *All relations of $\mathcal{I}_Z$ are in degree $\leq \tau + 2$ and there are a relations in degree $\tau + 2$.*

PROOF. (i) This is well known to algebraists: its root is the point of the Hilbert syzygy theorem that H_Z is the alternating sum of the Hilbert functions of the (shifted) free modules occuring in the minimal resolution of $\mathcal{I}_Z$, which Hilbert used to show that the Hilbert function is a polynomial in high degrees (here a constant s): see [Hi1], or also [Ei2, p. 45]). The claim follows from applying the third (r-th) order difference operator Δ^3, to the equality $H_Z = \sum H(R)_{-e_i} - \sum H(R)_{-d_i}$, and noting that for the shifted free module R_{-e}, we have $\Delta^3 H(R) = (\ldots 0, 1_e, 0, \ldots)$, with 1 in degree e.

(ii). The sequences $\Delta^2(H)$ and $\Delta^3(H)$ were calculated in (5.3.21). We have $\Delta^2(H)_{\tau+2} = 0$, so

$$1 = \sum_{1 \leq i \leq \tau+2} -\Delta^3(H)_i = \# \text{ generators } - \# \text{ relations in degree } \leq \tau + 2.$$

Since all generators are in degree $\leq \tau + 1$ by Theorem 1.69 and by the Hilbert-Burch theorem the number of relations is one less then the number of generators we see that all relations are in degrees $\leq \tau + 2$. The equality $\Delta^2(H)_{t+1} = -a$ yields similarly

$$1 = (\# \text{ generators } - \# \text{ relations in degree } \leq \tau + 1) - a.$$

This shows that $\mathcal{I}_Z$ has a generators in degree $\tau + 2$. $\qquad \square$

THEOREM 5.39. MORPHISM FROM $Gor_{sch}(T)$ TO $Hilb^H(\mathbb{P}^2)$, WHEN T CONTAINS THE SUBSEQUENCE $(s-a, s, s, s-a)$. *Suppose that T satisfies (5.3.18), so $T = \mathrm{Sym}(H, j)$, where H is the Hilbert function of some degree-s scheme Z_T with $t = \tau(Z_T)$, and $j = 2t + 1$. Let $Gor_{sch}(T)$ be the irreducible variety defined above. Then $p : f \mapsto \mathrm{Proj}(R/J_I)$, $I = \mathrm{Ann}(f)$, $J_I = (I_{\leq(t+1)})$ defines a morphism $Gor_{sch}(T) \rightarrow Hilb^H(\mathbb{P}^2)$ which has open, dense image and the fibers are open, dense subsets in $\mathbb{A}^s$. Furthermore the morphism p satisfies the properties (i)–(v) of Theorem 5.31. No $f \in Gor(T) - Gor_{sch}(T)$ has an annihilating scheme of degree $\leq s$.*

PROOF. Let $f \in Gor_{sch}(T)$. According to Lemma 5.37 if $v = 2u+1$ is the number of generators of $I = \mathrm{Ann}(f)$, then $d_{u+2} = t + 2 = (j + 3)/2$. Thus condition (5.3.16) is satisfied, so in the minimal resolution of I, determined by an alternating matrix Ψ, one has the zero $u \times u$ block as in (5.3.10). The proof of Theorem 5.39, concerning the morphism p, is like that of Theorem 5.31.

Suppose $f \in Gor(T)$ has an annihilating scheme Z of degree $\leq s$. We want to prove that this implies $f \in Gor_{sch}(T)$. By Lemma 5.3(D,E) we have $\deg(Z) = s$, $(\mathcal{I}_Z)_{\leq t+1} = \mathrm{Ann}(f)_{\leq t+1}$, so $H_Z = H$, $\tau(Z) = t$.

Considering a minimal resolution of $I = \text{Ann}(f)$, we see that the generators of I in degrees $\leq t+1$ are the generators of $\mathcal{I}_Z$, say $\phi_1, \ldots, \phi_{u+1}$. Each relation of I in degree $t+2$ has the form $\psi_{i1}\phi_1 + \psi_{i2}\phi_2 + \cdots$ where every $\deg \psi_{ik} > 0$ since the resolution is minimal, so no generators of I of degree $t+2$ enter into this linear combination. This implies that the relations of I and $\mathcal{I}_Z$ in degree $t+2$ are the same, so I has a relations in degree $t+2$ according to Lemma 5.38. Now $\Delta^3(T)_{t+2} = 0$, so I has a generators in degree $t+2$ by Theorem 5.25(i) which by definition means f belongs to $Gor_{\text{sch}}(T)$.   □

REMARK. Theorem 5.39 together with (5.3.17) yields the following formula for the dimension of $Gor_{\text{sch}}(T)$, T as in (5.3.18).

$$\dim Gor_{\text{sch}}(T) =$$

$$2s + 1 - \frac{1}{2}\left(\sum_{i \leq t}(T_i - T_{i-3})\Delta^3(T)_i + a^2 + (a+b)(b-2a)\right). \quad (5.3.30)$$

Comparing with (4.4.9), we see that the codimension of $Gor_{\text{sch}}(T)$ in $Gor(T)$ is $a(a-1)/2$. This is in accordance with a formula of M. Boij (cf. (4.4.19)).

5.3.4. Morphism: the case $T \supset (s-a, s, s-a)$.

Suppose now that the Gorenstein sequence T has the form

$$T = (1, 3, \ldots, s-a-b, s-a, s, s-a, \ldots, 3, 1), \quad (5.3.31)$$

$a \geq 1$, with $T_{t-1} = s-a, T_t = s$ so $j = 2t$, and

$$\Delta^2(T) = (\ldots, a \overset{t}{-} b, -2a, a-b, \ldots),$$

$$\Delta^3(T) = (1, \ldots, \overset{t+1}{-3a}+b, 3a-b, \ldots) \quad (5.3.32)$$

Here $H = (T_{\leq t}, \underline{s}) = (1, 3, \ldots, s-a-b, s-a, \overset{t}{s}, s, \ldots)$. That ΔH is an O-sequence (Theorem 5.25(i)) implies $b \geq a$.

As in the previous case we want first to calculate the resolution data for $I = \text{Ann}(f)$, where f is a divided power sum whose polar polyhedron Z is a general enough set of distinct points in $\mathbb{P}^2$ having Hilbert function H — thus, Z is parametrized by a general point of $Hilb^H \mathbb{P}^2$.

LEMMA 5.40. *Let $T = \text{Sym}(H, j)$ be as above. Let Z be a general enough set of s points in $\mathbb{P}^2$ with $H_Z = H$. Let $f = \sum_{u=1}^{s} L_u^{[j]}$ have polar polyhedron Z and let $I = \text{Ann}(f)$.*

i. If $b < 2a$ then $v_i(f) = \Delta^3(T)_i^-$ for every $0 < i \le j + 2$, here the generator degree sequence of $\mathrm{Ann}(f)$ is that of a general element in $Gor(T)$.

ii. If $b \ge 2a$ then $v_{t+1}(f) = a$, $v_{t+2}(f) = b - 2a$ and $v_i(f) = \Delta^3(T)_i^-$ for every $i \ne t + 2, t + 3$, $0 < i \le j + 2$.

PROOF. It is similar to the proof of Lemma 5.36, using $\Delta^3(H)_{t+1,t+2} = (b - 2a, a)$ from (5.3.21). $\qquad\square$

REMARK 5.41. One of the main points in Theorems 5.31 and 5.39 is that we find necessary and sufficient conditions for $f \in Gor(T)$ to have an annihilating scheme of degree s; and, moreover, when it exists, the graded ideal of such a scheme is generated by the apolar forms of degree $\le \tau + 1$, where $\tau = \min\{i \mid (H_f)_i = s\}$. In other words Z is unique and can be recovered explicitly from f (cf. Lemma 5.3(E), the case $j \ge 2\tau + 1$). In the case we consider now we have $\tau = t$, $j = 2\tau$, so in order to obtain similar results a natural restriction should be that $\mathcal{I}_Z$ is generated by $(\mathcal{I}_Z)_{\le t}$, since then $\mathcal{I}_Z$ would be generated by $\mathrm{Ann}(f)_{\le t}$ (cf. Lemma 5.3(E.iii)). If $Z \in Hilb^H \mathbb{P}^2$ is general then, according to Corollary 5.34, $v_{t+1}(\mathcal{I}_Z) = 2a - b$ if $b < 2a$ and $v_{t+1}(\mathcal{I}_Z) = 0$ if $b \ge 2a$. The latter condition is thus a natural restriction to obtain results similar to those of Theorems 5.31 and 5.39.

Let us assume $b \ge 2a$ and let D_a be the degree sequence defined in Lemma 5.40(ii). Let

$$U_a(T) = Gor_{D_a}(T) = \{f \in Gor(T) \mid v_{t+1}(f) = a, \text{ and}$$
$$v_i(f) = -(\Delta^3(T)_i^-) \text{ for all other } i, \ 0 < i \le j + 2\}. \quad (5.3.33)$$

For T satisfying (5.3.31), we let

$$Gor_a(T) = \{f \in Gor(T) \mid v_{t+1}(f) = a\}.$$

The set $Gor_a(T)$ is open in the closed subset of $Gor(T)$ determined by the condition $v_{t+1}(f) \ge a$, thus $Gor_a(T)$ is locally closed in $Gor(T)$, and it is the union of Betti-strata. That $Gor_a(T)$ is not closed in $Gor(T)$ follows from $\Delta^2(T)_{t+1} = -2a$: by the Conca-Valla Theorem 5.25(v)(g), there are $f \in Gor(T)$ with $v_{t+1}(f) = 2a$.

LEMMA 5.42. GENERATORS OF $\mathrm{Ann}(f)$, $f \in Gor_a(T)$. Let T be a Gorenstein sequence of the form (5.3.31) and suppose $b \ge 2a$. Consider the minimal resolution of $I = \mathrm{Ann}(f)$. Suppose $v_{t+1}(f) = a$. Then $v_{t+2}(f) = b - 2a$ and the number of relations in degrees $t + 1$, $t + 2$ is $w_{t+1}(f) = b - 2a$, $w_{t+2}(f) = a$ respectively. Let $D = (d_1, \dots, d_v)$, $v =$

$2u + 1$ be the generator degrees of I. Then

$$\nu(T) = d_1 \leq \cdots \leq d_{u+1} \leq t;$$

$$t + 1 = d_{u+2} = \cdots = d_{u+a+1} < \qquad (5.3.34)$$

$$< d_{u+a+2} = \cdots = d_{u+b-a+1} = t + 2 \leq \cdots \leq d_v,$$

where the segment $d_{u+a+2} = \cdots = d_{u+b-a+1} = t + 2$ enters only if $b - 2a \geq 1$. The variety $Gor_a(T)$ is irreducible and $U_a(T)$ is open, dense in $Gor_a(T)$. The ideal $J_I := (\phi_1, \ldots, \phi_{u+1}) = (I_{\leq t})$.

PROOF. The condition $v_{t+1}(f) = a$ means the sequence D has a segment $(t+1), \cdots, (t+1)$ of length a, which implies that the relation degree sequence E has a segment $(t+2), \cdots, (t+2)$ of length a, (since $e_i - d_i = j + 3 = 2t + 3$). Thus f has $w_{t+2}(f) = a$ relations in degree $t + 2$. We have from Theorem 5.25(i)

$$w_{t+2}(f) - v_{t+2}(f) = \Delta^3(T)_{t+2} = 3a - b.$$

Hence $v_{t+2}(f) = b - 2a$. Similarly $w_{t+1}(f) = b - 2a$. The proof of (5.3.34) is similar to that of Lemmas 5.29 and 5.37. Namely, for the sequence D_a we have from $\Delta^2(T)_t = a - b$ that

$$w + 1 := \sum_{0 < i \leq t} \Delta^3(T)_i^- = 1 + b - a + \sum_{t+3 \leq i \leq j+2} \Delta^3(T)_i^-. \qquad (5.3.35)$$

Noticing that for $f_0 \in U_a(T) = Gor_{D_a}(T)$ we have $b - a = v_{t+1}(f_0) + v_{t+2}(f_0)$ we conclude that (5.3.34) holds for every $f_0 \in U_a(T)$, and that $u = w$ from (5.3.35). Furthermore, if C_a is the sequence from Theorem 5.25(v) associated with f_0 we have

$$C_a = \begin{cases} C_{\min} \cup (1, -1)^{b-2a} & \text{if } b \in [2a, 3a] \\ C_{\min} \cup (1, -1)^a & \text{if } b \geq 3a. \end{cases} \qquad (5.3.36)$$

If $f \in Gor_a(T)$ is arbitrary, then as we showed above $v_{t+1}(f) = a$, $v_{t+2}(f) = b - 2a$, so if C is the sequence associated to f we have $C = C_a \cup (B', -B')$ where all integers in the nonincreasing sequence B' are ≥ 2. This shows (5.3.34) and with Theorem 5.25(v.f) proves that $Gor_a(T) = \overline{Gor_{D_a}(T)} \cap Gor_a(T)$, and $Gor_{D_a}(T)$ is open in $Gor_a(T)$. Hence $Gor_a(T)$ is irreducible. $\qquad \square$

REMARK 5.43. The key argument in Theorems 5.31 and 5.39, which prove the existence of a tight annihilating scheme, requires that the alternating $v \times v$ matrix Ψ, $v = 2u + 1$ which determines the minimal resolution of $I = \text{Ann}(f)$, has a zero block of size $u \times u$ as in (5.3.10). For general $f \in Gor_a(T)$ this condition is not satisfied unless $a = 1$. Indeed, the inequalities of (5.3.34) imply, using $e_i = 2t + 3 - d_i$

for $i \geq u+2$, that all entries in the $u \times u$ block with rows and columns $i, k = u + 2, \ldots, v$ are zero unless the entries are in the alternating $a \times a$ submatrix with rows and columns $i, k = u + 2, \ldots, u + a + 1$ which are of degree 1. The condition that this $a \times a$ block equals zero is equivalent to the following one.

Each relation of $I = \mathrm{Ann}(f)$ *of degree* $t + 2$

$$\psi_1 \phi_1 + \cdots + \psi_{u+1} \phi_{u+1} + \psi_{u+2} \phi_{u+2} + \cdots + \psi_{u+a+1} \phi_{u+a+1} \quad (5.3.37)$$

where $\deg \psi_i = t + 2 - \deg \phi_i$ *satisfies* $\psi_{u+2} = \cdots = \psi_{u+a+1} = 0$.

We define a subset of $Gor_a(T)$ as follows

$$Gor_{\mathrm{sch}}(T) = \{ f \in Gor(T) \mid v_{t+1}(f) = a, \quad \text{and}$$
$$I = \mathrm{Ann}(f) \quad \text{satisfies Condition (5.3.37)} \} \quad (5.3.38)$$

LEMMA 5.44. *Let T be the sequence (5.3.31) and suppose $b \geq 2a$. Let $f \in Gor_a(T)$. Then Condition (5.3.37) is equivalent to*

$$\mathrm{cod}_{R_{t+2}} R_2 \cdot I_t = s. \quad (5.3.39)$$

The subset $Gor_{\mathrm{sch}}(T)$ is closed in $Gor_a(T)$ and has codimension $\leq 3a(a-1)/2$. [10]

PROOF. Let ϕ_i, $1 \leq i \leq u+a+1$ be the generators of $I = \mathrm{Ann}(f)$ of degree $\leq t+1$, as in Lemma 5.42. Let $V = \langle \phi_{u+2}, \ldots, \phi_{u+a+1} \rangle$. Then $I_{t+1} = R_1 I_t \oplus V$ and Condition (5.3.37) is equivalent to

   a. $R_1 \otimes_k V \to I_{t+2}$ is an injection,
   b. $R_2 I_t \cap R_1 V = 0$.

These two conditions are equivalent to $R_2 I_t$ having codimension $3a$ in $R_1 I_{t+1}$. The codimension of $R_1 I_{t+1}$ in I_{t+2} is $v_{t+2}(f) = b - 2a$ by Lemma 5.42, so Condition (5.3.37) is equivalent to $\mathrm{cod}_{I_{t+2}} R_2 I_t = a+b$. Since $H(R/I)_{t+2} = s - a - b$ this is the same as (5.3.39).

Let $\mathbb{J}_i$ be the vector bundles over $Gor_a(T)$ with fibers $\mathbb{J}_i(f) = \mathrm{Ann}(f)_i$, $i = 1, \ldots, j-1$. Consider the multiplication map

$$\mu_2 : R_2 \otimes_k \mathbb{J}_t \longrightarrow \mathbb{J}_{t+2}.$$

The above considerations show that Conditions (a) and (b) are equivalent to the condition that the corank of μ_2 at the point f is the maximal possible one $a+b$. Thus $Gor_{\mathrm{sch}}(T)$ is the corank $\geq a+b$ determinantal locus of the bundle map μ_2, hence it is closed in $Gor_a(T)$.

For calculating the inequality for the codimension it suffices to intersect $Gor_{\mathrm{sch}}(T)$ with the open subset $\mathrm{U}_a(T)$. Here the generator

---

[10] See also Remark 5.47

degree sequence is fixed, and so is the type — degrees of the entries — of the alternating matrix Ψ giving the resolution of an ideal $I = \text{Ann}(f)$, $f \in U_a(T)$. Condition (5.3.37), which is the vanishing of the linear entries of the $a \times a$ alternating block, is the same as the vanishing of $3a(a-1)/2$ coefficients. This shows that the codimension of $Gor_{\text{sch}}(T) \cap U_a(T)$ in $U_a(T)$ is $\leq 3a(a-1)/2$. $\square$

Our next lemma implies that Condition (5.3.37) is satisfied by divided power sums of Hilbert function T.

LEMMA 5.45. *Let T be a Gorenstein sequence as in (5.3.31) and suppose $b \geq 2a$. Suppose $f \in Gor(T)$ has an annihilating scheme Z of degree s, such that $\mathcal{I}_Z$ is generated by $(\mathcal{I}_Z)_{\leq t}$. Then $f \in Gor_{\text{sch}}(T)$.*

PROOF. We have to prove that $v_{t+1}(f) = a$ and f satisfies (5.3.39). By hypothesis $\mathcal{I}_Z \subset \text{Ann}(f) = I$ and from Lemma 5.3(E) $(\mathcal{I}_Z)_{\leq t} = I_{\leq t}$. The assumption that $\mathcal{I}_Z$ is generated by degree $\leq t$ forms, equivalent to $v_{t+1}(\mathcal{I}_Z) = 0$, implies $R_1 I_t = (\mathcal{I}_Z)_{t+1}$. Now $H(R/\mathcal{I}_Z)_{t+1} = s$, $H(R/I)_{t+1} = s - a$, therefore $v_{t+1}(f) := \text{cod}_{I_{t+1}} R_1 I_t = a$. Next $R_2 I_t = R_2(\mathcal{I}_Z)_t = (\mathcal{I}_Z)_{t+2}$ and $H(R/\mathcal{I}_Z)_{t+2} = s$. Thus, Condition (5.3.39) holds. $\square$

Let $H = (1, 3, \ldots, s - a - b, s - a, \overset{t}{s}, s, \ldots)$, $b \geq 2a$. We denote by $(Hilb^H \mathbb{P}^2)_0$ the subset of $Hilb^H \mathbb{P}^2$ consisting of Z with $v_{t+1}(\mathcal{I}_Z) = 0$ and by $(Sm^H \mathbb{P}^2)_0$ its subset consisting of sets of s points. Both are open, dense in $Hilb^H \mathbb{P}^2$ (see Remark 5.41 and Corollary 5.34)

THEOREM 5.46. MORPHISM FROM $Gor_{\text{sch}}(T)$ TO $Hilb^H(\mathbb{P}^2)$, WHEN T CONTAINS THE SUBSEQUENCE $(s - a - b, s - a, s, s - a)$ AND $b \geq 2a$. *Suppose that T satisfies (5.3.31), so $T = \text{Sym}(H, j)$, where H is the Hilbert function of some degree-s scheme Z_T with $t = \tau(Z_T)$, and $j = 2t$. Suppose $b \geq 2a$. Let $Gor_{\text{sch}}(T)$ be the variety defined in (5.3.38) and Lemma 5.44. Then $p : f \mapsto \text{Proj}(R/J_I)$, $I = \text{Ann}(f)$, $J_I = (I_{\leq t})$ defines a morphism $p : Gor_{\text{sch}}(T) \to (Hilb^H \mathbb{P}^2)_0$ with open, dense image whose fibers are open, dense subsets in $\mathbb{A}^s$. The variety $Gor_{\text{sch}}(T)$ is irreducible. Properties (i),(ii) and (v) of Theorem 5.31 are satisfied and instead of Property (iv) we have: the image of p includes $(Sm^H \mathbb{P}^2)_0$. No $f \in Gor_a(T) - Gor_{\text{sch}}(T)$ has annihilating schemes of degree $\leq s$ and no $f \in Gor(T) - Gor_{\text{sch}}(T)$ has annihilating schemes of degree $\leq s$ whose graded ideal is generated by forms of degree $\leq t$.*

PROOF. Analogous to the proof of Theorem 5.31 — if $f \in Gor_{\text{sch}}(T)$ then the alternating matrix Ψ has a zero block as in (5.3.10). The only point that differs is the following. After proving that J_I is the saturated ideal of a zero dimensional scheme Z, we need to verify $(H_Z)_{t+1} = s$ in order to conclude $\deg Z = s$. We have $(\mathcal{I}_Z)_{t+1} = R_1(\mathcal{I}_Z)_t$ since J_I is generated by polynomials of degree $\leq t$. Since $I_t = (\mathcal{I}_Z)_t$ and $v_{t+1}(f) = a$ we have $\mathrm{cod}_{I_{t+1}} R_1(\mathcal{I}_Z)_t = a$. Furthermore $H(R/I)_{t+1} = s - a$, hence $\mathrm{cod}_{R_{t+1}} R_1(\mathcal{I}_Z)_t = s$, which proves $(H_Z)_{t+1} = s$.

That the image of p contains $(Sm^H\mathbb{P}^2)_0$ follows from Lemma 5.45 since every divided power sum f with polar polyhedron $Z \in (Sm^H\mathbb{P}^2)_0$ belongs to $Gor_{\mathrm{sch}}(T)$. The variety $Gor_{\mathrm{sch}}(T)$ is irreducible since its image and every fiber are irreducible.

Suppose $f \in Gor_a(T)$ has an annihilating scheme of degree $\leq s$. Then $\deg Z = s$ and $(\mathcal{I}_Z)_{\leq t} = I_{\leq t}$ where $I = \mathrm{Ann}(f)$ by Lemma 5.3 (D,E). That $v_{t+1}(f) = a$ implies $\bar{R}_1 I_t$ has codimension s in R_{t+1} (since $H(R/I)_{t+1} = s - a$), therefore $R_1(\mathcal{I}_Z)_t = (\mathcal{I}_Z)_{t+1}$, and we may apply Lemma 5.45, which also yields the last statement of the Theorem. $\square$

REMARK 5.47. CODIMENSION OF $Gor_{\mathrm{sch}}(T)$ IN $Gor(T)$. M. Boij announced in [**Bo4**] a formula (4.4.19) for the codimension of $Gor_D(T)$ in $Gor(T)$. Using Lemma 5.42, equation (5.3.36), that $v_{t+1} = a, v_{t+2} = b - 2a$ for $f \in Gor_a(T)$, and applying his formula we conclude that the codimension of $Gor_a(T)$ in $Gor(T)$ is $a(b - 2a)$. On the other hand the dimension of $Gor_{\mathrm{sch}}(T)$ is $s + \dim Hilb^H\mathbb{P}^2$ according to Theorem 5.46. Comparing formulas (4.4.10) and (5.3.17) we see that the codimension of $Gor_{\mathrm{sch}}(T)$ in $Gor_a(T)$ is $3a(a-1)/2$, so the inequality in Lemma 5.44 is in fact equality. We obtain

$$\mathrm{cod}_{Gor(T)} Gor_{\mathrm{sch}}(T) = a(b - 2a) + \frac{3}{2}a(a - 1).$$

Notice the case $a = 1$, $b = 2$ which is the unique one in which $Gor_{\mathrm{sch}}(T)$ is open in $Gor(T)$.

EXAMPLE. When $T = (1, 3, 6, 9, 10, 9, 6, 3, 1)$ where $t = 4$, $j = 8$, $a = 1$, $b = 3$, $C_{\min}(T) = (5, 3, 3)$, $C_a(T) = (5, 3, 3, 1, -1)$, $C_{\max}(T) = (5, 3, 3, 1, 1, -1, -1)$ corresponding to $D_{\min}(T) = (3, 4, 4)$, $D_a(T) = (3, 4, 4, 5, 6)$ and $D_{\max}(T) = (3, 4, 4, 5, 5, 6, 6)$. Here $D_{\max}(T)$ no longer satisfies $(\phi_1, \ldots, \phi_{u+1}) = I_{\leq t}$, with $v(I) = 2u + 1$, neither $(I_{\leq t+1})$ is generated by $u + 1 = 4$ elements. In fact we show in the next remark that there is no $f \in Gor_{D_{\max}}(T)$ with the property that we could choose the 7 generators of $I = \mathrm{Ann}(f)$ in such a way that the alternating 7×7 matrix Ψ has a 3×3 zero block as in (5.3.10). This is one reason for the narrow Definition of $Gor_{\mathrm{sch}}(T)$ in (5.3.38), rather than a broad definition as in (5.3.22).

REMARK 5.48. The aim of this remark is to show that the method exploited in this section is not applicable to the remaining cases when $j = 2t$, $T \supset (\ldots, s - a, s, s - a, \ldots)$, $a > 0$, namely when $b \geq 2a$ and $v_{t+1}(f) = a' > a$, or when $b < 2a$. In the latter case a general $f_0 \in Gor(T)$ has $v_{t+1}(f_0) = 3a - b > a$ by Lemma 5.40, so every $f \in Gor(T)$ has $v_{t+1}(f) = a' > a$ for some a'. Notice also that $v_{t+1}(f) > a$ whenever f has an annihilating scheme Z of degree s such that $\mathcal{I}_Z$ is not generated by $(\mathcal{I}_Z)_{\leq t}$. Indeed, one has $I_t = (\mathcal{I}_Z)_t$, $H(R/\mathcal{I}_Z)_{t+1} = s$ and $H(R/I)_{t+1} = s - a$, so $v_{t+1}(f) = \mathrm{cod}_{I_{t+1}} R_1 I_t = v_{t+1}(\mathcal{I}_Z) + a > a$.

We claim that for no element $f \in Gor(T)$ with $v_{t+1}(f) = a' > a$, can one choose the $v = 2u + 1$ generators of $I = \mathrm{Ann}(f)$ such that the alternating $v \times v$ matrix Ψ, giving the resolution of I, has a zero $u \times u$ block as in (5.3.10). Suppose by way of contradiction that such an f exists yielding a zero block in Ψ. Then, by the arguments of Theorem 5.31, the forms $\phi_1, \ldots, \phi_{u+1}$ generate the ideal $\mathcal{I}_Z$ of an annihilating scheme. We claim that Z has degree s. Indeed our assumption $v_{t+1}(f) = a'$ implies by the symmetry of generator-relations about $(j + 3)/2 = t + 1.5$, that I has $w_{t+2}(f) = a'$ relations in degree $t + 2$. Hence

$$v_{t+2}(f) = w_{t+2}(f) - \Delta^3(T)_{t+2} = a' - (3a - b) = b - 2a + a' - a.$$

Representing $v_{t+2}(f)$ as $(a' - a) + (b - 2a)$, using (5.3.35) and applying the arguments that follow it in the proof of Lemma 5.42 — adding (a'-a) to both v_{t+1} and v_{t+2} there, so now $u = w + (a' - a)$ — we conclude that the set $\{\phi_1, \ldots, \phi_{u+1}\}$ consists of all generators from the set $\{\phi_1, \ldots, \phi_v\}$ having degree $\leq t$ and of $a' - a$ generators of degree $t + 1$. Thus if $J_I = (\phi_1, \ldots, \phi_{u+1})$ we have $(J_I)_t = I_t$ and $(J_I)_{t+1}$ is of codimension a in I_{t+1}. Since $H(R/I)_{t,t+1} = (s, s - a)$ this implies by $J_I = \mathcal{I}_Z$ that $H(R/\mathcal{I}_Z)_{t,t+1} = (s, s)$. This implies by Theorem 1.69 that $\deg Z = s$. We have $I_{t+1} = (\mathcal{I}_Z)_{t+1} \oplus V$, where $V = \langle \phi_{u+2}, \ldots, \phi_{u+a+1} \rangle$. Arguments similar to those of Remark 5.43 may be applied, so the condition of the existence of a zero $u \times u$ block is equivalent to Condition (5.3.37). As in Lemma 5.44 the latter Condition is equivalent to

$$\mathrm{cod}_{R_1 I_{t+1}} R_1(\mathcal{I}_Z)_{t+1} = 3a. \tag{5.3.40}$$

Now $\mathrm{cod}_{I_{t+2}} R_1 I_{t+1} := v_{t+2}(f) = b - 2a + a' - a$ as shown above. From $H(R/I)_{t+2} = s - a - b$ we obtain $\mathrm{cod}_{R_{t+2}} R_1 I_{t+1} = s - 3a + a' - a$. So if (5.3.40) holds then $\mathrm{cod}_{R_{t+2}} R_1(\mathcal{I}_Z)_{t+1} = s + a' - a$ which is impossible since $H(R/\mathcal{I}_Z)_{t+2} = s$ and $(\mathcal{I}_Z)_{t+2} = R_1(\mathcal{I}_Z)_{t+1}$ by Theorem 1.69. This contradiction proves our claim of the non-existence of a zero $u \times u$ block when $v_{t+1}(f) > a$.

5.3.5. A dimension formula for the variety $Gor(T)$.

We first give a consequence of our results concerning a Lefschetz property, then study dimension. We say that the Gorenstein Artin algebra $A = R/I$ satisfies the weak Lefschetz property, if there is a linear element $g \in R_1$ such that the multiplication $m_g : A_i \to A_{i+1}$ has maximal rank for each i, $1 \le i \le j$. The following corollary of Theorem 5.31 generalizes a result of J. Watanabe for the case J_I is a complete intersection.

COROLLARY 5.49. WEAK LEFSHETZ PROPERTY. *If T contains (s, s, s), then every Gorenstein Artin algebra of Hilbert function T satisfies the weak Lefschetz condition (WL). If T contains (s, s), then every $A = R/\operatorname{Ann}(f)$, $f \in Gor_{\mathrm{sch}}(T)$ satisfies WL. If T satisfies (5.3.31), then every $f \in Gor_{\mathrm{sch}}(T)$ satisfies the weak Lefschetz property for degrees $i < t = j/2$, and degrees $i > t$.*

PROOF. That J_I is height two Cohen-Macaulay shows the weak Lefschetz condition for degrees i satisfying both $H_i = T_i$, $H_{i+1} = T_{i+1}$. It is well known that the duality then implies that $m_g : A_{j-i-1} \to A_{j-i}$ is surjective. This and Theorems 5.31, 5.39, and 5.46 imply the Corollary.   □

We now collect what we have shown concerning the relations between the dimensions of $Gor(T)$, $Gor_{\mathrm{sch}}(T)$ and $Hilb^H(\mathbb{P}^2)$, independently of the results announced by M. Boij on Betti strata — see in particular Corollary 5.34 above when $T \supset (s, s, s)$, and (5.3.30) when $T \supset (s - a, s, s, s - a)$. In the third case, $T \supset (s - a, s, s, s - a)$, recall that we gave in Remark 5.47 a precise formula for the codimension of $Gor_{\mathrm{sch}}(T)$ in $Gor(T)$, using Boij's result and our description of $Gor_{\mathrm{sch}}(T)$.

COROLLARY 5.50. DIMENSION OF $Gor(T)$ WHEN $r = 3$.

i. *When T is a Gorenstein sequence containing (s, s, s), so $T = \operatorname{Sym}(H, j)$, then the dimension of $Gor(T)$ satisfies*

$$\dim Gor(T) = s + \dim Hilb^H(\mathbb{P}^2) \qquad (5.3.41)$$

ii. *If T contains $(s - a, s, s, s - a)$, with $a \ge 1$, then*

$$\dim Gor_{\mathrm{sch}}(T) = s + \dim Hilb^H(\mathbb{P}^2),$$
$$\dim Gor(T) = \dim Gor_{\mathrm{sch}}(T) + a(a - 1)/2. \qquad (5.3.42)$$

In this case $Gor(T) = Gor_{\mathrm{sch}}(T)$ iff $a = 1$.

iii. *If T contains $(s - a - b, s - a, s, s - a, s - a - b)$ and $b \geq 2a$, and if $Gor_a(T) = \{f \in Gor(T) \mid v_{t+1}(I) = a\}$, then*

$$\dim Gor_{sch}(T) = s + \dim Hilb^H(\mathbb{P}^2),$$
$$\dim Gor(T) = \dim Gor_{sch}(T) + cod(Gor_a(T), Gor(T))$$
$$+ cod(Gor_{sch}(T), Gor_a(T)), \qquad (5.3.43)$$
$$cod(Gor_{sch}(T), Gor_a(T)) \leq \frac{3}{2}a(a-1).$$

PROOF. (i) and the first equalities in (ii), (iii) are immediate from Theorems 5.31, 5.39, 5.46. The second equality of (ii) uses the Conca-Valla dimension formula (see the Remark after Theorem 5.39). The inequality in (iii) is from Lemma 5.44.   □

Gerd Gotzmann determined the dimension of the Hilbert function strata $Hilb^H(\mathbb{P}^2) \subset \mathbf{Hilb}^n(\mathbb{P}^2)$, following work of M. A. Coppo and others in special cases. We state Gotzmann's result first in his original form, in terms of the dimension sequence $\varphi = (0, \ldots, \varphi_i, \ldots)$ where $\varphi_i = \dim_k I_i$ for an ideal $I = \mathcal{I}_Z$ in $R = k[x, y, z]$ defining a zero-dimensional scheme $Z \subset \mathbb{P}^2$. We denote by φ' the difference sequence $\varphi'_i = \varphi_i - \varphi_{i-1}$ and set $d_i = \varphi'_i - \varphi'_{i-1}$. We let the sequence H denote the corresponding Hilbert function of $R/\mathcal{I}_Z$, $S = \Delta H$ where $s_i = H_i - H_{i-1}$, and let $E = -\Delta S$ where $e_i = s_{i-1} - s_i$. Then, it is easy to see that

$$H_i = r_i - \varphi_i, \quad s_i = i + 1 - \varphi'_i, \quad e_i = d_i - 1. \qquad (5.3.44)$$

We here let $\nu = \nu(I) = order(I)$, and let $e = \sigma(Z) = \tau(Z) + 1 =$ regularity of $\mathcal{I}_Z$ (see Definition 1.68). We let $\mathbf{G}_S = \mathbf{G}(S)$ parametrize the graded ideals J in $k[x, y]$ of Hilbert function $H(R/J) = S$ (see also Chapter 8).

THEOREM 5.51. (G. Gotzmann, [**Got5**, Appendix 3], [**Got3**])[11]. DIMENSION OF $Hilb^H(\mathbb{P}^2)$.

i. *The dimension of $Hilb^H(\mathbb{P}^2)$ satisfies*

---

[11]The formulas are announced in [**Got3**], and shown in [**Got5**] in the form we adapt here

$$\dim Hilb^H(\mathbb{P}^2) = 1 - \varphi_\nu' + \sum_{i=\nu}^{e}[\varphi_i' - \varphi_{i-1}' - 1] \cdot \varphi_{i+1}'$$

$$= s_\nu - \nu + \sum_{i=\nu}^{e} e_i(i + 2 - s_{i+1})$$

$$= s_\nu + \sum_{i=\nu}^{e} e_i \cdot (i + 1 - s_{i+1}) \tag{5.3.45}$$

$$= \Delta(H)_\nu - \sum_{i=\nu}^{\tau+1} \Delta^2(H)_i(i + 1 - \Delta(H)_{i+1}).$$

ii. *The dimensions of $Hilb^H(\mathbb{P}^2)$ and of $\mathbf{G}_S$ satisfies,*

$$\dim Hilb^H(\mathbb{P}^2) = \dim \mathbf{G}_S + \sum_{i=\nu}^{e}(d_i - 1) \cdot \varphi_i'$$

$$= \sum_{i=\nu}^{e}(e_i + 1)(e_{i+1}) + \sum_{i=\nu}^{e} e_i \cdot (i + 1 - s_i). \tag{5.3.46}$$

REMARK. G. Gotzmann showed his dimension theorem, by reducing to the Artin case, where $\dim \mathbf{G}_S$ was already known [I1, Theorems 2.12, 3.13]. A different approach to the dimension theorem would be analogous to that of J. O. Kleppe in showing the smoothness/dimension of $Gor(T)$ [Kl2]. First, G. Ellingsrud showed that the deformations of codimension two CM schemes are unobstructed [El]; and J. O. Kleppe showed that the tangent space to $Hilb^H(\mathbb{P}^2)$ at a scheme Z with saturated ideal I is the zero-graded piece $\mathrm{Hom}(I, R/I)_0$ (see [Kl1]). So it would suffice to make a calculation of the tangent space at a single point, a "most lexicographic point" — such calculations appear in [ReeS], or may be done independently.

EXAMPLE 5.52. DIMENSION OF $Gor(T)$. Suppose $T = \mathrm{Sym}(H, j)$, $H = (1, 3, 6, 10, 12, \underline{13})$, $j \geq 12$. Then $S = \Delta H = (1, 2, 3, 4, 2, 1, \underline{0})$, $E = -\Delta S = (-1, -1, -1, -1, 2, 1, 1, \underline{0})$, $\nu = 4$, $e = 6$, and $\dim Gor(T) = 13 + \dim Hilb^H(\mathbb{P}^2) = 36$, since from (5.3.46)

$$\dim Hilb^H(\mathbb{P}^2) = [3(1) + 2(1)] + [2(5 - 2) + 1(6 - 1) + 1(7 - 0)] = 23.$$

The same number is obtained using (5.3.17). If $T' = \mathrm{Sym}(H', j)$, $j \geq 12$, $H' = (1, 3, 6, 10, \underline{13})$, then we have by either Theorem 4.1A or 4.5A, or Lemma 5.37, that $\dim Gor(T') = 39$. When $j = 12$, $T = (1, 3, 6, 10, 12, 13, 13, 13, 12, 10, 6, 3, 1)$ here.

EXAMPLE 5.53. DIMENSION OF $Gor_{sch}(T)$. Suppose $T=H(8,7,3)$ $= (1,3,6,8,8,6,3,1)$, $j = 7$. This is a sequence as in (5.3.18). Then since $D_{min}(T) = (3,3,4)$ (complete intersections) we have dim $CI(T) =$ $2(8) + 8 = 24$ so dim $Gor(T) = 1 + 24 = 25$ (as $CI(T)$ parametrizes ideals but $Gor(T)$ includes a choice of f). Here $t = 3$, $a = 2$, and $D_{sch}(T) := D_a = (3,3,4,5,5)$, with two extra generators in degree $t + 2 = 5$. We have $H = (1,3,6,\underline{8})$, the generic Hilbert function for 8 points in $\mathbb{P}^2$ so dim $Hilb^H(\mathbb{P}^2) = 16$ and dim $Gor_{sch}(T) =$ $8 + $ dim $Hilb^H(\mathbb{P}^2) = 24$ in accordance with (5.3.42). Here $C_{min}(T) =$ $(4,4,2)$, and $C_2(T) = C_{sch}(T) = (4,4,2,0,0)$. $C_{max}(T) = (4,4,2,2,0,$ $0,-2)$ corresponds to generator degrees $D_{max}(T) = (3,3,4,4,5,5,6)$ and $Gor_{sch}(T) = Gor(T, D_{sch}(T)) \cup Gor(T, D_{max}(T))$. The remaining $C \in \mathcal{L}(T)$ is $C = (4,4,2,2,-2)$ corresponding to $D = (3,3,4,4,6)$, generator degrees for a subfamily of DP-forms which do not have annihilating schemes of degree ≤ 8 (Theorem 5.39).

REMARK 5.54. The results of this section are related to those of A. Geramita, M. Pucci, and Y. Shin [**GPS**], and of T. Harima [**Hari3**], and M. Boij [**Bo1**], all of whom consider Gorenstein ideals I constructed in a similar manner from ideals $J = \mathcal{I}_Z$. The results here are new in showing there is a morphism from all of $Gor(T)$ to $Hilb^H(\mathbb{P}^2)$ when T contains (s, s, s), or in identifying the domain $Gor_{sch}(T)$ of such a morphism in terms of the resolution degrees of ideals when T contains (s, s).

The morphism from $Gor(T)$ to $Hilb^H(\mathbb{P}^2)$ allows also a comparison of the dimensions of the Betti strata of each of these varieties: strata where the minimal resolution degrees of the Gorenstein ideal I, or of the defining ideal $\mathcal{I}_Z$ are fixed. We develop this comparison in Section 5.5 below.

5.4. Power sum representations in three and more variables

In this section we reconsider the three main theorems 5.31, 5.39 and 5.46 of Section 5.3, focusing on Problem 0.1 from the Introduction – finding conditions for a form to be represented as a sum $f = L_1^j +$ $\cdots + L_s^j$. We also discuss a procedure for finding explicitly a power sum representation of a given form. We present the relevant results from a more elementary viewpoint, using only catalecticant matrices, and when necessary we reintroduce some of the notions from the previous section. As usual we will work with divided powers and arbitrary characteristic; recall that when $\text{char}(k) = 0$ or $\text{char}(k) > j$, then replacing $L^{[j]}$ by $\frac{1}{j!}L^j$ one obtains the analogous statements for forms and ordinary powers (see Appendix A).

Recall from Section 1.1 that for a non-zero DP-form f of degree j one defines $(H_f)_i = \operatorname{rk} Cat_f(i, j - i; r)$, $i = 0, \ldots, j$, with $(H_f)_0 = (H_f)_j = 1$. The minimum s for which an additive decomposition $f = L_1^{[j]} + \cdots + L_s^{[j]}$ is possible satisfies $s \geq \max\{(H_f)_i\}$; given f we look to find such a representation of minimum length in an explicit way.

Let us recall the classical solution for binary forms when $\operatorname{char}(k) = 0$ or $> j$ (see Section 1.3 for details). First, H_f is completely determined by j and $s = \max\{(H_f)_i\}$; and $s = (H_f)_t, t = \lfloor j/2 \rfloor$. If $j = 2t$ and $s = (H_f)_t$ has the maximum possible value $t + 1$, there is a one parameter family of length s additive decompositions. If $j = 2t + 1$ or $2t$ and $s = \max\{(H_f)_i\} < t + 1$, which is the case if and only if the sequence $T = H_f$ contains (s, s), then generically - for general enough f, with $H_f = T$, there is a unique additive decomposition of length s, which can be found explicitly as follows. Note that s is the smallest degree i for which $\operatorname{rk} Cat_f(j - i, i; 2) < i + 1 = r_i$, and that this rank is just $(H_f)_s = s$ (the corank is one). The Hankel matrix $Cat_f(j - s, s; 2)$ has type $(j - s + 1) \times (s + 1)$, so the linear system

$$Cat_f(j - s, s; 2)(Z_0, \ldots Z_s)^t = 0$$

has up to a constant multiple, a unique nonzero solution $(z_0, \ldots, z_s)$. One first decomposes into a product of linear factors the apolar polynomial

$$\phi = \sum_{i=0}^{s} z_i x^{s-i} y^i = \prod_{i=1}^{s}(b_i x - a_i y) \tag{5.4.1}$$

Choosing f general enough, such that $H_f = T$, assures that ϕ has no multiple roots — this is the only reason for choosing f general. Then one decomposes

$$f = c_1(a_1 X + b_1 Y)^j + \cdots + c_s(a_s X + b_s Y)^j \tag{5.4.2}$$

finding the constants c_i by expanding the powers and solving a system of linear equations. When f is arbitrary and ϕ has multiple roots then one should replace the additive decomposition (5.4.2) by a generalized additive decomposition as in (1.3.1)

When $r \geq 3$ there are two questions to discuss. First, when does there exist a length-s additive decomposition? Second, if it exists, how do we find it explicitly? The three theorems 5.31, 5.39 and 5.46 from Section 5.3 give a satisfactory answer to the first question when

$r = 3$ in most cases; the intrinsic conditions found may be considered an analog of the rank condition $(H_f)_t < t + 1$ for the Hankel matrices in the binary case. We postpone our discussion of this until Theorem 5.55 below. The second question is easier and we start by describing such a procedure for determining an additive decomposition for f, for arbitrary $r \geq 3$ based on Lemma 5.3.

As in the case $r = 2$ one considers the sequence $T = H_f$ and its maximum

$$s = \max\{T_i\} = \max\{\operatorname{rk} Cat_f(i, j - i; r) \mid i = 0, \dots, j\}.$$

As discussed in Section 5.1 instead of considering an apolar polynomial ϕ of minimum degree s and finding its decomposition (5.4.1) one looks for a zero dimensional scheme Z of degree s such that each polynomial ϕ from the ideal of Z is apolar to f, i.e. in the language of Section 5.1, the scheme Z is an annihilating scheme (so $\mathcal{I}_Z \subset \operatorname{Ann}(f)$); which is moreover tight: $\deg Z = s = \max\{(H_f)_i\}$. When Z happens to be smooth, i.e. Z is a set of s points, then f has an additive decomposition of length s with polar polyhedron Z (Corollary 5.4). So, if Z could be found explicitly from f and if it were smooth, then the coordinates of its s points are the coefficients of the linear forms in the additive decomposition

$$f = c_1 L_1^{[j]} + \cdots + c_s L_s^{[j]}.$$

Then the coefficients $c_i \in k$ can be found by expanding the divided powers and solving a system of linear equations. The analog of the generalized additive decomposition, which is of interest when Z happens not to be smooth, is not known at present and would be interesting to find.

Lemma 5.3 addresses the problem of finding explicitly the annihilating scheme Z of degree $s = \max\{(H_f)_i\}$ if it exists. Namely, one considers the minimum degree i for which $(H_f) = s$ and denotes it by τ. One denotes by ν the minimum number i such that $(H_f)_i = \operatorname{rk} Cat_f(i, j - i; r) < r_i = \binom{i+r-1}{r-1}$. For each $i \in [\nu, \tau + 1]$ one solves the linear system

$$Cat_f(j - i, i; r) Z^t = 0. \tag{5.4.3}$$

With each solution $(z_{i_1, \dots, i_r} \mid i_1 + \cdots + i_r = i)$ one associates an apolar polynomial $\phi = \sum_{i_1 + \cdots + i_r = i} z_{i_1, \dots, i_r} x_1^{i_1} \cdots x_r^{i_r}$. Denote by I_i the linear subspaces of $R_i = k[x_1, \dots, x_r]_i$ obtained in this way, consisting of all degree-i polynomials apolar to f. Then Lemma 5.3(D,E) says: provided an annihilating scheme Z of degree s exists one has

a. $\mathcal{I}_Z$ is generated by polynomials of degree $\leq \tau + 1$.

b. If $j \geq 2\tau + 1$ then $\mathcal{I}_Z$ is generated by the apolar polynomials from I_i for $\nu \leq i \leq \tau + 1$.

c. If $j = 2\tau$ and $\mathcal{I}_Z$ were generated by the polynomials of degree $\leq \tau$, then $\mathcal{I}_Z$ is generated by the apolar polynomials from I_i, $\nu \leq i \leq \tau$.

This shows in Cases (b) and (c) that the scheme Z is the solution, taking into account multiplicities, of a system of homogeneous polynomial equations. The system is obtained explicitly from f.

An economical way to obtain this system of equations is the following. One finds a fundamental system of solutions of (5.4.3) for $i = \nu$ consisting of $v_\nu(f)$ elements. Since $\mathrm{Ann}(f)$ is an ideal one has $R_1 I_\nu \subset I_{\nu+1}$ and one complements this subspace with a subspace having a basis of $v_{\nu+1}(f)$ elements. One repeats this procedure, complementing each $R_1 I_{i-1} \subset I_i$ by $v_i(f)$ linearly independent forms for every $\nu + 1 \leq i \leq \tau + 1$ in Case (b) and for every $\nu + 1 \leq i \leq \tau$ in Case (c). Let us denote the forms obtained in this way by $\phi_1, \ldots, \phi_{u+1}$ (the notation is in accordance with that of Theorems 5.31, 5.39 and 5.46). We obtain, under the assumptions of (b), (c) above, that if Z were smooth – a set of s points – then these s points are the solutions of the system

$$\phi_i(w) = 0, \qquad i = 1, \ldots, u + 1.$$

When Z is not smooth the system above still has s solutions counting multiplicities, i.e. $\deg \mathrm{Proj}\, R/(\phi_1, \ldots, \phi_{u+1}) = s$. In either case the sheaf of ideals of Z is $\mathcal{I}_Z = (\phi_1, \ldots, \phi_{u+1})$.

The next theorem — a corollary of Theorems 5.31, 5.39, 5.46 and Lemma 5.3 — gives when $r = 3$ necessary and sufficient conditions for the existence of an annihilating scheme of degree s.

THEOREM 5.55. *Suppose* $r = 3$. *Let* $f \in \mathcal{D}_j$, $f \neq 0$, *let* $T = (1, T_1, \ldots, T_i, \ldots, 1)$ *be the sequence of ranks* $T_i = (H_f)_i = \mathrm{rk}\, \mathrm{Cat}_f(i, j - i; 3)$. *Suppose* $T_1 = 3$.[12] *Let* $I = \mathrm{Ann}(f)$ *be the graded ideal of polynomials apolar to* f. *Let* $s = \max\{T_i\}$ *and let* $\tau = \min\{i \mid T_i = s\}$.

Case 1. Let $j = 2t$ *or* $2t + 1$. *If* $T \supset (s, s, s)$, *in particular* $s \leq r_{t-1} = \binom{t+1}{2}$, *then* f *has a zero-dimensional annihilating scheme* Z *of degree* s, *whose ideal* $\mathcal{I}_Z$ *is generated by the apolar forms to* f *of degree no greater than* t.

---

[12]The case $T_1 < 3$ is reduced to binary forms according to Lemma 1.22.

Case 2. *Let* $j = 2t+1$. *If* $T \supset (s-a, s, s, s-a)$, $a \geq 1$, $T_t = T_{t+1} = s$ *then* f *has a zero-dimensional annihilating scheme* Z *of degree* s *if and only if*

$$\mathrm{cod}_{R_{t+2}} R_1 I_{t+1} = s, \qquad (5.4.4)$$

in which case $\mathcal{I}_Z$ *is generated by the apolar forms to* f *of degree no greater than* $t+1$.

Case 3. *Let* $j = 2t$. *If* $T \supset (s - b, s - a, s, s - a)$, $a \geq 1, T_t = s$ *and* $b \geq 2a$, *then* f *has an zero-dimensional annihilating scheme* Z *of degree* s *whose ideal is generated by forms of degree* $\leq t$ *if and only if*

$$\mathrm{cod}_{R_{t+1}} R_1 I_t = s \quad and \quad \mathrm{cod}_{R_{t+2}} R_2 I_t = s. \qquad (5.4.5)$$

In each of the three cases Z *is unique and can be explicitly recovered from* f: *the graded ideal* $\mathcal{I}_Z$ *is generated by the apolar forms to* f *of degree* $\leq \tau + 1$ *(here* $\leq t$*) in Case 1, of degree* $\leq t + 1$ *in Case 2 and of degree* $\leq t$ *in Case 3. The annihilating scheme also satisfies the following property. It imposes* T_i *conditions on forms of degree* i *when* $i < \tau$ *and* s *conditions on forms of degree* i *when* $i \geq \tau$. *Fixing* T *let us denote by* $Gor(T)$ *the set of DP-forms* f *with* $T = H_f$. *In Case 2 or Case 3 denote by* $Gor_{\mathrm{sch}}(T)$ *the subsets of* $Gor(T)$ *satisfying conditions* (5.4.4) *or* (5.4.5) *respectively. Then these are irreducible algebraic varieties. If* f *is a general enough element of* $Gor(T)$ *in Case 1, or of* $Gor_{\mathrm{sch}}(T)$ *in Cases 2 or 3, then the unique length-s annihilating scheme* Z *of* f *is smooth, and* f *has an additive decomposition*

$$f = c_1 L_1^{[j]} + \cdots + c_s L_s^{[j]}$$

with polar polyhedron Z.

Let us consider as an example the sequence $T = H(s, j, 3) = (1, 3, \ldots, r_{\tau-1}, s, s, \ldots, 3, 1)$, $T_\tau = s$. In Case 1 we are in the set-up of Theorems 4.1A, 4.5A and we obtain a new proof of these theorems: $\overline{PS(s, j; 3)} = \overline{Gor(T)}$; but also we obtain in addition that every element in $Gor(T)$ has a degree-s annihilating scheme whose ideal is generated by apolar forms. Cases 2 and 3 yield the following supplement to Theorems 4.1B and 4.5B.

COROLLARY 5.56. *Suppose* $j = 2t$ *or* $2t + 1$ *and let* $r_{t-1} < s \leq r_t = \binom{t+2}{2}$.

a. *If* $j = 2t + 1$ *then* $\overline{PS(s, j; 3)} \cap Gor(T)$ *is the subset of* $Gor(T)$ *distinguished by the condition*

$$\mathrm{cod}_{R_{t+2}} R_1 \mathrm{Ann}(f)_{t+1} = s.$$

b. *If $j = 2t$ and $s \leq r_{t-1} + \frac{t}{2}$ then $\overline{PS(s,j;3)} \cap Gor(T)$ has as an open, dense subset the subset of $Gor(T)$ distinguished by the conditions*

$$\operatorname{cod}_{R_{t+1}} R_1 \operatorname{Ann}(f)_t = s, \quad \operatorname{cod}_{R_{t+2}} R_2 \operatorname{Ann}(f)_t = s.$$

We conclude this section by summarizing the calculations one needs to do in order to find explicitly a divided power sum representation of a given DP-form f

$$f = c_1 L_1^{[j]} + \cdots + c_s L_s^{[j]} \tag{5.4.6}$$

where no two L_i, L_k are proportional to each other, $c_i \neq 0$ and $s = \max\{\operatorname{rk} Cat_f(i, j-i; r)\}$, provided such an additive decomposition exists. Let $Z \subset \mathbb{P}(\mathcal{D}_1)$ be the polar polyhedron to be found.

Step 1. One finds the ranks of $Cat_f(i, j-i; r)$ for $1 \leq i \leq j/2$. If (5.4.6) holds this is a nondecreasing sequence, and the first differences are an O-sequence. One may stop the first time one gets equal ranks for two consecutive values $i = \tau, \tau+1$ and then $s = \operatorname{rk} Cat_f(\tau, j-\tau; r)$. This follows from $T = H_f = \operatorname{Sym}(H_Z, j)$ and Theorem 1.69. If the sequence of ranks for $1 \leq i \leq j/2$ is strictly increasing then in the case $j = 2t+1$ one lets $\tau = t$, $s = \operatorname{rk} Cat_f(t, t+1; r)$ and in the case $j = 2t$ one lets $\tau = t$, $s = \operatorname{rk} Cat_f(t, t; r)$.

Step 2. One calculates the apolar forms $\phi_1, \ldots \phi_{u+1}$ of degree $\leq \tau+1$ if $j \geq 2\tau+1$ and those of degree $\leq \tau$ if $j = 2\tau$ as described on page 185.

Step 3. One solves the system of homogeneous polynomial equations

$$\phi_i(x) = 0, \quad i = 1, \ldots, u+1 \tag{5.4.7}$$

and obtains a finite set of s points $Z \subset \mathbb{P}^{r-1}$. [13]

Step 4. The coordinates of Z are the coefficients of $L_1, \ldots, L_s$. One expands the divided powers in

$$f = c_1 L_1^{[j]} + \cdots + c_s L_s^{[j]} \tag{5.4.8}$$

and obtains a linear system for $c_1, \ldots, c_s$.

---

[13] Notice however that in the case $j = 2t$, $\tau = t$ one might get more then s solutions if $\mathcal{I}_Z$ were not generated by polynomials of degree $\leq t$.

REMARK. Notice that when $r = 3$ every DP-form f has the property that the sequence of ranks $\operatorname{rk} Cat_f(i, j-i; 3)$ is nondecreasing and is strictly increasing for $i \in [1, \tau]$. This follows from Theorem 5.25(i) and Theorem 1.69. This might be not the case when $r \geq 4$ (see [BeI, BoL]), but then failure to obtain the sequence of ranks as described in Step 1 means f cannot have a decomposition as in (5.4.6). When $r = 3$ Theorem 5.55 gives necessary and sufficient condition that the system (5.4.7) has s solutions in the scheme-theoretic sense – counting multiplicities, furthermore when f satisfies the conditions of this theorem, and is general enough, then the system has exactly s distinct solutions.

EXAMPLE 5.57. Suppose $r = 3, k = \mathbb{C}$, and $R = k[x, y, z], \mathcal{R} = k[X, Y, Z]$ are polynomial rings: here we will use the differentiation action of R on $\mathcal{R}$. In the first two cases below, $H_f = (1, 3, 5, 5, 3, 1)$, in the last, $H_f = (1, 3, 4, 4, 3, 1)$.

a. Let $f = X^5 + Y^5 + Z^5 + 30X^2YZ^2 + 20XY^3Z$. Since $Cat_f(i, j - i; 3)$ has rows determined by the $j - i$-order partials, we have $H_f = (1, 3, s, s, 3, 1)$, where $s = \operatorname{rk} Cat_f(3, 2; 3)$, which is the vector space dimension of the space $< x^2 \circ f, \dots z^2 \circ f >$; since $(xz - y^2) \circ f = 0$, the span of $R_2 \circ f$ is

$$x^2 \circ f = 20X^3 + 60YZ^2, xy \circ f = 60XZ^2 + 60Y^2Z,$$

$$xz \circ f = y^2 \circ f = 120XYZ + 20Y^3,$$

$$yz \circ f = 2X^2Z + 3XY^2, z^2 \circ f = 20Z^3 + 2X^2Y,$$

and, since these are linearly independent, we have (Step 1) $s = 5$. In calculating $R_3 \circ f$, one finds $R_1 \cdot \phi_1 \circ f = 0$ where $\phi_1 = (xz - y^2)$, and new annihilators $\phi_2 = x^3 - yz^2, \phi_3 = x^2y - z^3$ (Step 2). A check by computation, or by noting that f satisfies the hypothesis of Case 2 of Theorem 5.55, shows that ϕ_1, ϕ_2, ϕ_3 generate an ideal of Hilbert function $(1, 3, 5, 5, 5, \dots)$, so they determine a degree-5 scheme Z. Using a primary decomposition method, or other method, one finds (Step 3) that the solution to $\phi_1 = \phi_2 = \phi_3 = 0$ is the smooth scheme Z consisting of the five points

$$p_1 = x + \zeta y + \zeta^2 z, \ p_2 = \sigma(p_1), \dots, p_5 = \sigma^4 p_1,$$

where ζ is a primitive 5th root of 1, and $\sigma : \zeta \to \zeta^2$ generates $Aut(\mathbb{Q}(\zeta)/\mathbb{Q})$. Finally, one solves (Step 4)

$$f = c_1(X + \zeta Y + \zeta^2 Z)^5 \cdots + c_5(X + \zeta^3 Y + \zeta Z)^5,$$

obtaining each $c_i = 1/5$.

b. Let $f = X^2Y^2Z$, then $\mathrm{Ann}(f) = (z^2, x^3, y^3)$; here $\phi_1 = z^2, \phi_2, \phi_3$ generate the apolar ideal to f, so there is no tight (degree 5) annihilating scheme: here f is not even the limit of a parametrized family of forms that are a sum of 5 fifth powers (see Proposition 5.18).

c. Let $f = X^5 + X^4Y + X^4Z + X^3YZ$. Then, one finds from Steps 2 and 3 that $\mathrm{Ann}(f)_2 = (y^2, z^2)$, determines a degree-4 scheme concentrated at $(1, 0, 0)$: so there is no way to write f as a sum of 4 powers of linear forms, but f is the limit of such sums.

5.5. Betti strata of the punctual Hilbert scheme

In Corollary 5.34 we used the morphism $p : Gor(T) \to Hilb^H\mathbb{P}^2$ to the postulation Hilbert scheme from Theorem 5.31, to calculate the generator degrees of $\mathcal{I}_Z$ for a general $Z \in Hilb^H\mathbb{P}^2$, and to provide a simple formula for $\dim Hilb^H\mathbb{P}^2$ based on the Conca-Valla formula for $\dim Gor(T)$. In this section we go further and use this morphism and Theorem 5.31 to study the Betti strata of $Hilb^H\mathbb{P}^2$ – the subvarieties with fixed generator degrees.

According to the Hilbert-Burch theorem (see e.g. [**Ei2**, Theorem 20.15]) if Z is a zero dimensional subscheme of $\mathbb{P}^2$ then its graded ideal $J = \mathcal{I}_Z$ has a minimal resolution

$$0 \to \sum_{1 \leq i \leq u} R(-e_i) \xrightarrow{\Phi} \sum_{1 \leq i \leq u+1} R(-d_i) \xrightarrow{\lambda} J \to 0, \qquad (5.5.1)$$

where $d_1 \leq \cdots \leq d_{u+1}$, $e_1 \geq \cdots \geq e_u$. Also, the $(u+1) \times u$ matrix $\Phi = (\varphi_{ij})$ has entries homogeneous polynomials (possibly zero) of degrees $\deg \varphi_{ij} = e_i - d_j$ when $e_i - d_j > 0$, and the entries are 0 otherwise. The row-matrix λ has entries $\lambda_i = (-1)^i \det \Phi_i$ where Φ_i is the $u \times u$ matrix obtained by deleting the i-th row of Φ and $\sum_{k=1}^{u} e_k = \sum_{k=1}^{u+1} d_k$. The latter condition assures that $\det \Phi_i$ is a nonzero homogeneous polynomial of degree d_i for each $i = 1, \ldots, u + 1$ (cf. Theorem B.2). Conversely, given a matrix Φ as above, if the ideal $J = (\lambda_1, \ldots, \lambda_{u+1})$ has height (=grade) ≥ 2 then the complex (5.5.1) is exact and therefore J is the saturated ideal of a zero dimensional scheme (cf. the argument on page 163).

DEFINITION 5.58. To each nondecreasing sequence of positive integers $D = (d_1, \ldots, d_n)$ we associate a finite sequence of non-negative integers $\beta(D) = (\beta_i)$, $1 \leq i \leq d_n$, setting β_i to be the number of times

i enters in $d_1, \ldots, d_n$. Conversely given β_i one obtains a unique D such that $\beta(D) = (\beta_i)$, namely $D = \cdots \underbrace{i, \ldots, i}_{\beta_i} \cdots$.

Suppose the Hilbert function $H = H_Z$ is fixed. We assume $H_1 = 3$, since otherwise Z is a subscheme of $\mathbb{P}^1$ which is a trivial case. The sequence of generator degrees $D = (d_1, \ldots, d_{u+1})$ determines the sequence of relation degrees as follows. Let $\beta = \beta(D)$ be as above. Then using Lemma 5.38 we have for each $i \geq 1$

$$\#\{\text{relations of degree } i\} \;=\; \beta_i + \Delta^3(H)_i.$$

This determines $e_1, \ldots, e_u$.

Fixing H and D let us denote by $Hilb_D^H(\mathbb{P}^2)$ the subset of $Hilb^H \mathbb{P}^2$ consisting of schemes Z with sequence of generator degrees equal to D. We first give a preparatory Lemma, generalizing our earlier result concerning the local closedness of the generator-strata.

LEMMA 5.59. *Let $R = k[x_1, \ldots, x_r]$. Let $\mathbb{J}_i \to X$ be vector subbundles of the trivial bundle $R_i \times X$ for each $i \geq 1$, set $\mathbb{J}_0 = 0$ and suppose $R_1(\mathbb{J}_i)_x \subset (\mathbb{J}_{i+1})_x$ for each $i \geq 1$, in other words $\oplus_{i \geq 1} \mathbb{J}_i$ is a flat family of graded ideals in R. Then for each $i \geq 1$ the function $v_i(\mathbb{J}_x) := \mathrm{cod}_{(\mathbb{J}_i)_x} R_1(\mathbb{J}_{i-1})_x$ is upper semicontinuous.*

PROOF. Consider the multiplication bundle map $\mu : R_1 \otimes \mathbb{J}_{i-1} \to \mathbb{J}_i$. Then $v_i(\mathbb{J}_x) = \mathrm{rk}(\mathbb{J}_i) - \mathrm{rk}(\mu_x)$ and under specialization $\mathrm{rk}(\mu_x)$ could only drop. □

The possible sequences D for which $Hilb_D^H(\mathbb{P}^2) \neq \emptyset$ were found implicitly by R. Fröberg, and explicitly by G. Campanella [**Cam**].[14] Our goal is to compare his results with those of S. J. Diesel (Theorem 5.25(v.f)) and prove that the collection $\{Hilb_D^H(\mathbb{P}^2)\}$ of Betti strata forms a family of locally closed subsets of $Hilb^H \mathbb{P}^2$ which satisfies the frontier property (see page 34). We apply the formula (4.4.19) announced by M. Boij to obtain a formula for $\dim Hilb_D^H(\mathbb{P}^2)$. We derive these results from facts about $Gor(T)$, using Theorem 5.31.

THEOREM 5.60. *Let $Hilb_D^H(\mathbb{P}^2) \neq \emptyset$. Then*

  i. *$Hilb_D^H(\mathbb{P}^2)$ is an irreducible variety.*
  ii. (G. Campanella [**Cam**]). *$Hilb_D^H(\mathbb{P}^2)$ has an open, dense subset parameterizing smooth schemes.*

---

[14]R. Fröberg had shown that any Betti numbers for height two ideals in $k\{x, y\}$ are attainable by monomial ideals [**Fr**, Section 5]; but monomial ideals can be lifted to $k[x, y, z]$, preserving the Betti numbers by Hartshorne's connectedness methods ([**Har1**], see also [**GGR**]).

Proof. The essential part is Campanella's result [**Cam**, Corollary 4.2] that if $Hilb_D^H(\mathbb{P}^2) \neq \emptyset$, then there exists a set Z of distinct points with Hilbert function H and generator degrees D. Having this, (i) implies (ii).

The proof of (i) is similar to that of the corresponding statement about $Gor_D(T)$, that we discussed on page 152. Fixing the sequence of generator degrees D one fixes the size of the matrix Φ and the degrees of its entries in (5.5.1), hence $Hilb_D^H(\mathbb{P}^2)$ is an image of an irreducible variety, so it has an irreducible closure in $Hilb^H\mathbb{P}^2$. We want to prove that $Hilb_D^H(\mathbb{P}^2)$ is open in its closure. First we observe that the number of possible sequences D for $Hilb^H\mathbb{P}^2$ is finite. This follows from Theorem 1.69 since if $s = \max\{h_i\}$ and $\tau = \min\{i \mid h_i = s\}$ then for every $Z \in Hilb^H\mathbb{P}^2$ the number of degree-i generators $v_i(\mathcal{I}_Z) = 0$ for $i > \tau+1$. From Definition 5.19 of $Hilb^H\mathbb{P}^2$ it is evident that the graded ideals $\mathcal{I}_Z$, $Z \in Hilb^H\mathbb{P}^2$ fit in a direct sum of vector bundles $\oplus_{i \geq 1}\mathcal{J}_i$ over $Hilb^H\mathbb{P}^2$. Lemma 5.59 applies, so under specialization $v_i(\mathcal{I}_Z)$ could only jump. Reasoning by descending induction on $w \in \mathbb{Z}_+$ we conclude that each subset

$$\mathcal{Y}_w = \left\{ Z \in Hilb^H\mathbb{P}^2 \mid \sum_{i \geq 1}(\mathcal{I}_Z) \geq w \right\} \tag{5.5.2}$$

is closed in $Hilb^H\mathbb{P}^2$. If $\mathcal{X} = Hilb_D^H(\mathbb{P}^2)$ and $\overline{\mathcal{X}}$ is the closure in $Hilb^H\mathbb{P}^2$ then $\overline{\mathcal{X}} - \mathcal{X} = \overline{\mathcal{X}} \cap \mathcal{Y}_{\#D+1}$, so $\mathcal{X}$ is open in $\overline{\mathcal{X}}$. This proves that $Hilb_D^H(\mathbb{P}^2)$ is an irreducible variety.[15]   □

Let $H = (1, 3, \ldots, s - a, \overset{\tau}{s}, s, \ldots)$, $a \geq 1$ be a Hilbert sequence — a fixed sequence of integers that occurs as the Hilbert function of a zero-dimensional scheme Z in $\mathbb{P}^2$ (thus, the first difference $\Delta(H)$ is an O-sequence - see Theorem 5.21). Let $v = \min\left\{i \mid \binom{i+2}{2} > h_i\right\}$, $\tau = \min\{i \mid h_i = s\}$. Let us fix an integer $j \geq 2\tau+2$. We need to determine a generator sequence $\tilde{D}$ for a Gorenstein ideal, that will be related to the generator sequence D for an ideal $\mathcal{I}_Z$. With every sequence $D = (d_1, \ldots, d_{u+1})$, $v \leq d_1 \leq \cdots \leq d_{u+1} \leq \tau + 1$ let us associate a new sequence $\tilde{\beta}$ as follows. Consider $\beta = \beta(D)$ from Definition 5.58,

---

[15]See [**El**, Theorem 1] for a general result implying local closedness of the Betti strata of Hilbert schemes.

giving the number of generators of degree-i. Let

$$\tilde{\beta}_i = \begin{cases} \beta_i & \text{if } i \in [1, \tau + 1] \\ 0 & \text{if } i \in [\tau + 2, j - \tau] \\ a & \text{if } i = j + 1 - \tau \\ \Delta^3(H)_{j+3-i} + \beta_{j+3-i} & \text{if } i \in [j + 2 - \tau, j + 3 - \nu] \end{cases} \tag{5.5.3}$$

As in Definition 5.58, we associate to $\tilde{\beta} = (\tilde{\beta}_i)$ a sequence of degrees $\tilde{D}$.

LEMMA 5.61. *Let H, $j \geq 2\tau + 2$ be as above. Let $T = \mathrm{Sym}(H, j)$. Every sequence of generator degrees for $\mathrm{Gor}(T)$ has the form $\tilde{D}$ for some sequence D, and $\mathrm{Hilb}_D^H(\mathbb{P}^2) \neq \emptyset$ if and only if $\mathrm{Gor}_{\tilde{D}}(T) \neq \emptyset$. One has $\mathrm{Gor}_{\tilde{D}}(T) = p^{-1}\left(\mathrm{Hilb}_D^H(\mathbb{P}^2)\right)$ where $p : \mathrm{Gor}(T) \to \mathrm{Hilb}^H\mathbb{P}^2$ is the morphism from Theorem 5.31. The morphism p maps $\mathrm{Gor}_{\tilde{D}}(T)$ dominantly to $\mathrm{Hilb}_D^H(\mathbb{P}^2)$ and*

$$\dim \mathrm{Gor}_{\tilde{D}}(T) = s + \dim \mathrm{Hilb}_D^H(\mathbb{P}^2). \tag{5.5.4}$$

PROOF. All statements except the last one are clear from Lemma 5.29 and Theorem 5.31 since if $Z = p(f)$ then $(\mathcal{I}_Z)_{\leq \tau + 1} = \mathrm{Ann}(f)_{\leq \tau + 1}$. If X is a smooth scheme in $\mathrm{Hilb}_D^H(\mathbb{P}^2)$ then the fiber $p^{-1}(X)$ consists of divided power sums with polar polyhedron X and using Boij's Theorem B.12 we conclude $p^{-1}(X) \subset \mathrm{Gor}_{\tilde{D}}(T)$. By Theorem 5.60(ii) the smooth schemes form an open dense subset. This proves the dominance and (5.5.4). $\qquad\square$

COROLLARY 5.62. *Suppose $r = 3$ and T is a Gorenstein sequence which contains (s, s, s). Then for every possible sequence of generator degrees S as in Theorem 5.25(v.f), each general enough element $f \in \mathrm{Gor}_S(T)$ has an additive decomposition $f = \sum_{u=1}^s c_u L_u^{[j]}$.*

The possible sequences of generator degrees for a given H were calculated by G. Campanella in [Cam]. [16] Part of the next theorem rephrases his result in a way parallel to Diesel's Theorem 5.25(v.f). In fact Diesel's result implies Campanella's, as is evident from the arguments below, however such a proof is more complicated then the original one which is based on the study of monomial ideals in $k[x, y]$.

---

[16]Again, see also R. Fröberg's [Fr]; these generator degrees are the same as that for graded Artin quotients of $k[x, y]$ having Hilbert function ΔH, which were partially delimited by F. H. S. Macaulay [Mac1], and implicitly given in [I1, Theorem 4.3 and §4B]; the possible sequences of relation degrees in the Artin case is stated in [I3, Theorem 4.6C].

THEOREM 5.63. *Let* $H = (h_i)_{i \geq 0} = (1, 3, \ldots, s-a, s, s, \ldots)$, $a \geq 1$ *be the Hilbert function of a zero dimensional scheme in* $\mathbb{P}^2$, *let* $\nu = \min \left\{ i \mid \binom{i+2}{2} > h_i \right\}$, *let* $\tau = \min\{i \mid h_i = s\}$. *Then the possible sequences of generator degrees* $D = (d_1, \ldots, d_{u+1})$, $d_1 \leq \cdots \leq d_{u+1}$ *are described as follows.*

   i. *The sequence with minimum number of generators* $D_{\min}$, *occurring for* Z *in an open, dense subset of* $\mathrm{Hilb}^H \mathbb{P}^2$ *has*

$$\beta(D_{\min})_i := \nu_i(\mathcal{I}_Z) = \max\{0, -\Delta^3(H)_i\} := \Delta^3(H)_i^-$$

      generator degrees equal to i *for each* $i \geq 1$. *The minimum number of generators is*

$$\mu(H) = \sum_{i \geq 1} \Delta^3(H)_i^-.$$

   ii. *There is a unique sequence* $D_{\max}$ *with a maximal number of generators and* $\beta(D_{\max})$ *satisfies*

$$\beta(D_{\max})_i = \begin{cases} 1 - \Delta^2(H)_i & \text{if } i = \nu \\ \max\{0, -\Delta^2(H)_i\} & \text{if } i \in [\nu+1, \tau+1] \\ 0 & \text{otherwise.} \end{cases}$$

      The number of generators in the maximal sequence equals $\nu + 1$.

  iii. *One has* $D_{\max} = D_{\min} \cup G$ *where either* $G = \emptyset$ *or* G *is a nondecreasing sequence of positive integers. Then* D *is a sequence of generator degrees for a certain* $Z \in \mathrm{Hilb}^H \mathbb{P}^2$ *if and only if* $D = D_{\min} \cup G'$ *for some nondecreasing sequence* $G' \subset G$ *(including* $\emptyset$). *The number of possible sequences equals* $\prod_{i=\nu}^{\tau+1}(\beta_i(G) + 1)$.

The family of locally closed subsets $\{\mathrm{Hilb}_D^H(\mathbb{P}^2)\}$ *satisfies the frontier property:* $\mathrm{Hilb}_{D'}^H(\mathbb{P}^2)$ *is contained in the closure of* $\mathrm{Hilb}_D^H(\mathbb{P}^2)$ *if and only if an element of* $\mathrm{Hilb}_{D'}^H(\mathbb{P}^2)$ *is contained in this closure, and this holds if and only if* $D' \supset D$. *The dimension of* $\mathrm{Hilb}^H \mathbb{P}^2$ *is given by either of the formulas (5.3.17) or (5.3.45). For a given* D *let* β_i *be the number of times* i *enters in* D. *Then the codimension of* $\mathrm{Hilb}_D^H(\mathbb{P}^2)$ *in* $\mathrm{Hilb}^H \mathbb{P}^2$ *is given by* [17]

$$\mathrm{cod}\, \mathrm{Hilb}_D^H(\mathbb{P}^2) = \sum_{\nu \leq i \leq \tau+1} \beta_i(\beta_i + \Delta^3(H)_i) \qquad (5.5.5)$$

PROOF. Statements (i) – (iii) are equivalent to Campanella's description of possible D in [**Cam**]. The reader may see the case $D_{\max}$ treated in the proof of Theorem B.13.

---

[17]This statement depends on Boij's announcement (4.4.19) in [**Bo4**]

Let us fix $j \geq 2\tau + 2$. Let $T = \mathrm{Sym}(H, j)$. Suppose $D_2 \supset D_1$ and $D_2 = D_1 \cup G_1$. Then the associated sequences of generator degrees for Gorenstein ideals in $\mathbb{P}Gor(T)$ satisfy $\tilde{D}_2 = \tilde{D}_1 \cup (G_1, \sigma(G_1))$ where $\sigma(d) = j+3-d$. Passing to the sequences C_2, C_1 as in Theorem 5.25(v), we see that $C_2 = C_1 \cup (B_1, -B_1)$ where the correspondence between G_1 and B_1 is bijective, given by $d \mapsto j+3-2d$. Conversely, by Lemma 5.61 each $C_2 = C_1 \cup (B_1, -B_1)$ as in Theorem 5.25(v.f) is obtained as above from $D_2 = D_1 \cup G_1$.

Let us now show the frontier property. The proof of Theorem 5.60 shows that if Z' belongs to the closure of $Hilb_D^H(\mathbb{P}^2)$ then its sequence of generator degrees D' satisfies $D' \supset D$. Let $D' \supset D$. Then $\tilde{D}' \supset \tilde{D}$, so associating to them the sequences C', C as above and using Theorem 5.25(v.f) we conclude that $Gor_{\tilde{D}'}(T)$ is contained in the closure of $Gor_{\tilde{D}}(T)$. Using the dominance property from Lemma 5.61, we conclude that this implies $Hilb_{D'}^H(\mathbb{P}^2)$ is contained in the closure of $Hilb_D^H(\mathbb{P}^2)$.

According to (5.5.4) and Theorem 5.31 the codimension of $Hilb_D^H(\mathbb{P}^2)$ in $Hilb^H\mathbb{P}^2$ equals the codimension of $Gor_{\tilde{D}}(T)$ in $Gor(T)$. The latter was calculated by M. Boij in [**Bo4**] and equals (see (4.4.19) and notice that when $j \geq 2\tau + 2$ is odd the extra term in degree $(j+3)/2$ is zero)

$$\mathrm{cod}\, Gor_{\tilde{D}}(T) = \frac{1}{2} \sum_{i=1}^{j+2} \tilde{\beta}_i \tilde{\beta}_{j+3-i}$$

$$= \sum_{i=\nu}^{\tau+1} \beta_i (\beta_i + \Delta^3(H)_i) \qquad \text{by (5.5.3)}.$$

This proves (5.5.5).   □

EXAMPLE 5.64. We again consider $H = (1, 3, 6, 8, 9, 9, \dots)$ (cf. Example 5.35). One has

$$\Delta^2(H) = (1, 1, 1, -1, -1, -1, 0, 0, \dots)$$
$$\Delta^3(H) = (1, 0, 0, -2, \ 0, \ \ 0, 1, 0, \dots)$$

So $D_{\min} = (3, 3)$, $D_{\max} = (3, 3, 4, 5)$ and $D_{\max} = D_{\min} \cup (4, 5)$. One obtains 4 possible sequences of generator degrees: $D_{\min}$, $D_1 = (3, 3, 4)$, $D_2 = (3, 3, 5)$, and $D_{\max}$. One has $\dim Hilb^H\mathbb{P}^2 = 16$ since 8 general points $\{p_1, \dots, p_8\}$ in $\mathbb{P}^2$ determine a pencil of cubics whose base locus $\{p_1, \dots, p_8, p_9\}$ is a general element of $Hilb^H\mathbb{P}^2$ (see Example 5.35). According to (5.5.5) $\dim Hilb_{D_1}^H(\mathbb{P}^2) = \dim Hilb_{D_2}^H(\mathbb{P}^2) = 15$

and dim $Hilb_{D_{\max}}^H (\mathbb{P}^2) = 14$. We want to see what are the general configurations of points in the Betti strata associated with D_1, D_2 and $D_{\max}$. Let $Z \in Hilb^H \mathbb{P}^2$.

Case 1. The linear system $|(\mathcal{I}_Z)_3|$ has no fixed components. If ϕ_1, ϕ_2 is a basis of $(\mathcal{I}_Z)_3$, then they form a regular sequence, so the ideal $J = (\phi_1, \phi_2)$ has the Koszul resolution

$$0 \longrightarrow \mathcal{O}_{\mathbb{P}^2}(-6) \longrightarrow \mathcal{O}_{\mathbb{P}^2}(-3) \oplus \mathcal{O}_{\mathbb{P}^2}(-3) \longrightarrow J \longrightarrow 0.$$

Comparing with the resolution from Example 5.35 we see that $Z \in Hilb_{D_{\min}}^H (\mathbb{P}^2)$.

Case 2. The linear system $|(\mathcal{I}_Z)_3|$ has a fixed component which is an irreducible conic C with equation $F_2(x) = 0$. A general set Z of points of this type consists of 8 points on a conic $\{p_1, \ldots, p_8\} \subset C$ and $p_9 \notin C$ and $|(\mathcal{I}_Z)_3| = C + |\mathcal{O}_{\mathbb{P}^2}(1)(-p_9)|$. The number of parameters for such configurations is:

- $5 \cdot 2 = 10$ for 5 general points fixing a conic;
- 3 for the remaining points on the conic;
- $1 \cdot 2 = 2$ for the remaining point in $\mathbb{P}^2$.

The total is 15. We claim that for a general set of points Z of this type one has $R_1(\mathcal{I}_Z)_3 \subsetneq (\mathcal{I}_Z)_4$, thus $v_4(\mathcal{I}_Z) \geq 1$. From $H^1(\mathbb{P}^2, \mathcal{O}_{\mathbb{P}^2}(2)) = 0$ one has an exact sequence

$$0 \longrightarrow H^0(\mathcal{O}_{\mathbb{P}^2}(2)) \longrightarrow H^0(\mathcal{O}_{\mathbb{P}^2}(4)) \longrightarrow H^0(C, \mathcal{O}_C(4)) \longrightarrow 0.$$

So $|\mathcal{O}_{\mathbb{P}^2}(4)(-\sum_{i=1}^8 p_i)|$ has a base locus $\{p_1, \ldots, p_8\}$ and for a general point $p_9 \in \mathbb{P}^2$ there is an irreducible quartic ($F_4(x) = 0$) which contains $p_1, \ldots, p_9$. Since all polynomials from $R_1(\mathcal{I}_Z)_3$ are divisible by F_2 we see that $R_1(\mathcal{I}_Z)_3 \subsetneq (\mathcal{I}_Z)_4$. From the two sequences D with $\beta(D)_4 \geq 1$ only $D_1 = (3, 3, 4)$ corresponds to a stratum of dimension ≥ 15, and in fact dim $Hilb_{D_1}^H (\mathbb{P}^2) = 15$. This shows the general element of the stratum $Hilb_{D_1}^H (\mathbb{P}^2)$ consists of 8 points on a conic and a point outside the conic.

Case 3. The linear system $|(\mathcal{I}_Z)_3|$ has a fixed component which is a line L. Then $|(\mathcal{I}_Z)_3| - L$ is a pencil of conics with base locus of 4 points. A general set of points of this type consists of 5 general points on a line $\{p_1, \ldots, p_5\} \subset L$ and 4 general points outside the line. These 4 points are complete intersection of 2 conics with equations $F_2'(x) = 0$, $F_2''(x) = 0$. The number of parameters is $2 \cdot 2 + 3 + 4 \cdot 2 = 15$. If a quartic Q contains Z it should contain L since $(Q \cdot L) = 4$, so $Q - L = (F_3(x) = 0)$ where F_3 vanishes on the 4 points outside L. From Max Noether's $AF + BG$ Theorem one has $F_3 = AF_2' + BF_2''$, thus $(\mathcal{I}_Z)_4 = R_1(\mathcal{I}_Z)_3$. Similarly to Case 2 one shows $R_1(\mathcal{I}_Z)_4 \neq$

$(\mathcal{I}_Z)_5$, hence the degree sequence is $D_2 = (3,3,5)$ and the described configuration occurs for a general element of $Hilb^H_{D_2}(\mathbb{P}^2)$.

Case 4. It remains to find a configuration depending on 14 parameters with sequence $D_{\max} = (3,3,4,5)$. Such a configuration should be limit of both the configurations associated with D_1 and D_2 by the frontier property of Theorem 5.63. This suggests that we consider $\{p_1, \dots, p_5\} \subset L_1$, $\{p_6, p_7, p_8\} \subset L_2$, $p_9 \neq L_1 \cup L_2$. These sets of points depend on $(2 \cdot 2 + 3) + (2 \cdot 2 + 1) + 1 \cdot 2 = 14$ parameters. Being a limit of sets from Case 2 and Case 3 they satisfy $v_4(\mathcal{I}_Z) \geq 1$, $v_5(\mathcal{I}_Z) \geq 1$. Thus the only possible sequence of generator degrees is $(3,3,4,5) = D_{\max}$. This proves that the general element of $Hilb^H_{D_{\max}}(\mathbb{P}^2)$ consists of 5 points on a line, 3 points on another line and 1 point outside the two lines.

Notice that in this example we do not need (5.5.5) to conclude that $\dim Hilb^H_D(\mathbb{P}^2)$ equals 15 for $D = D_1$ or D_2 and 14 for $D = D_{\min}$. Indeed the dimensions are less or equal to these numbers by the frontier property, and by explicitly constructing families with the corresponding number of parameters we prove equality for the dimension numbers. Note, this is the configuration of points determined by the alignment character, a configuration which has $D_{\max}$; see [**GPS**].

REMARK 5.65. The irreducibility of the Betti strata of the punctual Hilbert scheme $Hilb^s(\mathbb{P}^n)$ for $n = 2$ does not extend to higher embedding dimensions n: for example the ideal $\mathcal{I}_Z$ of the punctual scheme Z consisting of 8 points in general enough position in $\mathbb{P}^4$ has the minimum possible Betti numbers associated to the Hilbert function $H = (1,5,8,8,\dots)$ (i.e., Z satisfies the "Minimum resolution conjecture"), according to a computation using "Macaulay". But the degree-8 nonsmoothable punctual scheme W concentrated at the origin of $\mathbb{A}^4$ determined by 7 general quadric forms (see Definition 6.20 and [**EmI1, I4**]) has the same Betti numbers.

Any Betti sequence possible for a set Z of smooth points in $\mathbb{P}^n$, with Hilbert function H occurs also for an Artin algebra with Hilbert function ΔH, since $R/\mathcal{I}_Z$ is Cohen-Macaualay, and has a nonzero-divisor. The converse is not true. That Betti numbers possible for a graded Artin quotient of $k[x,y,z]$ need not occur for a monomial ideal of Hilbert function T — even when $T = \Delta H$, where $H = H_s(r)$ is the Hilbert function of s points in generic position — was noted by A. Geramita, D. Gregory and L. Roberts in [**GGR**, Theorem 4.7]; this does not yet show there is a missing Betti stratum for smooth degree-s punctual schemes — for elements of $Hilb^H(\mathbb{P}^n) \subset Hilb^s(\mathbb{P}^n)$. However, such missing Betti strata are provided by the counterexamples to the minimal resolution conjecture (MRC) for points in $\mathbb{P}^n$.

The missing Betti number sequences are those for the minimal resolution of an Artin algebra $A = R/I$ defined by a suitable choice of generic homogeneous forms of a given degree t. Thus there are Hilbert functions of the form $H_s(r)$ (here $H_s(r)_i = \min(s, r_i)$), and a Betti number sequence β possible for an Artin algebra A having Hilbert function $H(A) = \Delta H_s(r)$, that cannot occur for a smooth zero-dimensional scheme of Hilbert function $H_s(r)$. An example is the sequence of Betti numbers for the Artin quotient $A = k[x_1, \ldots, x_6]/I$, where I is generated by 17 general enough quadratic forms, of Hilbert function $T = (1, 6, 4, 0) = \Delta H$, with $H = H_{11}(7) = (1, 7, 11, 11, \ldots)$: it is well known that the scheme Z consisting of 11 general enough points of $\mathbb{P}^6$ does not satisfy the minimal resolution conjecture, as the minimal resolution has an extra term $R(-5)^1, R = k[x_0, \ldots, x_6]$ in the third and fourth syzygies (see [EiP1, DGM]); a check using "Macaulay" showed that the Artin algebra A has the expected minimal resolution, without the extra terms. This example is related to a Fröberg Conjecture of [Fr], and to the following related conjecture shown to be equivalent by R. Fröberg (unpublished; see [I8, p.321]):

CONJECTURE. *The minimal resolution of an ideal $I(F)$ generated by s generic homogeneous forms $F = (f_1, \ldots, f_s)$ of given degrees D is the minimum one that is consistent with their (expected) Hilbert function: that is, the Koszul resolution up to the smallest degree where $(F)_i = R_i$ is possible.*

The questions of which Betti numbers occur for height two arithmetically Cohen-Macaulay (ACM) varieties that are nonsingular curves in $\mathbb{P}^3$, and which occur for smooth arithmetically Gorenstein (AG) curves in $\mathbb{P}^4$, have been solved, the former by T. Sauer [Sa], the latter by J. Herzog, N. V. Trung, and G. Valla in [HeTV] (we state the latter condition in Theorem 5.76 below, Section 5.7). See also [GMi1, GMi2, KMMNP]; for a summary of work on which Betti numbers are possible for smooth subvarieties of $\mathbb{P}^n$, that are ACM codimension two, or AG of codimension three , see [HeTV] and as well [Mig, §4.3].

5.6. The length of a form, and the closure of the locus $PS(s, j; 3)$ of power sums

We first collect the several definitions of length of a homogeneous polynomial used in the book, and show the connections between them.

We next formulate results and some examples concerning the closure of $PS(s, j; 3)$. The goal of this section is to determine if $f \in \overline{PS(s, j; 3)}$, using intrinsic data from f, or from its Gorenstein ideal $\mathrm{Ann}(f)$.

We now define the several lengths used in the book. Recall from 1.32 that the length $\ell(f)$ of a binary form is the length of a generalized additive decomposition. By Lemma 1.38 this length is just the smallest s such that $f \in \overline{PS(s, j; r)}$, $r = 2$, and we use the latter for $r \geq 3$ as our natural definition of length. We recall also the two other lengths of a form already defined, and add another.

DEFINITION 5.66. LENGTHS OF A FORM. If $f \in \mathcal{D}_j$,

i. The *length $\ell(f)$* is the smallest nonnegative integer s such that $f \in \overline{PS(s, j; r)}$.

ii. The differential length $\ell\mathrm{diff}(f) = \max\{(H_f)_i\}$, the maximum rank of any of the catalecticants $Cat_f(i, j - i; r), 1 \leq i \leq j - 1$.

iii. The scheme length $\ell\mathrm{sch}(f)$ is the smallest degree of a zero-dimensional subscheme Z of $\mathbb{P}^n, n = r - 1$, such that $\mathcal{I}_Z \subset \mathrm{Ann}(f)$.

iv. The smoothable scheme length $\ell\mathrm{schsm}(f)$ of f is the smallest degree of a smoothable zero-dimensional subscheme of $\mathbb{P}^n$ satisfying $\mathcal{I}_Z \subset \mathrm{Ann}(f)$ (see Definition 5.16 for *smoothable* scheme).

We now compare the above lengths.

PROPOSITION 5.67. *The above lengths are finite, and satisfy,*

$$\ell\mathrm{diff}(f) \leq \ell(f) \leq \ell\mathrm{schsm}(f), \text{ and}$$
$$\ell\mathrm{sch}(f) \leq \ell\mathrm{schsm}(f). \tag{5.6.1}$$

PROOF. For the first inequality, $f \in PS(s, j; r)$ implies $f = L_1^{[j]} + \cdots + L_s^{[j]}$, whence $R_i \circ f \in \langle L_1^{[j-i]}, \ldots, L_s^{[j-i]} \rangle$, of dimension no greater than s, so $f \in PS(s, j; r)$ implies $ldiff(f) \leq s$. This inequality carries over to the closure of $PS(s, j; r)$, showing $\ell\mathrm{diff}(f) \leq \ell(f)$. The second inequality $\ell(f) \leq \ell\mathrm{schsm}(f)$ is Lemma 5.17. The last one is immediate from the definitons. Since every $f \in D_j$ satisfies, $f \in PS(r_j, j; r)$, with $r_j = \dim_k R_j$, there is a smooth annihilating scheme of degree no greater than r_j; this shows that each of the lengths is finite. $\square$

For a finer comparison between the lengths, and further results concerning the uniqueness of the annihilating scheme, see Section 6.1.

THEOREM 5.68. *Suppose that $f \in \mathcal{D}_j$. If f has a tight smoothable annihilating scheme Z of degree s, then $\ell(f) = s$.*

PROOF. Since $f \in \overline{PS(s,j;r)}$ implies $\ell\mathrm{diff}(f) \leq s$ (see 5.67 for a proof), the hypothesis implies $f \in \overline{PS(s,j;r)} - \overline{PS(s-1,j;r)}$, which is the definition of $\ell(f) = s$. $\qquad\square$

Since any zero-dimensional subscheme of $\mathbb{P}^2$ is smoothable, by Fogarty's result [**Fog**], we have

COROLLARY 5.69. *If $r = 3$ and f has an annihilating scheme of degree s, then $f \in \overline{PS(s,j;3)}$ and $\ell(f) \leq s$. If f has a tight annihilating scheme of degree s, then $\ell(f) = s$.*

PROOF. This is immediate from $\ell\mathrm{diff}(f) \leq \ell\mathrm{schsm}(f)$ in equation (5.6.1), the definitions, Fogarty's result, and Lemma 5.17 or Theorem 5.68. $\qquad\square$

We now determine the Hilbert functions that occur for $f \in \overline{PS(s,j;3)}$.

PROPOSITION 5.70. *The Zariski closure $\overline{PS(s,j;3)}$ meets each $Gor(T)$ such that $T \leq H(s,j;3)$, or, equivalently, such that $\max\{t_i\} \leq s$.*

PROOF. The Buchsbaum-Eisenbud structure theorem implies that any codimension three graded Gorenstein Artin algebra of socle degree $j = 2t$ or $2t+1$ has Hilbert function T satisfying, $(\Delta T_{\leq t}, 0, 0, \dots)$ is an O-sequence (see Proposition 3.3 of [**BE2**] or Theorem 4.2 of [**St1**] as well as Theorem 5.25). Suppose that $T \leq H(s,j,3)$. By Lemma 7.1, there are points f_p of $Gor(T)$ in the closure of $PS(s,j;3)$. $\qquad\square$

We can now answer in most cases Problem 2.21 (see also Problem 0.2) asking for intrinsic criteria to decide if $f \in \overline{PS(s,j;3)}$. Recall that if D is a sequence of generator degrees, $Gor_D(T) \subset Gor(T)$ is the subset of f such that $\mathrm{Ann}(f)$ has generator degrees D. Recall that $Gor_{\mathrm{sch}}(T)$ is defined in (5.3.22) for T containing $(s-a, s, s, s-a)$, and $Gor_{\mathrm{sch}}(T)$ is defined in (5.3.38) for T containing $(s-a, s, s-a)$.

THEOREM 5.71. WHAT IS $\overline{PS(s,j;3)} \cap Gor(T)$?

i. *If T contains (s,s,s), $\overline{PS(s,j;3)} \cap Gor(T) = Gor(T)$.*

ii. *Suppose $j = 2t+1$. If T contains $(s-a, s, s, s-a)$, $a \geq 1$, then $\overline{PS(s,j;3)} \cap Gor(T) \supset Gor_{\mathrm{sch}}(T)$. If furthermore $s \leq r_{t-1}$ then*

$$\overline{PS(s,j;3)} \cap Gor(T) = Gor_{\mathrm{sch}}(T)$$
$$= \{f \in Gor(T) \mid v_{t+2}(f) \geq a\}$$
$$= \{f \in Gor(T) \mid v_{t+2}(f) = a\}. \qquad (5.6.2)$$

iii. *If T contains $(s - a - b, s - a, s, s - a)$, $a \geq 1$, and $b \geq 2a$, then*

$$\overline{PS(s, j; 3)} \cap Gor(T) \supset Gor_{sch}(T) =$$
$$= \{f \in Gor(T) \mid v_{t+1}(f) = a, \, \mathrm{cod}_{R_{t+2}} R_2 \cdot I_t = s\}. \qquad (5.6.3)$$

Proof. The inclusion of $Gor(T)$, $Gor_{sch}(T)$, and $\overline{Gor_{sch}(T)}$, respectively, in the closure $\overline{PS(s, j; 3)}$ is immediate from the smoothability of 0-dimensional schemes in $\mathbb{P}^2$, Lemma 5.17, and Theorems 5.31, 5.39, and 5.46, respectively. Alternatively the inclusion also follows from the irreducibility of these varieties and the fact that their general elements are divided power sums of length s, as proved in those theorems. The equivalence $v_{t+2}(I) \geq a$ with $v_{t+2}(I) = a$ in (ii) is from Lemma 5.37. The opposite inclusion in (ii) follows from the equality $\overline{PS(s, j; 3)} = \overline{Gor(T')}$, $T' = H(s, j, 3)$ and Theorem 7.9 below applied to T and T', since when $s \leq r_{t-1}$ we have $T'_{t+2} = s = T_{t+2} + a$. $\square$

Question.

A. Is there equality also in (ii) when $s > r_{t-1}$?

B. Is it true that in (iii) one has $\overline{PS(s, j; 3)} \cap Gor(T) = \overline{Gor_{sch}(T)}$, the closure in $Gor(T)$, and if not, how to describe intrinsically the closure of $PS(s, j; 3)$?

Example 5.72. Closure of $PS(6, 6; 3)$. The only sequences $T \leq H(6, 6; 3)$ for which there might be a question are those for which both $T \not\supseteq (s, s, s)$, $s \leq 6$ and $t_1 = 3$ (since $t_1 = 2$ is by Lemma 1.22 essentially the binary case for which $\ell(f) = \ell\mathrm{diff}(f)$). There are two such T that are Gorenstein sequences, $T_1 = (1, 3, 5, 6, 5, 3, 1)$ and $T_2 = (1, 3, 4, 5, 4, 3, 1)$. The possible degree sequences for T_1 are $D_{\min} = (2, 3, 4)$, corresponding to complete intersections, and $D_{\max} = (2, 3, 4, 4, 5)$. We are in the situation of (5.3.31) with $j = 6 = 2t$, $a = 1$, $b = 2$. If $f \in Gor_{D_{\min}}(T_1)$ then $v_4(f) = 1 = a$, and Condition 5.3.37 is automatically satisfied since $a = 1$, thus $Gor_{D_{\min}}(T_1) = Gor_{sch}(T_1)$ (see (5.3.38)). By Theorem 5.46 every general enough element of $Gor_{D_{\min}}(T_1)$ is a divided power sum of 6 linear forms, thus $Gor_{D_{\min}}(T_1) \subset \overline{PS(6, 5; 3)}$ and consequently $Gor(T_1) \subset \overline{PS(6, 5; 3)}$.

Theorem 5.71 does not answer whether $Gor(T_2)$ is in the closure; this is a case $a = 1$, $b = 1$ out of the range covered by Theorem 5.46 (see also Remark 5.48). Notice this is the same $Gor(T_2)$, that occured in Example 3.6. Recall that if $f \in Gor(T_2)$, then the ideal $J = \mathrm{Ann}(f)$ satisfies $J_2 = h \cdot V$ where $V \in R_1$ is a two-dimensional subspace. We may suppose without loss of generality that $J_2 = <xy, xz>$, whence, following the argument of Example 3.6, $f \in \langle R_4 J_2 \rangle^\perp = \langle Y^{[6]}, Y^{[5]}Z, \ldots, Z^{[6]}, X^{[6]} \rangle$. It follows that $\ell \mathrm{schsm}(f) \leq 4 + 1 = 5$, since any form in $\mathcal{D}'_6$, (here $\mathcal{D}'$ is the divided power ring of Y, Z), is in the closure of $PS(4, 6; 2)$, by Theorem 1.44, and the presence of $X^{[6]}$ adds one. We conclude that $\overline{Gor(T_2)} \subset \overline{PS(5, 6; 3)}$! Thus,

$$\overline{PS(6, 6; 3)} = \bigcup_{T \leq H(6,6,3)} \overline{Gor(T)}.$$

REMARK 5.73. We'd like to thank P. Rao for noting that a morphism p from a component C of $Gor(T)$ to $\mathbf{Hilb}^s \mathbb{P}^{r-1}$ does not in general extend to a morphism $\tilde{p}$ from the closure of C in $\mathbb{P}(\mathcal{D}_j)$. For example, if $r = 3, R = k[x, y, z]$, there is a morphism p from $Gor(T), T = (1, 3, 3, 3, 1)$ to $\mathbf{Hilb}^3 \mathbb{P}^2$ given by $p: f \to \mathrm{Proj}(R/Sat(Ann(f)_3)$ (see Theorem 5.31). We have

$$\overline{Gor(T)} - Gor(T) = Gor(T_1) \cup Gor(T_2) \cup Gor(T_3),$$

$$T_1 = (1, 2, 3, 2, 1), T_2 = (1, 2, 2, 2, 1), \text{ and } T_3 = (1, 1, 1, 1, 1).$$

The DP-form $g = X^{[4]} + Y^{[4]} + (X + Y)^{[4]} \in Gor(T_1)$ has many additive decompositions of length three, lying in the divided power ring in X, Y (compare g of Example 1.34). Each such decomposition $L = L_1^{[4]} + L_2^{[4]} + L_3^{[4]}$ is the limit of a curve C_L of decompositions, $L(t) = L_1(t)^{[4]} + L_2(t)^{[4]} + L_3(t)^{[4]}, t \in C_L \subset Gor(T)$ where $L_1(t), L_2(t), L_3(t), t \neq o$ are three linearly independent linear forms: for example $X^{[4]} + Y^{[4]} + (X + Y)^{[4]} = \lim_{t \to 0} X^{[4]} + Y^{[4]} + (X + Y + tZ)^{[4]}$; and the image $p(L(t)) = p_1(t)) \cup p_2(t)) \cup p_3(t)) \in \mathbf{Hilb}^3(\mathbb{P}^2)$. Thus the above morphism cannot be extended to g: different ways to approach g along curves in $Gor(T)$ will lead to different limit annihilating schemes.

A similar argument can be applied to the DP-form $X^{[4]} \in Gor(T_3)$: one looks for a distinguished annihilating degree-3 scheme concentrated at the point $p_0 = (1, 0, 0) \in \mathbb{P}^2$: there is one such candidate $\mathrm{Proj}(R/m_{p_0}^2)$, which works for most degenerations, however an approach to $X^{[4]}$ from DP-forms $L_1^{[4]} + L_2^{[4]} + L_3^{[4]}$ whose corresponding sets of three points are on an approximate parabola, will determine a curvilinear length 3 scheme such as $\mathrm{Proj}(R/(y, z^2))$ at p_0.

5.7. Codimension three Gorenstein schemes in $\mathbb{P}^n$

Hitherto we have dealt with the relation between a zero-dimensional "annihilating scheme" Z in $\mathbb{P}^n = \mathrm{Proj}(R)$ and an ideal $I \subset R$ defining an Artinian Gorenstein quotient. There is a similar relation between an arithmetically Gorenstein dimension-d scheme W in $\mathbb{P}^n$ and an arithmetically Cohen-Macaulay (ACM) dimension $d+1$ subscheme Z of $\mathbb{P}^n$. Recall that $\deg(Z)$ denotes the degree of Z. If $H = H(W) = H(R/\mathcal{I}_W)$, then we define the k-th difference function inductively by $\Delta^k H(i) = \Delta^{k-1}H(i) - \Delta^{k-1}H(i-1)$. In particular, if W is a d-dimensional Gorenstein subscheme of $\mathbb{P}^n$, we let $T(W) = \Delta^{d+1}H(R/\mathcal{I}_W)$ be the Gorenstein sequence arising from W: we call T the h-vector of W. In the rest of this Section, by *Gorenstein scheme* W we will mean an arithmetically Gorenstein subscheme of $\mathbb{P}^n$. Its defining ideal $\mathcal{I}_W$ and likewise the defining ideal $\mathcal{I}_Z$ of the ACM scheme Z are the saturated ideals of R associated to W, Z respectively. By *codimension* of an ideal I of R we will mean its height, the dimension of the localization R_I (see [Ei1, p. 225]).

DEFINITION 5.74. TIGHT SUBSCHEME. The d-dimensional Gorenstein subscheme W of $\mathbb{P}^n$ is contained as a tight codimension one subscheme in the ACM subscheme Z of dimension $d+1$ if the following two conditions are satisfied:

   i. $\mathcal{I}_Z \subset \mathcal{I}_W$,
   ii. $\deg(Z) = \max_i\{T(W)_i\}$.

Note that if W is a Gorenstein codimension one subscheme of the ACM variety $Z \subset \mathbb{P}^n$, and if L is a general $(d+1)$-dimensional subspace of R_1, then the triple (L, W, Z) determines a ring $R' = R/(L)$, a codimension one ideal $\mathcal{I}_{Z'} = (\mathcal{I}_Z + (L))/(L)$ of R' (equivalently, the dimension $\dim(R'/\mathcal{I}_{Z'}) = 1$), and an ideal $\mathcal{I}_{W'} = (\mathcal{I}_W + (L))/(L)$ of R' defining a Gorenstein Artin quotient, such that $Z' = \text{Proj}(R'/\mathcal{I}_{Z'}) \subset \text{Proj}(R') = \mathbb{P}^{n-d-1}$ is a dimension zero subscheme of $\mathbb{P}^{n-d-1}$, and the degree $\deg(Z) = \deg(Z')$. Furthermore $\mathcal{I}_{W'}$ is M'-primary, where M' is the maximal graded ideal of R', so $R'/\mathcal{I}_{W'}$ is a graded Gorenstein Artin algebra of Hilbert function $H(R'/\mathcal{I}_W') = T(W)$, the h-vector of W. We let $j =$ the socle degree of $R'/\mathcal{I}_{W'}$. Then we have

LEMMA 5.75. TIGHT SUBSCHEME AND ANNIHILATING SCHEME. *The Gorenstein scheme W is a tight subscheme of the ACM scheme Z of dimension $d+1$ in $\mathbb{P}^n$ iff W is a codimension one subscheme of Z, and if the triple (L, W, Z), where L is a general $d+1$-dimensional subspace of R_1, determines as above a ring $R' = R/(L)$, an ideal $\mathcal{I}_{Z'} = (\mathcal{I}_Z + (L))/(L)$ of R', and an M'-primary Gorenstein ideal $\mathcal{I}_{W'} = (\mathcal{I}_W + (L))/(L)$ of R' having socle degree j, that satisfy one of the following two equivalent conditions*
   i. *The zero-dimensional scheme $Z' = \text{Proj}(R'/\mathcal{I}_{Z'}) \subset \mathbb{P}^{n-d-1}$ defined by $\mathcal{I}_{Z'}$ is a tight annihilating scheme of $f \in (\mathcal{I}_{W'})^{\perp}_j$;*

ii. $\deg(Z) = \deg(Z') = \max_i\{T_i\}$, where $T = H(R'/\mathcal{I}_{W'}) = T(W)$.

NOTE ON PROOF. The Lemma is immediate from the existence and properties of minimal reductions of Cohen-Macaulay ideals (see [**BruH**, §4.5]), and Definition 5.74.   □

REMARK. It follows from Lemma 5.75 that to obtain a dimension d Gorenstein scheme W that cannot be a tight codimension one subscheme, it suffices to lift a Gorenstein Artin algebra A not having a tight annihilating scheme. Such Artin algebras abound: see Examples 5.8, 5.9, or pick any Gorenstein Artin algebra with non-unimodal Hilbert function. However, the lifting problem is nontrivial. For instance it is not obvious that there is an irreducible and generically smooth lifting W of Spec A. As an example, J. Herzog, N. V. Trung, and G. Valla showed [**HeTV**] that there are strong restrictions on the generator degrees and the H-vector in order for a resolution skeleton (Betti numbers) of a height three Gorenstein ideal to occur for a smooth arithmetically Gorenstein curve in $\mathbb{P}^4$. It is not hard to see that in particular, it is a consequence of their conditions that in order for there to be such a lifting, H must not contain the subsequence (s, s, s)!

We explain this briefly, referring to our work in Section 5.3.2. Recall that $v = \#$ generators, suppose $v = 2u + 1$, and we will denote by U the degree matrix $U = \deg \Psi$, associated to the alternating matrix Ψ in the Buchsbaum-Eisenbud minimal resolution of a Gorenstein algebra (see (B.2.3) of Theorem B.2). J. Herzog, N. V. Trung, and G. Valla show that always $u_{ik} > 0$ when $i + k = 2u + 3 \ (= v + 2)$, and further,

THEOREM 5.76. [**HeTV**, Theorem 1.2]. *A matrix of integers U as above, is the degree matrix of a smooth arithmetically Gorenstein curve in $\mathbb{P}^4$ iff $u_{ik} > 0$ whenever $i + k = 2u + 4 \ (= v + 3)$.*

We claim that H must not contain the subsequence (s, s, s) in order for the condition of the Theorem to be satisfied. Clearly, $u_{ik} = (c_i + c_k)/2$, where C is the difference sequence $C = E - D$ of generator minus relation degrees. Considering first the sequence C_{min} from (5.3.7), we see that $v = \mu(T) = 2w + 1$, and if $i = k = w + 2$, we have $i + k = v + 3$, but $2c_{w+2} = 2(2t + 1 - j) < 0$ since $j \geq 2t + 2$. From the proof of Lemma 5.29, we see that other sequences C associated to $T \supset (s, s, s)$ must satisfy $C = C_{min} + B$, where B consists of pairs $\alpha, -\alpha$ located on either side of $2t + 1 - j$, hence if now $v = 2u + 1$, again $c_{u+2} = 2t + 1 - j$, so $2c_{u+2} < 0$, contradicting the condition of the above Theorem. This shows our claim.

A. Geramita and J. Migliore in [**GMi2**] study which Betti numbers are possible for graded height three Gorenstein ideals in $R = k[x_1, \ldots, x_4]$: they show that any minimal resolution possible for a height three Gorenstein ideal actually occurs for a Gorenstein "stick figure" — a collection of lines having transversal intersections — in $\mathbb{P}^4$ (ibid. Theorem 3.1).

They also ask, for which zero dimensional Gorenstein subschemes W of $\mathbb{P}^3$ does the initial portion of the ideal $\mathcal{I}_W$ define an ACM curve in $\mathbb{P}^3$ [**GMi2**, Remark 2.3]. The following Theorem answers their question in most cases. If W is a d-dimensional Gorenstein subscheme of $\mathbb{P}^n$, we let $T(W) = \Delta^{d+1} H(R/\mathcal{I}_W)$ be the Gorenstein sequence arising from W (the h-vector of W): we assume it has socle degree j, and we let $s = \max_i\{T_i\}$, $\tau = \min\{i \mid T_i = s\}$. Recall from page 167 that $v_i(I) = \#$ of generators of I having degree i.

THEOREM 5.77. TIGHT SUBSCHEMES IN CODIMENSION 3. *Let W be a codimension 3, Gorenstein subscheme of $\mathbb{P}^{d+3}$, defined by $\mathcal{I}_W$ (a saturated ideal). Then*

  a. *If T contains (s, s, s), (here $j \geq 2\tau + 2$), then there exists a unique codimension 2 ACM subscheme Z in $\mathbb{P}^{d+3}$ such that W is a tight subscheme of Z. Here the saturated ideal $\mathcal{I}_Z$ equals $((\mathcal{I}_W)_{\leq \tau+1})$.*

  b. *If T contains $(s - a, s, s, s - a)$, $a \geq 1$ (here $j = 2t + 1, t = \tau$), then there is a codimension 2 ACM subscheme Z containing W as a tight codimension one subscheme iff $v_{t+2}(I) = a$, the maximum possible value. If so, it is unique and $\mathcal{I}_Z = ((\mathcal{I}_W)_{\leq t+1})$.*

  c. *If T contains $(s - a - b, s - a, s, s - a)$, $a \geq 1$ (here $j = 2t, t = \tau$) then there is a codimension 2 ACM subscheme Z containing W as a tight codimension one subscheme if $\mathcal{I}_W$ can be generated by the Pfaffians of an alternating matrix Ψ, appearing in the Buchsbaum-Eisenbud resolution (B.2.3), such that a of the relations in degree $t + 1$ have zero coefficients on a generators of degree $t + 1$. If also $b \geq 2a$, and $v_{t+1}(I) = a$, then Z is unique and $\mathcal{I}_Z = ((\mathcal{I}_W)_{\leq t})$. Furthermore, if T contains $(s - a - b, s - a, s, s - a)$, then W is contained tightly in a codimension 2 ACM subscheme Z only if for a general $d + 1$-dimensional linear space $L \subset R_1$, the quotient $R'/\mathcal{I}_{W'} = (R/(L))/((\mathcal{I}_W + (L))/(L))$ lies in $Gor_{sch}(T)$, as defined in (5.3.38).*

PROOF. Exactly analogous to the proofs of Theorem 5.31, Theorem 5.39, and Theorem 5.46. Note that for (a) and (b), taking hyperplane sections in $\mathbb{A}^{n+1}$, one gets the necessary inequalities for the generators degrees in the $r = 3$ case, that enter into the hypotheses for the respective Theorems, in order to obtain zero blocks. Also note that almost all steps in the proof of the crucial Claim in Theorem 5.31 are independent of the dimension of R/I. The one exception is eliminating the first alternative after (5.3.14), namely

1. There exists a triple i, j, k, such that $\det B_{ij,k} \notin (g, \phi_k)$.

Here as before, I consists of zero divisors of the class of $\det B_{ij,k}$ in $\overline{R} = R/(g, \phi_k)$. As before $\overline{R}$ has projective dimension no greate than two. The D. Buchsbaum and D. Eisenbud criterion for exactness [**BE1**] and the D. Buchsbaum and M. Auslander formula (5.3.15) imply that the codimension (= height) of associated primes to an R module N can be no greater than the projective dimension of N: see, for example, [**Ei2**, Corollary 20.14], and note that $\operatorname{cod} P = depth\, P\ (= grade(P))$ for primes in the polynomial ring R.

The second part of (c) uses Lemma 5.75 and Theorem 5.46.   □

REMARK. Note that (c) leaves vague what happens when there is no matrix Ψ satisfying the sufficient condition that there is a certain a by a zero block.

PROBLEM 5.78. The d-dimensional tight subscheme problem in $\mathbb{P}^n$ is to determine for a given d-dimensional Gorenstein scheme W whether there is an ACM scheme Z on which W is a tight codimension one scheme. The problem we studied earlier — determining which homogeneous polynomials f have a tight annihilating scheme — can be regarded as a special case, provided one is willing to accept when $\dim Z = 0$, that the Gorenstein Artin algebra $A = R/I$ defines a "ghost scheme" W of dimension $d = -1$. Lemma 5.75 and Theorem 5.77 illustrate the strong connection between the tight subscheme problem and the main theme of this Section, the study of annihilating schemes of a form.

For other work on Gorenstein schemes and liaison, see the comprehensive book by J. Migliore, which includes an analysis and review of the literature, and the references therein [**Mig**]; and as well the articles by S. L. Kleiman and B. Ulrich, and by J .O. Kleppe et al [**KleU, KMMNP**].

CHAPTER 6

Forms and Zero-Dimensional Schemes, II: Annihilating Schemes and Reducible $Gor(T)$

The previous chapter was mainly devoted to the case of 3 variables. We here further study the apolarity relation between 0-dimensional schemes in $\mathbb{P}^{r-1}$ and DP-forms when $r \geq 3$. One result of our study is an example of a variety $Gor(T) \subset \mathbb{A}(\mathcal{D}_j)$ with at least two irreducible components, when $r \geq 7$, $r \neq 8$. One of the components parameterizes DP-forms f with length $2r$ additive decompositions $f = L_1^{[j]} + \cdots + L_{2r}^{[j]}$ where the annihilating scheme – the points corresponding to $L_1, \ldots, L_{2r}$ – are "self-associated point sets" of $\mathbb{P}^{r-1}$; the other component parameterizes DP-forms f that are not in the closure of $PS(2r, j; r)$ because the annihilating schemes are nonsmoothable "thick points" (Lemmas 6.16, 6.23, Theorem 6.26). These examples have Hilbert function $T = T(j, r)$, where

$$T(j, r) = (1, r, 2r - 1, 2r, \ldots, 2r, 2r - 1, r, 1). \qquad (6.0.1)$$

We give similar examples, with more complicated sequences T, for $r = 5, 6$.

Recall that Z is throughout a degree-s punctual subscheme of $\mathbb{P}^{r-1}$ (thus, $\ell(\mathcal{O}_Z) = s$), and that $\mathcal{I}_Z \subset R$ is its saturated homogeneous defining ideal, so $Z = \mathrm{Proj}(R/\mathcal{I}_Z)$. We denote by $\tau(Z)$ the integer $\tau(Z) = \tau(\mathcal{I}_Z) = \min\{i \mid H(R/\mathcal{I}_Z)_i = s\}$, and by $\sigma(Z) = \tau(Z) + 1$ the Castelnuovo regularity of Z (Definition 1.68). Finally, $\mathrm{Sym}(H_Z, j)$ denotes the symmetrization of the initial portion of H_Z about $j/2$ (see Equation (5.0.1)).

QUESTION. Suppose that f is a general element of $(\mathcal{I}_Z)_j^{\perp}$. Is $H_f = \mathrm{Sym}(H_Z, j)$? In particular, is there equality when Z is locally Gorenstein? If there is equality, can we recover Z from f?

Example 5.10 above shows that when $Z \subset \mathbb{P}^2$ is an arbitrary, non-locally Gorenstein zero-dimensional scheme, then even if $f \in (\mathcal{I}_Z)_j^{\perp}$ is general, the Hilbert function H_f may not equal $\mathrm{Sym}(H_Z, j)$. To obtain equality we will require an assumption even more restrictive than

Z locally Gorenstein (Lemma 6.1). However, we report on a recent extension of Lemma 6.1 to the locally Gorenstein case in Section 6.4.

OUTLINE BY SECTION:

In Section 6.1, we show a key result: if Z is smooth or "conic" we can recover Z from a general element f annihilated by Z, if f has large enough degree (Lemma 6.1). We then give several technical lemmas connecting further f and the tangent space T_f to $Gor(T)$ with Z and $\mathcal{I}_Z^2$. Lemma 6.6 relates closures of components of the punctual Hilbert scheme to closures of components of $Gor(T)$. We apply this to compare the closure of $PS(s, j; r)$, with the variety parameterizing DP-forms $f \in Gor(T)$ having smoothable annihilating schemes (Proposition 6.7).

In Section 6.2 we study the components of $Gor(T)$, $T = T(j, r)$. We begin with some results concerning self-associated point sets Z in $\mathbb{P}^n$: these are sets of $2n + 2$ points in linear general position imposing only $2n + 1$ conditions on quadrics. Those DP-forms f having self-associated points sets as annihilating schemes, are an open dense subset of a component of $Gor(T)$ if $3 \leq r \leq 9$ and $6 \leq j$, or if $3 \leq r$ and $8 \leq j$ (Lemma 6.16). We then consider certain nonsmoothable conic "compressed" schemes concentrated at a single point of $\mathbb{P}^n$: these determine a second component of $Gor(T)$ (Lemma 6.21, Example 6.23). In Theorem 6.26 we show that if $7 \leq r \leq 13$, $r \neq 8$ and $6 \leq j$ or if $9 \leq r$ and $8 \leq j$, these components of $Gor(T)$ are distinct.

In Section 6.3 we extend these results to certain other sequences T. We find examples of reducible $Gor(T)$ when $r = 5, 6$.

In Section 6.4 we report first on an extension of the main Lemma 6.1, by the first author and Y. Cho, and as well on some consequences for finding Gorenstein sequences T for $r \geq 5$ for which $Gor(T)$ has several irreducible components [**ChoI1, ChoI2**]. We also report on examples of reducible $Gor(T)$ found by M. Boij, for arbitrary $r \geq 4$ [**Bo2**].

6.1. Uniqueness of the annihilating scheme; closure of $PS(s, j; r)$

The scheme Z concentrated at the point $p = (1, 0, ..0)$ of $\mathbb{P}^{r-1}$ is conic if Z is defined by a homogeneous ideal $\mathcal{I}_{p,Z}$ in the local ring $\mathcal{O}_p = k\{x_2, \ldots, x_r\}$. We let $\mathcal{D}'$ be the divided power ring on the variables $X_2, \ldots, X_r$, dual to $R' = k[x_2, \ldots, x_r]$. Recall that $H_Z = H(R/\mathcal{I}_Z)$; if Z is conic Gorenstein determined by $g \in \mathcal{D}'_r$, then H_Z is related to $H'_g = H(R'/\operatorname{Ann}(g))$ by $(H_Z)_i = \sum_{u \leq i} (H'_g)_u$.

LEMMA 6.1. HILBERT FUNCTION H_F OF A GENERAL DP-FORM $f \in (\mathcal{I}_Z)_j{}^\perp$, AND UNIQUENESS OF THE ANNIHILATING SCHEME. *Suppose that Z is a degree-s zero-dimensional subscheme of $\mathbb{P}^{r-1}$ let $\tau = \tau(Z)$ and let $j \geq 2\tau$. If either*

   i. *Z is Gorenstein, concentrated at a single point p, and is conic,*
      or
  ii. *Z is smooth,*

Then

   a. *If f is a sufficiently general element of $(\mathcal{I}_Z)_j{}^\perp$, we have $H_f = \operatorname{Sym}(H_Z, j)$, $\ell\operatorname{diff}_k(f) = s$, and Z is a tight annihilating scheme of f.*

If Z is any degree-s zero-dimensional subscheme of $\mathbb{P}^{r-1}$, if $j \geq 2\tau$, $\tau = \tau(Z)$, and if $f \in (\mathcal{I}_Z)_j{}^\perp$ satisfies $H_f = \operatorname{Sym}(H_Z, j)$, then

   b. *If $Y \subset \mathbb{P}^{r-1}$ is a zero-dimensional subscheme of degree $\ell(Y) \leq s$ such that $\mathcal{I}_Y \subset \operatorname{Ann}(f)$, then $\ell(Y) = s$ and $(\mathcal{I}_Y)_i = (\mathcal{I}_Z)_i$ for $i \leq j - \tau$.*
   c. *Let $\tau = \tau(Z)$. If also $j \geq 2\tau + 1$, or $j = 2\tau$ and $(\mathcal{I}_Z)_\tau$ generates the sheaf $\check{\mathcal{I}}_Z(\tau)$, then Z is the unique tight annihilating scheme of f.*
   d. *Let $\nu = \min\{i \mid (H_f)_i \neq r_i\}$, and suppose that $i \leq j - (\tau - \nu)$, then $(\operatorname{Ann}(f)^2)_i = (\mathcal{I}_Z^2)_i$.*

PROOF OF (a). Suppose first that Z is Gorenstein conic, concentrated at the point $p = (1, 0, \ldots, 0)$ of $\mathbb{P}^{r-1}$. The defining ideal $\mathcal{I}_{p,Z}$ in the local ring $\mathcal{O}_p = k\{x_2, \ldots, x_r\}$ is graded Gorenstein, so we may regard it as a graded ideal of R'. By Lemma 2.12, it is the annihilator of a DP-form $g \in \mathcal{D}'$ in the duality between R' and $\mathcal{D}'$. One has $(\mathcal{I}_Z)_i = \sum\limits_{0 \leq u \leq i} x_1^{i-u}(\mathcal{I}_{p,Z})_u$ for every $i \geq 1$, and this is clearly a saturated ideal. Since $H(\mathcal{O}_p / \mathcal{I}_{p,Z}) = \Delta H(R / \mathcal{I}_Z)$, we have $\deg(g) = \tau(Z)$. Let $f = X_1^{[j-\tau]} \cdot g \in \mathcal{D}$. Then

$$
\begin{aligned}
\mathcal{A}_i(f) = R_{j-i} \circ f &= \sum_{\max(0, \tau-i) \leq v \leq \min(\tau, j-i)} X_1^{[i+v-\tau]}(R'_v \circ g) \\
&= \sum_{\max(0, i-(j-\tau)) \leq u \leq \min(\tau, i)} X_1^{[i-u]} \cdot \mathcal{A}'_u(g),
\end{aligned}
\tag{6.1.1}
$$

a direct sum. It follows from (4.4) that $H_f = \operatorname{Sym}(H_Z, j)$. Since H_f is upper semicontinuous, and attains its upper bound $\operatorname{Sym}(H_Z, j)$ for a particular DP-form f, $H_f = \operatorname{Sym}(H_Z, j)$ for a general DP-form f.

When Z is smooth, sufficiently general $f \in (\mathcal{I}_Z)_j{}^{\perp}$ may be made more precise — suffices that f has additive decomposition with polar polyhedron Z. Then the statement follows from Corollary 5.4.

PROOF OF (b) AND (c). These are included in Lemma 5.3(E).

PROOF OF (d). This is a straightforward consequence of Lemma 5.3(Eii). $\qquad\square$

LEMMA 6.2. GENERATION OF THE IDEAL OF A CONIC GOREN-STEIN SCHEME. *Let $r \geq 3$, $\tau \geq 2$, suppose $g \in \mathcal{D}'_\tau$, and let Z be the conic, Gorenstein subscheme of $\mathbb{P}^{r-1}$ concentrated in $(1, 0, \ldots, 0)$ determined by $\mathrm{Ann}(g) \subset R'$ as in Lemma 6.1. Suppose further that g is not a divided power of an element of $\mathcal{D}'_1$ (so $g = L^{[\tau]}$ has no solution $L \in \mathcal{D}'_1$). Then $\mathcal{I}_Z \subset R$ is generated by $(\mathcal{I}_Z)_{\leq \tau}$.*

PROOF OF LEMMA. We have

$$(\mathcal{I}_Z)_i = \sum_{0 \leq u \leq i} x_1^{i-u}(\mathrm{Ann}(g))_u, \qquad (6.1.2)$$

where $\mathrm{Ann}(g)$ is in R'. Since $\tau(Z) = \tau = \deg(g)$ as in the proof of Lemma 6.1(a), we have by Theorem 1.69 that the regularity $\sigma(Z) = \deg(g) + 1$, so we only need to show that the map

$$m : \mathrm{Ann}(g)_\tau \otimes R'_1 \to R'_{\tau+1} \qquad (6.1.3)$$

is surjective. If m should not be surjective, then, since $\mathrm{cod}(\mathrm{Ann}(g)_\tau)$ in R_τ is one, a special case of Macaulay's Hilbert function inequality (Corollary C.7 of Theorem C.2 in Appendix C) shows that the image has codimension one in $R'_{\tau+1}$. In this case, let $J = (\mathrm{Ann}(g)_{\leq \tau})$ be the ideal of R' generated by the elements of degree $\leq \tau$ in $\mathrm{Ann}(g)$. A special case of Gotzmann's persistence theorem [**Got1**] (see Proposition C.32(2)) shows that in $\mathbb{P}^{r-2} = \mathrm{Proj}(R')$ the ideal $(\mathrm{Ann}(g)_{\leq \tau})$ is saturated, and determines the scheme of a point $p = (a_2 : \cdots : a_r) \in \mathbb{P}^{r-2}$. Then $g = (a_2 X_2 + \cdots + a_r X_r)^{[\tau]}$ by the Apolarity Lemma 1.15. This completes the proof.

If $r \geq 3$ and $g \in \mathcal{D}'_\tau$ is general (this is in fact the case we need below), one can obtain the surjectivity of (6.1.3) as a consequence of a result of M. Hochster and D. Laksov [**HoL**]; or, for g a sum of at least $\dim_k R_{\tau+1}/r$ powers, from the Alexander-Hirschowitz Theorem 1.66, via Lemma 2.2, as in the proof of Theorem 1.61, p. 59. $\qquad\square$

EXAMPLE 6.3. If $r = 3$, and $g = X_2 X_3^{[2]} \in \mathcal{D}'$ then $\mathrm{Ann}(g) = (x_2^2, x_3^3)$ in R', and $Z = Spec(R'/\mathrm{Ann}(g))$ has degree 6. The saturated ideal $\mathcal{I}_Z = (x_2^2, x_3^3) \cdot R$ has Hilbert function $H_Z = H(R/\mathcal{I}_Z) = (1, 3, 5, 6, 6, \ldots)$, and is generated by $(\mathcal{I}_Z)_{\leq 3}$. If instead we choose

$g' = X_3^{[3]}$, then $\text{Ann}(g') = (x_2, x_3^4)$ in R', with a generator of degree 4; here Z' has degree 4, and $H_{Z'} = (1, 2, 3, 4, 4, \ldots)$.

Recall that $R' = k[x_2, \ldots, x_r]$; we let $r'_u = \dim_k R'_u = \binom{r+u-2}{u}$.

Lemma 6.4. Hilbert function of $R/\text{Ann}(F)^2$, when Z_F is a conic Gorenstein scheme. *Suppose that $g \in \mathcal{D}'_r$, and let $F = X_1^{[n]}g$. Let $I = \text{Ann}(g) \cap R' \subset R'$ and let $Z = Z_g$ be the scheme at $p = (1, 0, \ldots, 0)$ defined by I (so $I = \mathcal{I}_{Z,p}$ and $I_Z = I \cdot R$). Let $\deg(Z) = \dim_k(R'/I)$ denote the degree of Z, let $\nu(I) = \nu$ denote the order of I, and suppose that $n \geq \tau = \tau(Z)$. Let $J = \text{Ann}(F) \subset R$. Then J satisfies $J = (I_Z, x_1^{n+1})$, the Hilbert function $H(R/J) = \text{Sym}(H_Z, n + \tau)$, and*

i. $J_i = \displaystyle\sum_{\max(\nu, i-n) \leq u \leq i} x_1^{i-u} I_u \oplus \sum_{u \leq i-(n+1)} x_1^{i-u} R'_u$

ii. $J^2 = (I_Z^2, x_1^{n+1} I_Z, x_1^{2n+2})$; *if also* $i \leq \min\{2\nu + n, 2n + 1\}$
$(J^2)_i = (I_Z^2)_i \oplus x_1^{n+1} \cdot (I_Z)_{i-n-1}$.

iii. *If* $i \leq \min\{2\nu + n, 2n + 1\}$ *then*
$H(R/J^2)_i = \displaystyle\sum_{u \leq i} H(R'/I^2)_u - \sum_{\nu \leq u \leq i-n-1} (r'_u - H(R'/I)_u)$

iv. *If* $\tau \leq 2\nu$, *then*
$\dim_k T_F = \displaystyle\sum_{u \leq \tau + n} H(R'/I^2)_u - \sum_{\nu \leq u \leq \tau-1} (r'_u - H(R'/I)_u)$

v. *If* $\tau \leq 2\nu$, *then* $\dim_k T_F = \displaystyle\sum_{\tau \leq u \leq \tau+n} H(R'/I^2)_u + \deg(Z) - 1$.

Furthermore, if $n \geq \tau$, then Z is a tight annihilating scheme for F. If $n = \tau$ and g is not the power of an element in $\mathcal{D}'_1$, or if $n > \tau$, then Z is the unique tight annihilating scheme of F.

Proof. That $H(R/J) = \text{Sym}(H, n+\tau)$ was shown in Lemma 6.1, and follows from (i). The equation (i) is immediate from $J = (I, x_1^{n+1})R = (I_Z, x_1^{n+1})$ and (6.1.1). Also, (ii) and (iii) follow from (i). Theorem 3.9, Lemma 6.1 and (iii) for the degree $i = n + \tau$ imply (iv). Under the assumption $\tau \leq 2\nu$, the sum on the right of (iv) is

$$\sum_{u \leq \tau-1} r'_u + \sum_{\tau \leq u \leq \tau+n} H(R'/I^2)_u - \sum_{\nu \leq u \leq \tau-1} r'_u + \sum_{\nu \leq u \leq \tau-1} H(R'/I)_u.$$

Since $\sum_{u \leq \nu-1} r'_u + \displaystyle\sum_{\nu \leq u \leq \tau-1} H(R'/I)_u = \deg(Z) - 1$, we have (iv) is equivalent to (v). The last statements are immediate from (i), Lemma 6.1, and Lemma 6.2. $\qquad\square$

EXAMPLE 6.5. As in Example 6.3, we suppose $r = 3$ and $g = X_2 X_3^{[2]} \in \mathcal{D}'$, where $\tau(Z) = 3$. Now let $F = X_1^{[3]} g = X_1^{[3]} X_2 X_3^{[2]} \in \mathcal{D}$ of degree $2\tau(Z)$; then $J = \text{Ann}(F) = (x_1^4, x_2^2, x_3^3) \subset R$, of Hilbert function $T = H(R/J) = (1,3,5,6,5,3,1) = Sym(H_Z, 6)$. The scheme $Z = Proj(R/(x_2^2, x_3^3))$ is the unique tight annihilating scheme of F, as $\mathcal{I}_Z$ satisfies the further condition of being generated in degrees $\leq \tau(Z)$ (Lemma 6.1(c)). Here $H(R'/(x_2^2, x_3^3)) = (1,2,3,4,4,3,1)$. The tangent space T_F at $F \in \mathbf{Gor}(T)$ has dimension

$$\dim_k T_F = \dim_k (R/J^2)_6 = \dim_k R_6 / \left(x_1^4 x_2^2, x_2^4, x_2^2 x_3^3, x_3^6 \right)_6 = 17$$
$$= (4 + 4 + 3 + 1 + \deg Z - 1),$$

as in Lemma 6.4(v).

We need a lemma relating the closure of a subvariety $\mathcal{H}$ of the punctual Hilbert scheme, to the closure of the corresponding subvariety $\mathcal{C}$ of $Gor(T)$. Before stating it we introduce some additional notation. Let $\mathcal{H} \subset \mathbf{Hilb}^s \mathbb{P}^{r-1}$ be an irreducible subvariety of the Hilbert scheme of zero dimensional degree-s subschemes of $\mathbb{P}^{r-1}$. Let $Z \in \mathcal{H}$ be general, $\tau = \tau(Z)$, $j \geq 2\tau$ and suppose that for general $f \in (\mathcal{I}_Z)_j^\perp$ one has $\ell \text{diff}_k(f) = s$. Then Z is a tight annihilating scheme of f, and $H_f = T = Sym(H_Z, j)$ (Lemma 5.3(E)). One obtains an irreducible sublocus $\mathcal{C}$ of $Gor(T)$ of dimension $\dim \mathcal{C} \leq \dim \mathcal{H} + s$. By Lemma 5.3 if either $j \geq 2\tau + 1$ or if $j = 2\tau$ and for general $Z \in \mathcal{H}$ the linear system $|\mathcal{O}_{\mathbb{P}^{r-1}}(\tau)(-Z)|$ has a finite number of base points then $\dim \mathcal{C} = \dim \mathcal{H} + s$.

LEMMA 6.6. RELATION BETWEEN CLOSURES IN THE HILBERT SCHEME AND IN $Gor(T)$. Let $\mathcal{H}, \mathcal{C}$ be as above. Let $g \in \mathcal{D}_j$ and suppose that $\ell \text{diff}_k(g) = s$ and $g \in \overline{\mathcal{C}}$. Suppose g has a tight annihilating scheme X with $\tau(X) = \tau'$. Suppose that either $j \geq 2\tau' + 1$, or $j = 2\tau'$ and the space $(\mathcal{I}_X)_{\tau'}$ generates the sheaf $\tilde{\mathcal{I}}_X(\tau')$ (cf. Lemma 5.3 E(iii)). Then $X \in \overline{\mathcal{H}}$.

PROOF. Let $\nu' = j - \tau'$. Then by Lemma 5.3 $\text{Ann}(g)_{\leq \nu'} = (\mathcal{I}_X)_{\leq \nu'}$. Since $g \in \overline{\mathcal{C}}$ one can find a family of DP-forms $\{f(t) \mid t \in W\}$, where W is a smooth affine curve with distinguished point $o \in W$, $g = f(o)$ and $f(t) \in \mathcal{C}$ for $t \neq o$ with a corresponding family of annihilating schemes $\{Z(t) \mid t \in W - \{o\}\}$, $Z(t) \in \mathcal{H}$ (cf. Section 5.2). Furthermore, since $\mathcal{H}$ and $\mathcal{C}$ are irreducible, fixing any open dense subsets of $\mathcal{H}$ and $\mathcal{C}$ one can choose the families so that for $t \neq o$, $f(t)$, $Z(t)$ belong to these sets (so are "general"). Now $\mathbf{Hilb}^s \mathbb{P}^{r-1}$ is proper, so there is a limit scheme $Z(o) \in \overline{\mathcal{H}}$. We wish to prove that $X = Z(o)$. The

Hilbert function T' of $R/\operatorname{Ann}(g)$ satisfies $T' \leq T$ and both attain the maximum value s. Thus from Lemma 5.3 one has

$$j/2 \leq \nu' \leq \nu = j - \tau.$$

Let $I(t) = \operatorname{Ann}(f(t))_{\nu'}$ for $t \neq o$. This is a subspace in $R_{\nu'}$ of codimension s. Since W is smooth there is a limit subspace $I(o) \in Grass(r_{\nu'} - s, R_{\nu'})$. Now, from Lemma 5.3 $\operatorname{Ann}(f(t))_{\nu'} = (\mathcal{I}_{Z(t)})_{\nu'}$, thus $I(o) \subset (\mathcal{I}_{Z(o)})_{\nu'}$ according to Lemma 5.12(i). On the other hand

$$I(o) = \lim_{t \to o} \operatorname{Ann}(f(t))_{\nu'} = \operatorname{Ann}(g)_{\nu'} = (\mathcal{I}_X)_{\nu'}. \tag{6.1.4}$$

Now, our hypothesis assumes that the conditions of Lemma 5.3E(iii) are satisfied for X and g. By its proof we know that the saturation of the graded ideal $R \cdot (\mathcal{I}_X)_{\nu'}$ coincides with $\mathcal{I}_X$. However from $I(o) \subset (\mathcal{I}_{Z(o)})_{\nu'}$ and (6.1.4) this saturation is contained in $\mathcal{I}_{Z(o)}$. Thus $\mathcal{I}_X \subset \mathcal{I}_{Z(o)}$, which implies $X = Z(o)$ since these are 0-dimensional schemes of equal degree s.   □

A DP-form $f \in PS(s, j; r)$ iff it has a smooth degree-s annihilating scheme (Lemma 5.3 B). Our next proposition relates DP-forms from the border $\overline{PS(s, j; r)} - PS(s, j; r)$ with smoothable punctual schemes (see Definition 5.16).

PROPOSITION 6.7. CLOSURE OF $PS(s, j; r)$. Let $f \in \mathcal{D}_j$.
A. If f has a smoothable annihilating scheme Z of degree s, then $f \in \overline{PS(s, j; r)}$.
B. Let $\ell\mathrm{diff}_k(f) = s$ and suppose $f \in \overline{PS(s, j; r)}$. Let Z be a tight annihilating scheme of f, let $\tau(Z) = \tau$ and suppose that either $j \geq 2\tau + 1$ or $j = 2\tau$ and the space $(\mathcal{I}_Z)_\tau$ generates the sheaf $\tilde{\mathcal{I}}_Z(\tau)$ Then Z is smoothable.
C. Suppose $f \in \overline{PS(s, j; r)}$ and let $s \leq j + 1$. Then the converse to (A) holds: f has a smoothable annihilating scheme of degree s.

PROOF OF (A). This is Lemma 5.17.
PROOF OF (B). This is immediate from Lemma 6.6 where $\mathcal{H} \subset$ Hilb$^s \mathbb{P}^{r-1}$ parameterizes the smooth, degree-s schemes.
PROOF OF (C). As in the proof of Lemma 6.6 we can find a family of DP-forms $\{f(t) \mid t \in W\}$, $f(o) = f$ and a flat family of punctual degree-s schemes $\{Z(t) \mid t \in W\}$, such that for $t \neq o$ $Z(t)$ is a smooth annihilating scheme of $f(t)$. We may also assume the curve is chosen so that when $t \neq o$, $H_{f(t)} = H(s, j, r)$ and

$$H(R/\mathcal{I}_{Z(t)}) = H = (1, \ldots, r_{\tau-1}, \overset{\tau}{s}, \overset{\tau+1}{s}, \ldots)$$

where τ is a fixed number, equal to the order of each $\mathrm{Ann}(f(t))$. We want to prove that $Z(o)$ is an annihilating scheme of $f(o) = f$. Let $\mathcal{I}_0$ be the limit ideal, $\mathcal{I}_o = \lim_{t \to o} \mathcal{I}_{Z(t)}$. According to Lemma 5.12 we have $\mathcal{I}_o \subset \mathcal{I}_{Z(o)}$ and clearly $H(R/\mathcal{I}_o) = H$. According to Theorem 1.69 $\tau(Z(o)) \leq s - 1$. So, our hypothesis $s \leq j + 1$ implies $\tau(Z(o)) \leq j$. From the inclusion $\mathcal{I}_o \subset \mathcal{I}_{Z(o)}$ and the form of the Hilbert function of $R/\mathcal{I}_o$ we conclude that $(\mathcal{I}_o)_i = (\mathcal{I}_{Z(o)})_i$ for every $i \geq \tau(Z(o))$, in particular for $i = j$. Now, $(\mathcal{I}_{Z(t)})_j \subset (\mathrm{Ann}\, f(t))_j$ for $t \neq o$ implies $(\mathcal{I}_o)_j \subset (\mathrm{Ann}(f))_j$, thus $(\mathcal{I}_{Z(o)})_j \subset (\mathrm{Ann}(f))_j$. From Lemma 2.15 we conclude that $\mathcal{I}_{Z(o)} \subset \mathrm{Ann}(f)$. This proves that $Z(o)$ is an annihilating scheme of f; it is of degree s and by construction it is smoothable. $\quad\square$

PROBLEM 6.8. Is it true that every $f \in \overline{PS(s, j; r)}$ has an annihilating scheme of degree $\leq s$.

NOTE (and Warning). We recall from Remark 5.73 that a map from a component C of $Gor(T)$ to the Hilbert scheme $\mathbf{Hilb}^s(\mathbb{P}^{r-1})$ generally does not extend to a map from the closure of C in $\overline{Gor(T)}$.

6.2. Varieties $Gor(T)$, $T = T(j, r)$, with several components

Recall that $T(j, r) = (1, r, 2r - 1, 2r, \ldots, 2r, 2r - 1, r, 1)$. We first study DP-forms whose annihilating scheme is a set of "self-associated points" in $\mathbb{P}^{r-1}$ (pages 214–220). We next study DP-forms whose annihilating scheme is a conic compressed Gorenstein scheme, concentrated at a single point $(1, 0, \ldots, 0)$ of $\mathbb{P}^{r-1}$ (pages 220–223). We show that these two kinds of DP-forms determine distinct irreducible components of $Gor(T)$ for $T = T(j, r)$ when (j, r) are suitable (Theorem 6.26). This result generalizes to certain other $Gor(T)$, where $T = \mathrm{Sym}(H, j)$, and H is the Hilbert function of a "generic" compressed algebra (Theorem 6.27, Corollaries 6.28, 6.29).

Forms having self-associated points sets as annihilating schemes. We will say that a subset $Z \subset \mathbb{P}^{r-1}$ consisting of $2r$ points in linear general position is "self-associated" if $\dim_k(R/\mathcal{I}_Z)_2 = 2r - 1$, i.e. Z imposes one less condition on quadrics than a generic set of $2r$ points. These have been studied by A. Coble [Co2, Co1] and more recently by V. Shokurov, I. Dolgachev and D. Ortland, A. Geramita and F. Orecchia (special case of $2r$ points on a rational normal curve), and D. Eisenbud and S. Popescu, but key information, such as the regularity degree and possible Hilbert functions of $\mathcal{I}_Z^2$ appear to be unknown (see [Sho, DO, GO, EiP2], and Example 6.17).

EXAMPLE 6.9. SCHEME CONSISTING OF $2r$ POINTS ON A RATIO-
NAL NORMAL CURVE. Let $r \geq 3$. Suppose that Z is a nonsingular,
reduced scheme consisting of $2r$ general points lying on a rational normal curve $(1 : t : t^2 : \cdots : t^{r-1})$ of $\mathbb{P}^{r-1}$. Then the Hilbert function

$$H_Z = (1, r, 2r - 1, 2r, 2r, \ldots), \tag{6.2.1}$$

and Z is an arithmetically Gorenstein scheme with $\tau(\mathcal{I}_Z) = 3$ (see the
Claim p. 131 of [GO]), as well as the discussion below).

EXAMPLE 6.10. HYPERPLANE SECTIONS OF CANONICAL CURVES.
Another example of reduced schemes Z consisting of $2r$ points with
Hilbert function H_Z as above in (6.2.1) is given by the general hyperplane sections of canonical curves C of genus $r + 1$ in $\mathbb{P}^r$ (see [GriH]
or [ACGH]). If the curve is either trigonal or isomorphic to a plane
quintic, then the intersection of quadrics which contain C is a rational normal scroll, respectively a Veronese surface, and the general
hyperplane section consists of $2r$ points in general position lying on a
rational normal curve. This latter is a particular case of the one in
Example 6.9.

The ordered sets of $2r$ points in $\mathbb{P}^{r-1}$ which are in linear general position and which impose $2r - 1$ conditions on quadrics, thus
whose Hilbert function satisfies (6.2.1), are classically known as *self-
associated point sets* (see [DO, Ch. III, Section 3]). We shall here use
the same notation also for unordered point sets. According to [DO,
p.51], the self-associated point sets in $\mathbb{P}^{r-1}$ form an irreducible variety
W of dimension

$$\dim(W) = r(r - 1)/2 + (r - 1)(r + 1). \tag{6.2.2}$$

Let W_1 be the subvariety of those self-associated point sets that lie on
some rational normal curve. We need a lemma.

LEMMA 6.11. GENERATION OF THE IDEALS OF SELF-ASSOCIATED
POINTS SETS. *For every general* $Z \in W$ *the ideal* $\mathcal{I}_Z$ *is generated by*
$(\mathcal{I}_Z)_2$. *For every general* $Z \in W_1$ *the ideal* $\mathcal{I}_Z$ *is generated by* $(\mathcal{I}_Z)_{\leq 3}$.
In each case Z *is arithmetically Gorenstein.*

PROOF. The condition that $\mathcal{I}_Z$ is generated in degree 2, respectively, in degrees no greater than 3, is an open condition, so it suffices to
show that it is not empty. We choose a smooth canonical curve $C \subset \mathbb{P}^r$
of genus $r + 1$. Consider a general hyperplane section $Z = C \cap H$. The
curve C is projectively normal [ACGH], so by a lemma in [S-D], or
[Is1] one has that the restriction map

$$r_H : \oplus_{n \geq 0} H^0(\mathbb{P}^r, \mathcal{O}(n)) \to \oplus_{n \geq 0} H^0(H, \mathcal{O}_H(n))$$

sends $\mathcal{I}_C$ surjectively onto $\mathcal{I}_Z$, furthermore this surjection is an isomorphism in degree 2. Now, using the Enriques-Petri theorem [**ACGH**] we conclude that if we choose the canonical curve C of genus $r+1 \geq 4$ to be neither trigonal nor isomorphic to a plane quintic, then $\mathcal{I}_C$ is generated by $(\mathcal{I}_C)_2$, thus the ideal of $Z \in W$ is generated in degree 2. Similarly, if we choose C to be trigonal, then $\mathcal{I}_C$ is generated by $(\mathcal{I}_C) \leq 3$, thus for the hyperplane section $Z \in W_1$ the ideal $\mathcal{I}_Z$ is generated by $(\mathcal{I}_Z) \leq 3$. That a general $Z \in W$ is Gorenstein results from Z being Gorenstein in a special case, as $Z \in W_1$, or Z arising from a suitable cyclic polytope. $\square$

REMARK 6.12. Shokurov proved in [**Sho**] by a direct analysis of point sets that $\mathcal{I}_Z$ is generated by $(\mathcal{I}_Z)_2$ for every set Z of $2r$ points in general position in $\mathbb{P}^{r-1}$ which impose $2r - 1$ conditions on quadrics, and do not lie on a rational normal curve.

COROLLARY 6.13. UNIQUENESS OF SELF-ASSOCIATED ANNIHILATING SCHEME. *Suppose that $j \geq 6$, and that $Z \in W$ is a self-associated point set. Then a general element F of $((\mathcal{I}_Z)_j)^\perp$ satisfies, $H_F = T(j,r)$, the sequence in (6.0.1). If $Z \in W - W_1$ is general, or if $Z \in W_1$ is general, then Z is the unique minimum degree scheme with $\mathcal{I}_Z \subset \mathrm{Ann}(F)$.*

PROOF. Immediate from Lemma 6.1a,c and Lemma 6.11. $\square$

EXAMPLE 6.14. TANGENT SPACE FOR A FORM WHOSE ANNIHILATING SCHEME IS $2r$ POINTS ON A RATIONAL NORMAL CURVE. Suppose $F \in ((\mathcal{I}_Z)_j)^\perp$, $Z \in W_1$. We calculated the Hilbert function $H = H(R/(\mathrm{Ann}(F))^2$, using "Macaulay" for the seven pairs $(j,r) = (6,r)$, $3 \leq r \leq 9$, where $F \in ((\mathcal{I}_Z)_j)^\perp$, $Z \in W_1$ is $2r$ special points (see Example 6.17), and $H(R/J) = H(6,r)$. We found,

$$H(R/(\mathrm{Ann}(F))^2) =$$

$$= \begin{cases} (1, 5, 15, 35, 50, 51, \mathbf{44}, 26, 12, 5, 1) & \text{when } r = 5, \\ (1, 6, 21, 56, 77, 74, \mathbf{62}, 32, 14, 6, 1) & \text{when } r = 6, \\ (1, 7, 28, 84, 110, 101, \mathbf{83}, 38, 16, 7, 1) & \text{when } r = 7, \\ (1, 8, 36, 120, 149, 132, \mathbf{107}, 44, 18, 8, 1) & \text{when } r = 8, \\ (1, 9, 45, 165, 194, 167, \mathbf{134}, 50, 20, 9, 1) & \text{when } r = 9. \end{cases}$$

By Theorem 3.9 we have (the formula is valid for $3 \leq r \leq 9$)

$$\dim_k T_F = \begin{cases} 44 & \text{if } r = 5 \\ 62 & \text{if } r = 6 \\ 83 & \text{if } r = 7 \\ 107 & \text{if } r = 8 \\ 134 & \text{if } r = 9 \end{cases} = 2r^2 - \binom{r-1}{2}. \qquad (6.2.3)$$

EXAMPLE 6.15. DIMENSION OF THE VARIETY OF FORMS WITH ANNIHILATING SCHEME $2r$ POINTS ON A RATIONAL NORMAL CURVE. The variety $C_1 \subset Gor(T)$, $T = T(j,r)$ parameterizing such forms F arising from rational normal curves in $\mathbb{P}^{r-1}$ is dominated by a variety $\tilde{C}_1$ which is fibred over the variety RNC parameterizing rational normal curves X in $\mathbb{P}^{r-1}$, by a fibre of dimension $4r$. To see this, choose a general set Z of $2r$ points on X, then $(\mathcal{I}_Z)_j^{\perp}$ is of dimension $2r$; the fiber consists of the pairs (Z, F), $F \in (\mathcal{I}_Z)_j^{\perp})$, so we have $\dim(\tilde{C}_1) = \dim(RNC) + 4r$. By Corollary 6.13, if $j \geq 6$ then a general $F \in (\mathcal{I}_Z)_j^{\perp}$ has a unique annihilating scheme of degree $2r$ (namely Z), thus the map $\tilde{C}_1 \to C_1$ is of finite degree (is birational if $char(k) = 0$), so $\dim(\tilde{C}_1) = \dim(C_1)$. Further, $\dim(RNC) = r^2 - 4$, since a rational normal curve is determined by an r-tuple of homogeneous polynomials of degree $r - 1$ in 2 variables up to the action of GL_2. So, we obtain for $r \geq 3$, $j \geq 6$ that

$$\dim(C_1) = (r^2 - 4) + 2r + 2r = r^2 + 4r - 4. \qquad (6.2.4)$$

LEMMA 6.16. DIMENSION OF THE VARIETY PARAMETERIZING DP-FORMS HAVING ANNIHILATING SCHEME A SELF-ASSOCIATED POINT SET. *Suppose that $3 \leq r \leq 9$ and $j \geq 6$, or that $r \geq 3$ and $j \geq 8$, and let $T = T(j,r) = (1, r, 2r - 1, 2r, \ldots, 2r, 2r - 1, r, 1)$. Consider the constructible set $C \subset Gor(T)$, $T = T(j,r)$ which parameterizes all $F = L_1^{[j]} + \cdots + L_{2r}^{[j]}$, such that the point set $Z = \{P_1, \ldots, P_{2r}\} \subset \mathbb{P}^{r-1}$ which corresponds to the hyperplanes $\{L_i(x) = 0\}$ is self-associated. Then C has dimension*

$$\dim C = \dim W + 2r = 2r^2 - \binom{r-1}{2}. \qquad (6.2.5)$$

Furthermore, $\overline{C}$ is an irreducible component of $\overline{Gor(T)}$. The dimension of the tangent space T_f to $Gor(T)$ at a general point f of C is the same as $\dim C$, and $\mathbf{Gor}(T)$ is generically smooth along C. If $r \geq 3$ and $j \geq 6$, then C lies in an irreducible component of $\mathbf{Gor}(T)$ having at least the dimension given in (6.2.5).

PROOF. The dimension result (6.2.5) uses the calculation of Dolgachev-Ortland for $\dim(W)$ (see [DO] and (6.2.2)) as well as the fact that $2r$ points in linear general position impose independent conditions on the forms of degree $j \geq 3$, thus $\dim_k(\mathcal{I}_Z)_j^\perp = 2r$. Let $f \in C$ be a general point and let $I = \mathrm{Ann}(f)$. We will show that the dimension of the tangent space T_f satisfies

$$\dim_k T_f \leq 2r^2 - \dim_k(I_2) = 2r^2 - \binom{r-1}{2}.$$

Let $Z \in W$ be the annihilating scheme of f; from Lemma 6.11 and Corollary 6.13, $\mathcal{I}_Z = (\mathcal{I}_Z)_{\leq 2} = (I_2)$, and Z is uniquely determined by f. Choose $h \in I_{j-2} - (\mathcal{I}_Z)_{j-2}$. We claim that

  i. Since $f \in (\mathcal{I}_Z)_j^\perp$ is general, we have $h(P_i) \neq 0$ for each $i = 1, \ldots, 2r$.
  ii. $\mathcal{I}_Z^{(2)} \cap R_2 = 0$ (i.e there are no degree 2 polynomials vanishing to order 2 at all of the points P_i).
  iii. If $3 \leq r \leq 10$ and $j \geq 6$, or if $r \geq 3$ and $j \geq 8$, then $\dim_k(R/\mathcal{I}_Z^2)_j \leq 2r^2$.

It follows from (i) and (ii) that $(\mathcal{I}_Z^{(2)})_j \cap h \cdot I_2 = (\mathcal{I}_Z^{(2)})_j \cap h \cdot (\mathcal{I}_Z)_2 = 0$ hence,

$$\dim_k(R/I^2)_j \leq \dim_k(R/\mathcal{I}_Z^2)_j - \dim_k I_2 \leq 2r^2 - \binom{r-1}{2}.$$

Proof of (i). By assumption $Z = \{P_1, \ldots, P_{2r}\} \subset \mathbb{P}^{r-1}$ is a self-associated set whose points are in linear general position. Let $\{L_1, \ldots, L_{2r}\}$ be the corresponding linear forms in $\mathcal{D}$. Since any subset of $2r-1$ points of Z imposes independent conditions on quadrics, it follows from Lemma 1.15 that there is a unique linear relation between $L_i^{[2]}$ (up to multiplication by a constant) and it is of the form

$$c_1 L_1^{[2]} + \cdots + c_{2r} L_{2r}^{[2]} = 0$$

with all $c_i \neq 0$. Now let $g = a_1 L_1^{[j]} + \cdots + a_{2r} L_{2r}^{[j]}$ be an element of C with all $a_i \neq 0$, $g \in (\mathcal{I}_Z)_j^\perp$, clearly f is such element. Then

$$h \circ g = \sum a_i h(L_i) L_i^{[2]} = 0$$

and since at least one $h(P_k) \neq 0$ it follows that $h(L_i) = c_i/a_i \neq 0$ for each i.

Proof of (ii). For a quadric the set of singular points forms a proper linear subspace, so it cannot contain all P_i.

Proof of (iii). When $j \geq 8 = 2\sigma(Z)$, Theorem 1.70 implies that $(\mathcal{I}_Z)_j^2 = (\mathcal{I}_Z)_j^{(2)}$, and $(\mathcal{I}_Z)^{(2)}$ is regular in degree 8. Since the multiplicity of $(\mathcal{I}_Z)^{(2)}$ is $2r^2$, this proves (iii) when $j \geq 8$. If $3 \leq r \leq 9$ and $j = 6$, the result follows from (6.2.3). When $r = 10$ and $j = 6$, we calculated $H(R/\mathcal{I}_Z^2)$ (See Example 6.18), finding $\dim_k(R/\mathcal{I}_Z^2)_6 = 2r^2$. By Theorem 1.70 (ii), and Lemma 6.11 if Z is general in W then $\mathcal{I}_Z^{(2)}$ is regular in degree 6, and of course $\mathcal{I}_Z^2$ is generated by degree 4. Thus equality of $\mathcal{I}_Z^{(2)}$ and $\mathcal{I}_Z^2$ in degree 6 will imply equality in higher degrees. It follows that for Z general in W

$$\dim_k(R/\mathcal{I}_Z^2)_6 = 2r^2 \Rightarrow \text{ for } i \geq 6, \ (\mathcal{I}_Z^2)_i = (\mathcal{I}_Z^{(2)})_i.$$

This and either Example 6.14 ($3 \leq r \leq 9$) or Example 6.17 below ($3 \leq r \leq 10$) complete the proof of (iii) and the lemma. $\qquad\square$

REMARK. In the proof of the above lemma we used a result from [DO] which gives $\dim(W)$. When $j = 6$, $3 \leq r \leq 9$, one can obtain in a more elementary way that $\overline{C}$ is the unique irreducible component of $\overline{Gor(T)}$ which contains C_1, and it satisfies the properties of Lemma 6.11. Namely, the locus in $\operatorname{Sym}^{2r}(\mathbb{P}^{r-1})$ of $2r$ points which impose only $2r - 1$ conditions on quadrics is the corank 1 locus of a $2r \times r_2$ matrix. Thus any of its components has codimension at most $r_2 - (2r - 1)$ in $\operatorname{Sym}^{2r}(\mathbb{P}^{r-1})$, so has dimension at least $2r(r-1) - (r_2 - (2r - 1)) = 3r^2/2 - r/2 - 1$. This yields possibly several components of $\overline{Gor(T)}$ which contain C_1 and are of dimension at least $3r^2/2 - r/2 - 1 + 2r = 2r^2 - \binom{r-1}{2}$. The calculation of the tangent spaces to $Gor(T)$ in (6.2.3) now gives the desired result.

EXAMPLE 6.17. HILBERT FUNCTIONS FOR $\mathcal{I}_Z^2$ AND $\mathcal{I}_Z^{(2)}$ WHEN Z IS $2r$ POINTS ON A *RNC*. We used Macaulay to calculate $H(R/\mathcal{I}_Z^2)$ and $H(R/\mathcal{I}_Z^{(2)})$ for the scheme $Z = \{q_0, q_1, q_{-1}, \ldots, q_{r+1}\}$, $q_i = (1 : i : i^2 : \cdots : i^n)$ on $\mathbb{P}^n$, $n = r - 1$, for $3 \leq n \leq 9$. We tabulate the first differences $SQ(r) = \Delta H(R/\mathcal{I}_Z^2)$ and $SYMSQ(r) = \Delta H(R/\mathcal{I}_Z^{(2)})$ below. The values are consistent with the Conjecture 6.18 below.

When $n = 9$, the ideal $\mathcal{I}_Z$ has 8 degree-3 generators (in addition to the 36 in degree 2); Mats Boij points out that this is a maximum possible number, and that using an ideal of points arising from a suitable polytope, one should find a Gorenstein set of points $S \in W_1$ with the same Hilbert function but with a single degree-3 generator. Thus, it is natural to expect that $H(R/\mathcal{I}_Z^2)$ would be extremal for the above choice.

degree i	0	1	2	3	4	5	6
$SQ(5)$	1	4	10	20	15	1	-1
$SYMSQ(5)$	1	4	10	19	13	3	0
$SQ(6)$	1	5	15	35	21	-3	-2
$SYMSQ(6)$	1	5	15	31	17	3	0
$SQ(7)$	1	6	21	56	26	-9	-3
$SYMSQ(7)$	1	6	21	46	21	3	0
$SQ(8)$	1	7	28	84	29	-17	-4
$SYMSQ(8)$	1	7	28	64	25	3	0
$SQ(9)$	1	8	36	120	29	-27	-5
$SYMSQ(9)$	1	8	36	85	29	3	0
$SQ(10)$	1	9	45	165	25	-39	-6
$SYMSQ(10)$	1	9	45	109	33	3	0

TABLE 6.1. Difference Hilbert functions $SQ(r)$ and $SYMSQ(r)$ for $2r$ special points Z on a RNC in $\mathbb{P}^{r-1}$, calculated by "Macaulay". See Example 6.17.

CONJECTURE 6.18. *For a scheme Y consisting of $2n + 2$ points on a rational normal curve in $\mathbb{P}^n$, the first differences for $SQ(r)$ and $SYMSQ(r)$ are no greater in absolute value than the following values, which occur for the scheme Z of Example 6.17.*

$$SQ(r) = \left(1, n, \binom{n+1}{2}, \binom{n+2}{3}, f(r), 3 - (n-3)(n-2), 3 - n\right)$$

where $f(r) = 2r^2 - \binom{n+3}{3} + (n-1)(n-3) - 3.$

$$SYMSQ(r) = \left(1, n, \binom{n+1}{2}, 2r^2 - \binom{n+2}{2} - 4n, 4n - 3, 3\right).$$

NOTE. Calculations by M. Boij show that for $4 \le n \le 7$, a general self-associated point scheme $Z \subset \mathbb{P}^n$ satisfies $\Delta H(R/\mathcal{I}_Z^2)$ non-negative.

QUESTION 6.19. Let Z be a degree-s punctual subscheme of $\mathbb{P}^{r-1}$. Let $\tau = \tau(Z)$, $H = H(R/\mathcal{I}_Z)$, $T = \mathrm{Sym}(H, 2\tau)$, let $j \ge 2\tau$ and set $T' = \mathrm{Sym}(H, j)$. Suppose that if f is general in $(\mathcal{I}_Z)_{2\tau}^{\perp}$, and F general in $(\mathcal{I}_Z)_j^{\perp}$, then $f \in Gor(T)$ and $F \in Gor(T')$. Then is $\dim_k T_f = \dim_k T_F$? The answer is "yes" if Z is conic Gorenstein, arising from a local ideal $I, \tau \le 2\nu(I_Z)$, and $I^2 \supset R'_{2\tau}$ (Lemma 6.4(v)), and "yes" if H satisfies the condition, $i \le \tau \Rightarrow H_i = \min(r_i, s)$ (Lemma 6.1d and Theorem 1.70).

Forms with non-smoothable annihilating schemes.

DEFINITION 6.20. NONSMOOTHABLE SCHEME. We say that a degree-s punctual scheme is *nonsmoothable* if there is no (flat) deformation to s distinct points; an Artin algebra is nonsmoothable if it has no deformation to a direct sum $k \oplus \cdots \oplus k$. We say an Artin algebra is *generic nonsmoothable* if it is nonsmoothable and lies on a single irreducible component of the variety parameterizing dimension n commutative algebras, parameterized by their structure constants (see [EmI1]). Recall from Definition 3.11 and Proposition 3.12 that a Gorenstein Artin algebra $A = R/I$ is compressed if it has the maximum possible Hilbert function $H(A) = H(j, r)$, $|\ H(j, r, i) = \min(r_i, r_{j-i})$, given the socle degree j, and embedding dimension r.

LEMMA 6.21. NONSMOOTHABLE GORENSTEIN POINT SCHEMES. *Suppose that* $6 \leq r' \leq 12$, *but* $r' \neq 7$, *and that* g *is a general element of* $\mathcal{D}'_3$ *(we use here the notation of the proof of Lemma 6.1). Suppose further that* char $k = 0$. *Let* $B_g = R'/\operatorname{Ann}(g)$. *Then* $H(B_g) = (1, r', r', 1) = H(3, r')$, *and*

$$H(R'/(\operatorname{Ann}(g))^2) = \left(1, r', \binom{r'+1}{2}, \binom{r'+2}{3}, r', 0\right). \qquad (6.2.6)$$

Furthermore, the scheme Z *concentrated at* $(1, 0, \ldots, 0) \in \mathbb{P}^{r-1}$, *given by* $Z = \operatorname{Spec}(B_g)$ *is generic nonsmoothable (see above). In particular* B_g *has no deformations except to local Artinian Gorenstein algebras* B' *with* $H(B') = H(B_g)$.

PROOF. We have checked (6.2.6) using the "Macaulay" algebra program for $r' = 6$ and $8 \leq r' \leq 12$, and specific DP-forms g. This check we made in characteristic 31,991, and several others, but the calculation is certainly not characteristic free. Thus, we consider Equation (6.2.6) valid in characteristic zero, though it must also be valid in almost all characteristics – modulus one's trust of the computer calculation. Henceforth in this section, we will assume that char $k = 0$, or that (6.2.6) has been checked if used, in the specific characteristic needed.

The small tangent space argument of J. Emsalem and the first author [EmI1], along with finding the dimension of the variety parameterizing Gorenstein algebras (nongraded) of Hilbert function $(1, r', r', 1)$ shows that the algebra B_g is generic nonsmoothable. For a discussion see [EmI1] (but note the error in calculation there for $r' = 4, 5$). A similar result occurs for many compressed algebras of any odd socle degree (Conjecture 6.30). $\qquad \square$

LEMMA 6.22. TANGENT SPACE $\mathcal{T}_G$ TO **Gor**(T) WHEN G HAS COMPRESSED ANNIHILATING SCHEME. *Let $r' = r-1$. Let $G = X_1^{[n]}g$, where $g \in \mathcal{D}'_\tau$ of odd degree $\tau = j'$, $n \geq \tau$ and g satisfies,*

   i. $H(R'/\text{Ann}(g)) = T(j', r-1)$ *(Ann(g) is compressed Gorenstein).*

   ii. $H(R'/(\text{Ann}(g)^2) = (1, r', r'_2, \ldots, r'_{j'}, r', 0)$.

Let Z denote the annihilating scheme $Z = \text{Proj}(R/(\text{Ann}(g)))$, concentrated at $(1, 0, \ldots, 0) \in \mathbb{P}^{r-1}$. Then $\deg(Z) = \dim_k R'/\text{Ann}(g)$; if $\tau = 3$, then $\deg(Z) = 2r$. The tangent space $\mathcal{T}_G$ to **Gor**(T), $T = \text{Sym}(H_Z, n + j')$ *satisfies*

$$\dim_k \mathcal{T}_G = \deg(Z) - 1 + r'_{j'} + r' \quad \text{(for any } \tau \geq 3\text{)}$$

$$= \binom{r+1}{3} + 3r - 2 \quad \text{(if } \tau = 3\text{)}.$$

PROOF. Immediate from Lemma 6.4(v) and (6.2.6).   □

LEMMA 6.23. FORMS G WHOSE ANNIHILATING SCHEME IS A COMPRESSED GORENSTEIN POINT SCHEME WHEN $r \geq 7$. *Suppose that $Z \subset \mathbb{P}^{r-1}$ is a Gorenstein scheme of Hilbert function $H = H(R/\mathcal{I}_Z) = (1, r, 2r-1, 2r, 2r, \ldots)$ concentrated at $(1, 0, \ldots, 0) \in \mathbb{P}^{r-1}$ as in Lemma 6.1. Let $j \geq 6$, suppose that $G \in (\mathcal{I}_Z)_j^\perp$, is general enough and let $J = \text{Ann}(G)$, $A_G = R/J$, and denote by $\mathcal{T}_G$ the tangent space to* **Gor**(T), $T = T(j, r)$. *Then*

   i. $H(A_G) = T(j, r)$

   ii. *Let $C_2 \subset \text{Gor}(T)$ parametrize all DP-forms $G \in \text{Gor}(T)$ such that G has as annihilating scheme some degree $2r$ compressed scheme Z as in Lemma 6.1. Then*

$$\dim(C_2) = \binom{r+1}{3} + 3r - 2. \tag{6.2.7}$$

   iii. *Suppose $Z = \text{Spec}(B_g)$ satisfies (6.2.6), e.g. if the conditions of Lemma 6.21 are satisfied. Then*

$$\dim_k(\mathcal{T}_G) = \binom{r+1}{3} + 3r - 2,$$

C_2 is a component of $\text{Gor}(T)$, and **Gor**(T) *is generically smooth along C_2.*

PROOF. By Lemma 6.1(a) the algebra $A_G = R/\text{Ann}(G)$ has Hilbert function $H(A_G) = \text{Sym}(H; j) = T(j, r)$. By Lemma 6.22, if G has the special form $G = X_1^{[j-\tau]}g$, $g \in \mathcal{D}'_3$, then $\mathcal{T}_G$ has the dimension given in

r	3	4	5	6	7	8	9	10	11	12	13
$\dim_k \mathcal{T}_G$	17	29	44	62	75	107	145	193	251	320	401
$\dim(C_2) =$		20	33	51	75	106	145	193	251	320	401

TABLE 6.2. Dimensions of tangent spaces $\mathcal{T}_G$ at points of $Gor(T)$, $T = H(6, r)$, and the value of $\dim(C_2)$. (See Lemma 6.23)

(iii): it follows that for a general $G \in (\mathcal{I}_Z)_j^\perp$, $\dim_k \mathcal{T}_G$ is no greater. Thus (iii) will follow from (ii). By the uniqueness of Z given G (see Lemma 6.1c and Lemma 6.2), the variety $C_2 \subset Gor(T)$ is fibred over a subset $P \subset \mathbf{Hilb}^{2r} \mathbb{P}^{r-1}$ parameterizing the schemes Z_G; the fiber is an open dense subset in the affine space $\mathbb{A}^{2r}$ parameterizing $G \in (\mathcal{I}_{Z_p})_j^\perp$. The variety P itself is fibred over $\mathbb{P}^{r-1}$, with fibre an open subset of $\mathrm{Proj}(\mathcal{O}_p)_3 \cong \mathbb{P}^{r'_3 - 1}$, so we have $\dim(C_2) = r'_3 - 1 + r' + 2r = r'_3 + 3r - 2$, which is (6.2.7). □

EXAMPLE 6.24. TANGENT SPACE TO $\mathbf{Gor}(T)$, AT POINTS WITH COMPRESSED ANNIHILATING SCHEME. For pairs (j, r) with $j = 6$ and $3 \leq r \leq 13$ we calculated using "Macaulay" the Hilbert functions $H(R/(\mathrm{Ann}(G))^2)$ when $X_1^{[3]} g$ and the tangent spaces $\mathcal{T}_G$ to $\mathbf{Gor}(T)$, $T = H(6, r)$, finding[1]

$$H(R/(\mathrm{Ann}(G))^2) =$$

$$= \begin{cases} (1, 5, 15, 35, 49, 50, \mathbf{44}, 25, 10, 5, 1) & \text{when } r = 5, \\ (1, 6, 21, 56, 72, 72, \mathbf{62}, 28, 11, 6, 1) & \text{when } r = 6, \\ (1, 7, 28, 84, 90, 90, \mathbf{75}, 20, 13, 7, 1) & \text{when } r = 7, \\ (1, 8, 36, 120, 128, 128, \mathbf{107}, 24, 15, 8, 1) & \text{when } r = 8. \end{cases} \quad (6.2.8)$$

REMARK 6.25. We included the cases $r = 5, 6$ in both Examples 6.14 and 6.24 because of the surprising equality in the dimensions 44, 62 of the tangent spaces. The Hilbert functions in (6.2.8) are determined by those for R'/J' and R'/J'^2, $J' = \mathrm{Ann}(g) \cap R'$ (Lemma 6.4).

Recall that $T(j, r) = (1, r, 2r - 1, 2r, \ldots, 2r, 2r - 1, 1)$. Let $T = T(j, r)$, $j \geq 6$, and let $C_1 = C_1(T)$ denote the subvariety of $Gor(T)$ arising from *smooth* annihilating schemes Z consisting of a self-associated set of $2r$ points as in Lemma 6.16. We let $C_2 = C_2(T)$ denote the subvariety of $Gor(T)$ parameterizing DP-forms f as in Lemma 6.23.

---

[1] We used the "random" matrix request in "Macaulay" to generate the form g.

THEOREM 6.26. $\overline{Gor(T)}$ HAS TWO IRREDUCIBLE COMPONENTS WHEN $r \geq 7$, $r \neq 8$. *Suppose* char$(k) = 0$ *and* $j \geq 8$, $r \geq 9$, *or* $j \geq 6$, $7 \leq r \leq 13$, $r \neq 8$, *and* $T = T(j, r)$ *(see* (6.0.1)*). Then* $C_1(T)$ *(smooth annihilating schemes) and* $C_2(T)$ *(conic Gorenstein annihilating schemes) are contained in two distinct irreducible components of* $\overline{Gor(T)}$, *whose dimensions are bounded below by* $\dim C_1(T) = 2r^2 - \binom{r-1}{2}$, *and* $\dim C_2(T) = \binom{r+1}{3} + 3r - 2$ *respectively. When* $j \geq 8$, $r \geq 3$, *or* $j \geq 6$, $3 \leq r \leq 10$, *then* $\overline{C_1(T)}$ *is a component of* $\overline{Gor(T)}$, *and* **Gor**(T) *is generically smooth along* $C_1(T)$. *When* $j \geq 6$, $7 \leq r \leq 13$, $r \neq 8$, *then* $\overline{C_2(T)}$ *is a component of* $Gor(T)$, *and* **Gor**(T) *is generically smooth along* $C_2(T)$.

PROOF. Let $j \geq 8$, $r \geq 9$. Then according to Lemma 6.16, $\overline{C_1(T)}$ is an irreducible component of $\overline{Gor(T)}$ of dimension $2r^2 - \binom{r-1}{2}$. By Lemma 6.23, $C_2(T)$ has dimension $\binom{r+1}{3} + 3r - 2$ which is greater than $\dim C_1(T)$, thus $C_2(T)$ is contained in a component of $\overline{Gor(T)}$ different from $\overline{C_1(T)}$.

Now, let $7 \leq r \leq 13$, $r \neq 8$. Lemma 6.23 (see also Table 6.2) shows that $\overline{C_2(T)}$ is an irreducible component of $\overline{Gor(T)}$. It is obtained as in the paragraph preceding Lemma 6.6 from $\mathcal{H}_2 \subset \mathbf{Hilb}^s \mathbb{P}^{r-1}$, where $\mathcal{H}_2$ is described in Lemma 6.23. If $f \in C_1(T)$ is a sufficiently general element and Z its annihilating degree-s scheme, then $\mathcal{I}_Z = ((\mathcal{I}_Z)_{\leq 3})$ as we proved in Lemma 6.11. Thus if we suppose that $f \in \overline{C_2(T)}$, then Lemma 6.6 shows that $Z \in \overline{\mathcal{H}_2}$. This is absurd since the schemes of $\overline{\mathcal{H}_2}$ have support consisting of one point, while Z consists of $2r$ distinct points. This proves the theorem. $\square$

6.3. Other reducible varieties $Gor(T)$

Recall that R', $\mathcal{D}'$ are the rings analogous to R, $\mathcal{D}$, but in $r' = r - 1$ variables (so $R' = k[x_2, \ldots, x_r]$). Recall from Definition 3.11 that a compressed Gorenstein Artin algebra $A = R'/J$ has the maximal Hilbert function given (j', r'); by Proposition 3.12 this Hilbert function is $T(j', r')_i = \max(r'_i, r'_{j'-i})$. Suppose that $A' = R'/J$ is Gorenstein of Hilbert function $T(j', r')$, with j' odd. We say that A' has *small tangent space* if

$$H(R'/J^2)_{j'+1, j'+2} = (r', 0). \tag{6.3.1}$$

THEOREM 6.27. REDUCIBLE $Gor(T)$. *Let* $T = \mathrm{Sym}(H, j)$, *where* $\Delta H = T(j', r')$, *and* $j \geq 2j'$. *Suppose that for general* $g \in \mathcal{D}'_{j'}$ *the graded compressed algebra* $B_g = R'/\mathrm{Ann}(g)$ *of Hilbert function*

$T(j', r')$ satisfies the small tangent space condition (6.3.1). Let $\mathcal{H} \subset$ $\mathbf{Hilb}^s \mathbb{P}^{r-1}$, $s = \Sigma_i T(j', r')_i$ be an irreducible subvariety which consists of subschemes Z of $\mathbb{P}^{r-1}$, concentrated at a point $p \in \mathbb{P}^{r-1}$, isomorphic to $\mathrm{Spec}(B_g)$ where g belongs to a certain Zariski open subset of $\mathcal{D}_{j'}$. Then $\mathcal{H}$ yields an irreducible component $\overline{C_2(T)}$ of $\overline{Gor(T)}$ via the construction of Lemma 6.1 with $Z \in \mathcal{H}$, $f \in (\mathcal{I}_Z)_j{}^\perp$. Furthermore, $\overline{Gor(T)}$ has at least two irreducible components, another component $\overline{C_1(T)}$ parameterizes DP-forms f with smoothable annihilating scheme.

PROOF. This is proved similarly to Theorem 6.26 using Lemma 6.6. By Lemmas 6.1 and 6.4, for a general $f \in (\mathcal{I}_Z)_j{}^\perp$, $Z \in \mathcal{H}$, we have $\dim_k \mathcal{T}_f = r'_{j'} + r' + \deg(Z) - 1$. But $\dim(\mathcal{H}) = r'_{j'} - 1 + (r') + \deg(Z)$ (choice of g, choice of point p, and choice of $f \in (\mathcal{I}_Z)_j{}^\perp$). The equality of these numbers shows that $\overline{C_2(T)}$ is an irreducible component of $\overline{Gor(T)}$. We now only need to construct a DP-form $f \in Gor(T)$ with a smooth annihilating scheme of degree s. By a result of P. Maroscia ([**Mar**, Theorem 1.8], see also A. Geramita, P .Maroscia, and L. Roberts' [**GMR**, Theorem 3.3], and Theorem 5.21(A) above) there is a finite subset S of $\mathbb{P}^{r-1}$ with Hilbert function H. Indeed, the hypothesis of his theorem requires that ΔH be an O-sequence. But we assume that $\Delta H = T(j', r')$, the Hilbert function of the compressed algebra B_g, so this condition holds. Now $\tau(S) = j'$ and by Lemma 6.1, a general $f \in (\mathcal{I}_S)_j{}^\perp$ satisfies the required properties. □

COROLLARY 6.28. REDUCIBLE $\overline{Gor(T)}$. Let $\mathrm{char}(k) = 0$. When $r = 6$, $j \geq 10$, and the sequence

$$\overset{j-9 \text{ times}}{T = (1, 6, 21, 36, 41, \overbrace{42 \ldots 42}, 41, 36, 21, 6, 1)}$$

then $\overline{Gor(T)}$ has at least two irreducible components $\overline{C}_1$ (smooth $Z(f)$) and $\overline{C}_2$ (nonsmoothable $Z(f)$). Here $T = \mathrm{Sym}(H, j)$ and $\Delta H_{\leq j/2} = T(5, 5)_{\leq j/2}$.

COROLLARY 6.29. REDUCIBLE $\overline{Gor(T)}$. Let $\mathrm{char}(k) = 0$. If $r = 5$, $j \geq 30$, $\Delta H_{\leq j/2} = T(15, 4)_{\leq j/2}$, and $T = \mathrm{Sym}(H, j)$, then $\overline{Gor(T)}$ has at least two irreducible components $\overline{C}_1$ and $\overline{C}_2$. Here

$$H_f = (1, 5, 15, 35, 70, 126, 210, 330, 450, 534,$$

$$\overset{d-29 \text{ times}}{590, 625, 645, 655, 659, \overbrace{660 \ldots 660}, 659, \ldots, 1)}$$

$(\Delta H_f)_{\leq 15} = T(15, 4) =$

$$(1, 4, 10, 20, 35, 56, 84, 120, 120, 84, 56, 35, 20, 10, 4, 1)$$

PROOF OF COROLLARIES 6.28, 6.29. We have verified (6.3.1) by computer for the two cases. Thus

a. when $(j', r') = (5, 5)$ the general compressed Gorenstein Artin algebra $R'/\operatorname{Ann}(g)$ of Hilbert function $T(5, 5) = (1, 5, 15, 15, 5, 1)$, dimension 42, is generic nonsmoothable. Here the tangent space dimension $= \dim_k (R/I^2)_j = 172$, agreeing with the dimension calculation $\dim C_2 = r'_5 - 1 + 5 + n = 125 + 5 + 42 = 172$.

b. When $(j', r') = (15, 4)$ a general compressed Gorenstein Artin algebra $R'/\operatorname{Ann}(g)$ of Hilbert function $T(15, 4)$, vector space dimension 660, is generic nonsmoothable. □

The following Conjecture generalizes the examples of nonsmoothable generic Gorenstein Artin algebras used above. In each case it would suffice to show that $R'/(\operatorname{Ann}(g))^2$ satisfies (6.3.1). small tangent space!to parameter space of compressed algebras—textbf

CONJECTURE 6.30. GENERICITY OF CERTAIN COMPRESSED ALGEBRAS. *Suppose that j is an odd integer. For a general $g \in \mathcal{D}_j$, the graded compressed Gorenstein Artin algebra $B_g = R/\operatorname{Ann}(g)$ having Hilbert function $T(r, j)$ satisfies the small tangent space condition (6.3.1) when $r \geq 6$ and $j \geq 3$ (except for the pair $(j, r) = (3, 7)$, when $r = 5$ and $j \geq 5$, and when $r = 4$ and $j \geq 15$.*

REMARK. Since Gorenstein singularities concentrated at a point of $\mathbb{P}^3$ are smoothable (see [Kl.H]), the method we have used above to show the reducibility of $Gor(T)$ does not apply to $r = 4$. However, M. Boij has recently shown that when $T = (1, 4, 7, 10, 13, 13, 10, 7, 4, 1)$, then $Gor(T)$ has two irreducible components, each of dimension 38: one component parameterizes forms whose annihilating scheme consists of 13 points lying on a conic; the other parameterizes forms f such that $f \in (I_L{}^2)^\perp$, where I_L is the ideal of a line. See also his Theorem 6.42 below, which gives other examples. We give a similar example in Appendix C (Example C.38).

6.4. Locally Gorenstein annihilating schemes

In this Section we report on work of Y. Cho and the first author, concerning forms having locally Gorenstein annihilating schemes, and also on the work of M. Boij in which he extends the relation between annihilating schemes and parameter spaces $Gor(T)$ to higher dimension schemes.

We first state an improvement of the Lemma 6.1a of Section 6.1, from [Cho1], showing that we can recover the annihilating scheme Z from a sufficiently general form F of high-enough degree, that is annihilated by Z, provided Z is locally Gorenstein (Theorem 6.31). We next consider degree-s schemes Z concentrated at a single point $p \in \mathbb{P}^n$, and defined by a general non-homogeneous compressed algebra quotient of the local ring at p. Then the global Hilbert function is $H_Z = H_s(r)$, where $H_s(r)_i = \min\{r_i, s\}$ (Theorem 6.34 (i)), which is the same as that for a scheme consisting of s general distinct points of $\mathbb{P}^{r-1}$ (Lemma 1.19). The symmetrization $\mathrm{Sym}(H_Z, j)$ is $T = H(s, j, r)$, the Hilbert function of a generic element of $PS(s, j, r)$. Also, for j large enough, a general $F \in ((\mathcal{I}_Z)_j)^\perp$ has Z as tight annihilating scheme, and the Hilbert function $H_F = H(s, j, r)$. (Theorem 6.34(i)).

The subscheme $\mathbf{C}_{SM}(T)$ of $\mathbf{Gor}(T)$ parametrizes forms in $PS(s, j; r)$, and was studied in Theorem 4.10A: there we showed that it is generically smooth for j large enough so that $T \supset (s, s, s)$. We now determine the dimension, and in some cases the dimension of the tangent space, of the subscheme $\mathbf{C}_{COMP}(T)$ in $\mathbf{Gor}(T)$ parametrizing forms having annihilating schemes Z, that are general enough compressed (Theorem 6.34(ii),(iii)). This result allows the construction of many examples of Gorenstein sequences $T = H(s, j, r)$, such that $Gor(T)$ has at least two irreducible components, when $r \geq 5$ (Theorem 6.34(iv),(v); Corollaries 6.36 and 6.37). We generalize, by giving the Hilbert function H_Z for compressed locally Gorenstein Z whose local annihilator f has a given number of nonzero terms (Theorem 6.39)). However, the dimension of the tangent spaces to the analogous varieties $\mathbf{C}_{SM}(T)$, and $\mathbf{C}_{COMP}(T), T = \mathrm{Sym}(H_Z, j)$ is still open for this more general class of Gorenstein sequences T, so less is known about the irreducible components of $Gor(T)$ (see Remark 6.41).

Finally we report on constructions of M. Boij, involving his extension of "annihilating schemes" to higher dimension, and his more relative context [Bo2] (Theorem 6.42). In particular, Boij gave the first examples in embedding dimension 4 for which $Gor(T)$ has several irreducible components (Example 6.43).

A zero-dimensional subscheme $Z \subset \mathbb{P}^n$ is *locally Gorenstein* if its structure sheaf $\mathcal{O}_Z$ has the property that for each point $z \in Z$ the local ring $\mathcal{O}_{Z,z}$ is Gorenstein. We will sometimes call such schemes *Gorenstein* for short, reserving the term *arithmetically Gorenstein* for the case $R/\mathcal{I}_Z$ is Gorenstein.

First, we state the result of Y. Cho and the first author, showing that Lemma 6.1(a) extends to the case Z is locally Gorenstein.

We need to introduce some notation. We let $\alpha(Z)$ denote the maximum socle degree of an irreducible component of Z; and we denote by $\beta(Z) = \tau(Z) + \max\{\alpha(Z), \tau(Z)\}$. If Z is either smooth, or concentrated at a single point and "conic", then $\beta(Z) = 2\tau(Z)$.

THEOREM 6.31. [Chol1] LOCALLY GORENSTEIN ANNIHILATING SCHEMES. *Let Z be a (locally) Gorenstein zero-dimensional subscheme of $\mathbb{P}^n$. If $j \geq \beta(Z)$, and F is a general enough element of $(\mathcal{I}_Z)_j^{\perp}$, then $H(R/\operatorname{Ann}(F)) = \operatorname{Sym}(H_Z, j)$, and for i satisfying $\tau(Z) \leq i \leq j - \tau(Z)$ we have $\operatorname{Ann}(F)_i = (I_Z)_i$.*

PROOF IDEA. The proof depends on a study of inverse systems for locally Gorenstein ideals, and in particular of the relation between the local inverse system to $(\mathcal{I}_Z)_p$ at a point $p \in \mathbb{P}^n$, and the global inverse system to $\mathcal{I}_Z$.   □

REMARK. It is well known that when Z is a zero-dimensional Gorenstein subscheme of $\mathbb{P}^n$, the global Hilbert function $H_Z = H(R/\mathcal{I}_Z)$ is not determined by the local Hilbert functions $H(\mathcal{O}_p/(\mathcal{I}_Z)_p)$ at the points p in the support of Z; this is true even when Z is concentrated at a single point $p \in \mathbb{P}^n$ (see Example 6.38). This indeterminacy is in contrast to the case Z local and "conic", where the global Hilbert function H_Z is simply the sum-function of the local Hilbert function (Lemma 6.1, Theorem 6.26).

We used Lemma 6.1, in the study of $Gor(T), T = Sym(H_Z, j)$. The extension of its first part in Theorem 6.31 allows study of $Gor(T)$ for a larger class of Gorenstein sequences T- provided we can control the global Hilbert function H_Z, and can understand sufficiently well the subscheme $\mathbf{Hilb}_{LG}^H(\mathbb{P}^n)$ of the postulation scheme $\mathbf{Hilb}^H(\mathbb{P}^n)$ parameterizing zero-dimensional locally Gorenstein subschemes of $\mathbb{P}^n$ having Hilbert function (postulation) $H_Z = H$.[2]

Thus, large subvarieties $U \subset \mathbf{Hilb}_{LG}^H(\mathbb{P}^n)$ parametrizing such locally Gorenstein subschemes should lead to parameter spaces $Gor(T)$ having large subvarieties fibred over U; if U parametrizes compressed schemes $Z_u \mid u \in U$ concentrated at a single point, and satisfying a suitable small tangent space condition (6.3.1), then the resulting subvariety C_U of $Gor(T)$ lying over U should be an irreducible component (Theorem 6.34(iv)).

---

[2]Since being locally Gorenstein is an open condition on the subscheme $Z \subset \mathbb{P}^n$) – as invertibility of the relative dualizing sheaf is an open condition – there is a natural induced subscheme structure on $\mathbf{Hilb}_{LG}^H(\mathbb{P}^n)$.

Alternatively, as we shall see below, to show $Gor(T)$ has several components, we need only show that there is an irreducble component $C_{SM}(T)$ whose annihilating schemes are smooth, and that there is another subvariety C of $Gor(T)$ lying over compressed locally Gorenstein annihilating schemes, and having higher dimension (Theorem 6.34(v)).

We now define *compressed Gorenstein Artin algebra* in the nonhomogeneous case (see [**EmI1, I3, EliI, I6**]). This extends the earlier Definition 3.11, which is the case A homogeneous. We write $H(t, n)$ below instead of $H(j, r)$ because we consider compressed Gorenstein schemes Z concentrated at a single point of $\mathbb{P}^n$.

DEFINITION 6.32. We denote by $R' = k\{x_1, \ldots, x_n\}, n = r - 1$ the completed local ring at a point $p \in \mathbb{P}^n$. If $A = R'/I$, we denote by $A^* = Gr_m(A)$ the associated graded algebra, with respect to the maximal ideal m, of A. The Hilbert function $H(A)$ is by definition $H(A) = H(A^*)$. The Artin quotient A is by definition *Gorenstein* if $\dim_k(0 : m) = 1$. The Gorenstein Artin algebra $A = R'/I$ of socle degree t is *compressed*, if it has Hilbert function $H(A) = H(A^*) = H(t, n)$, satisfying $H(t, n)_i = \min(n_i, n_{t-i})$, where $n_i = \binom{n+i-1}{i} = \dim_k R'_i$. This is the maximum Hilbert function possible for an Gorenstein Artin quotient of R' having socle degree t.

We say a scheme Z concentrated at p is "compressed" Gorenstein if its local ring at p satisfies $\mathcal{O}_p/(\mathcal{I}_Z)_p$ is Gorenstein, of Hilbert function $H' = H(t, n)$ for some t, n.

If $p \in \mathbb{P}^n$, we denote by $CompAlg_p(H') \subset Hilb^s(\mathbb{P}^n), s =| H' |$ the parameter space for – nonhomogeneous in general – compressed schemes at p having local Hilbert function $H' = H(t, n)$.

LEMMA 6.33. ([**EmI1, FL, I3, I6**]) *The variety* $CompAlg_p(H(t, n))$ *is irreducible, of dimension (recall $n_i = \binom{n+i-1}{i}$)*

$$\dim CompAlg_p(H(t, n)) = \sum_{i=\lceil t/2 \rceil}^{t} (n_i - H(t, n)_i). \qquad (6.4.1)$$

What is the global Hilbert function H_Z for a compressed Gorenstein Z, defined by a nonhomogeneous ideal of $\mathcal{O}_p$? The full answer is subtle (see Theorem 6.39). However, for Z parametrized by an open dense subset U_{Comp} of $CompAlg_p(H(t, n))$ the global Hilbert function $H_Z = H(R/\mathcal{I}_Z)$ satisfies, $H_Z = H_s(r)$, where $H_s(r)$ is defined by

$$H_s(r)_i = \min(r_i, s). \qquad (6.4.2)$$

This is also the Hilbert function H_S for the smooth subscheme S consisting of s general points of $\mathbb{P}^n$ (see Lemma 1.19). Recall that $\tau_{s,r} = \min\{i \mid s \le r_i = dim_k R_i\}$ (Equation 3.2.6). Note that if $j \ge 2 \cdot \tau_{s,r}$, then evidently

$$\text{Sym}(H_s(r), j) = H(s, j, r),$$

the Hilbert function of forms $F \in PS(s, j; r)$, that are generic sums of s degree-j powers of linear forms. For $T = H(s, j, r)$, we denote by $C_{SM}(T)$ the subscheme of $\mathbf{Gor}(T)$ whose closed points parametrize forms F with smooth tight annihilating scheme Z, satisfying $H_Z = H_s(r)$. Thus, $C_{SM}(T)$ contains an open dense subset of $PS(s, j; r)$. If $s = \mid H(t, n) \mid = \sum_{i=0}^{t} H(t, n)_i$, we denote by $\mathbf{C}_{COMP}(T)$ the subscheme of $\mathbf{Gor}(T), T = H(s, j, r)$ whose closed points parametrize forms F having a compressed Gorenstein tight annihilating scheme Z_f, concentrated at a some point $p \in \mathbb{P}^n$, and satisfying $H_Z = H_s(r)$. By Theorem 6.31 and Lemma 6.1, if Z is such a compressed scheme, if $j \ge t + \tau_{s,r}$, and F is a general element of $((\mathcal{I}_Z)_j)^{\perp}$, then $H_F = H(s, j, r)$. This is part (i) of the following Theorem. Recall from (6.3.1) that such a scheme Z is said to satisfy the *small tangent space* condition if its local defining ideal $J \subset R'$ satisfies $H(R'/J^2)_{j'+1, j'+2} = (r', 0)$.

THEOREM 6.34. [ChoI2] *Suppose $t \ge 3$, and fix (t, n), $n = r - 1$, and let $s = \mid H(t, n) \mid$. Suppose further that $j \ge 2\tau_{s,r}$, and let $T = H(s, j, r)$.*

  i. *Assume $j \ge t + \tau_{r,s}$ and let $p \in \mathbb{P}^n$. Then $C_{COMP}(T)$ is nonempty; there is a dense open subset U of $CompAlg_p(H(t, n))$ such that if $Z \in U$, and F is a general element of $((\mathcal{I}_Z)_j)^{\perp}$, then $H_F = H(s, j, r)$, and Z is a tight annihilating scheme of F.*

  ii. *Assume $j \ge 2\tau_{r,s}+1$. The dimension of the subfamily $C_{COMP}(T)$ of $Gor(T)$, if it is nonempty, satisfies*

$$\dim(\mathbf{C}_{COMP}(T)) = \dim(CompAlg_p(H(t, n))) + n + s. \qquad (6.4.3)$$

  iii. *Assume $j \ge 2\tau_{r,s} + 1$ and that $F \in \mathbf{C}_{COMP}(T)$ has tight annihilating scheme Z satisfying the small tangent space condition (6.3.1). The tangent space T_F to $\mathbf{Gor}(T)$ at $F \in \mathbf{C}_{COMP}(T)$ satisfies*

$$\dim_k(T_F) = H(R/\mathcal{I}_Z^2)_j = \dim(\mathbf{C}_{COMP}(T)). \qquad (6.4.4)$$

  iv. *Let Z be a compressed local Gorenstein scheme with small tangent space, and suppose $j \ge 2\tau_{s,r} + 1$ (equivalently suppose $T \supset (s, s)$). If $\overline{C_{COMP}(T)}$ is nonempty, then it is an irreducible, generically smooth component of $Gor(T)$.*

v. *If $r \geq 5$, if $j \geq \max\{6, , 2\tau_{s,r} + 2\}$ and if also $C_{COMP}(T)$ is nonempty, and $\dim(CompAlg_p(H(t,n))) \geq ns - n$, then $Gor(T)$ has at least two irreducible components, the closure of $C_{SM}(T) = PS(s,j;r) \cap Gor(T)$, which is generically smooth, and a second component containing $C_{COMP}(T)$.*

PROOF. Parts (i) and (ii) are immediate from Lemma 6.1, Theorem 6.31, and Equation 6.4.2 for the global Hilbert function of a general enough compressed Z. The calculation for (ii), is,

$$\dim(C_{COMP}(T)) = \dim(CompAlg_p(H(t,n))) + \dim(\mathbb{P}^n) + s.$$

We omit the proof of (iii), which involves showing that the homogenization of the square J^2 of the local defining ideal of Z, agrees with I_Z^2 in degrees $i \geq 2\tau_{r,s} + 1$ (they need not agree in lower degrees). The part (iv) is immediate from the definitions and (iii). As for (v), by Lemma 4.12, since $T \supset (s,s,s)$, and $j \geq 6$, the tangent space to $C_{SM}(T)$ has dimension rs; also, the dimension of the subvariety $C_{SM}(T)$ satisfies $\dim C_{SM}(T) = sn + s = sr$, thus $\overline{C_{SM}(T)}$ is an irreducible component of $Gor(T)$ (see Theorem 4.10A). The assumption $\dim(CompAlg_p(H(t,n))) \geq ns - n$ implies by (ii) that the irreducible variety $C_{COMP}(T)$ has dimension at least rs, so must lie on a different component of $Gor(T)$ than $C_{SM}(T)$. □

REMARK 6.35. The restriction $r \geq 5$ is needed in (v) so that the condition $\dim(CompAlg_p(H(t,n))) \geq ns - n$ can be satisfied – in $n = 3$ variables every locally Gorenstein zero-dimensional scheme is smoothable. At the time of writing, it is open whether the hypothesis of Theorem 6.34(i) might be improved to simply $j \geq 2 \cdot \tau_{s,r}$ in place of $j \geq t + \tau_{s,r}$.

We can use Theorem 6.34 to construct examples of reducible $Gor(T)$ in two ways: first, either by showing Z has small tangent space (requiring usually a computer verification for each j – see Conjecture 6.30), or, second, by choosing (t,n) so that $\dim(CompAlg_p(H(t,n))) \geq n(s-1)$, and then taking j as in (v), a much simpler criterion.

We now give the first such examples of reducible $Gor(T)$, in embedding dimensions 5, 6, ... that use the second, dimension approach above. These are analogs of the examples in Corollaries 6.28 and 6.29, and some of those from Theorem 6.26 earlier, but now the Hilbert function is $T = H(s,j,r)$. Note that the minimal degree j for $r = 5$ is now smaller than in Corollary 6.29. We will understand by $PS(s,j;r)$ and $\overline{PS(s,j;r)}$ below the intersection of these families with $Gor(T), T = H(s,j,r)$.

COROLLARY 6.36. *When $r \geq 5$, then $Gor(T), T = H(s, j, r)$ has at least two irreducible components, one of which, $\overline{PS(s, j; r)}$, is generically smooth, in the following cases:*

   i. *For $r = 5$, $T = H(252, j, 5)$, $j \geq 17$.*
   ii. *For $r = 6$, $T = H(42, j, 6)$, $j \geq 8$.*
   iii. *For $r = 7$, $T = H(56, j, 7)$, $j \geq 8$.*
   iv. *For $r = 8$, $T = H(72, j, 8)$, $j \geq 8$*
   v. *For $r \geq 9$, $T = H(2r, j, r)$, $j \geq 6$.*

PROOF. When $r = 5$, let $(t, n) = (11, 4)$. Then

$$H(11, 4) = (1, 4, 10, 20, 35, 56, 56, 35, 20, 10, 4, 1),$$

$s = 252$, $\tau_{s,r} = 6$, and $\dim(CompAlg_p(H(11, 4))) = 1113$, larger than $4s - 4$, the dimension needed to apply Theorem 6.34(v)

When $r = 6$, let $(t, n) = (5, 5)$, $H(5, 5) = (1, 5, 15, 15, 5, 1)$, $s = 42$, and $\tau_{s,r} = 3$, so by Theorem 6.34(i), the subvariety $C_{COMP}(T)$ is nonempty for $j \geq 8$. By Equation (6.4.1), the dimension

$$\dim(CompAlg_p(H(5, 5))) = (35 - 15) + (70 - 5) + (126 - 1) = 210,$$

more than the $5(s - 1)$ required by Theorem 6.34(v). The next two cases are calculated similarly, from $H(5, n), n = 6, 7$. For $r = 9$, so $n = 8$, we have $H(3, 8) = (1, 8, 8, 1), \tau_{18,4} = 2$, and

$$\dim(CompAlg_p(H(3, 8))) = (36 - 8) + (120 - 1) = 147,$$

so $T = (1, 9, 18, \ldots, 18, 9, 1)$, and $\dim C_{COMP}(T) = 147 + 8 + 18 = 173$, larger than the dimension $rs = 2r^2$, here 162 of $C_{SM}(T)$.   □

When $r = 8$ and $T = H(2r, j, r) = (1, 8, 16, \ldots, 16, 8, 1)$ we have $(t, n) = (3, 7)$; the dimensions of $C_{COMP}(T)$ and $PS(16, j, 8)$ are then 127 and 128, respectively, so it is plausible that if $j \geq 6$ there are at least two components of $Gor(T)$, but a further argument is needed. This parallels also the exceptional case to the small tangent space argument of Lemma 6.21 and Theorem 6.26, so Theorem 6.34(iv) does not help.

However, the analogous case for $r = 7$ can be handled by the small tangent space argument of Theorem 6.34(iv).

COROLLARY 6.37. *Let $r = 7$, $T = H(14, j, 7), j \geq 5$, so $T = (1, 7, 14, \ldots, 14, 7, 1)$. Then $\overline{C_{COMP}(T)}$ and $\overline{PS(14, j, 7)}$ are generically smooth components of $Gor(T)$, of dimensions 90 and 98, respectively.*

PROOF. The claim for $\overline{C_{COMP}(T)}$ follows from Theorem 6.34(iv), and equation (6.2.6) of Lemma 6.21 for $r' = 6$. That $\overline{PS(14, j, 7)}$ is generically smooth follows when $j \geq 6$ from Theorem 6.34(v), and when $j = 5$ from Lemma 6.1(a),(d), and a calculation in "Macaualy" that $\dim_k(R/\mathcal{I}_Z^2)_5 = 98$ for Z the subscheme of $\mathbb{P}^6$ defined by 14 generic points.   □

REMARK. Note that if instead we take a "conic" X, defined by a general form $f_3 \in \mathcal{D}_p'$, the local dual module, then the global Hilbert function $H_X = (1, 7, 13, 14, 14, \ldots)$, and one finds as in Theorem 6.26 that $Gor(T'), T' = (1, 7, 13, 14, 13, 7, 1)$ has several irreducible components.

Before stating the generalization of Equation (6.4.2), we give a simple example to illustrate why the Hilbert function H_Z might depend on the structure of f.

EXAMPLE 6.38. Suppose that $\mathcal{D}'$ is the divided power ring in two variables, dual to the completed local ring $R' = k\{y, z\}$ of the point $p = (1, 0, 0) \in \mathbb{P}^2$. Consider first the "conic" case. Let $f = Y^{[3]} + Z^{[3]} \in \mathcal{D}'$, itself homogeneous, and let $Z = Z_f = \mathrm{Spec}(\mathcal{O}_p/\mathrm{Ann}(f))$. Then Z has local Hilbert function $H' = H_f = (1, 2, 2, 1)$ at p, and degree 6. It is defined globally by the ideal $\mathcal{I}_Z = (yz, y^3 - z^3, y^4, z^4)$, and it has (global) Hilbert function $H_Z = H(R/\mathcal{I}_Z) = (1, 3, 5, 6, \ldots)$: thus, Z does not cut out independent conditions on quadrics.

Next consider the nonhomogeneous form $g = Y^{[3]} + Z^{[3]} + YZ \in \mathcal{D}'$, whose local defining ideal is $\mathcal{J}_p = (yz - y^3, y^3 - z^3, y^4, z^4)$, and let X be the corresponding scheme, which also has local Hilbert function $H' = (1, 2, 2, 1)$. Homogenizing $\mathcal{J}_p$ using the variable x, one obtains $\mathcal{I}_X = \mathcal{J} = (xyz - y^3, y^3 - z^3, y^4, z^4)$, having global Hilbert function $H_X = (1, 3, 6, 6, \ldots)$: thus, X cuts out independent conditions on quadrics. Here $H' = (1, 2, 2, 1) = H(3, 2)$ is compressed (although that is not critical for the example). The nonhomogeneous ideal $J_p = \mathrm{Ann}(g)$ has an order-two element $yz - y^3$ that homogenizes to degree 3 instead of 2. This increases the global Hilbert function H_X by one in degree 2, over H_Z, provided the term f_2 added to $f = f_3$ is general enough: here $f_2 = YZ$ is indeed sufficiently general for the global Hilbert function to change.

The above example suggests that the number $v + 1$ of nonzero terms $f_t + \ldots + f_{t-v}$ in the compressed "dual generator" $f \in \mathcal{D}_p'$ for $(\mathcal{I}_Z)_p$, might determine the Hilbert function H_Z that arises, and thus its symmetrization T. This suggestion turns out to be true, provided we assume that $f_t, \ldots f_{t-v}$ are sufficiently general (Theorem 6.39).

We now give the generalization of the determination of the global Hilbert functions H_Z in Equation (6.4.2) and Example 6.38, by Y. Cho and the first author. Let $T' = H(t, n)$, a compressed Hilbert function in $n = r - 1$ variables, and let $s = \sum_{i=1}^{t} H(t, n)_i$. We denote by $\mathcal{D}'$ the ring of divided powers in n variables

THEOREM 6.39. [ChoI2] *Suppose* $t/2 \geq v \geq 0$ *and suppose that* $f \in \mathcal{D}'$ *satisfies* $f = f_t + f_{t-1} + \ldots + f_{t-v}$, *where* $f_t \in \mathcal{D}'_t, \ldots, f_{t-v} \in \mathcal{D}'_{t-v}$ *are sufficiently general. Let* Z *denote the degree-s scheme concentrated at a point* $p = (1, 0, \ldots, 0) \in \mathbb{P}^n$, *defined by* $(\mathcal{I}_Z)_p = \mathrm{Ann}(f) \subset \mathcal{O}_p \cong k\{x_1, \ldots, x_n\}$, *the local ring at* p. *Then the global Hilbert function* $H = H_Z$ *satisfies,*

$$(H_Z)_i = \begin{cases} \min\{r_i, \sum_{u=0}^{i+v} h_u\}, & \text{for all } 0 \leq i \leq t - v, \\ \min\{r_i, s\} & \text{for } i \geq t - v, \end{cases} \qquad (6.4.5)$$

where $h_u = H(t, n)_u$. *Furthermore, if* $j \geq t + \tau(Z)$ *then a general member* $F \in ((\mathcal{I}_Z)_j)^{\perp}$ *satisfies,* $H_f = \mathrm{Sym}(H_Z, j)$.

EXAMPLE 6.40. Consider the local compressed Hilbert function $H' = H(5, 4) = (1, 4, 10, 10, 4, 1)$, of length 30. We give H_Z for differing numbers of terms in $f \in \mathcal{D}'$ determining Z concentrated at p, having local Hilbert function H'.

 i. If $f = f_5$, then $H_Z = (1, 5, 15, 25, 29, 30, \ldots)$.
 ii. If $f = f_5 + f_4$, then $H_Z = (1, 5, 15, 29, 30, \cdots)$.
 iii. If $f = f_5 + f_4 + f_3$, then $H_Z = (1, 5, 15, 30, \ldots)$.

To construct the sequences simply, note that the first is the sum sequence of H'; the second is

$$\min\left((5, 15, 25, 29, 30, \ldots), (1, 5, 15, 35, 70, \ldots, \dim_k R_i, \ldots)\right).$$

The third is

$$\min\left((15, 25, 29, 30, \ldots), (1, 5, 15, 35, 70, \ldots)\right).$$

By Theorems 6.31 and 6.39, in the second case, a general enough form $F \in ((\mathcal{I}_Z)_9)^{\perp}$ determines an element of

$$Gor(T_2), T_2 = (1, 5, 15, 29, 30, 30, 29, 15, 5, 1)$$

having Z as tight annihilating scheme. In the third case, the analogous Hilbert function is $T_3 = (1, 5, 15, 30, 30, 30, 15, 5, 1) = H(30, 8, 5)$.

REMARK 6.41. For $n \geq 4$ the subscheme $\mathbf{Hilb}_{LG}^H(\mathbb{P}^n)$ of the Hilbert scheme $\mathbf{Hilb}^s\mathbb{P}^n$ can have a very rich component structure, hinted at by Conjecture 6.30. "Elementary" components (concentrated at one point) combine to determine more complex components of higher degree, parameterizing generic non-local Gorenstein Artin algebras (see [**I4, I5**]). When a component U of $\mathbf{Hilb}_{LG}^H(\mathbb{P}^n)$ has been identified, arguments analogous to that of Theorems 6.26 and 6.27 may show that when $r \geq 5$, then $Gor(T)$, $T = \mathrm{Sym}(H,j)$, has at least two components, one containing a subvariety $C_U(T)$, lying over U, and another $\overline{C_{SM}(T)}$ lying over a family of smooth points in $\mathbb{P}^n$. We illustrated this above in Theorem 6.34 (iv),(v) for $T = H(s,j,r)$. However to employ this technique, one needs either to understand well the tangent space to points of $\mathbf{Hilb}_{SM}^H(\mathbb{P}^n)$ — to show that $C_{SM}(T)$ is an irreducible component of $Gor(T)$ — or one needs to employ some small tangent space argument to show that $C_U(T)$ is a component. One needs to rule out there being a large component C, perhaps all of $Gor(T)$, whose general element does not have a tight annihilating scheme, but which specializes to both $C_{SM}(T)$ and $C_U(T)$. By choosing a very large $j \geq 2s + 1$, Proposition C.33 of Appendix C applies, and $\dim C_U(T) \geq \dim C_{SM}(T)$ would suffice to show that $Gor(T)$ has several components. We have little knowledge of the tangent space to $\mathbf{Hilb}_{SM}^H(\mathbb{P}^n)$ when $n \geq 3$ — a generalization of Lemma 4.12 to determine $H(R/I_Z^2)_j$ for "general" sets Z of smooth points with given Hilbert function H would be needed. For further discussion see [**ChoI2**].

We now state M. Boij's main result from [**Bo2**], where he shows that when $r \geq 4$ there are Hilbert functions T with arbitrarily long subsequences of indices i such that $T_i = s$, and for which $\mathbb{P}Gor(T)$ has at least two irreducible components. We let $n = r - 1$ and denote by $B_{n,s,j}$ the Gorenstein sequence satisfying

$$(B_{n,s,j})_d = \min\{nd + 1, s, n(j - d) + 1\}. \tag{6.4.6}$$

We denote by $\mathcal{F}_{n,s,j}$ the subset of $\mathbb{P}Gor(B_{n,s,j})$ parameterizing forms $F \in \mathcal{D}_j$ with annihilating scheme s points lying on a rational normal curve in $\mathbb{P}^n$. We denote by $\mathcal{G}_{n,j}$ the subset of $\mathbb{P}Gor(B_{n,j+2,j})$ parameterizing $F \in \mathcal{D}_j$ such that $\mathrm{Ann}(F) \supset I_W{}^2$, where W is some line in $\mathbb{P}^n$. M. Boij first shows that if $r \geq 4$ and if $j = 5$, or $j \geq 7$ then the Gorenstein algebras parameterized by $\mathcal{G}_{n,j}$ do not satisfy a weak Lefschetz property, in the sense of Corollary 5.49 (see [**Bo2**, Proposition 3.5]). Gorenstein Artin algebras not satisfying weak Lefschetz had been found previously in embedding dimension four by H. Sekiguchi and in other cases by J. Watanabe (see [I6, p.67]).

THEOREM 6.42. [Bo2, Theorem 4.1]. *Assume that $n \geq 3$ and $j \geq \max\{11, 3n - 2\}$. Then the parameter space $\mathbb{P}Gor(B_{n,j+2,j})$ is reducible, with one component C_1 containing $\mathcal{F}_{n,j+2,j}$, and another component containing $\mathcal{G}_{n,j}$.*

EXAMPLE 6.43. M. Boij's first example in embedding dimension $r = 4$ for which $\mathbb{P}Gor(T)$ has several irreducible components is

$$T = B_{3,13,11} = (1, 4, 7, 10, 13, 13, 13, 13, 10, 7, 4, 1),$$

where the two components described above both have dimension 37. There may be further components. He has shown also that if $T = (1, 4, 6, 6, \ldots, 4, 1)$, then $Gor(T)$ has several irreducible components.

These examples of M. Boij show that Theorem 5.31 does not extend to $r \geq 4$: if $F \in \mathcal{G}_{n,j}$ one cannot as in Theorem 5.31 recover a zero-dimensional tight annihilating scheme from $Ann(F), F \in Gor(T)$ even though T may contain an arbitrarily long constant subsequence. Note however, that if one fixes the constant s, then by Corollary C.33 if $T_i = T_{i+1} = s$, for a value $i \geq s$, then $F \in Gor(T)$ determines a zero-dimensional annihilating scheme.

REMARK. M. Boij's approach more generally is to study Artinian Gorenstein quotients of R/J, where J is an ideal of R, usually geometrically defined, and where $Z = Proj(R/J)$ is not restricted to being a zero-dimensional scheme. Different geometrical structures for Z may in this way lead to different components of $Gor(T)$. When $J \neq 0$ the structure of the family of $Gor_{R/J}(T)$ may be quite complicated, having several irreducible components even when $r = 3$. A. Conca made some calculations of this in special cases.

For any $F \in Gor(T), T$ arbitrary, we may construct a sequence of generalized annihilating schemes of F in $\mathbb{P}^n$, using the "ancestor ideal" partial saturations defined earlier (see Equation 2.3.3): let $Z_i(F) = Proj(R/\overline{I}_i), I = Ann(F)$. Then we have

$$Z_1 \supset Z_2 \supset \ldots Z_i \supset Z_{i+1} \ldots \supset Z_j \supset Z_{j+1} = \varnothing. \qquad (6.4.7)$$

The study of this sequence should lead to further connections between the geometry of schemes and subvarieties of $\mathbb{P}^n$ and the parameter spaces $Gor(T)$.

CHAPTER 7

Connectedness and Components of the Determinantal Locus $\mathbb{P}V_s(u, v; r)$

We denote by $\mathbb{P}V_s(u, v; r)$ the projective algebraic set in $\mathbb{P}(\mathcal{D}_j)$ associated to the affine set $V_s(u, v; r)$, defined by the $(s + 1) \times (s + 1)$ determinantal minors of the catalecticant matrix $Cat_F(u, v; r)$. When $\mathrm{char}(k) = 0$ or $\mathrm{char}(k) > j$ we show that $\mathbb{P}V_s(u, v; r)$ is connected in Section 7.1. We also prove connectedness in some cases when the characteristic is arbitrary. We give a sufficient condition for the closure $\overline{PS(s, j; r)}$ to intersect $Gor(T)$ (Lemma 7.1). We then give in Section 7.2 an obstruction to deforming from $Gor(T)$ to $Gor(T')$ (Theorem 7.9), and we use it to show that $V_s(t, t; 3)$ has at least $[t/4] + 1$ irreducible components (Theorem 7.15).

When $r = 3$ we give a necessary condition for a Gorenstein sequence T satisfying $T \leq H(s, j, 3)$ to satisfy the condition that $Gor(T) \subset \overline{PS(s, j; 3)}$ (Corollary 7.12); we also give an example of a Gorenstein sequence $T \leq H(s, j, 3)$ for which $Gor(T) \not\subset \overline{PS(s, j; 3)}$ (Example 7.13). These results for $r = 3$ are related to, but complement those of Section 5.6, where we determined the intersection $Gor(T) \cap \overline{PS(s, j; 3)}$ in most cases (Theorem 5.71). In Section 7.3 we apply our results to study the $s-$ multisecant variety $Sec_s(v_j(\mathbb{P}^n))$ of the j-th Veronese embedding of $\mathbb{P}^n$: it is the projectivization of the closure of $\overline{PS(s, j; n + 1)}$.

7.1. Connectedness of $\mathbb{P}V_s(u, v; r)$

First, we give a criterion for $Gor(T) \cap \overline{PS(s, j; r)}$ to be nonempty for arbitrary r (Lemma 7.1). We then apply the criterion to $r = 3$, giving a result stronger than connectedness in that case (Theorem 7.3). We next show that if r is arbitrary and k is algebraically closed of characteristic zero, or greater than j, then each irreducible component of the projective algebraic set $\mathbb{P}V_s(u, v; r)$ contains $\mathbb{P}PS(1, j; r)$ (Theorem 7.6). Theorem 7.8 is about square catalecticants in arbitrary characteristic. Theorem 7.6 implies the connectedness results in

Theorems 7.3 and 7.8 except if $\text{char}(k) = p \leq j$; we include the other results in order to allow the reader to compare these approaches. Also each of the connectedness results Theorems 7.3, 7.6 and 7.8 applies to cases not covered by the other two.

Recall that T is a Gorenstein sequence if it occurs as the Hilbert function of an Artinian Gorenstein algebra; $\Delta T = (1, \delta_1, \delta_2, \dots)$ is the first difference sequence $\delta_i = t_i - t_{i-1}$. Recall that a sequence $(1, r', \dots)$ is an O-sequence if it occurs as the Hilbert function of a graded commutative algebra quotient of $R' = k[x_1, \dots, x_{r'}]$ (see Corollary C.3).

LEMMA 7.1. A SUFFICIENT CONDITION FOR T SUCH THAT $\text{Gor}(T) \cap \overline{PS(s, j; r)}$ IS NONEMPTY. *Let $j = 2t$ or $2t + 1$, $s \leq \dim_k R_t$, and for the Gorenstein sequence $T = (T_0, T_1, \dots, T_j)$ of socle degree j let $\Delta T = (1, \delta_1, \dots)$ be its first difference sequence. If T satisfies both $T \leq H(s, j, r)$, and the sequence $(1, \delta_1, \dots, \delta_t, 0, 0, \dots)$ is an O-sequence, then $\text{Gor}(T) \cap \overline{PS(s, j; r)}$ is nonempty.*

PROOF. We let H_Z denote the sequence satisfying

$$(H_Z)_i = \begin{cases} T_i & \text{for } i \leq t \\ T_t & \text{for } i \geq t. \end{cases}$$

Then $\Delta H_Z = (1, \delta_1, \dots, \delta_t, 0, 0, \dots)$ is an O-sequence by hypothesis, and $T = \text{Sym}(H_Z, j)$. By the result of P. Maroscia [**Mar**, Theorem 1.8] or A. Geramita, P. Maroscia, and G. Davis, [**GMR**, Theorem 3.3] there is a smooth subscheme S of $\mathbb{P}^{r-1}$ with the same Hilbert function $H_S = H_Z$, in particular $\dim S = 0, \deg S = s' = \max\{T_i\} \leq s$. Since $\delta_i = T_i - T_{i-1} \geq 0$, for $i \leq t$ by the assumption that ΔT is a O-sequence, we have that the sequence $\{T_i\}_{i \leq t}$ is nondecreasing, so in fact $s' = T_t$.

By Lemma 6.1(a) every sufficiently general DP-form $f \in (\mathcal{I}_S)_j^{\perp}$ satisfies $H_f = \text{Sym}(H_Z, j) = T$. Thus, $f \in \text{Gor}(T)$. If $S = \{p(1), \dots, p(s')\}$, $s' = T_t$ and $L_S = \{L_1, \dots, L_{s'}\}$, $L_i = L_{p(i)}$ is the corresponding subset of $\mathcal{D}_1$ then $f \in \langle L_1^{[j]}, \dots, L_{s'}^{[j]} \rangle$. We now find a deformation $f_w|_{w \in X}$, X an affine, smooth curve, $0 \in X$, such that $f = f_0$ and $f_w|_{w \neq 0}$ is in the open subset U of $PS(s, j; r)$ for which $H_{f_w} = H(s, j, r)$. It follows that $f \in \overline{PS(s, j; r)} \subset \overline{\text{Gor}(H(s, j, r))}$. $\square$

REMARK 7.2. Notice that this Lemma gives a sufficient condition for T to be the Hilbert function H_f for a DP-form $f \in \overline{PS(s, j; r)}$: in other words, for H_f to be the sequence of ranks of the catalecticant matrices $\text{Cat}_f(i, j - i; r)$, $0 \leq i \leq j$ for a special form f. This is one of the questions in Problem 0.2. The condition that $(\Delta T_{\leq t}, 0, \ldots)$ be an O-sequence is of combinatorial character, and we refer to Appendix C and the references therein for more details.

When $r = 3$ the condition $\Delta T_{\leq t}$ be an O-sequence is satisfied by all Gorenstein sequences T (a consequence of the Buchsbaum-Eisenbud structure theorem – see Theorem 5.25). Thus, when $r = 3$, Lemma 7.1 leads to a solution to the above problem of characterizing H_f for forms on the boundary $\overline{PS(s, j; 3)} - PS(s, j; 3)$, as shown in Proposition 5.70. Since when $r = 5$ there are non-unimodal Gorenstein sequences T (having several maxima), whose first differences $\Delta T_{\leq t}, 0, \ldots$ is not a O-sequence [**BeI, BoL**], and since it is an open question whether such T exist when $r = 4$, Lemma 7.1 does not completely resolve the question under consideration from Problem 0.2 when $r \geq 4$.

We showed in Proposition 5.70 that $\overline{PS(s, j; 3)}$ meets every $Gor(T)$, such that $T \leq H(s, j, 3)$; and we determined portions of the intersection $\overline{PS(s, j; 3)} \cap Gor(T)$ in Theorem 5.71. We now derive another consequence of these results.

Since $V_s(u, v; r)$ is the affine cone over $\mathbb{P}V_s(u, v; r)$, the connectedness of $\mathbb{P}V_s(u, v; r)$ is equivalent to the connectedness of $V_s(u, v; r) - \{0\}$.

THEOREM 7.3. CONNECTEDNESS OF $V_s(t, j - t; 3) - \{0\}$ AND $U_s(t, j - t; 3)$. *When* $r = 3$, $j = 2t$ *or* $2t + 1$ *and* $s \leq \dim_k R_t$, *then* $V_s(t, j - t; 3) - \{0\}$, $U_s(t, j - t; 3)$ *and* $\mathbb{P}V_s(t, j - t; 3)$ *are each connected, and the closure* $\overline{PS(s, j; 3)}$ *meets every irreducible component of* $V_s(t, j - t; 3) - \{0\}$.

PROOF. Since Gorenstein sequences in height three are unimodal (see Theorem 5.25) if $Gor(T)$ intersects $V_s(t, j - t; 3) - \{0\}$ then $T \leq H(s, j, 3)$ and consequently $Gor(T) \subset V_s(t, j - t; 3)$. Therefore by Diesel's Theorem 1.72 the irreducible components of $V_s(t, j - t; 3)$ are closures $\overline{Gor(T_i)}$ for certain Gorenstein sequences $T_i \leq H(s, j, 3)$. Let $T = T_i$ be one such sequence. By Theorem 5.25(i), $(\Delta(T)_{\leq t}, 0, 0, \ldots)$ is an O-sequence, so by Lemma 7.1, there are points f_p of $Gor(T)$ in the closure of $PS(s, j; 3)$. This shows that the irreducible variety $\overline{PS(s, j; 3)}$, which is contained in $V_s(t, j - t; 3)$, intersects each irreducible component $\overline{Gor(T_i)}$ in a non-zero point (in $Gor(T_i)$). This completes the proof of connectedness for $V_s(t, j - t; 3) - \{0\}$, and its projectivization. The set $U_s(t, j - t; 3)$ meets $PS(s, j; 3)$ in an open dense set of $PS(s, j; 3)$: this and the foregoing argument suffice to show the connectedness of $U_s(t, j - t; 3)$. $\square$

REMARK 7.4. The same proof as above yields for arbitrary r the connectedness of $\overline{W} - \{0\}$ where W is the set

$$W = \bigcup_{\substack{T \mid \max\{T_i\} \leq s \text{ and} \\ (\Delta T)_{\leq j/2}, 0, \dots) \text{ is an } O\text{-sequence}}} Gor(T)$$

REMARK 7.5. It is conjectured that when $r = 4$, each Gorenstein sequence $T = (1, 4, \dots, h_j)$, $j = 2t$ or $2t + 1$ satisfies, $(\Delta T_{\leq t}, 0, 0, \dots)$ is an O-sequence. The analogous statement is false when $r = 5$, so Lemma 7.1 does not imply $\mathbb{P}V_s(t, t; r)$ is connected when $r \geq 5$. (See § 5A of [I6], especially the Remark p. 67 for an account of what is known about Gorenstein sequences, and for further references). Note that it is an open question whether $Gor(T)$ is connected when $r \geq 4$. However, the connectedness of $\overline{Gor(T)} - \{0\}$ is shown below.

Recall that $\mathcal{H}(j, r)$ is the set of all Gorenstein sequences for degree-j DP-forms in $\mathcal{D}$. An upper triangular matrix has ones on the main diagonal, and zeroes below.

THEOREM 7.6. CONNECTEDNESS OF THE CLOSURES OF $\mathbb{P}Gor(T)$ AND OF THE CATALECTICANT LOCI. *Suppose that k is algebraically closed and* $\mathrm{char}(k) = 0$ *or* $\mathrm{char}(k) = p > j$; *suppose that $T \in \mathcal{H}(j, r)$ is a nonzero Gorenstein sequence of socle degree j. Then any irreducible component of the Zariski closure $\overline{\mathbb{P}Gor(T)}$ contains the projectivization $\mathbb{P}PS(1, j; r) \cong \mathbb{P}^{r-1}$ of $PS(1, j; r)$. Likewise, if $u + v = j$ and $s \geq 1$, any irreducible component of $\mathbb{P}V_s(u, v; r)$ contains $\mathbb{P}PS(1, j; r)$.*

PROOF. Under the hypotheses, we may use the polynomial ring $\mathcal{R}$ instead of $\mathcal{D}$ and the usual partial differentiation action of R on $\mathcal{R}$ (see Lemma A.12). The uppertriangular subgroup $\mathcal{G} \subset PGl(r)$ in the projective linear group, acts on the projective algebraic set $\overline{\mathbb{P}Gor(T)}$ or $\mathbb{P}V_s(u, v; r)$ via the usual action of $PGl(r)$ on f: if $g \in PGl(r)$ then $g(f) = f(g^{-1}X_1, \dots, g^{-1}X_r)$. Since $\mathcal{G}$ is an affine, connected, solvable algebraic group, and the sets acted upon are projective algebraic sets, the Borel fixed point theorem ([Bor, 10.4, p. 242]) shows that the closure of each orbit of $\mathcal{G}$ must contain a $\mathcal{G}$-fixed point. Since the sets are $\mathcal{G}$-invariant, each irreducible component C of $\overline{\mathbb{P}Gor(T)}$ or of $\mathbb{P}V_s(u, v; r)$ must contain a fixed point under the action, which can only be $\langle X_1^j \rangle$. Since C is also $PGl(r)$ invariant, it contains $\mathbb{P}PS(1, j; r)$, the $PGl(r)$ orbit of $\langle X_1^j \rangle$. $\square$

REMARK 7.7. When $r = 3$, $u = v$, and $2 \leq s$, Theorem 7.3 shows that $\mathbb{P}V_s(t, t; 3) - \mathbb{P}PS(1, j; 3)$ is still connected. Note that when

$\text{char}(k) = p$, $0 \neq p \leq j$, we could use the contraction action of R on the divided power ring $\mathcal{D}$, and show similarly that the closure of each $\mathcal{G}$ orbit contains a fixed point under the action. However, the fixed point set of $\mathbb{P}(\mathcal{D}_j)$ under $\mathcal{G}$ may no longer be $\langle X_1^{[j]} \rangle$, and it may not be connnected.

We now show a connectedness result for arbitrary characteristic. Recall that if $s \leq \dim_k R_t$ we let $s^\vee = \dim_k R_t - s$.

THEOREM 7.8. CONNECTEDNESS OF SOME $\mathbb{P}V_s(t, t; r)$ IN ARBITRARY CHARACTERISTIC. *If k is algebraically closed and s satisfies $\dim_k R_{2t} \geq \binom{s^\vee + 1}{2} + 2$ the loci $\mathbb{P}V_s(t, t; r)$ and $V_s(t, t; r) - \{0\}$ are connected. The hypothesis is satisfied when $r = 3$ and $s^\vee \leq 2t$, and in particular when $r = 3$ and $s \geq \dim_k R_{t-1}$.*

PROOF. We let $j = 2t$, $n = \dim_k R_t = \binom{t+r-1}{r-1}$, and $N = \dim_k R_j = \binom{j+r-1}{r-1}$. We let SM_n denote the vector space of symmetric matrices, and $\mathbb{P}(SM_n)$ the associated projective space. We let $\Delta_s \subset \mathbb{P}(SM_n)$ denote the sublocus of symmetric matrices of rank no greater than s, up to k^*-multiple; this is known to be irreducible, and of codimension $\binom{s^\vee + 1}{2}$ (Proposition 1.23 for $j = 2$). There is an injective catalecticant map

$$Cat : \mathbb{P}(\mathcal{D}_j) = \mathbb{P}^{N-1} \to \mathbb{P}(SM_n).$$

and $\mathbb{P}V_s(t, t; r)$ is mapped to $Cat(\mathbb{P}(\mathcal{D}_j)) \cap \Delta_s$. By our assumption,

$$\dim \mathbb{P}(\mathcal{D}_j) + \dim \Delta_s \geq \dim \mathbb{P}(SM_n) + 1.$$

Thus, according to the Fulton-Hansen Theorem [**FuH**], $\mathbb{P}V_s(t, t; r)$ is a connected locus of dimension greater or equal 1. □

7.2. The irreducible components of $V_s(u, v; r)$

We give as preliminary result, a necessary condition for f to be in the closure of $Gor(T')$. We suppose throughout, unless otherwise stated that T is a Gorenstein sequence of socle degree j.

THEOREM 7.9. NECESSARY CONDITION FOR f TO BE IN THE ZARISKI CLOSURE OF $Gor(T')$. *For arbitrary r, if $T \leq T'$ are Gorenstein sequences, if u satisfies $1 \leq u \leq j - 1$, and $T_u < T'_u$ but $T_{u-1} = T'_{u-1}$ then an element f of $Gor(T)$ can be in the Zariski closure of $Gor(T')$ only if for $J = \text{Ann}(f)$*

$$\dim_k(J_u / R_1 J_{u-1}) \geq T'_u - T_u, \tag{7.2.1}$$

In other words, J must have at least $(T'_u - T_u)$ linearly independent generators in degree u.

PROOF. Let $f_w|_{w \in X}$, X an affine smooth curve, $o \in X$, be a deformation of $f = f_o$ such that when $w \neq o$, $f_w \in Gor(T')$. Let $J = \text{Ann}(f_o)$, and for $w \neq o$ let $I_w = \text{Ann}(f_w)$. Let $d_i = \dim_k R_i - T_u$. We define

$$I_o = \lim_{w \to o} I_w, \quad (I_o)_i = \lim_{w \to o} (I_w)_i \text{ in } Grass(d_i, R_i).$$

in the sense of limit of ideals having the same Hilbert function. Thus, the Hilbert function of $H(R/I_o)$ is T', but if $T \neq T'$, then I_o is no longer Gorenstein. We have $I_o \subset J = \text{Ann}(f_o)$. This and the assumption that $T_{u-1} = T'_{u-1}$ implies that $(I_o)_{u-1} = J_{u-1}$. Thus

$$R_1 J_{u-1} = R_1(I_o)_{u-1} \subset (I_o)_u.$$

It follows that the number of degree-u generators of J satisfies,

$$\dim_k(J_u/R_1 J_{u-1}) \geq \dim_k J_u - \dim_k(I_o)_u = T'_u - T_u.$$

This completes the proof.   □

COROLLARY 7.10. SUFFICIENT CONDITION FOR THE DETERMINANTAL LOCUS TO BE REDUCIBLE. *Let $j = 2t$ or $j = 2t + 1$ and let $s \leq \dim_k R_t$. Suppose there is a Gorenstein sequence $T = (1, r, \ldots, 1)$ with $T_t = s$ and $T \neq H(s,j,r)$, but satisfying $T \leq H(s,j,r)$ termwise. Suppose there is an element $f \in Gor(T)$ such that $J = \text{Ann}(f)$ is generated by elements of degree less or equal t. Then $V_s(t, j - t; r)$ has at least two irreducible components.*

PROOF. Let $T' = H(s,j,r)$. Our hypothesis implies that $T \leq T'$, $T \neq T'$. If $s \geq \dim_k R_{t-1}$ then $Gor(T')$ is a Zariski open dense subset of some union of irreducible components of $V_s(t, j-t; r)$, since then T' is the maximum Gorenstein sequence compatible with $V_s(t, j - t; r)$. If $s \leq \dim_k R_{t-1}$ then by Theorem 4.10A, $\overline{PS(s,j;r)}$ is an irreducible component of both $\overline{Gor(T')}$ and $V_s(t, j - t; r)$. Theorem 7.9 applied with the minimal $u > t$ such that $T_u < T'_u$ shows that f cannot be in the closure of $Gor(T')$. This shows that there exists an irreducible component of $V_s(t, j - t; r)$ which does not contain f. This proves the Corollary.   □

EXAMPLE 7.11. DETERMINANTAL LOCUS OF A SQUARE CATALECTICANT WITH AT LEAST TWO IRREDUCIBLE COMPONENTS. $V_{10}(4, 4; 3)$ has at least two irreducible components $Gor(T_1)$, $T_1 = H(10, 8, 3)$, and $Gor(T_2)$, $T_2 = (1, 3, 6, 9, 10, 9, 6, 3, 1)$. The general element f of

$Gor(T_2)$ determines a complete intersection $I = \text{Ann}(f) = (h_1, h_2, h_3)$ with generator degrees $(3, 4, 4)$. The projectivization $\mathbb{P}Gor(T_2)$ is fibred over the Grassmanian $Grass(1, R_3) = Grass(1, 10)$ parametrizing h_1, by $Grass(2, R_4/h_1 R_1) \cong Grass(2, 12)$ parametrizing the choice of $\langle h_2, h_3 \rangle$ from the vector space $\langle R_4/xh_1, yh_1, zh_1 \rangle$. Thus,

$$\dim \mathbb{P}Gor(T_2) = \sum_{\deg h_i} t_i = (9 + 10 + 10) = 29$$

and $Gor(T_2) \subset \mathbb{A}^8$ has dimension 30. By Theorem 4.1A, $Gor(T_1)$ also has dimension 30, so neither $Gor(T_1)$ nor $Gor(T_2)$ can be entirely in the closure of the other. It is easy to see that the tangent space T_f at a general point f of $Gor(T)$, $T = T_1$ or $T = T_2$, also has dimension 30 (by Theorem 4.1A for T_1, and by Theorem 4.19 for T_2, or by Kleppe's Smoothness Theorem 4.21). Corollary 7.10 implies that $Gor(T_1)$ and $Gor(T_2)$ determine different irreducible components of $V_{10}(4, 4; 3)$, independent of the dimension calculation above.

We give several corollaries of Theorem 7.9 that depend on Theorem 5.25 from Section 5.3, that determines the possible locations of generators and relations for $I = \text{Ann}(f)$, $f \in Gor(T)$, $r = 3$. Recall that we let r_i denote $\dim_k R_i$.

COROLLARY 7.12. CRITERION FOR $f \in Gor(T)$ TO BE OUTSIDE THE CLOSURE OF THE LENGTH-s POWER SUMS, WHEN $r = 3$. *If $r = 3$, $j = 2t$, $s \leq \dim_k R_{t-1}$, and T is a Gorenstein sequence such that $T \leq H(s, j, r)$, $T_t = s$, if $f \in Gor(T)$ and $\mu_i = \mu_i(\text{Ann}(f))$ is the number of degree-i generators of $\text{Ann}(f)$, then $\mu_{t+1}(f) < s - T_{t+1}$, implies that $f \notin \overline{Gor(H(s, j, 3))} = \overline{PS(s, j; 3)}$. If*

$$-\Delta^3(T)_{t+1} < s - T_{t+1}$$

then a general DP-form $f \in Gor(T)$ is not in the closure $\overline{PS(s, j; 3)}$.

PROOF. Immediate from Theorem 7.9 ($T' = H(s, j, 3)$), Theorem 4.1A and Theorem 5.25, implying $\mu_{t+1}(f) < s - T_{t+1}$ for a general $f \in Gor(T)$. □

EXAMPLE 7.13. FORM NOT IN THE CLOSURE OF LENGTH s POWER SUMS. If $r = 3$, $j = 2t$, $t \geq 4$, $s = \dim_k R_{t-1}$,

$$T = (1, 3, \ldots, r_{t-2}, s-1, s, s-1, r_{t-2}, \ldots, 1), \tag{7.2.2}$$

then the general element $f \in Gor(T)$ is not in the closure of $PS(s, j; 3)$, as T satisfies the hypotheses of Corollary 7.12. Indeed, $\Delta^3(T)_{t+1} = t - 4$, so by Theorem 5.25, $\mu_{t+1}(f) = 0$, for f general.

When $(t, s) = (4, 10)$ in (5.2), the form $f = X^2Y^3Z^3 \in Gor(T)$ with $\mathrm{Ann}(f) = (x^3, y^4, z^4)$, is not in $\overline{PS(10, 8; 3)}$. By the discussion of Example 5.8, the form f has no degree 10 annihilating scheme.

When $(t, s) = (5, 15)$ in (7.2.2), the form $f = X^2Y^8 + X^4Y^3Z^3 \in Gor(T)$ with $\mathrm{Ann}(f) = (z^4, x^5, y^4z, y^5 - x^2z^3, x^3y^4)$ is not in $\overline{PS(15, 10; 3)}$.

COROLLARY 7.14. SEQUENCE OF INCOMPARABLE VARIETIES $Gor(T)$. *Suppose $r = 3$, $j = 2t$, $a \geq 0$, $b \geq 2a + 1$, and $t \geq 4b - a$, and define*

$$T(a, b, j) = (1, 3, \ldots, r_{t-3}, r_{t-2} - a, r_{t-1} - b, r_{t-1},$$
$$r_{t-1} - b, r_{t-2} - a, \ldots, 3, 1) \quad (7.2.3)$$

of socle degree $2t$. Then the general point of $Gor(T(a, b, j))$ does not belong to the closure of $Gor(T(a', b', j))$ when $b' < b$.

PROOF. The second difference sequence $\Delta^2 T$ is symmetric, and is

$$\Delta^2(T) = \ldots, 1, 1, 1 - a, 1 - b + 2a, 2b - a - t, -2b, 2b - a - t, \ldots$$

The condition for T to be a Gorenstein sequence when $r = 3$ is equivalent to $\Delta^2 T_i \leq 0$ for $v(T) \leq i < t$. For $T(a, b, j)$ this is equivalent to $b \geq 2a + 1$ and $2b \leq a + t$. The relevant portion of the third difference sequence $\Delta^3(T)$ is

$$\Delta^3(T) = \ldots, 0, 0, -a, 3a - b, 3b - 3a - t - 1, -4b + a + t, 4b - a - t, \ldots,$$

where $-a$ is in degree $t - 2$, and $a + t - 4b$ is in degree $t + 1$. By Theorem 5.25 the condition $t \geq 4b - a$ assures that when f is general in $Gor(T(a, b, j))$, there are no generators for I_f in degree $t + 1$. This and Theorem 7.9 imply the Corollary. $\qquad \square$

From Corollary 7.14 we obtain,

THEOREM 7.15. DETERMINANTAL LOCI HAVING MANY COMPONENTS, WHEN $r = 3$. *When $s = s_0 = \dim_k R_{t-1}$, and $j = 2t$, the algebraic set $U_s(t, t; 3)$ (and $V_s(t, t; 3)$) has at least $[t/4] + 1$ irreducible components of the form $\overline{Gor(T(0, b, j))}$, $0 \leq b \leq [t/4]$. There are at most $2^{t+1} - 1$ components.*

PROOF. The upper bound is from Diesel's Theorem 1.72 showing $Gor(T)$ is irreducible, and her Lemma B.17 (see Appendix B) counting the number of Gorenstein sequences of socle degree j. $\qquad \square$

Corollary 7.16. Scheme length $\ell\mathrm{sch}(f) > \ell\mathrm{diff}_k(f) = s$, when $r = 3$. *Suppose that* $j = 2t$, $s = \dim_k R_{t-1}$ *and* f *is a general element of* $Gor(T(0, b, j))$, $1 \le b \le [t/4]$. *Then* f *has no tight annihilating scheme* S, *(S of degree* s, $f \in (\mathcal{I}_S)_j^{\perp}$).

Proof. Any punctual subscheme of $\mathbb{P}^2$ is smoothable. Proposition 6.7 shows that if S is smoothable, and if f is in $(\mathcal{I}_S)_j^{\perp}$, then $f \in \overline{PS(s, j; 3)}$. Theorem 4.1A shows that $\overline{PS(s, j; 3)} = \overline{Gor(H(s, j, 3))}$ when $s = \dim_k R_{t-1}$. But $H(s, j, 3) = T(0, 0, j)$ and Theorem 7.15 shows that $Gor(T(0, b, j)) \not\subset \overline{Gor(T(0, 0, j))}$ if $1 \le b \le [t/4]$. $\qquad\square$

Remark 7.17. This result may be compared with Examples 5.8, and 5.9: where $\ell\mathrm{sch}(f) > \ell\mathrm{diff}_k(f)$ for forms f such that $\mathrm{Ann}(f)$ is a complete intersection ideal and f is general enough.

7.3. Multisecant varieties of the Veronese variety

Let $\mathbb{P}^{r-1} = \mathbb{P}(V)$, V an r-dimensional vector space. Let $n = r - 1$, $N = r_j - 1 = \dim_k R_j - 1$, and consider the j-th Veronese embedding

$$v_j : \mathbb{P}^n \to \mathbb{P}^N = \mathbb{P}(\mathrm{Sym}^j(V^*)^*) = \mathbb{P}(\mathcal{D}_j).$$

The s-secant variety to the Veronese variety $Sec_s(v_j(\mathbb{P}^n))$ is by definition the closure of the union of the subspaces of $\mathbb{P}^N$ spanned by s points of $v_j(\mathbb{P}^n)$. If $s < \dim_k \mathcal{D}_j = \binom{j+n}{n}$ this is the same as the closure of the union of the s-secant $(s-1)$-planes to $v_j(\mathbb{P}^n)$. According to Corollary A.10 (cf. Definition 1.12) we have $Sec_s(v_j(\mathbb{P}^n)) = \overline{PS(s, j; n+1)}$. In this section we for convenience reformulate some of the results about the variety $PS(s, j; n+1)$ in terms of the s-secant varieties to the Veronese varieties. If $f \in \mathcal{D}_j$ we denote by $\overline{f}$ its class *mod* k^* in $\mathbb{P}(\mathcal{D}_j)$. We refer to Theorem 1.56 for the case $n = 1$ of multisecant varieties of a rational normal curve.

Theorem 7.18. Multisecant varieties of the Veronese varieties. *Suppose* $n \ge 2$ *and* $j \ge 3$. *Let* $N = \binom{j+n}{n} - 1$ *and let* $Sec_s(v_j(\mathbb{P}^n))$ *be the* s-secant variety to the Veronese variety $v_j(\mathbb{P}^n) \subset \mathbb{P}^N$

A. *Suppose* $\mathrm{char}(k) \nmid j$.[1] *Then*

$$\dim Sec_s(v_j(\mathbb{P}^n)) = \min(s(n+1) - 1, N)$$

---

[1]This restriction on $\mathrm{char}(k)$ may be removed if s satisfies the conditions of Theorem 2.6(i)

except for the triples $(s,j,n) = (5,4,2),(9,4,3),(14,4,4)$ *and* $(7,3,4)$, *when the dimension is one less and equals* $N-1$. *In other words* $Sec_s(v_j(\mathbb{P}^n))$ *is a deficient multisecant variety only in the four exceptional cases listed above, in which case the deficiency is* 1 *(cf.* [**Ha**]*).*

B. *Suppose* $s \leq \binom{t+n}{n} - n - 1$ *when* $j = 2t$, *or* $s \leq \binom{t+n}{n}$ *when* $j = 2t + 1$. *Then a general point* $\bar{f} \in Sec_s(v_j(\mathbb{P}^n))$ *belongs to a unique* s-*secant* $(s-1)$-*plane.*

C. *Assume* $j = 2t$ *or* $2t + 1$ *and* $s \leq \binom{t-1+n}{n}$. *Let* J_{s+1} *be the homogeneous ideal generated by the* $(s+1) \times (s+1)$ *minors of all generic catalecticant matrices* $Cat_F(i, j-i; n+1)$, $1 \leq i \leq j/2$ $(J_{s+1} = I(\mathbf{Gor}_{\leq}(T))$, $T = H(s,j,n+1))$. *Then the homogeneous ideal of* $Sec_s(v_j(\mathbb{P}^n))$ *is equal to one of the isolated primary components of* J_{s+1}. *If* $n = 2$ *this prime ideal equals the ideal of* $\overline{Gor(T)}$, $T = H(s,j,3)$.

D. *Suppose that* $n = 2$, *and let* $H_f = T$ *be the sequence of ranks of the catalecticant matrices* $Cat_f(i, j-i; 3)$, $0 \leq i \leq j$ *(cf. Definition 1.9). Then* $\bar{f}$ *belongs to* $Sec_s(v_j(\mathbb{P}^n))$ *in the following cases.*

  i. T *contains a subsequence* (s,s,s).
  ii. *If* T *contains a subsequence* $(s-a, s, s, s-a)$ *with* $a \geq 1$, *and* $j = 2t+1$, *then* $\bar{f} \in Sec_s(v_j(\mathbb{P}^n))$ *if and only if* $\mathrm{Ann}(f)$ *has* a *generators of degree* $t+2$.
  iii. *If* T *contains* $(s-a-b, s-a, s, s-a)$, *so* $T_t = s$, $j = 2t$, *then* $\bar{f} \in Sec_s(v_j(\mathbb{P}^n))$ *if* $f \in Gor_{\mathrm{sch}}(T)$ *as defined in* (5.3.38).

E. *Suppose that* f *has a zero-dimensional smoothable degree-s annihilating scheme (see Definition 5.1). Then* $\bar{f}$ *belongs to* $Sec_s(v_j(\mathbb{P}^n))$. *If* $s \leq j+1$ *then the converse holds.*

F. *For* $\bar{f} \in Sec_s(v_j(\mathbb{P}^n))$ *suppose* $s = \ell\mathrm{diff}(f)$ *and suppose* f *has a tight annihilating scheme* Z *of length* s. *Then the scheme* Z *is smoothable provided either that* $j \geq 2\tau(Z) + 1$ *or that* $j = 2\tau(Z)$ *and* $\mathcal{I}_Z$ *is generated by* $(\mathcal{I}_Z)_{\leq \tau(Z)}$.

PROOF. Part (A) is Theorem 1.61. Part (B) is Theorem 2.6(ii). Part (C) is from Theorem 4.10A and, when $n = 2$, Diesel's Theorem 1.72. The fact that the primary component with radical $I(\overline{PS(s,j;n+1)})$ is in fact a prime ideal, and hence equals the homogeneous ideal of $Sec_s(v_j(\mathbb{P}^n))$, is a consequence of $\mathbf{Gor}(T)$ being smooth along a dense open subset of $PS(s,j;r)$. Part (D) is shown in Theorem 5.71 and Parts (E) and (F) are from Proposition 6.7. $\square$

REMARK 7.19. A. Geramita, M. Pucci, and Y. Shin have shown that when $n = 2$, $s \leq r_{t-1} = \binom{t+1}{2}$, $T = H(s,j,3)$, then $\mathbb{P}\mathbf{Gor}(T)$

is a smooth scheme. This yields that $\mathbb{P}Gor(T)$ is a smooth subset of $Sec_s(v_j(\mathbb{P}^2))$ and moreover the ideal sheaf of $Sec_s(v_j(\mathbb{P}^2))$ is locally generated by the $(s+1) \times (s+1)$ minors of all catalecticant matrices $Cat_F(i, j - i; r)$, $1 \le i \le j/2$, at the points of $\mathbb{P}Gor(T)$. Their proof uses Theorem 3.9 and their calculation of $(I^2)_j$ for an arbitrary graded Gorenstein ideal of Hilbert function T.

REMARK 7.20. DEGREE OF $Sec_s(v_j(\mathbb{P}^n))$. Suppose $\mathrm{char}(k) = 0$. We gave above the classical formulas for the degree of $Sec_s(v_j(\mathbb{P}^n))$ when $j = 2$ (Theorem 1.26) and when $n = 1$ (Theorem 1.56). Another interesting case is $Sec_3(v_3(\mathbb{P}^2)) = \mathbb{P}\overline{PS(3, 3; 3)}$. This is a hypersurface in $\mathbb{P}^9$ of degree 4. Its equation is the Aronhold invariant (see e.g. [DK, p. 250] for its explicit form). It is not connected with catalecticant matrices, as a generic degree 3 form f in $\mathcal{D}, r = 3$ has $H_f = (1, 3, 3, 1)$, but cannot be written as a sum of 3 cubes of linear forms.

The degree of $Sec_s(v_j(\mathbb{P}^2))$ was calculated by G. Ellingsrud and S. A. Strømme for $2 \le s \le 8$, $j \ge s - 1$, $(s, j) \ne (5, 4)$ in [ElS]. More precisely, under these assumptions, Ellingsrud and Strømme obtain a formula for the product $p(s, j, 2) \cdot \deg Sec_s(v_j(\mathbb{P}^2))$ where $p(s, j, n)$ is the number of ways a general form $f \in PS(s, j; n + 1)$ can be represented as a sum of s powers of linear forms. We claim that if $j \ge s - 1$ and $(s, j) \ne (5, 4)$ and $n = 2$ that $p(s, j, n) = 1$. One easily checks that if $n = 2$ and $j \ge 5$, or $n \ge 3$ and $j \ge 3$, then the condition $j \ge s - 1$ implies the condition of our Theorem 2.6(ii), so $p(s, j, n) = 1$. Now let $n = 2$. The case $s = 2$ is trivial and the case $s = 3$, $j \ge 3$ is dealt with in [ElS]. It remains to consider the case $(s, j) = (4, 4)$. Here for a general form of type $f = L_1^4 + \cdots + L_4^4$ the set Z of 4 points in the dual plane which correspond to $\{L_i\}$ equals exactly the base locus of the linear system $|\mathrm{Ann}(f)_2|$ (cf. the proof of Theorem 2.6), thus $p(4, 4, 2) = 1$.

So, the formulas in [ElS] yield $\deg Sec_s(v_j(\mathbb{P}^2))$ within the above limits. For each fixed s, $2 \le s \le 8$, this is a polynomial in j of degree $2s$. Its dominant term equals $j^{2s}/s!$. We refer to [ElS] for the explicit formulas. For instance

$$2 \deg Sec_2(v_j(\mathbb{P}^n)) = j^4 - 10j^2 + 15j - 6.$$

$$6 \deg Sec_3(v_j(\mathbb{P}^n)) = j^6 - 30j^4 + 45j^3 + 206j^2 - 576j + 384.$$

Closures of the Variety $Gor(T)$, and the Parameter Space $G(T)$ of Graded Algebras

If T is a Gorenstein sequence we let $\mathbf{G}(T)$ denote the scheme parametrizing the family of all graded ideals in R of Hilbert function $H(R/I) = T$; this is constructed by J. O. Kleppe from the functor $GRADALG_T$, in [Kl1], and is used by him in his study of $\mathbb{P}\mathbf{Gor}(T)$ [Kl2] (see Theorem 4.21 above). He shows that the tangent space $T_{I,\mathbf{G}(T)}$ to $\mathbf{G}(T)$ at the closed point $I = \text{Ann}(f), f \in \mathcal{D}_j$ and the tangent space $T_{f,\mathbb{P}\mathbf{Gor}(T)}$ at the point corresponding to f in $\mathbb{P}\mathbf{Gor}(T)$ are equal, and satisfy

$$T_{I,\mathbf{G}(T)} = Hom_R(I, A)_0 \cong T_{f,\mathbb{P}\mathbf{Gor}(T)} \qquad (8.0.1)$$

We restrict ourselves to the closed points of $\mathbf{G}(T)$ and $\mathbb{P}\mathbf{Gor}(T)$ in the discussion below. Let $d = \min\{i \mid t_i < \dim_k R_i\}$, and $j = \max\{i \mid t_i \neq 0\}$. We set $BigGrass(T) = \prod_{d \leq i \leq j} Grass(r_i - t_i, R_i)$; and define the scheme $G(T)$ as the reduced subscheme of $\mathbf{G}(T)$. We define an inclusion

$$i : G(T) \rightarrow BigGrass(T) \quad \text{by} \quad i(I) = (I_d, \dots, I_j). \qquad (8.0.2)$$

and identify $G(T)$ with its image in $BigGrass(T)$. There is a natural projection,

$$\pi : G(T) \rightarrow \mathbb{P}\overline{Gor(T)} \subset \mathbb{P}(\mathcal{D}_j), \quad A = R/I \rightarrow \pi(A) = I_j^\perp.$$

Notice that if $I_j^\perp = \langle f \rangle$, then $\text{Ann}(f) = \bar{I}_j$, where $\bar{I}_j$ is the ancestor ideal of I_j defined in (2.3.3), and it satisfies $\bar{I}_j \supset I$. The natural inclusion

$$\iota : \mathbb{P}Gor(T) \rightarrow G(T), \quad \iota(\langle f \rangle) = \text{Ann}(f),$$

with image the Gorenstein ideals in all graded ideals of Hilbert function T, is a section of π. When $r = 2$ the scheme $\mathbf{G}(T)$ may be constructed similarly to the construction of $G(T)$ in [I1, Theorems 2.9,3.4], by glueing affine space patches. Also, when $r = 2$, $\mathbf{G}(T)$ is a natural desingularization of $\mathbb{P}\overline{Gor(T)}$ (Theorem 8.1). However, when $r = 3$, $\mathbf{G}(T)$ is in general singular (Lemma 8.3).

Recall from Theorems 1.44 and 1.54 that if $A = R/I$ is an Artinian Gorenstein algebra when $r = 2$, then I is a CI, $I = (g,h)$, $\deg g + \deg h = j + 2$; if $s = \deg g \leq \deg h$, $H(A) = H(s,j,2)$ (this is the sequence T of Theorem 1.44) with $s \leq [j/2] + 1$. For simplicity, we assume for the rest of the section that k is algebraically closed, and that char $k = 0$ or char $k > j$. The first two and last parts of the next Theorem are proven in Section 1.3: for parts (a) and (b) see Theorems 1.44, 1.45, and Remark 1.46; for part (d) see also Lemma 1.38. We outline an alternative proof of the first two parts below.

THEOREM 8.1. THE MORPHISM $\mathbf{G}(T) \to \mathbb{P}\overline{\mathbf{Gor}(T)}$ WHEN $r = 2$.

  a. *If $T = H(s,j,2)$, then*
  $$\overline{Gor(T)} = \bigcup_{s' \leq s} Gor(T'), \quad T' = H(s',j,2)$$
  is identical to $V_s(t, j - t; 2)$, $t = [j/2]$, the determinantal variety of the Hankel matrix $Cat_F(t, j - t; 2)$.

  b. *Except when $(s,j) = (t + 1, 2t)$, the variety $\mathbb{P}Gor(T)$ is fibred over $\mathbb{P}^s = \mathrm{Sym}^s(\mathbb{P}^1)$ by fibres that are open sets in $\mathbb{P}^{s-1}$. When $(s,j) = (t+1, 2t)$, then $\mathbb{P}Gor(T)$ is an open subvariety of $Grass(2, R_{t+1})$.*

  c. *The morphism $\mathbf{G}(T) \to \mathbb{P}\overline{\mathbf{Gor}(T)}$ is a desingularization, whose fibre over $f \in \mathbb{P}Gor(T')$, $T' = H(s',j,2)$ is the family of graded ideals $I \in G(T)$ satisfying $I \subset \mathrm{Ann}(f)$.*

  d. $\overline{Gor(T)} = \overline{PS(s,j;2)}$, *and is the family of generalized power sums, or generalized additive decompositions of length $\leq s$ (see Definiton 1.30 or Lemma 5.3 C).*

PROOF OUTLINE. See the Appendix to [I2] for references: the key point is the irreducibility and nonsingularity of $\mathbf{G}(T)$. For this see [I1, Theorems 2.9, 3.13]; however those results are stated for the reduced scheme structure on $G(T)$, and must be extended to the scheme $\mathbf{G}(T)$. The subvariety $CI(T)$ of $G(T)$ parametrizing complete intersections is an open subset, so must be dense. To complete the proof of the frontier property (a), one now may observe that $\pi(G(T)) = \mathbb{P}\overline{Gor(T)}$, or use the properties of the determinantal varieties $V_s(t, j - t; 2)$ of the Hankel matrix as in Section 1.3 (see also [Ei1]). The statement (b) is immediate from the fact, $\mathrm{Ann}(f) = (g, h)$ is a CI when $r = 2$; so $\mathbb{P}^s = \mathrm{Sym}^s(\mathbb{P}^1)$ parametrizes the degree-s zero-divisors $|g|$, while the fiber parametrizes the choices of $h \in R_{j+2-s} \bmod (g) \cap R_{j+2-s}$. The statement (c) requires first the smoothness of $\mathbf{G}(T)$ and the definition of π; then a construction shows that there exist ideals $I \in G(T)$ satisfying $I \subset \mathrm{Ann}(f)$ – this is the hardest step. The proof of (d) is that of Section 1.3. $\qquad\qquad\qquad\qquad\qquad\qquad\qquad\qquad\qquad\qquad$ $\square$

REMARK 8.2. THE SET $PS(s, j; 2)$ AND THE HILBERT SCHEME, s SMALL. When $r = 2$, s is "small" in the sense of Chapters 4, 6 when $s \leq \dim_k R_{t-1} = t$, and $j = 2t$ or $2t + 1$. Thus s is small unless $(s, j) = (t + 1, 2t)$. If s is small, and $\mathrm{Ann}(f) = (g, h)$, $\deg(g) = s$, then the annihilating scheme Z_f is $|g|$, the zero-divisor of g in $\mathbb{P}^1$. If $T = H(s, j, 2)$, then $Gor(T)$ is fibred over $\mathbb{P}^s = \mathrm{Sym}^s(\mathbb{P}^1) = \mathbf{Hilb}^s(P^1)$.

The fibre of $G(T)$ over $|g| \in \mathbb{P}^s$ is essentially independent of j, being the affine space $\mathbb{A}(R_j/(gR_{j-s}))$. However, the variety $G(T)$ desingularizing $\mathbb{P}\overline{Gor(T)}$ is a bundle over $\mathbf{Hilb}^s(\mathbb{P}^1)$ whose structure depends on j. J. Yameogo and the first author have shown that although the Betti numbers of $G(T)$ are independent of j, when s is fixed and "small", the cohomology ring $H^*(G(T))$ does depend nontrivially on j. They also study a simpler desingularization by $G(T_{s,j})$, $T_{s,j} = (1, 2, \ldots, s, s, \ldots s, 1)$ of socle degree j, which has the advantage that the cohomology ring structure of $G(T_{s,j})$ is known [**IY**].

When $T = H(s, j, 3)$, $Gor(T)$ is also a desingularization of $G(T)$, whose cohomology ring structure is unknown. A. King and C. H. Walter have further studied the varieties $G(T)$, when $r = 3$, and shown that, for any O-sequence T, the inclusion i of (8.0.2) induces a surjection i^* of homology rings.

The following Lemma gives an Example developed by J. Yameogo, showing that $\mathbf{G}(T)$ is no longer nonsingular when $r = 3$. This prevents an analog to Theorem 8.2 when $r = 3$, based on the properties of the scheme parametrizing graded ideals. In the Lemma, the Hilbert function of R/I is $T = (1, 3, 4, 3, 1)$, but that of the Gorenstein ideal $\mathrm{Ann}(Z^4)$ is $T_0 = (1, 1, 1, 1, 1)$, the most singular Hilbert function for elements in $\mathbb{P}\overline{Gor(T)}$.

LEMMA 8.3. (With J. Yameogo). SINGULAR $\mathbf{G}(T)$ WHEN $r = 3$. Let $T = (1, 3, 4, 3, 1) = H(4, 4, 3)$, and let the ideal $I = \mathrm{Ann}(V)$, $V = \langle Z^4, X^3, Y^3, XY \rangle$. Then

a. $I = (xz, yz, x^2y, xy^2, x^4, y^4, z^5)$, and
   $\pi(I) = \mathrm{Ann}(Z^4) = (x, y, z^5) \in \mathbb{P}\overline{Gor(T)}$.

b. $\iota(\mathbb{P}Gor(T))$ is a smooth, open subscheme of $\mathbf{G}(T)$, whose closure contains I. The dimensions $\dim(\mathbb{P}Gor(T)) = \dim(G(T)) = 11$.

c. $\mathbf{G}(T)$ is singular at the point z_I parametrizing I.

d. A generic point $J = \mathrm{Ann}(f)$ of $\iota(\mathbb{P}Gor(T))$ is a CI: $\mathrm{Ann}(f) = (g_1, g_2, h)$, where $\deg(g_1) = \deg(g_2) = 2$, $\deg(h) = 3$.

e. $PS(2,2;3)$ *is open dense in* $Gor(T)$ *and parametrizes forms* $f \in$
$\mathcal{R}_4$ *having a degree-4 smooth annihilating scheme* Z: *such that*
$\mathrm{Ann}(f)_2 = \langle g_1, g_2 \rangle$, *and* $(g_1 = g_2 = 0)$ *consists of four points.*

f. *The fibre of* $G(T)$ *over an element* $J = \mathrm{Ann}(f)$ *of* $\mathbb{P}\overline{Gor}(T)$
consists of the graded ideals $I \in G(T)$ *contained in* J.

PROOF OF (b). If $J = \mathrm{Ann}(f) \in \mathbb{P}\mathbf{Gor}(T)$, the tangent space
$\mathcal{T}_J$ to $\mathbf{G}(T)$ at $\iota(J)$ is $Hom_0(J, R/J) \cong (J/J^2)_j$, where $j = $ socle
degree$(R/J) = 4$. Since $\dim_k(J_2) = 2$, also $\dim_k(J_4^2) = 3$, and
$\dim_k((J_4^2)^{\perp}) = 12$, so $\dim_k((J/J^2)_4) = 11$. Since $\dim(G(T)) = 11$,
$\iota(\mathbb{P}\mathbf{Gor}(T))$ is a smooth subscheme of $\mathbf{G}(T)$.

The subfamily $I(t) = \{\mathrm{Ann}(f_t) \mid t \neq 0, t \in k\}$ of $\mathbb{P}\mathbf{Gor}(T) \subset G(T)$

$$f_t = (Z^4 + t(X^4 + Y^4 + tX^2Y^2),$$

$$\mathrm{Ann}(f_t) = (xz, yz, xy^2 - tx^3, x^2y - ty^3, x^4 - tz^4)$$

evidently has limit $I(0) = I = \mathrm{Ann}(V)$.

PROOF OF (c). We write a degree-zero deformation of I as

$$xz + \alpha_1 x^2 + \beta_1 xy + \chi_1 y^2 + \delta_1 z^2, \ \ yz + \alpha_2 x^2 + \beta_2 xy + \chi_2 y^2 + \delta_2 z^2,$$

$$x^2y + \varepsilon_1 x^3 + \phi_1 y^3 + \varphi_1 z^3, \ \ xy^2 + \varepsilon_2 x^3 + \phi_2 y^3 + \varphi_2 z^3,$$

$$x^4 + \gamma_1 z^4, \ \ y^4 + \gamma_2 z^4.$$

The relations

$$y\underline{xz} - x\underline{yz} = 0 \quad \mathrm{mod}\ I: \ \chi_1 = \alpha_2 = 0,$$

$$x^2\underline{yz} - z\underline{x^2y} = 0 \quad \mathrm{mod}\ I: \ \varphi_1 = 0$$

$$y^2\underline{xz} - z\underline{y^2x} = 0 \quad \mathrm{mod}\ I: \ \varphi_2 = 0,$$

leave a dimension 12 tangent space $Hom_0(I, A)$ at the special point z_I.
Thus, $\mathbf{G}(T)$ is singular at z_I.

PROOF OF (d), (e), (f). These are straightforward. □

REMARK 8.4. We believe that $G(T) = \iota(\mathbb{P}\overline{Gor}(T))$, here, but have
not verified this. If so, then $V_4(2,2;3) = \overline{PS(4,4;3)}$, as one can show
$\pi(G(T)) = V_4(2,2;3)$ in this case. Note that $\mathbf{V}_4(2,2;3)$ was known to
be irreducible, by calculation.

For larger $j = 2t$, $t \geq 4$, Example 7.13 shows that if $T = H(s, j, 3)$,
$s = \dim_k R_{t-1}$, then either $G(T)$ fails to be irreducible, or there are no
ideals of Hilbert function T contained in the general Gorenstein ideal
of Hilbert function T', where T' is the sequence of (7.2.2).

It is natural to ask, which graded ideals of Hilbert function T are
in the closure $\overline{\iota(Gor(T))}$? This is open in general when $r \geq 3$. The

study of the $Sl(r+1)$ action on $\overline{Gor(T)}$ and $G(T)$, would be of interest: see J. Weyman's articles [**We3, We4**], for such a study for varieties defined by the multiple root loci of binary forms; see D. Bayer's and I. Morrison's [**BaMo**] or also E. Viehweg's [**Vi**, Chapter 7] for a study of the $Sl(r+1)$ action on the Hilbert scheme $\textbf{Hilb}^Q(\mathbb{P}^n)$.

There is a study of the decomposition of $G(T)$ into subvarieties determined by monomial ideals when $r = 2$, in [**IY**]. The work of R. Notari and M. L. Spreafico [**NotSp**], is related to that of D. Bayer, and concerns a decomposition of the Hilbert scheme into locally closed subschemes determined by order ideals in the sense of Appendix C. This work suggests that it might be fruitful to study the decomposition of $\textbf{G}(T)$ into such locally closed subschemes in the more difficult case of $r \geq 3$.

CHAPTER 9

Questions and Problems

A. Defining equations for $\overline{PS(s,j;r)}$. Several of our results, such as Theorems 4.1A, 4.5A, 4.10A, 5.55, 5.71, were focused on Problem 0.1 of finding intrinsic conditions on a homogeneous form f (or a DP-form in arbitrary characteristic) such that $f \in \overline{PS(s,j;r)}$. A natural approach – considering rank conditions on the catalecticant matrices associated with f – may be applied when $s \leq \binom{t+r-2}{r-1}$ (this is $\dim_k R_{t-1}$) for $\deg(f) = j = 2t$ or $2t+1$, according to Theorem 4.10A, but our method is insufficient outside this range for dimension reasons. When $r \geq 3$ an approach using rank conditions alone is insufficient as well, except possibly for very small s, due to Lemma 7.1 and Example 7.13, so some further conditions are necessary.

The case of ternary forms, $r = 3$, seems approachable due to the Buchsbaum-Eisenbud structure theorem for height 3 Gorenstein algebras. Our Theorem 5.71 is an example of such an application. and in many cases for the sequence of ranks of the catalecticants H_f this theorem decides when $f \in \overline{PS(s,j;3)}$. At present it is still an open and interesting problem to find a complete list of intrinsic conditions on a ternary form which are necessary and sufficient for the form to belong to $\overline{PS(s,j;3)}$, and to translate these conditions into polynomial equations on the coefficients of the form.

It is an open problem whether $f \in \overline{PS(s,j;r)}$ implies f has a smoothable annihilating scheme of degree $\leq s$. Some partial results were given in Proposition 6.7.

In our Theorem 5.55 we gave necessary and sufficient conditions, when $r = 3$, for $f \in Gor(T)$ to have an annihilating scheme[1] of degree $s = \max\{T_i\}$ in case $\deg(f)$ is odd and for most of the possible sequences T when $\deg(f)$ is even. It would be interesting to find such conditions also in the remaining cases, which our approach failed to cover.

---

[1] Any zero dimensional subscheme of $\mathbb{P}^2$ is smoothable by Fogarty's result [**Fog**]

The second part of Problem 0.1 - finding generators of the ideal of the affine variety $\overline{PS(s, j; r)}$ - is much more difficult and, apart from some sporadic examples when $\overline{PS(s, j; r)}$ is a hypersurface (see e.g. Corollary 2.3), these generators are known only when $j = 2$ or $r = 2$ or $s \leq 2$ – they are minors of appropriate catalecticant matrices (see the discussion in Section 1.2). Here the next case $s = 3$ seems approachable.

B. Squares of vanishing ideals at s general points of $\mathbb{P}^{r-1}$. We have studied primarily the Hilbert functions $T = H(s, j, r)$ that occur as H_f for $f = L_1^{[j]} + \cdots + L_s^{[j]}$ a sum of divided powers of general enough linear forms. We have studied the relation between the closures of $Gor(T)$, $T = H(s, j, r)$ and of the variety $PS(s, j; r)$, primarily when $s \leq s_0 = \dim_k R_{t-1}$ if $j = 2t$ or $2t + 1$. Understanding these closures in the remaining cases $s > s_0$ requires as a first step resolving the portion of Conjectures 3.20, 3.23 concerning the tangent space to **Gor**(T) (equivalently to $\mathbf{V}_s(t, j - t; r)$) at sufficiently general points $f \in PS(s, j; r)$. The tangent space portion of these two Conjectures are equivalent together to the simply stated Conjecture 3.25 concerning the Hilbert function of the square $\mathcal{I}_Z^2$, Z a set of s general points of $\mathbb{P}^{r-1}$, $r \geq 4$.

C. The dimension of $Sec_s(v_j(\mathbb{P}^n))$. We do not know the dimension of the multisecant variety of the Veronese variety $Sec_s(v_j(\mathbb{P}^n))$ when char$(k)|j$ and s is beyond the limits of Theorem 2.6(i) (cf. Section 7.3).

D. Cohen-Macaulayness property. A. Geramita posed the following question in [**G1**, p.55]. Are the multisecant varieties $Sec_s(v_j(\mathbb{P}^n))$ arithmetically Cohen-Macaulay: equivalently, are $\overline{PS(s, j; n + 1)}$ Cohen-Macaulay affine varieties? This is known to be true when either $j = 2$, or $n = 1$, or $s \leq 2$ (see Sections 1.2 and 1.3). In view of the importance of the Cohen-Macaulay property when proving that certain polynomials generate the ideal of an affine variety (see e.g. the proof of Theorems 1.45 and 3.14) it would be interesting to know for which Gorenstein sequences T the schemes **Gor**$_\leq(T)$ and $\overline{\mathbf{Gor}(T)}$ are Cohen-Macaulay.

E. The boundary of $\overline{PS(s, j; r)}$, of $\overline{Gor(T)}$, and of $\overline{Hilb^H(\mathbb{P}^{r-1})}$. In the case of binary forms there is a nice description of the boundary $\overline{PS(s, j; 2)} - PS(s, j; 2)$ given by generalized additive decompositions (GAD, see Section 1.3). An interesting open question is to find an analog of GAD's, so as to describe in some explicit way the boundary

$\overline{PS(s,j;r)} - PS(s,j;r)$ when $r \geq 3$. Our study of annihilating schemes in Chapters 5 and 6, particularly the case $r = 3$, might be a starting point.

For more general Gorenstein sequences T, the study of $\overline{Gor(T)}$ is just begun when $r \geq 3$: these varieties do not satisfy a frontier condition, even when $r = 3$ (see Chapter 7).

The related question (see Section 5.3) of whether the postulation strata $Hilb^H(\mathbb{P}^2)$ of the punctual Hilbert scheme $Hilb^s(\mathbb{P}^2)$ — which are irreducible and smooth by [Got3] — satisfy a frontier condition, is apparently (and surprisingly) open, although almost certainly they do not. There is a stratification of $Hilb^s(\mathbb{P}^2)$ which evidently satisfies a frontier condition — by the lowest degree of a curve containing the punctual scheme; this is the stratification by the order of H. It is related to the Brill-Noether stratification of $Hilb^s(\mathbb{P}^2)$ by strata $W_{d,i}$: $H_Z(d) \leq i$ (see [HRW, Rah]). These strata are complicated unions of the postulation strata.

When $r \geq 4$ the postulation Hilbert scheme strata themselves are in general reducible, and likewise $Gor(T)$, as we noted in Chapter 6 (see also Remark 5.65); their closures are even less well understood than when $r = 3$.

F. Determination of Gorenstein sequences. When is a symmetric sequence T equal to $H(A_f)$ for some DP-form f? This is an open problem when $r \geq 4$. When $r = 4$, does every Gorenstein sequence T satisfy $T = \text{Sym}(H, j)$, with H the Hilbert function of a 0-dimensional scheme in $\mathbb{P}^3$? The analogous statement is no longer true in 5 variables, where there are non-unimodal Gorenstein sequences. See [BeI, BoL, Bo3], and §5 of [I6] for a discussion; articles by J. Watanabe and H. Ikeda (né Sekiguchi) [Wa2, Sek, Ik1, Ik2] discuss related questions concerning the subtler Lefschetz-Stanley properties.

What are the possible Hilbert functions H_f for $f \in \overline{PS(s,j;r)}$? This question is related to the discussion in (A) above and the answer is known when $r \leq 3$ (see Theorem 1.54 and Proposition 5.70).

G. What is the relation between $Gor_{\text{sch}}(T)$ and $Gor(T)$? When $r = 4$ the problems of determining the Gorenstein sequences and deciding which $Gor(T)$ are irreducible are salient next steps. If $r \geq 4$, and T contains a length-k subsequence $(s, s, \ldots, s)$, $s = \max(T_i)$, let $H = (t_0, \ldots t_{\lfloor j/2 \rfloor}, s, \ldots, \underline{s})$, namely $T_{\leq j/2}$ extended by s, and let $Gor_{\text{sch}}(T)$ denote the subfamily of $f \in Gor(T)$ such that $J = \text{Ann}(I)_{\leq (j+k)/2}$ defines a zero-dimensional scheme $Z \subset \mathbb{P}^{r-1}$ of Hilbert function H. Is $Gor_{sch}(T)$ the union of irreducible components of $Gor(T)$, each corresponding to a related component of $Hilb^H(\mathbb{P}^{r-1})$? This is true - for a single component - in M. Boij's examples of reducible $Gor(T)$ when $r = 4$.

It is certainly true for $r > 3$, and likely when $r = 3$ that the variety $Hilb^H_{LGor}(\mathbb{P}^r)$ parametrizing locally Gorenstein zero-dimensional schemes of given Hilbert function H has many irreducible components in general, in contrast to the known irreducibility of $Hilb^H(\mathbb{P}^2)$ [Got3]. Is it possible to construct such components using the methods of Chapter 6? That is, one begins with components of $Hilb^s(\mathbb{P}^{r-1})$, using them to construct $Gor(T)$ with several components, in r-variables. Regarding these algebras as defining schemes concentrated at a point of $\mathbb{P}^r$, will some of these determine components of $Hilb^H(\mathbb{P}^r)$, where $\Delta(H) = T$? This requires a study of the obstruction to deforming elements of $Gor(T)$ concentrated at one point of $\mathbb{P}^n$, to other locally Gorenstein schemes having the same global Hilbert function.

It is also not clear whether the constructions of Chapter 6 or of M. Boij could lead to new families of elementary "generic" Gorenstein schemes of $\mathbb{P}^r$ - a family $U_C, C \subset Gor(T)$ of schemes concentrated at a point p of $\mathbb{P}^r$ (so C is fibred over $\mathbb{P}^r$), having specified local Hilbert function T, whose members have no deformation within $\textbf{Hilb}^N(\mathbb{P}^r), N = |T|$, that are outside of U_C (see by analogy Definition 6.20).

H. Points in special position as annihilating schemes and varieties $Gor(T)$ with several irreducible components. How is the geometry of sets of points, or punctual schemes in $\mathbb{P}^{r-1}$, related to the irreducible components of $Gor(T)$? There has been much study of sets of smooth points in special position in projective space: those lying on rational curves, self-associated sets, those lying on arrangements of lines in special position: see, for just a few examples [**GPS, Hari1, Hari2, GHS1, GHS2**], and much work of the Italian school of geometers; and the references cited in our Section 6.2 on self-associated sets.

When a length-s zero-dimensional scheme Z in $\mathbb{P}^{r-1}$ is either smooth, or Gorenstein "conic" of Hilbert function $H(R/\mathcal{I}_Z) = H$, or now also locally Gorenstein (see Section 6.4), then the method of Section 6 can be used with $T = \text{Sym}(H, j), j \geq \tau(Z)$, to find homogeneous polynomials $f \in Gor(T)$ for which Ann $f \supset \mathcal{I}_Z$. Such forms f are also in the closure of $PS(s, j; r)$, provided the original scheme Z is either smooth or smoothable (Proposition 6.7). To find forms f satisfying

$\ell\text{diff}(f) \leq s$, but which are not in the closure of $PS(s, j; r)$, where $s \leq s_0 = r_{t-1}, t = \lfloor j/2 \rfloor$, appears to require the use of nonsmoothable locally Gorenstein schemes Z, if a method relying on annihilating schemes is to be used. This we studied in Chapter 6, Sections 6.2,6.3, where the examples of reducible $Gor(T)$ we found were based on nonsmoothable compressed local Gorenstein schemes in $\mathbb{P}^{r-1}$ of odd socle degree — cases where we could calculate that there are no nontrivial negative degree deformations ("small tangent space"). We report on some examples of reducible $Gor(T)$ (for different T) in Section 6.4, that do not require a small tangent space argument. There are many nonsmoothable local Gorenstein schemes Z, that do have nontrivial negative deformations (do not satisfy a "small tangent space" condition): for example, some that are compressed of even socle degree: do they determine irreducible components of $Gor(T), T = \text{Sym}(H_Z, j)$?

EXAMPLE 9.1. COMPRESSED NONSMOOTHABLE GORENSTEIN GRADED ALGEBRAS OF EVEN SOCLE DEGREE: If $r' \geq 11$ and $g \in \mathcal{R}'_4$ is general, then the Gorenstein ideal $\text{Ann}(g)$, of Hilbert function

$$T(4, r) = \left(1, r', \binom{r'+1}{2}, r', 1\right)$$

of length n' is nonsmoothable, since $n'r' < r'_4 - 1$, the dimension of the family of $g \in \mathcal{R}'_4$. Let $Z = \text{Spec}(R'/\text{Ann}(g))$ at $(1, 0, \ldots, 0)$, and set $r = r' + 1$. By the proof of Lemma 6.1(a), $H = H(R/\mathcal{I}_Z)$ is the sum function of $T(4, r')$, so $H = (1, r, r_2, r_2 + r - 1, r_2 + r, r_2 + r, \ldots) = H(R/\mathcal{I}_Z)$. Let $j \geq 8$, and finally set $T = \text{Sym}(H, j)$. Then $Gor(T)$ is a candidate for having several irreducible components, corresponding to forms $f \in \mathcal{R}_j$ having smoothable, or nonsmoothable annihilating scheme, respectively. However, the small tangent space argument of Theorem 6.27 does not apply, since $\dim \text{Hom}_R(I, A)_{-1} = \dim_k(\text{Ann}(g)/\text{Ann}(g)^2)_{j+1} = r'_{j+1}$, implying there are the maximum possible number of degree -1 tangents, in contrast to the r' required (see (6.3.1)). One needs another argument to show that the two putative components are not in the closure of a component whose general point has no length $n(T)$ annihilating scheme.

The first such example begins with $H(R'/\text{Ann}(g)) = T(4, 11) = (1, 11, 66, 11, 1)$, so $H(R/\mathcal{I}_Z) = (1, 12, 78, 89, 90, 90, \ldots)$ and $T = (1, 12, 78, 89, 90, 89, 78, 12, 1)$.

M. Boij has shown that when $j' = 4$, r' arbitrary, the smooth Gorenstein punctual subschemes with such Hilbert functions $H(R/\mathcal{I}_Z)$ are "4-self-associated" in a sense generalizing that of Section 6.2. This may give a route for showing that the forms $f \in \mathcal{R}_8$ with $H_f = T$, and having a smooth annihilating scheme would form a component of $Gor(T)$.

There has been little attempt yet to look at at components of $Gor(T)$ that might arise from sets of several nonsmoothable points in special position in $\mathbb{P}^n$.

I. Closure $\overline{Gor(T)}$, its singular locus and desingularization. A natural problem is to determine the singular locus of the scheme $\mathbf{Gor}(T)$ and of the Zariski closure $\overline{\mathbf{Gor}(T)} \subset \mathbb{A}(\mathcal{R}_j)$. When $r = 3$ Kleppe's Theorem 4.21 shows that $\mathbf{Gor}(T)$ is nonsingular. Comparing with the binary case one may ask whether $\mathbf{Gor}(T)$ is the nonsingular locus of $\overline{\mathbf{Gor}(T)}$.

Is there a natural desingularization of either the closure $\overline{Gor(T)}$ in $\mathbb{A}(\mathcal{R}_j)$, or of the larger closure in $G(T)$, the family of all graded ideals in R of Hilbert function T? When $r = 2$, there are several natural desingularizations that are semismall in the sense of Goresky-MacPherson: the variety X of Theorem 1.45, and as well $GradAlg(T)$ (see [**IY**]). Even then, when $r = 2$, the rational homology ring or homotopy structure of $\overline{Gor(T)}$ or $V_s(u, v; 2)$ is only partially understood, despite a good understanding of the homology ring of (several) desingularizations (see [**IY, GoS**]). B. Shapiro has noted that the Hankel varieties – having semismall desingularizations – are exactly at a border where some theorems, which connect intesection homology with singular homology, do not apply.

When $r = 3$ and s is small enough ($s \leq j/2$), there is the desingularization of $Gor(T), T = H(s, j, 3)$ by the universal bundle over the Hilbert scheme $Hilb^s(\mathbb{P}^2)$, used by G. Ellingsrud and S. A. Strömme to determine the degree of the multisecant variety [**ElS**] (see also Remark 7.20). This connection deserves further exploration.

J. Subloci of $Gor(T)$ depending on the number of generators of $\mathrm{Ann}(f)$, $f \in Gor(T)$. On each irreducible component C of $Gor(T)$ there is a generic number of generators for the family of Gorenstein ideals parametrized by C: this is just the minimum number that occurs on C, by Lemma 5.23. What is the codimension of the locus in C where there are more than the generic number of generators? M. Boij has announced a solution when $r = 3$ (see Theorem 4.34). When $r > 3$, what are the possible Betti strata of $Gor(T)$? Unlike the case $r = 3$, the Betti strata need not be irreducible (see Section 5.5 and Remark 5.65); what are their components, and the dimensions of the components? Do the irreducible components of the Betti strata have some geometric explanation?

K. Minimal resolution of $R/\operatorname{Ann}(f)$, $f = L_1^j + \cdots + L_s^j$, as j varies. Suppose that $f = \sum_{1 \leq i \leq s} L_i^j$, where the degree j may vary. How does the minimal resolution of $A = R/\operatorname{Ann}(f)$ over R change with j? The point of this question is that there should be a "stable" part of the resolution skeleton, that depends predictably for large j only on the triple (s, j, r). In the middle degrees $\delta \leq i \leq j - \delta$, $|\delta| < \min\{u \mid \dim_k R_u \geq s\}$, we have $\dim_k(A_i) = s$. By Lemma 1.19, $A_i = R_i/I_i$ where $I_i = (\mathcal{I}_P)_i$, the degree-i piece of the vanishing ideal $\mathcal{I}_P$ at the set P of points $p_1, \ldots, p_s$ in $\mathbb{P}^{r-1}$ corresponding to the forms $L_1, \ldots, L_s$. Thus, only at the smallest and highest degrees can I_i depend on the degree j. The minimal R-resolution of the algebra A should vary predictably with j. This behavior, when it occurs, is certainly related to results of M. Boij and others [**Bo1**] (see also Appendix B.3).

K. Ranestad and F.-O. Schreyer have studied when $j = 3$ the minimal R-resolution of $R/\operatorname{Ann}(f)$, and its dependence on invariants such as the length s of the form [**RS**].

L. Finding conditions on a form f so that $\operatorname{Ann}(f)$ is a complete intersection. B. Angéniol and M. Lejeune-Jalabert have studied the residue map when $I = \operatorname{Ann}(f) = (g_1, \ldots, g_r)$ is a complete intersection [**AnL**]. It is well known that the Jacobian of the CI generators gives an element of the socle of $A = R/I$ [**ScSt**]. In principle, these results should be enough to determine the generator f of the inverse system of I, given the CI generators $(g_1, \ldots, g_r)$: perhaps one should merely adjust the coefficients of the Jacobian in a suitable way. However, this has not been written down simply. It may be that this problem has no solution beyond simply finding I_j, j the socle degree and defining f by $\langle f \rangle = (I_j)^{\perp}$. R. Buchweitz has remarked that tracing out the residue map leads to the $\langle f \rangle = (I_j)^{\perp}$ construction.

The converse problem, determining direct conditions on a form f so that $\operatorname{Ann}(f)$ is a complete intersection is quite open.

M. Connections with geometry. K. Ranestad and F.-O. Schreyer, C. Peskine, and others have studied connections between the Hilbert function H_f and the study of curves and surfaces (see, for example [**RS, Schr, IlR**]). B. Hassett and Y. Tschinkel have studied a relation between $Gor(T)$, as well as parameter spaces $G(T)$ of more general Artin algebras, and the classification of certain equivariant compactifications of $\mathbb{G}_a^n$, the direct product of the additive group [**HasT**]. In a different direction, D. Eisenbud and H. Levine have for a (not necessarily graded) Artinian Gorenstein algebra A over the reals studied decompositions $A = W \oplus W^* \oplus D$ where W is a maximal isotropic subspace with respect to an inner product on A. When A is a complete intersection, $A = \mathbb{R}\{x_1, \ldots, x_r\}/(f_1, \ldots, f_r)$, determined by a C^∞ map germ $f = (f_1, \ldots, f_r)$, they show that the dimension of D is the degree of f (see [**EiL**, **Te**]). When $A = A_F = A/\operatorname{Ann}(F)$ is graded and $H(A) = (1, r, 1)$, this degree is the signature of F, which depends on determining the number of positive and negative coefficients in a minimal representation of F as a sum of squares of linear forms. When A_F is graded and $H(A)$ is arbitrary, the dimension of D may give a finer, geometric invariant related to the length of F: one can decompose the family $Gor_{\mathbb{R}}(T)$ of length-s degree-j homogeneous polynomials F of Hilbert function $H_F = T$ into strata defined by the degree, and study these strata.

N. Connections with topology. Most graded Artinian Gorenstein rings occur as the cohomology rings of a manifold (see [**I3**, Section 5A] for a discussion). V. Puppe has used "generic" Gorenstein algebras to construct manifolds having no circle action [**Puc**]. Can the Gorenstein algebras that are generic points of irreducible components of $Gor(T)$ be similarly used to construct interesting manifolds?

The Gorenstein algebras we have considered are "standard": they have generators in a single degree, being quotients of R. Usually the cohomology ring of a manifold has generators of different degrees. It could be of interest to see which of the results concerning standard Gorenstein algebras would extend to the more general setting.

O. Connections with singularity theory Many authors – including J. Damon, A. Dimca and C. Gibson, T. Gaffney [**Gaf**], C. T. C. Wall [**Wall**], have studied the classification of finite map germs, and their deformations (for further references see A. duPlessis and C. T. C. Wall's [**DuPW**]). An isomorphism class of Artinian algebras determines a right-left equivalence class finite mapping germs. Thus, obstructions to deforming Artinian Gorenstein algebras – in the wider class of nonhomogeneous Artinian Gorenstein algebras – have relevence to singularity theory. Artinian Gorenstein algebras that are generic points of components of $Gor(T)$ should give singularities of interest.

P. Limit behavior of the catalecticant for large j, r. Suppose first that $r = 3$. The result of Y. Cho and B. Jung [**ChoJ2**] (see Section 4.4) shows the existence of large dimensional irreducible components $Gor(T)$ of the catalecticant variety $V_s(t, t; 3)$, when $s \approx r_{t-1}$. As j increases, what shape Hilbert function T corresponds to a largest component? For example, taking $s = r_{t-1}$, determine a function

$g : [0, 1] \longrightarrow [0, 1]$ such that $T_i = \lfloor g(i/2t) \cdot r_{t-1} \rfloor$ yields a component asymptotically of highest dimension. Given the dimension formulas reported in Section 4.4, this is an entirely combinatorial problem. A similar question could be asked for $r > 3$, except that a further aspect would be to identify the large dimensional components of each $Gor(T)$.

Next we fix $j = u + v$, and let $r \to \infty$. The number of distinct entries of $Cat_F(u, v; r)$, satisfies

$$\#\{\text{distinct entries}\}/\#\{\text{entries}\} = \dim_k(R_j)/(\dim_k R_u \cdot \dim_k R_v).$$

As r increases and the pair of integers (u, v) is fixed, it is easy to see that the above ratio *decreases* from the value $(j + 1)/(u + 1)(v + 1)$ when $r = 2$, to the limit value

$$c_{u,v} = \binom{j}{u}^{-1}.$$

The density of repeated entries of $Cat_F(u, v; r)$ *increases* to the limit value $1 - c_{u,v}$ as r increases. Is there a limit behavior as $r \to \infty$ of the determinantal varieties $V_s(u, v; r), s = \lfloor \alpha \cdot r_u \rfloor$, where $0 < \alpha < 1$ is fixed? Determining such behavior could include finding the dimension, and the number of irreducible components.

Q. Problems posed by A. Geramita on the structure of catalecticant ideals. A. Geramita in **[G2]** proposes several further problems on the structure of catalecticant ideals. We briefly report on these, with some pointers to where in the book we discuss related issues; recall that $I_{s+1}(Cat_F(u, j-u; r))$ denotes the ideal of $s+1 \times s+1$ minors, defining the rank-s locus $V_s(u, j - u; r)$.

Q1. When is the ideal $I_{s+1}(Cat_F(u, j - u; r))$ a prime ideal?

As A. Geramita notes, when $r \geq 3$ there are many counterexamples, arising from cases where we know that the determinantal locus has several irreducible components; the first, for $r = 3$, appeared in S. Diesel's article **[Di]** (see our Example 3.6). Chapter 6 contains our study of counterexamples when $r \geq 4$, for which even the smaller variety $Gor(T)$ is reducible. We studied the reducibility of the determinantal locus also in Section 7.2.

That the determinantal ideals of $Cat_F(1, j - 1; r)$ are prime for all s less than r when $j \geq 2$ was settled for char $k = 0$ by O. Porras's Theorem **[Po**, Section 4] (see our Theorem 1.28; note that the variety $V_s(1, j - 1; r)$ is clearly irreducible). The case $j = 2$ is classical (see Theorem 1.26 and the references there). Also, primeness of the catalecticant ideals in the case $r = 2$ of binary forms is classical (see Theorem 1.56 and Corollary 1.58). M. Pucci showed that the ideal

$I_2(Cat_F(u, j - u; r))$ for $1 \leq u \leq j - 1$ is independent of u, and that it is the (prime) ideal defining the Veronese variety, apparently in arbitrary characteristic [**Puc**].[2] Another positive case is that of corank 1, that is, $I_{s+1}(Cat_F(u, j - u; r))$ is a prime ideal when $s = dim_k R_u - 1$, for $1 \leq u \leq j - u$ (see Theorem 3.14).

Q2,3. Is $I_{s+1}(Cat_F(u, j - u; r))$ a radical ideal? Saturated?

The first remains a very open question, apart from the cases when the catalecticant ideal is known to be prime. A. Geramita states in [**G2**] that M. Deery has shown that the catalecticant ideals are saturated.

Q4. Is Macaulay's theorem on the growth of Hilbert functions reflected in catalecticant ideals?

Roughly, A. Geramita asks, if certain natural inclusion relations among the radicals of catalecticant ideals that arise from Macaulay's Hilbert function inequalities (see Corollary C.4), actually extend to relations among the catalecticant ideals themselves. He gives two "simple test cases":

Q5a. Is it true for $j \geq 3$ that

$$I_3(Cat_F(2, j - 2; r)) = I_3(Cat_F(u, j - u; r))$$

for all u with $2 \leq u \leq j - 2$?

Q5b. Is it true for $j \geq 3$ that

$$I_3(Cat_F(1, j - 1; r)) \subsetneqq I_3(Cat_F(2, j - 2; r))?$$

A. Geramita conjectures, by analogy with M. Pucci's result, that the ideals $I_3(Cat_F(u, j - u; r)), 2 \leq u \leq j - 2$, should each be the ideal defining the secant line (chordal) variety to the Veronese variety. The case $j = 3$ is known (see Theorem 1.27). V. Kanev showed in [**Ka**] that the two ideals of Q5b, together generate the ideal of the secant line variety $\overline{PS(2, j; r)}$ (see page 21).

See [**G2**] for further discussion of Q1-Q5.

---

[2]There is no mention of characteristic in [**Puc**], whose Theorem 5.4 states that every 2-minor of $Cat(1, j - 1; r)$ is either itself a 2-minor or the sum of two 2-minors of $Cat_F(u, j - u; r)$, leading to the independence result, Corollary 5.5.

APPENDIX A

Divided Power Rings and Polynomial Rings

The aim of this appendix is to introduce the divided power rings and prove the few elementary facts about them we use. For a more complete, coordinate free exposition the reader is referred to [**Ei2**, Section A2.4]. Let k be a field of arbitrary characteristic. Let $R = k[x_1, \ldots, x_r] = \bigoplus_{j \geq 0} R_j$. Let $\mathcal{D}$ be the graded dual of R, i.e.

$$\mathcal{D} = \bigoplus_{j \geq 0} Hom_k(R_j, k) = \bigoplus_{j \geq 0} \mathcal{D}_j$$

We consider the vector space R_1 with basis $x_1, \ldots, x_r$ and the left action of $GL_r(k)$ on R_1 defined by $Ax_i = \sum_{j=1}^{r} A_{ji} x_j$. Since $R = \bigoplus_{j \geq 0} Sym^j R_1$ this action extends to an action of $GL_r(k)$ on R. By duality this action induces a left action of $GL_r(k)$ on $\bigoplus_{j \geq 0} \mathcal{D}_j$. We denote by $x^U = x_1^{u_1} \cdots x_r^{u_r}, |U| = u_1 + \cdots + u_r = j$ the standard monomial basis of R_j. Let $X_1, \ldots, X_r$ be the basis of $\mathcal{D}_1$ dual to the basis $x_1, \ldots, x_r$.

DEFINITION A.1. We denote by

$$X^{[U]} = X_1^{[u_1]} \cdots X_r^{[u_r]} \tag{A.0.1}$$

the basis of $\mathcal{D}_j$ dual to the basis $\{x^U : |U| = j\}$. We call these elements divided power monomials, or for short, DP-monomials. We call the elements of $\mathcal{D}_j$ divided power forms, or for short, DP-forms and the elements of $\mathcal{D}$ divided power polynomials, or DP-polynomials. We extend the definition of $X^{[U]}$ to multidegrees $U = (u_1, \ldots, u_r)$ with negative components by letting $X^{[U]} = 0$ if $u_i < 0$ for some i.

DEFINITION A.2. We define for every i, j a contraction map

$$R_i \times \mathcal{D}_j \longrightarrow \mathcal{D}_{j-i}$$

as follows. For $\phi \in R_i$ and $f \in \mathcal{D}_j$ let $\phi \circ f \in \mathcal{D}_{j-i}$ be 0 if $j < i$ and be the functional

$$\langle \psi, \phi \circ f \rangle = \langle \psi \phi, f \rangle \quad \text{for} \quad \psi \in R_{j-i} \tag{A.0.2}$$

One extends these maps by linearity to a contraction map $R \times \mathcal{D} \to \mathcal{D}$.

PROPOSITION A.3. *The contraction map has the following properties*

   i. *It is $GL_r(k)$-equivariant*

   ii. *One has*

$$x_1^{u_1} \cdots x_r^{u_r} \circ X_1^{[j_1]} \cdots X_r^{[j_r]} \;=\; X_1^{[j_1-u_1]} \cdots X_r^{[j_r-u_r]} \qquad (A.0.3)$$

   iii. *It is a left action of R on $\mathcal{D}$.*

   iv. *The R-module $\mathcal{D}$ is isomorphic to Macaulay's inverse system, the R-module $T = k[x_1^{-1}, \ldots, x_r^{-1}]$ (see e.g. [Ei2, p.526]), by the isomorphism sending $X_1^{[j_1]} \cdots X_r^{[j_r]}$ to $x_1^{-j_1} \cdots x_r^{-j_r}$*

PROOF. (i). The equality $A(\phi \circ f) = (A\phi) \circ (Af)$ for $A \in GL_r(k)$ is clear from the definition of contraction (A.0.2).

(ii). This is obvious.

(iii). Using formula (A.0.3) one verifies the equality $(\phi\psi) \circ f = \phi \circ (\psi \circ f)$.

(iv). This is immediate from (A.0.3). $\qquad\qquad\square$

REMARK A.4. We have not used so far that k is a field. The same construction holds for every commutative ring k, e.g. $k = \mathbb{Z}$. In the latter case we denote by $R_{\mathbb{Z}}$ the ring $\mathbb{Z}[x_1, \ldots, x_r]$ and by $\mathcal{D}_{\mathbb{Z}}$ the corresponding $R_{\mathbb{Z}}$-module constructed above.

EXAMPLE A.5. Suppose k is a field of characteristic 0. Let $\mathcal{R} = k[X_1, \ldots, X_r]$. One considers the differentiation action of $R = k[x_1, \ldots, x_r]$ on $\mathcal{R}$ given by

$$\phi \circ f \;=\; \phi\left(\frac{\partial}{\partial X_1}, \ldots, \frac{\partial}{\partial X_r}\right) f$$

It is $GL_r(k)$-equivariant and yields a duality between R_j and $\mathcal{R}_j$ for every $j \geq 0$. The basis of $\mathcal{R}_j$ dual to the monomial basis x^U, $|U| = j$ is

$$X^{[U]} \;=\; \frac{1}{u_1! \cdots u_r!} X_1^{u_1} \cdots X_r^{u_r} \qquad (A.0.4)$$

So, in this case the R-module $\mathcal{D}$ is isomorphic to the polynomial ring $k[X_1, \ldots, X_r]$ with divided power monomials given by (A.0.4). This isomorphism is compatible with the action of $GL_r(k)$.

We now define a ring structure on $\mathcal{D}$. It is modeled on the characteristic 0 case. One defines multiplication of monomials by the equality

$$X^{[U]} \cdot X^{[V]} \;=\; \frac{(U+V)!}{U! V!} X^{[U+V]} \qquad (A.0.5)$$

where $\frac{(U+V)!}{U!V!} = \frac{(u_1+v_1)!\cdots(u_r+v_r)!}{u_1!\cdots u_r!v_1!\cdots v_r!}$. This is extended by linearity and gives a structure of a k-algebra on $\mathcal{D}$.

EXAMPLE A.6. If $r = 1$, char $k = 0$, we may view $X^{[n]} \in \mathcal{D}$ as the operator $\frac{\partial}{n!\partial x^n}$ on R. By (A.0.5), $X^{[2]} \cdot X^{[2]} = \frac{(2+2)!}{2!2!}X^{[4]} = 6X^{[4]}$. Accordingly, we have

$$(X^{[2]} \cdot (X^{[2]}) \circ x^4 = \left(\frac{\partial}{2\partial x^2}\frac{\partial}{2\partial x^2}\right) \circ x^4 = 6 = 6X^{[4]} \circ x^4.$$

Thus, the multiplication introduced allows us to view $\mathcal{D}$ as the ring of (higher order) partial differential operators on R, when char $k = 0$. We do not carry this viewpoint further (in which $\mathcal{D}$ acts on R), since we need primarily the action of R on the inverse system $\mathcal{D}$. However, we will use the properties of the multiplication in $\mathcal{D}$.

PROPOSITION A.7. *The multiplication introduced above is commutative and associative. It is equivariant with respect to the action of* $GL_r(k)$:

$$A(fg) = (Af)(Ag) \tag{A.0.6}$$

PROOF. It suffices to check commutativity and associativity for monomials, which one verifies immediately by the definition (A.0.5). For proving equivariance it again suffices to check equality (A.0.6) for monomials $f = X^{[U]}, g = X^{[V]}$. Furthermore, since $GL_r(k)$ is generated by the subgroup of diagonal matrices and the transvections $t_{ij} = E + E_{ij}, i \neq j$ it is enough to prove (A.0.6) for A a diagonal matrix or a transvection. From the definition of $X^{[U]}$ it is clear that for $A = Diag(c_1, \ldots, c_r)$ one has $A(X^{[U]}) = c_1^{-u_1} \cdots c_r^{-u_r} X^{[U]}$, hence (A.0.6) holds for diagonal matrices. For transvections it suffices to check that the formula holds in $\mathcal{D}_{\mathbb{Z}}$ (see Remark A.4). As we saw in Example A.5 $\mathcal{D}_{\mathbb{Z}} \otimes \mathbb{Q} \cong \mathbb{Q}[X_1, \ldots, X_r]$. This isomorphism preserves the multiplication by (A.0.4) and (A.0.5). So, (A.0.6) reduces to a standard fact for the action of $GL_r(\mathbb{Q})$ on polynomials with rational coefficients. $\square$

DEFINITION A.8. Let $L = a_1 X_1 + \cdots + a_r X_r \in \mathcal{D}_1$. The *divided power* $L^{[j]}$ is defined as

$$L^{[j]} = \sum_{j_1+\cdots+j_r=j} a_1^{j_1} \cdots a_r^{j_r} X_1^{[j_1]} \cdots X_r^{[j_r]}$$

PROPOSITION A.9. *Let* $L \in \mathcal{D}_1$ *be as above. Then*

i. $L^j = (j!)L^{[j]}$

ii. *The element $L^{[j]} \in \mathcal{D}_j = \mathrm{Hom}_k(R_j, k)$ is equal to the functional defined by $\langle \phi, L^{[j]} \rangle = \phi(a)$.*

iii. *$A(L^{[j]}) = (A(L))^{[j]}$ for every $A \in GL_r(k)$.*

iv. *$L^{[i]} \cdot L^{[j]} = \frac{(i+j)!}{i! \cdot j!} L^{[i+j]}$.*

PROOF. Part (i) is easily proved by induction on j. Part (ii) is immediate from the definition of $L^{[j]}$.

Part (iii). Let $a = (a_1, \ldots, a_r)$ and let $B = ({}^tA)^{-1}$. Then the row vector of coordinates of $A(L)$ is ${}^t(B({}^ta))$. Recall that we defined the action of $GL_r(k)$ on R_j requiring that $\langle \cdot, \cdot \rangle$ is $GL_r(k)$-invariant on $R_1 \times \mathcal{D}_1$ and then extending to $R_j = Sym^j(R_1)$. This implies that $(A\phi)(a) = \phi({}^t(B^{-1}({}^ta)))$. Now, for every $\phi \in R_j$ we have

$$\langle \phi, A(L^{[j]}) \rangle = \langle A^{-1}\phi, L^{[j]} \rangle = (A^{-1}\phi)(a)$$
$$= \phi({}^t(B({}^ta))) = \langle \phi, (A(L))^{[j]} \rangle$$

Therefore $A(L^{[j]}) = (A(L))^{[j]}$.

Part (iv). This is by definition if $L = X_1$. The general case is reduced to this using Part (iii). □

COROLLARY A.10. COORDINATE-FREE DESCRIPTION OF THE VE-RONESE MAP. *Let $V = \mathcal{D}_1$. Then the Veronese map $v_j : \mathbb{P}(V) \to \mathbb{P}H^0(\mathcal{O}_{\mathbb{P}(V)}(j))^*$ can be identified with the map $\mathbb{P}(\mathcal{D}_1) \to \mathbb{P}(\mathcal{D}_j)$ given by $L \mapsto L^{[j]}$.*

PROOF. We have $V^* \cong R_1$ and $H^0(\mathcal{O}_{\mathbb{P}(V)}(j)) \cong Sym^j R_1 = R_j$. Thus the target space of v_j is $\mathbb{P}(\mathcal{D}_j)$. The identification of v_j with the map $L \mapsto L^{[j]}$ follows from Proposition A.9(ii). □

REMARK A.11. The definitions of divided powers $L^{[j]}$ and the product in $\mathcal{D}$ shows that the right-hand side in the definition of the DP-monomials $X^{[U]}$ (see (A.0.1)) is not just a symbolic expression, but has the meaning of a product of divided powers of X_i

We conclude the appendix by comparing divided power rings and polynomial rings in char$(k) > 0$.

PROPOSITION A.12. *Let $\mathcal{R} = k[Y_1, \ldots, Y_r]$ and let $\mathcal{D}$ be the divided power ring. Let $R = k[x_1, \ldots, x_r]$ act on $\mathcal{D}$ by contraction and on $\mathcal{R}$ by differentiation (see Example A.5). Then*

i. *There is an isomorphism of k-algebras*

$$\varphi : \mathcal{D}/ \oplus_{n \geq j+1} \mathcal{D}_n \longrightarrow \mathcal{R}/ \oplus_{n \geq j+1} \mathcal{R}_n$$

given on monomials by

$$\varphi(X_1^{[u_1]} \cdots X_r^{[u_r]}) \;=\; \frac{1}{u_1! \cdots u_r!} Y_1^{u_1} \cdots Y_r^{u_r} \tag{A.0.7}$$

and sending $L^{[j]}$ to $\frac{1}{j!}L^j$.

ii. *There is an isomorphism of R-modules*

$$\varphi' : \oplus_{i \leq j} \mathcal{D}_i \longrightarrow \oplus_{i \leq j} \mathcal{R}_i$$

given by the same formula as in (A.0.7)

iii. *Suppose* char$(k) = 0$. *Then* (A.0.7) *defines an isomorphism $\varphi : \mathcal{D} \to \mathcal{R}$ which commutes with the action of R*

PROOF. By Proposition A.9(i) the homomorphism $\vartheta : k[Y_1, \dots , Y_r] \to \mathcal{D}$ sending $Y_1^{u_1} \cdots Y_r^{u_r}$ to $X_1^{u_1} \cdots X_r^{u_r}$ is an isomorphism if char$(k) = 0$, or an isomorphism in degrees $\leq j$ if char$(k) > j$. The statement about $L^{[j]}$ follows from Proposition A.9(i). Part (ii) is clear from formula (A.0.3) $\hspace{3cm} \square$

It follows that if char$(k) > j$, the highest degree of a polynomial we use, or if char$(k) = 0$, we may interchange in some cases the R-algebras $\mathcal{D}$ and $\mathcal{R}$.

APPENDIX B

Height Three Gorenstein Ideals

We collect a number of facts that we use concerning height three Gorenstein ideals, and we give proofs for several of them.

In Section B.1 we prove an expansion formula for Pfaffians of alternating matrices with zero diagonal blocks, analogous to the Laplace formula for determinants. We use a particular case in the proofs of the main theorems of Section 5.3. We also state an interesting expansion formula for Pfaffians due to H. Srinivasan [**Sr1**].

In Section B.2 we first state the Buchsbaum-Eisenbud structure theorem, in the special case for graded Gorenstein ideals. We then state a variation of a Kustin-Ulrich theorem concerning the resolution of I^a when I is a general enough Pfaffian ideal [**KusU**], in the case $a = 2$ (Theorem B.3). We use this result to bound above the dimension of the tangent space to $\mathbf{Gor}(T)$, when $r = 3$ (Theorem 4.5B).

In Section B.3 we state several very useful results of M. Boij, connecting the coordinate ring of an ideal defining sets of points in $\mathbb{P}^n$ and a related Gorenstein Artin algebras; these culminate in a result determining the minimal resolution of the latter in terms of that for the ideal of sets of points (Proposition B.10).

In Section B.4 we first state and prove a nice criterion of A. Conca and G. Valla for determining the maximal Betti numbers consistent with a height three Gorenstein sequence T, directly in terms of the second difference $\Delta^2(T)$ (Theorem B.13). We use this result in alternative proofs of Lemmas 5.29 and 5.37. A. Conca and G. Valla state and prove a far more general version in [**CoV1**]. We prove the Conca-Valla criterion as a convenience, and also to illustrate some methods that we have not greatly stressed in the main text.

Finally we state a result of S. J. Diesel concerning the number of Gorenstein sequences of height three, having given order (of defining ideal) and socle degree. We use it in the proof of Theorem 7.15.

B.1. Pfaffian formulas

We use the definition of the Pfaffian of an alternating matrix according to Bourbaki [**Bou**]. Let S_{2n} be the group of permutations of $\{1, 2, \ldots, 2n\}$. Let $H \subset S_{2n}$ be the subgroup consisting of permutations which transform every subset $\{2k - 1, 2k\}$ into a subset $\{2\ell - 1, 2\ell\}$. Clearly H is a semidirect product of S_n with $(\mathbb{Z}_2)^n$ (this is the Weyl group $W(\mathbb{C}_n)$). Let A be a commutative ring with identity and let $X = (x_{ij})$ be a $2n \times 2n$ alternating matrix with $x_{ij} \in A$, $x_{ji} = -x_{ij}$ and $x_{ii} = 0$ for $i, j = 1, \ldots, 2n$. The Pfaffian of X is

$$\mathrm{Pf}(X) = \sum_{\overline{\sigma} \in S_{2n}/H} \mathrm{sgn}(\sigma) x_{\sigma(1)\sigma(2)} x_{\sigma(3)\sigma(4)} \cdots x_{\sigma(2n-1)\sigma(2n)}$$

$$= \frac{1}{n! 2^n} \sum_{\sigma \in S_{2n}} \mathrm{sign}(\sigma) x_{\sigma(1)\sigma(2)} \cdots x_{\sigma(2n-1)\sigma(2n)} \qquad (\text{B.1.1})$$

$$\text{when } \mathrm{char}(k) = 0.$$

In this way the Pfaffian is normalized so that $\mathrm{Pf}(X_0) = 1$, if $X_0 : x_{ij} = j - i$ if $| j - i | = 1$, and $x_{ij} = 0$ otherwise. The Pfaffians have the following two basic properties [**Bou**, Ch. IX, §5]. If Y is an arbitrary $2n \times 2n$ matrix with coefficients in A then

$$\mathrm{Det}(X) = \mathrm{Pf}(X)^2 \quad \text{and} \quad \mathrm{Pf}(YXY^t) = \mathrm{Det}(Y)\mathrm{Pf}(X) \qquad (\text{B.1.2})$$

Suppose $A = K$ is a field of characteristic 0. Let $\omega = \sum_{i,j=1}^{2n} x_{ij} e_i \wedge e_j$ be the corresponding bivector. Then

$$\overset{n}{\wedge} \omega = 2^n n! \, \mathrm{Pf}(X) e_1 \wedge e_2 \wedge \cdots \wedge e_{2n-1} \wedge e_{2n}.$$

Let us fix k columns of X with numbers $j_1 < \cdots < j_k$. For every subset $I \subset \{1, \ldots, 2n\}$, $\#I = k$, $I = \{i_1, \ldots, i_k\}$, $i_1 < \cdots < i_k$ we denote by X_J^I the submatrix consisting of the entries in the rows $i_1, \ldots, i_k$ and the columns $j_1, \ldots, j_k$. If $I \cap J = \emptyset$ we denote by $X_{I,J}$ the alternating $(2n - 2k) \times (2n - 2k)$ submatrix of X obtained by deleting the rows and columns with numbers $i_1, \ldots, i_k, j_1 \cdots j_k$.

LEMMA B.1. EXPANSION OF PFAFFIANS BY MINORS. *Let A be a commutative ring with identity and let $X = (x_{ij})$ be a $2n \times 2n$ alternating matrix with coefficients in A. Let $1 \leq j_1 < \cdots < j_k \leq 2n$ where $k \leq n$. Suppose the $k \times k$ block $(x_{j_\alpha j_\beta})_{1 \leq \alpha, \beta \leq k}$ equals 0. Then,* *expanding along k columns,*

$$\mathrm{Pf}(X) = \tag{B.1.3}$$

$$\sum_{\substack{\#I = k, \\ I \cap J = \emptyset}} (-1)^{\sum_{\alpha=1}^{k}(i_\alpha + j_\alpha) - k + \mathrm{inv}(i_1, j_1, \ldots, i_k, j_k)} \mathrm{Det}\left(X_J^I\right) \mathrm{Pf}\left(X_{I,J}\right)$$

where $\mathrm{inv}(i_1, j_1, \ldots, i_k, j_k)$ *is the number of inversions in* $(i_1, j_1, \ldots, i_k, j_k)$. *If the rows* $\{i_1, \ldots, i_k\} = I$ *are fixed and one assumes the* $k \times k$ *block* $\left(x_{i_\alpha i_\beta}\right) = 0$ *then expanding along the* k *rows,*

$$\mathrm{Pf}(X) = \tag{B.1.4}$$

$$\sum_{\substack{\#J = k, \\ I \cap J = \emptyset}} (-1)^{\sum_{\alpha=1}^{k}(i_\alpha + j_\alpha) - k + \mathrm{inv}(i_1, j_1, \ldots, i_k, j_k)} \mathrm{Det}\left(X_J^I\right) \mathrm{Pf}\left(X_{I,J}\right)$$

PROOF. We give the proof for expansion along columns. The formula for expansion along rows is proved similarly. Both sides of the equality (B.1.3) are polynomials in x_{ij} with integer coefficients, so it suffices to prove it for the ring $A = \mathbb{Z}[x_{ij}]_{i<j}$, and furthermore we may replace A by the field of fractions K of characteristic 0. We separate the summands of $\omega = \sum x_{ij} e_i \wedge e_j$ which do not contain indices $j_1, \ldots, j_k$ and using the assumption of the 0-block we obtain

$$\omega = 2\varphi + \omega', \quad \text{where}$$

$$\varphi = \sum_{i \notin J} \sum_{j \in J} x_{ij} e_i \wedge e_j$$

$$\omega' = \sum_{u,v \notin J} x_{uv} e_u \wedge e_v.$$

We know $\overset{n}{\wedge}\omega$ is proportional to $e_1 \wedge \cdots \wedge e_{2n}$. Expanding $\overset{n}{\wedge}(2\varphi + \omega')$ we see that there is only one summand which contains $e_{j_1} \wedge \cdots \wedge e_{j_k}$ as a factor, and thus might be nonzero:

$$\overset{n}{\wedge}\omega = \binom{n}{k} 2^k (\overset{k}{\wedge}\varphi) \wedge (\overset{n-k}{\wedge}\omega'). \tag{B.1.5}$$

Now,

$$\overset{k}{\wedge} \varphi =$$

$$= k! \sum_{I \cap J = \emptyset} \sum_{\sigma \in S_k} x_{\sigma(i_1)j_1} \cdots x_{\sigma(i_k)j_k} e_{\sigma(i_1)} \wedge e_{j_1} \wedge \cdots \wedge e_{\sigma(i_k)} \wedge e_{j_k}$$

$$= k! \sum_{I \cap J = \emptyset} \left(\sum_{\sigma \in S_k} \mathrm{sgn}(\sigma) x_{\sigma(i_1)j_1} \cdots x_{\sigma(i_k)j_k} \right) e_{i_1} \wedge e_{j_1} \wedge \cdots \wedge e_{i_k} \wedge e_{j_k}$$

$$= k! \sum_{I \cap J = \emptyset} \mathrm{Det}\left(X_J^I \right) e_{i_1} \wedge e_{j_1} \wedge \cdots \wedge e_{i_k} \wedge e_{j_k}. \tag{B.1.6}$$

Each of the wedge products

$$e_{i_1} \wedge e_{j_1} \wedge \cdots \wedge e_{i_k} \wedge e_{j_k} \wedge \left(\overset{n-k}{\wedge} \omega' \right) \tag{B.1.7}$$

cancels summands of $\overset{n-k}{\wedge} \omega'$ which contain either e_i, $i \in I$ or e_j, $j \in J$. So we can consider $\omega_{I,J} = \sum_{u,v \notin I \cup J} x_{uv} e_u \wedge e_v$, Then (B.1.7) equals

$$e_{i_1} \wedge e_{j_1} \wedge \cdots \wedge e_{i_k} \wedge e_{j_k} \wedge \left(\overset{n-k}{\wedge} \omega_{I,J} \right) = \tag{B.1.8}$$

$$= 2^{n-k}(n-k)! \, \mathrm{Pf}\left(X_{I,J} \right) \wedge_{\alpha=1}^{k} \left(e_{i_\alpha} \wedge e_{j_\alpha} \right) \wedge \left(\wedge_{u \notin (I \cup J)} e_u \right).$$

It only remains to apply a standard lemma which counts the number of inversions of the permutation $i_1, j_1, \ldots, i_k, j_k, u_1, \ldots, u_{2n-2k}$ with $u_1 < u_2 < \cdots < u_{2n-2k}$ and says it equals $\mathrm{inv}(i_1, j_1, \ldots i_k, j_k) + \sum_{\alpha=1}^{k}(i_\alpha + j_\alpha) - \frac{2k(2k+1)}{2}$ [Kur, I. §5]. This yields that the wedge product in (B.1.8) equals

$$(-1)^{\mathrm{inv}(i_1,j_1,\ldots,i_k,j_k)+\sum_{\alpha=1}^{k}(i_\alpha+j_\alpha)-k(2k+1)} e_1 \wedge \cdots \wedge e_{2n}.$$

Combining (B.1.5), (B.1.6) and (B.1.8) we obtain the expansion formula (B.1.3) □

Let us consider some particular cases of the formulas (B.1.3) and (B.1.4). If we fix the i-th row, then $x_{ii} = 0$ since X is alternating matrix and one obtains expansion of $\mathrm{Pf}(X)$ along a row:

$$\mathrm{Pf}(X) = \sum_{j=1}^{2n} (-1)^{i+j-1+\mathrm{inv}(i,j)} x_{ij} \, \mathrm{Pf}(X_{ij}) \tag{B.1.9}$$

where $\mathrm{inv}(i,j) = 0$ if $i < j$ and $\mathrm{inv}(i,j) = 1$ if $i > j$.

Suppose X has the form

$$X = \begin{pmatrix} A & B \\ C & 0 \end{pmatrix}$$

where A, B and C are square $n \times n$ matrices, A is alternating, $C = -B^t$. Then the condition of the lemma is satisfied with $J = \{n+1, \ldots, 2n\}$ and one obtains only one summand with $I = \{1, \ldots, n\}$, $X_J^I = B$. One has $\mathrm{inv}(1, n+1, \ldots, n, 2n) = (n-1)+(n-2)+\cdots+1 = n(n-1)/2$. So

$$\mathrm{Pf}(X) = (-1)^{\frac{n(n-1)}{2}} \mathrm{Det}(B) \qquad (\mathrm{B.1.10})$$

Another case that we need is

$$X = \begin{pmatrix} A & B \\ C & D = 0 \end{pmatrix}$$

where A is an alternating $(n+1) \times (n+1)$ matrix, B is an $(n+1) \times (n-1)$ matrix, $C = -B^t$ and D is a 0-block of size $(n-1) \times (n-1)$. Let us expand along the last $n-1$ columns. Then $J = \{n+2, \ldots, 2n\}$, the possible I are $I = \{1, 2, \ldots, n+1\} - \{i, j\}$, $1 \leq i < j \leq n+1$. The matrix $X_{I,J} = \begin{pmatrix} 0 & x_{ij} \\ -x_{ij} & 0 \end{pmatrix}$. We let $B_{ij} = X_J^I$, this is the matrix obtained from B deleting rows i and j. Applying (B.1.3) we obtain

$$\mathrm{Pf}(X) = \sum_{1 \leq i < j \leq n+1} (-1)^{i+j+\frac{n(n-3)}{2}} x_{ij} \mathrm{Det}(B_{ij}) \qquad (\mathrm{B.1.11})$$

We conclude this section by stating an interesting formula due to H. Srinivasan which may be used for an alternative proof of (B.1.11). We need first some more notation. Let X be an alternating matrix of order $2n$. If $S = (i_1, \ldots, i_q)$ denotes a length q increasing sequence of integers, with $1 \leq i_t \leq 2n$ for all t, then we denote by

$$X(S) = \begin{pmatrix} x_{i_1, i_1} & \cdot & x_{i_1, i_p} \\ \cdot & \cdot & \cdot \\ x_{i_p, i_1} & \cdots & x_{i_p, i_p} \end{pmatrix}$$

the submatrix using the rows and columns specified by S. We denote the Pfaffian $\mathrm{Pf}(X(S))$ by $\mathrm{Pf}(S)$. Srinivasan's formula [Sr1, Corollary 3.2] is the following. If I, J are increasing subsequences of S, and

$|I| = p, |J| = q$ satisfy $p < q$, then

$$\text{Pf}(I, J) = \sum_{t=0}^{q-2} (-1)^{(t-q)/2+1} \sum_{\substack{W \subset J \\ |W|=t}} \text{sign}\, \sigma \; \text{Pf}(I, W) \, \text{Pf}(J/W),$$

(B.1.12)

where σ is the permutation such that $\sigma(W, J/W) = J$.

B.2. Resolutions of height 3 Gorenstein ideals and their squares

We henceforth suppose for simplicity that $R = k[x_1, \ldots, x_r]$ is the polynomial ring over a field k, and denote by m_R the maximal ideal $m_R = (x_1, \ldots, x_r)$. Let us recall some standard notions from commutative algebra. If I is an ideal of R we denote by $V(I)$ the set of primes in the support of R/I: the primes containing I, or occurring in a primary decomposition of I. If M is an R-module the $grade\,(I, M)$ is the length of any maximal M-regular sequence in I; the grade of I is $grade\,(I, R)$, and satisfies (see [**Ei2**, Lemma 18.1], [**BruH**, Prop. 1.2.10])

$$grade\,(I) = \min_{\mathcal{P} \in V(I)} grade\,(I_{\mathcal{P}}, R_{\mathcal{P}}). \qquad (B.2.1)$$

The height (also called codimension) of I is, in the special case we consider, the codimension in $\mathbb{A}^r$ of the algebraic variety defined by I. In general, for a Noetherian ring R, the height of a prime ideal $\mathcal{P}$ is the maximum length k of a chain $\mathcal{P} = \mathcal{P}_0 \supset \mathcal{P}_1 \supset \ldots \supset \mathcal{P}_k$ of prime ideals of R containing $\mathcal{P}$, or, equivalently, the dimension of $R_{\mathcal{P}}$. The height of I satisfies (see [**BruH**, Prop. 1.2.14])

$$height\,(I) = \min_{\mathcal{P} \in V(I)} height\,(\mathcal{P});$$

$$height\,(I) \geq grade\,(I, R). \qquad (B.2.2)$$

Since $R = k[x_1, \ldots, x_r]$ is Cohen-Macaulay the grade and height of each ideal are equal (see [**BruH**, Cor. 2.1.4], [**Ei2**, Thm. 18.7]).

We refer to [**Ei2**, 21.11] for the definition of a (possibly non-Artinian) Gorenstein algebra R/I. The ideal I is by definition Gorenstein if R/I is Gorenstein. The socle degree of a Gorenstein algebra of dimension greater than zero is that of its Artinian reduction (see page 67).

We can now state the Buchsbaum-Eisenbud structure theorem, in the special case for graded Gorenstein ideals. The main applications of this theorem in the book are for the case $R = k[x_1, x_2, x_3]$, when

$f \in \mathcal{D}_j$ is a DP-form, and $I = \mathrm{Ann}(f)$. Then R/I is Artinian, so the height (= grade) of I is 3, and R/I is Gorenstein by Lemma 2.14. Note that if the algebra is to be nonzero, the degrees d_i of the generators are at least one, and the relation degrees e_i are at least two; the degree $j = d - 3$ below is the socle degree of R/I.

THEOREM B.2. (D. Buchsbaum and D. Eisenbud [**BE2, BE3**]) STRUCTURE OF HEIGHT THREE GORENSTEIN IDEALS, GRADED CASE. *Let $g \geq 3$ be an odd integer, and $d_1 \leq \ldots \leq d_g$ be a sequence of positive integers; set $d = \frac{2}{g-1}(d_1 + \cdots + d_g)$ and suppose this is an integer, let $e_i = d - d_i$, and $j = d - 3$, and we suppose $1 \leq d_1, d_g \leq j + 1$ (so $e_i \geq 2$).*

Let Ψ be an alternating $g \times g$ matrix with entries from the ring R, such that the entry ψ_{ij} is homogeneous of degree $e_i - d_j$ if $e_i > d_j$ and zero otherwise (so the entries belong to the maximal ideal m_R). Let Ψ_i be the $(g-1) \times (g-1)$ alternating matrix obtained by deleting the i-th row and column of Ψ. Then $\mathrm{Pf}(\Psi_i)$ is homogeneous of degree d_i. Let I be the ideal $\mathrm{Pf}(\Psi)$ generated by $\mathrm{Pf}(\Psi_i)$, $i = 1, \ldots, g$. Then I has grade (height) ≤ 3 in R. If I has grade 3, then I is a graded Gorenstein ideal of height three, and the socle degree of R/I is $j = d - 3$.

Let λ be the column vector with entries $\lambda_i = (-1)^i \mathrm{Pf}(\Psi_i)$.

  i. *Suppose I has the maximal possible grade 3. Then I has minimal resolution*

$$0 \to R(-d) \xrightarrow{\lambda} \sum_{1 \leq i \leq g} R(-e_i) \xrightarrow{\Psi} \sum_{1 \leq i \leq g} R(-d_i) \xrightarrow{\lambda^T} I \to 0. \quad \text{(B.2.3)}$$

$$0 \to R(-d) \to G^* \to G \to I \to 0. \quad \text{(B.2.4)}$$

  ii. *Conversely, if $I \neq R$ is a height three graded Gorenstein ideal of R, there is an alternating matrix Ψ as above, such that $I = \mathrm{Pf}(\Psi)$.*

NOTE ON PROOF. That the minimal number of generators of a height three Gorenstein ideal is odd, was shown by J. Watanabe [**Wa1**]. The structure theorem above was proven in [**BE2**]; the original version of their paper took a more naive approach, and appeared somewhat later in [**BE3**]. See also [**BruH**, §3.4] for an excellent brief account. A typo in the description of conditions on the sequences (D, E) that are possible for generator-relation degrees in [**BE2**] is corrected in [**Di, HeTV, Hari3**] (see also Theorem 5.25(v).)

We only give the proof of two simple facts. First that $\mathrm{Pf}(\Psi_i)$ is homogeneous of degree d_i, and second that (B.2.3) is a complex.

Acting by the group $G_m = k^*$ on $R = k[x_1, \ldots, x_r]$ multiplying each x_i by tx_i we want to prove that $t \circ \text{Pf}(\Psi_i) = t^{d_i} \text{Pf}(\Psi_i)$. Each entry ψ_{uv} of Ψ is multiplied by $t^{e_u - d_v}$ (including the cases $e_u \leq d_v$ when $\psi_{ij} = 0$ by assumption). So $\text{Det}(\Psi_i)$ is multiplied by

$$t^{\sum_{u \neq i} e_u - \sum_{v \neq i} d_v}.$$

Now, by assumption

$$(g-1)d = \sum_{u \neq i}(e_u + d_u) = 2(d_1 + \cdots + d_g).$$

This yields $\sum_{u \neq i} e_u - \sum_{v \neq i} d_v = 2d_i$. Thus $\text{Det}(\Psi_i)$ is multiplied by t^{2d_i} and $\text{Pf}(\Psi_i)$ by $\pm t^{d_i}$. The sign should be $+$ since $\text{Pf}(\Psi_i)$ is a polynomial in the coefficients of Ψ_i. This proves $\text{Pf}(\Psi_i)$ is homogeneous of degree d_i.

Now, let us prove that (B.2.3) is a complex. For every $1 \leq i \leq g$ consider the following alternating $(g+1) \times (g+1)$ matrix.

$$X = \begin{pmatrix} 0 & \psi_{i1} \cdots \psi_{ig} \\ \hline -\psi_{i1} & \\ \vdots & \Psi \\ -\psi_{ig} & \end{pmatrix}$$

Subtracting the $(i+1)$-th row of X from the first we see that $\text{Det}(X)$ is a multiple of $\text{Det}(\Psi)$, thus equals 0. Hence $\text{Pf}(X) = 0$. Expanding along the first row according to (B.1.9) we obtain

$$\sum_{j=1}^{g}(-1)^j \psi_{ij} \text{Pf}(\Psi_j) = 0.$$

So if we let λ be the column matrix with entries $\lambda_j = (-1)^j \text{Pf}(\Psi_j)$, we obtain $\Psi\lambda = 0$. Transposing we get $\lambda^T\Psi^T = -\lambda^T\Psi = 0$. This proves that (B.2.3) is a complex. $\square$

We next give a result that is an application of the Kustin-Ulrich theorem concerning powers of a Gorenstein height three ideal [**KusU**, Theorem 6.17]. The source is a letter from A. Kustin[1], to whom we are grateful. Using the result, we give an example showing that the tangent space of a component $\text{Gor}(T)$ of the catalecticant variety $\mathbf{V}_s(u, u; 3)$ may be larger for "smaller" T than for the maximum $T = H(s, j; 3), j = 2u$, such that $\text{Gor}(T)$ lies in $\mathbf{V}_s(u, u; 3)$.

---

[1]Letter of A. Kustin, October 1993.

Theorem B.3 (A. Kustin and B. Ulrich). Minimal resolution of I^2, for I a height 3 Artinian Gorenstein ideal.

Under the same hypotheses as Theorem B.2(i), I^2 has minimal resolution

$$0 \to \sum_{1 \le i < j \le g} R(-(e_i + e_j)) \longrightarrow \sum_{\substack{1 \le i,j \le g \\ (i,j) \ne (1,1)}} R(-(d_i + e_j)) \overset{\eta}{\longrightarrow}$$

$$\overset{\eta}{\longrightarrow} \sum_{1 \le i \le j \le g} R(-(d_i + d_j)) \to I^2 \to 0,$$

$$0 \to \Lambda^2 G^* \to G \otimes G^*/\eta \longrightarrow S^2 G \to I^2 \to 0.$$

Here $\eta = b_1 \otimes b_1^* + \cdots + b_g \otimes b_g^*$, where $b_1, \dots b_g$ is a basis of G, and $b_1^*, \dots, b_g^*$ is the dual basis of G^*.

Proof. A. Kustin refers to parts c. and d. of Theorem 6.17 in [KusU], with $q = 2$, $r = g - 2$. He observes that Ψ satisfies SPC_{g-2} because, from page 52 of [KusU], the grade of the Pfaffian ideal $PF_{g-1}(X) = 3 \ge g - (g-1) + 2$ (here $X = \Psi$); furthermore, $g - 1$ is the only even integer t in the range $r + 1 \le t \le g - 1$. The hypothesis SPC_{g-2} implies the others needed. □

Example B.4. Tangent spaces for points of a square catalecticant. The determinantal locus $V_{120}(15, 15; 3)$ contains $\mathbf{Gor}(T_1)$, $T_1 = H(120, 30, 3)$, and $\mathbf{Gor}(T_2)$, $T_2 = (1, 3, \dots, 105, 115, 120, 115, 105, \dots, 3, 1)$. Let f_1 be a general point of $\mathbf{Gor}(T_1)$ and f_2 a general point of $\mathbf{Gor}(T_2)$. We claim that $\dim_k T_{f_1} = 360$, and $\dim_k T_{f_2} = 365$, while $T_2 \le T_1$. In the light of J. O. Kleppe's smoothness result (see [Kl2] or Theorem 4.21), these calculations of tangent spaces give the dimensions of $\mathbf{Gor}(T_1)$ and $\mathbf{Gor}(T_2)$. Further such examples of large components of $V_s(t, t; 3)$ have been found by Y. Cho and B. Jung — see [ChoJ2] and also Example 4.28 in Chapter 4.

Proof of claim. The third difference sequence of T_2 contains the sequence $(-5, -1, -5, 5, 1, 5)$ in degrees $(t - 1, \dots, t + 4)$, $t = 15$. We truncate the Kustin-Ulrich resolution, using degrees no greater than $j = 2t = 30$. It becomes,

$$0 \to 0 \to 0 \to \mathrm{Sym}^2(G)_{\le j} \to (I^2)_{\le j} \to 0,$$

which yields

$$0 \to R_2^{15} \oplus R_1^5 \oplus R_0^1 \oplus R_0^{25} \to (I^2)_j \to 0.$$

Thus, the dimension of $(I^2)_j = 6 \cdot 15 + 3 \cdot 5 + 26 = 131$, out of $N = \dim_k R_{30} = \binom{32}{2} = 496$; thus $\dim_k T_{f_2} = 365$, as claimed. $\square$

B.3. Resolutions of annihilating ideals of power sums

In this section we state some important results of M. Boij from [Bo1] connecting the minimal resolutions for ideals defining point sets X in $\mathbb{P}^n$ and certain Gorenstein algebras A, for which X is an annihilating scheme. In particular he studies the case X smooth, but his main results apply to X any tight annihilating scheme for f, when $\deg(f)$ is sufficiently large (Corollaries B.9 and B.11). Some related results appear in the articles of M. Kreuzer, T. Harima, and others [Kr, Hari3, GPS]. We use his result concerning smooth X in the second half of Section 5.3, for Gorenstein sequences T not containing (s, s, s) (see Lemma 5.36).

We let $r \geq 2$ and let $X = \{p_1, \ldots, p_s\}$ be a set of s distinct points of $\mathbb{P}^{r-1}$; we denote by $A(X)$ or the quotient ring $R/\mathcal{I}(X)$ where as usual $R = k[x_1, \ldots, x_r]$.

LEMMA B.5. [Bo1, Proposition 2.3] *Let V be any codimension 1 subspace of $A(X)_c$, where $c \geq \tau(X)$. Then there are constants $\lambda_1, \lambda_2, \ldots, \lambda_s \in k$ not all zero, such that*

$$V = \{g \in A(X)_c \mid \sum_{i=1}^{s} \lambda_i g(P_i) = 0\}. \tag{B.3.1}$$

If B is a graded algebra, and $V \subset B_c$ we denote by $\overline{V}$ the largest ideal of B such that $\overline{V} \cap (B_c) = (V)$, the ideal generated by V (this $\overline{V}$ is a relative version of the ancestor ideal of Equation (2.3.3)). We have $\overline{V} = \oplus_{i \geq 0}(V : B)_i$ where

$$(V : B)_i = \{a \in B_d \mid ab \in V, \text{ for all } b \in B_{c-d}\}.$$

The next result is related to our Lemma 6.1(a), in the case (ii); but is more precise, in specifying the condition needed on λ (thus, on $f = V^{\perp}$), for $A = A_f$ to have the symmetrized Hilbert function of $H(R/\mathcal{I}_Z)$.

LEMMA B.6. [Bo1, Proposition 2.4] *Assume $c \geq 2\tau(X) - 1$ and let V be the hyperplane in $A(X)_c$ defined by $\lambda_1, \lambda_2, \ldots, \lambda_s \in k$ where for each i, $\lambda_i \neq 0$. Let $A = A(X)/\overline{V}$ be the Gorenstein Artin algebra quotient determined by V. Then the Hilbert function of A is*

$Sym(H(A(X)), j)$: *thus*

$$H(A)_i = \begin{cases} H(A(X))_i, & \text{if } 0 \le i \le c/2 \\ H(A(X))_{c-i} & \text{if } c/2 \le i \le c. \end{cases} \qquad (B.3.2)$$

THEOREM B.7. [**Bo1**, Theorem 3.5] *Assume that* $c \ge 2\tau(X) - 1$ *and* V *be a hyperplane of* $A(X)_c$ *as in Lemma B.6. Then* $J = \overline{V}$ *is isomorphic to the canonical module* ω_X.

Suppose that B is a graded Cohen-Macaulay algebra of Krull dimension d. We denote by $\tau(B)$ the integer satisfying, $\tau(B) + r - d$ is the highest degree of $\mathrm{Tor}^R_{r-d}(B, k)$. When B is a Gorenstein Artin algebra of socle degree c, then $\tau(B) = c$; when $B = A(X)$ for X a punctual scheme, then $\tau(B) = \tau(X)$. Given the ideal J of B we denote by $A = B/J$ the quotient. We denote by ω_B the canonical module of B (see [**BruH**, §3.6]). M. Boij shows

THEOREM B.8. [**Bo1**]*[Theorem 3.3]. Let* I *be a homogeneous ideal in* R, *let* $B = R/I$ *be a Cohen-Macaulay algebra of dimension* d, *and let* $J \subset B$ *be a homogeneous ideal of initial degree at least* $\tau(B) + 2$ *such that* $A = B/J$ *is Gorenstein of dimension* $d - 1$. *Then there is an isomorphism* $J \to \mathrm{Ext}^{r-d}_R(B, R) = \omega_B$, *which is homogeneous of degree* $-\tau(A) - r + d - 1$.

Note that the condition on degree in the following Corollary is equivalent to $f \in Gor(T)$, where $T \supset (s, s)$, and $s = \deg(X)$. We do not suppose that X is smooth.

COROLLARY B.9. *Suppose that* X *is a zero-dimensional subscheme of* $\mathbb{P}^n$, *that is a tight annihilating scheme of* $f \in \mathcal{D}_j$, *in the sense of Definition 5.1. Suppose further that* $j \ge 2\tau(X) + 1$. *Then* $J = \mathrm{Ann}(f)/\mathcal{I}_X$ *is isomorphic to the canonical module of* $R/\mathcal{I}_X$.

PROOF. Here $d = 1$. The hypothesis that X is a tight annihilating scheme of f implies that $H(R/\mathrm{Ann}(f)) = Sym(H_X, j)$; that $j \ge 2\tau(X) + 1$ guarantees that the initial degree of J is at least $\tau(X) + 2$, satisfying the hypothesis of Theorem B.8.   □

We now state M. Boij's result whose special case $B = A(X), d = 1$ connects the minimal resolution of the Gorenstein ideal $\overline{V}$ with that of $\mathcal{I}_X$. We suppose $B = R/I$ is a dimension-d Cohen-Macaulay ring, and denote by $[\mathrm{Tor}^R_h(B, k)]_i$ the vector space dimension of the degree-i summand of the h-th module in the minimal resolution of B as R-module. So for $h = 1$ this is the number of generators of degree i of I, for $h = 2$ this is the number of relations of degree i among the generators of I etc. Using the exact sequence of Tor associated to $0 \to J \to B \to A \to 0$, Boij shows

PROPOSITION B.10. [**Bo1**, Proposition 3.7] *Let* $B = R/I$ *be a Cohen-Macaulay algebra of dimension* d *and let* $J \subset B$ *be an ideal of initial degree at least* $\tau(B) + 1$ *isomorphic to the canonical module* ω_B, *and defining the quotient* $A = B/J$. *Then*

$$[\text{Tor}_h^R(A, k)]_i = [\text{Tor}_h^R(B, k)]_i \oplus [\text{Tor}_{r-d+1-h}^R(B, k)]_{\tau(A)+r-d+1-i},$$
(B.3.3)

for $h = 0, 1, 2, \ldots r - d$ *and* $i \in \mathbf{Z}$.

COROLLARY B.11. *Suppose that* X *is a zero-dimensional subscheme of* $\mathbb{P}^{r-1}$, *that is a tight annihilating scheme for* $f \in \mathcal{D}_j$, *and that* $j \geq 2\tau(X) + 1$. *Then the minimal resolution for* $A = R/\text{Ann}(f)$ *satisfies* (B.3.3), *where* $B = R/\mathcal{I}_X$ *and* $d = 1$.

We now suppose that $B = A(X)$ where X a set of s points $p_1, \ldots, p_s$ in $\mathbb{P}^2$, (so $r = 3$) and that $L_1, \ldots, L_s$ are the corresponding linear forms in $\mathcal{D}_1$, determined up to nonzero-constant multiple. We consider a divided power sum $f = \sum_{u=1}^s \lambda_u L_u^{[j]}$ where each $\lambda_u \neq 0$, we let $I = \text{Ann}(f)$. We consider a minimal resolution of I, and denote by v_i the number of generators of I in degree i, and by w_i the number of relations among the generators in degree i; we let $v_i(\mathcal{I}_X)$ and $w_i(\mathcal{I}_X)$ denote the corresponding numbers for $\mathcal{I}(X)$.

COROLLARY B.12. *Let* $B = A(X) = R/\mathcal{I}_X$, *where* $X = \{p_1, \ldots, p_s\}$ *is a subset of* $\mathbb{P}^2$, *suppose* $j \geq 2\tau(X)$, *and let* $f \in (\mathcal{I}_X)_j^\perp$ *satisfy* $f = \sum_{u=1}^s \lambda_u L_u^{[j]}$, *where each* $\lambda_u \neq 0$. *Let* $I = \text{Ann}(f)$. *Then we have*

$$\begin{aligned} v_i &= v_i(\mathcal{I}_X) + w_{j+3-i}(\mathcal{I}_X) \\ w_i &= w_i(\mathcal{I}_X) + v_{j+3-i}(\mathcal{I}_X). \end{aligned}$$
(B.3.4)

PROOF. By the Apolarity Lemma 1.15 the space V considered in Lemma B.5, Lemma B.6 and Theorem B.7 is, setting $c = j$, the hyperplane $V = \text{Ann}(f)_j/(\mathcal{I}_X)_j \subset A(X)_j$ and $\overline{V} = \text{Ann}(f)/\mathcal{I}_X \subset A(X)$ (cf. Lemma 2.15). Thus the condition on f being a length-s sum of divided powers is equivalent to the condition on V of Lemma B.6 and Theorem B.7 where $c = j$. The condition on j implies by Theorem B.7 that $J = I/\mathcal{I}(X)$ is isomorphic to the canonical module, and also by (B.3.2) that J has initial degree at least $\tau(X) + 1$. Thus Proposition B.10 applies with $r = 3, d = 1$. $\square$

B.4. Maximum Betti numbers, given T

In this section we first state and prove the Conca-Valla Generator Theorem, that shows how to find the degrees of a maximal generating set of a Gorenstein ideal I having Hilbert function $H(R/I) = T$, directly from T. A different way of finding the — same — maximal generating set from T had been given by Diesel [**Di**, §3], and by T. Harima [**Hari3**]. A. Conca and G. Valla show a rather more general result in [**CoV1**]. We then state and prove a result of S. J. Diesel that counts the height three Gorenstein sequences T having given order and socle degree, by associating them with certain partitions $P(T)$: these partitions are determined directly from the maximum set of generator degrees for T, or, more simply, from the first difference sequence of T (Lemma B.15).

The proof we give next of the Conca-Valla Lemma follows the approach of several authors including M. Boij, and A. Geramita, M. Pucci, and Y. S. Shin to related questions [**Bo1, GPS**]. We include it largely to give an introduction to their methods, which is to compare generators for ideals defining smooth point subschemes of $\mathbb{P}^2$ and ideals $I \in \mathbb{P}Gor(T)$.[2] We assume the first part of Equation (5.3.2), giving the maximum number of generators $v = 2\nu(T) + 1$ for $D_{\max}(T)$, a result of S. J. Diesel and also of T. Harima.

We do not assume the main results of Section 5.3, such as Theorem 5.31: these are shown independently of the Conca-Valla result, and so could give an alternate — but harder — route to the proof. The results of Section 5.3, show more, namely, that there is a mapping from the parameter space of ideals $I = \mathrm{Ann}(f), f \in Gor(T)$ (Theorem 5.31), or $f \in Gor_{\mathrm{sch}}(T)$ (Theorem 5.39) to $\mathbf{Hilb}^s(\mathbb{P}^2)$, rather than a map for certain ideals I lying over smooth points in $\mathbb{P}^2$, as here.

We now state the Conca-Valla Theorem. Recall from Theorem 5.25 that, given a height three Gorenstein sequence T, we denote by $D_{\max}(T), E_{\max}(T)$, respectively, the (nondecreasing) maximum ordered set of generator degrees consistent with T (or, nonincreasing maximum ordered set of relation degrees); and we let $C_{\max}(T) = E_{\max}(T) - D_{\max}(T)$. Recall that $\Delta(T)_i = T_i - T_{i-1}$, and that j is the socle degree of T; also $\nu(T) = \min\{i|T_i \neq r_i\}$ is the order of T, which we will denote by m in the proof, and in the following discussion.

---

[2]This proof is by the first author, and is his sole responsibility

THEOREM B.13. *(A. Conca and G. Valla* [CoV1]*) The number* $D_{\max}(T)_i$ *of generators of degree i in $D_{\max}(T)$ satisfies,*

$$
D_{\max}(T)_i = \begin{cases} -\Delta^2(T)_i\,, & \text{if } \nu(T) < i \leq j+2-\nu(T); \\ 1-\Delta^2(T)_i\,, & \text{if } i = \nu(T); \\ 0\,, & \text{otherwise} - \text{if } \Delta^2(T)_i = 1. \end{cases} \tag{B.4.1}
$$

PROOF. Recall first from Definition 4.30 the sequence $S = S(T)$, namely $S = \Delta(T)_{\leq j/2}$, augmented by zeroes in higher degrees,

$$
S(T) = (1, 2, \ldots, \nu(T), h_\nu, \ldots, h_t, 0, \ldots), \quad \text{with}
$$

$$
h_{\nu(T)-1} = \nu \geq h_\nu \geq \ldots h_{[j/2]}. \tag{B.4.2}
$$

By Theorem 5.25(i), $S = (S(0), S(1), \ldots, 0)$ is the Hilbert function of an Artin quotient of $k[x, y]$. We order the degree-i monomials in x, y lexicographically, $x^i > x^{i-1}y > \ldots > y^i$, and we now recall the definition in this case of the *lex-initial* monomial ideal M_S of Hilbert function $H(R/M_S) = S$ (see (4.4.15), and also Appendix C.1, Definition C.1). Let $m = \nu(T)$, $t = \min\{i \mid T_i = s\}$, and for $m \leq i \leq t$ let $\mu_i = x^{S(i)}y^{i-S(i)}$. If $i \geq m$, then $(M_S)_i$ is the span of the first $i + 1 - S(i)$ monomials in x, y in the lexicographic order, so,

$$
(M_S)_i = \langle x^i, x^{i-1}y, \ldots, x^{S(i)}y^{i-S(i)} \rangle.
$$

If μ is a monomial other than y^i, in the following formula we let $\mu' = y\mu : x$ denote the successor monomial to μ.

Evidently, a minimal generating set G for M_S satisfies

$$
G_i = \begin{cases} \{x^m, \ldots, \mu_m\} & \text{of length } 1 - \Delta(S)_m, \text{ if } i = m, \\ (y\mu_{i-1})', \ldots, \mu_i\} & \text{of length } -\Delta(S)_i \text{ if } m < i \leq t+1 \\ & \text{and } \Delta(S)_i < 0, \\ \emptyset & \text{in degree } i, \text{ otherwise.} \end{cases}
$$

$$\tag{B.4.3}$$

The total number of generators of M_S is

$$
1 + \sum_{i=m}^{t+1} \Delta(S)_i^- = m + 1.
$$

It is easy to check that for $m \leq i \leq t+1$ there are

$$-(\Delta S)_i \text{ degree } i + 1 \text{ relations among the generators of } S, \tag{B.4.4}$$

and none other, for a total of m relations: this is one less than the number of generators of M_S, as required by the Hilbert-Burch theorem [Ei2, Theorem 20.15].

By a result of A. Geramita, D. Gregory and L. Roberts [GGR], a refinement of a special case of [**Har1**], any Artin algebra C defined by a monomial ideal in a polynomial ring can be smoothed to a zero-dimensional subscheme X, whose coordinate ring $\mathcal{O}_Z = R/\mathcal{I}_X$ has a linear non-zero divisor, such that the quotient is C. It follows that A and $\mathcal{O}_Z$ have identical Betti information — degrees and numbers of generators, relation — and that $H(R/\mathcal{I}_Z) = H(T)$, the sum function of S. We apply this construction to M_S, finding a smooth scheme $X = X_S$.

Choosing a general element $f \in (\mathcal{I}_X)_j{}^\perp$, we find that X is a tight annihilating scheme of f (Lemma 6.1). By Proposition B.10 the minimal resolution for $I = \mathrm{Ann}(f)$ consists of that for $\mathcal{I}_X$, patched to its dual. In particular, we have

CLAIM B.14. A minimal generating set for I consists of two parts:

(i). First, the $m + 1$ generators of $\mathcal{I}_X \subset I$, in degrees $\leq t + 1$;

(ii). Second, m generators, beginning in degree $i = j + 1 - t$,

corresponding by $v_i = w_{j+3-i}(\mathcal{I}_X)$ of (B.3.4) to the relations of $\mathcal{I}_X$.

In either case, the number of such generators of degree i is given by the equation in (B.4.1).

PROOF OF CLAIM. Both (i) and (ii) follow directly from Corollary B.12 and Equation B.4.3. Since by (B.4.4) there are $-\Delta S_i$ relations of degree $i + 1$ for M_S, and hence for $\mathcal{I}_X$, for $m \leq i \leq t + 1$, it follows by (B.3.4) that there are $-\Delta S_i$ corresponding generators for I in degree $j + 3 - (i + 1) = j + 2 - i$. But $\Delta^2(T)_{j+2-i} = \Delta^2(T)_i$. If $T \supset (s, s)$ then we have

$$(\Delta^2 T)_i = (\Delta S)_i \text{ for } i \leq t + 1. \tag{B.4.5}$$

This completes the proof of the Claim if $T \supset (s, s)$. When T does not contain (s, s), then there is there is equality in (B.4.5) for $i \leq t$, but $(\Delta^2 T)_{t+1} = 2(\Delta S)_{t+1}$: in this case $j = 2t + 2$, and the two parts of the generating set overlap in degree $t + 1$. This completes the proof of the Claim. $\square$

Since there are $2m + 1 = 2\nu(T) + 1$ generators for I, and this is the maximum number, given T from [**Di**, Theorem 3.3] (see (5.3.2)), the proof of the Claim completes the proof of Theorem B.13. $\square$

A second way to show Theorem B.13 is to follow Diesel's determination of $C_{\max}(T)$ and $D_{\max}(T)$ via saturation in [**Di**]), and then compare the partition $P(T)$ derived from $D_{\max}(T)$ with the monomial ideal M_S (see

p. 123 and also the last paragraph of the proof of Lemma B.15 below). We leave this to the interested reader.

We now state a result of of S. J. Diesel, which is used in Theorem 7.15 ([**Di**, Proposition 3.9]). Let

$$D_{\max}(T) = (d_1, \ldots, d_{2m+1}), \quad d_1 \leq d_2 \leq \cdots \leq d_{2m+1},$$

denote the maximum set of generator degrees for T, (see Theorems 7.15 and 5.25). We denote by $P(T)$, and $Q(T)$ the partitions (e.g., nonincreasing sequence of integers, possibly zero),

$$P(T) = (d_2 - m, \ldots, d_{m+1} - m);$$
$$Q(T) = (d_2 - m, \ldots, d_{2m+1} - m). \tag{B.4.6}$$

The partition $P(T)$ is simply related to the "alignment character" of T, by equation (4.4.16). Given a pair of integers, $(m, j), j \geq 2m$, we let $M = M(m, j)$ denote the block-partition (whose Ferrer's graph is a rectangle) having $2m$ equal parts, each of length $m' = j - 2m + 2$, and $B = B(m, j)$ denote the block partition with m parts, each of length $t + 1 - m$, where $t = [j/2]$. The first parts of the following Lemma are from [**Di**]; the last statement is from [**I9**, Section III]. Recall that the sequence S is defined as $\Delta(T)_{\leq t}$ in (B.4.2). We regard $S_m \geq \cdots \geq S_t$ as a partition (possibly with zero parts) having $t + 1 - m$ parts, each of length no greater than m; the dual partition is obtained as usual by switching rows and columns of its Ferrer's graph.

LEMMA B.15. *The partition $Q(T) = P(T) \cup P^c(T)$, where $P^c(T)$ denotes the complement of P in an m by m' block. Also,*

$$C_{\max}(T) = (m' + 1, m' + 1 - 2Q(T)_1, \ldots, m' + 1 - 2Q(T)_{2m}). \tag{B.4.7}$$

The partition $P(T)$ is the dual partition to the partition $S_m, S_{m+1}, \ldots, S_t$.

PROOF IDEA. Diesel's proof of the first statement relies on a study of $C_{\max}(T)$, as the unique saturation of $C_{min}(T)$ obtained by adding successive pairs $(a, -a)$ until the condition $C_i + C_{v+2-i} = 2$ of (5.3.2) of Theorem 5.25(v.f) is satisfied for $C_{\max}(T)$. She then shows that the maximum number of generators must be $v = 2m + 1$. Alternatively, once we have identified the generating set $D_{\max}$ as in Claim B.14 above (this requires knowing $v(T) = 2m + 1$, for which we relied on [**Di**, Theorem 3.3]), we obtain that $Q(T) = P(T) \cup P^c(T)$ from (B.4.4). Then equation (B.4.7) arises from the basic equation $d_i + e_i = d = j + 3$ of Theorem B.2: the i-th row of M has length m'; it is partitioned into $d_{i+1} - m$, the i-th part P_i of P, and a complement P_i^c of length $m' - (d_{i+1} - m) = j - 2m + 2 - (d_{i+1} - m) = e_{i+1} - m - 1$. So for $i \geq 1$, we have

$$(C_{\max})_{i+1} = e_{i+1} - d_{i+1} = [m' - (d_{i+1} - m + (m+1))] - d_{i+1}$$
$$= m' + 1 - 2Q(T)_{i+1}.$$

That $P(T)$ is the dual partition of $(S_m, S_{m+1}, \ldots, S_t)$ follows from Claim B.14, and the description of the generators of M_S in (B.4.3): from these and the definition of $P(T)$ we have that for $m \leq i \leq t$, there are $-\Delta(S)_i = S_{i-1} - S_i$ parts of $P(T)$ having length $i - m$. This shows that $P(T)$ is indeed the dual partition as claimed. $\qquad\square$

EXAMPLE B.16. If $T = (1, 3, 6, 8, 9, 10, 9, 8, 6, 3, 1)$ then $S(T) = (1, 2, 3, 2, 1, 1)$, $m = 3, t = 5, j = 10, m' = 6$, $P(T) = (2, 1, 1)^{\vee} = (3, 1, 0)$, and $P^c(T) = (6 - 3, 6 - 1, 6 - 0) = (3, 5, 6)$. So $Q(T) = P(T) \cup P^c(T) = (0, 1, 3, 3, 5, 6)$, $D_{\max}(T) = (3, 3, 4, 6, 6, 8, 9)$ (including an initial degree $3 = m$), and $M_S = (x^3, x^2 y, x y^3, y^6)$. From (B.4.7) we have $C_{\max}(T) = (7, 7, 5, 1, 1, -3, -5)$. From Theorem 5.25(ii), equation (5.3.1), $D_{\min}(T) = (3, 3, 6, 6, 8)$ is read from $\Delta^3(T)$; and $C_{\min}(T) = (7, 7, 1, 1, -3)$ is obtained by removing the $(5, -5)$ pair from $C_{\max}(T)$.

THEOREM B.17. [Di, §3.4] GORENSTEIN SEQUENCES WHEN $r = 3$, AND PARTITIONS. *The Gorenstein sequences T of order $\nu(T) = m$ and socle degree $j(T) = j$ correspond one to one via $T \to P(T)$ to the set $\mathcal{P}(B)$ of partitions of integers having no greater than m nonzero parts, each part no greater than $t + 1 - m$. This is the set of partitions whose Ferrer's graph fits in a $m \times (t + 1 - m)$ block B. There is likewise a one to one correspondence $T \to Q(T)$ between such Gorenstein sequences T and the set $SCP(M)$ of pairs $(Q, M - Q), Q = M - Q$ of self-complementary partitions inside the partition $M = M(m, j)$.*

The number of such sequences T is the same as the number of elements of $\mathcal{P}(B)$, namely,

$$\binom{[j/2] + 1}{m}.$$

When $r = 3$, and $j = 2t$ or $j = 2t + 1$, there are a total of $2^{t+1} - 1$ Gorenstein sequences of socle degree j.

PROOF. Evidently $D_{\max}(T)$, hence T, is determined by the triple $(P(T), m, j)$. Since $P(T)$ is the dual partition to $(S_m, \ldots S_t)$, which may be chosen arbitrarily in $B^{\vee}$, the rectangular partition with $t+1-m$ parts of length no greater than $m = \nu(T)$, by Theorem 5.25(i.), it follows that $P(T)$ runs through all partitions that fit into B. This establishes the one-to-one map from the particular set of Gorenstein sequences to $\mathcal{P}(B)$. The set $\mathcal{SCP}(T)$ consists of partitions having the form $Q = P \cup P^c, P \in \mathcal{P}(B)$, so is also in one-to-one correspondence with the Gorenstein sequences of order m and socle degree j. The count of such partitions in $\mathcal{P}(B)$ is well known; we assume that $m \geq 1$, so the total number of height-three Gorenstein sequences of socle degree j is $2^{t+1} - 1$.   □

EXAMPLE B.18. If $\nu(T) = 3$ and $j = 8$, B has 3 rows of length 2, and there are $\binom{4+1}{3} = 10$ partitions in $\mathcal{P}(B)$, from $(2,2,2), (2,2,1), \ldots$ to $(0,0,0)$. A simple way to reconstitute T from $P(T) = (2,2,2)$, is to form $P^\vee(T) = (3,3)$, then $S(T) = (1,2,\ldots,m,P^\vee(T))$, here $S(T) = (1,2,3,3,3)$ then $T_{\leq t}$ as the sum function $\int S(T)$, here $(1,3,6,9,12)$, and symmetrize about $j/2$ to obtain $T = (1,3,6,9,12,9,6,3,1)$.

For further discussion of the partition $P(T)$, and its relation with the alignment character of T used by A. Geramita and others see pages 122-124.

APPENDIX C

The Gotzmann Theorems and the Hilbert Scheme

by Anthony Iarrobino and Steven L. Kleiman

There are several excellent sources for results on O-sequences, and for the Macaulay Theorem below describing the set of possible Hilbert functions (Theorem C.2) [**Sp, St1, BruH, St3, Gr2**]. There are as well several sources for the Gotzmann Persistence Theorem (Theorem C.17) [**Got1, Gr1, Gr2, BruH**][1]. However, the relationship of the Persistence Theorem to the Hilbert scheme is less well known, though it was an integral part of Gotzmann's work. We give here a brief summary of these results. In particular, we show that the Castelnuovo–Mumford m-regularity $\sigma(P)$ of a Hilbert polynomial is the same as a combinatorial invariant $\varphi(P)$, the Gotzmann number (Proposition C.24). We then state Gotzmann's Theorem describing the Hilbert scheme (Theorem C.29), and discuss its relationship to previous work of A. Grothendieck and D. Mumford. We also derive as a consequence of the Gotzmann Theorem and an argument of Grothendieck, that the Hilbert scheme is given locally by simply described determinantal conditions (Proposition C.30), which were conjectured by D. Bayer [**Ba**, Chap.VI].

In Section C.5 we apply this Theorem and a related result of G. Gotzmann from [**Got4**] to certain $Gor(T)$, in a manner similar to the applications in an article by A. Bigatti, A. Geramita and J. Migliore [**BGM**]. This leads to a better understanding of the structure of some $Gor(T)$ when T has a subsequence of maximal growth, and to a new example of $Gor(T)$ having several irreducible components when $r = 4$ (Proposition C.33 and Example C.38).

---

[1]M. Green proved the theorem when the ground ring is a field in [**Gr1**]. See also [**BruH**, 1998 ed. §4.3] and [**Gr2**] for this case

C.1. Order sequences and Macaulay's Theorem on Hilbert functions

Let $R = k[x_1, \ldots, x_r]$ be a polynomial ring over an arbitrary field k. Given a monomial $m = x_1^{\alpha_1} \cdot \cdots \cdot x_r^{\alpha_r}$, set $exp(m) = (\alpha_1, \ldots \alpha_r)$ in $\mathbb{N}^r$. Recall that, in the *lex order* of monomials, $m_1 > m_2$ if the first nonzero entry of $(exp(m_1) - exp(m_2))$ is positive. For example, $x_1^3 > x_1^2 x_2 > x_1^2 x_3 > x_1 x_2^2$ is in lex order.

DEFINITION C.1. LEX-INITIAL IDEAL AND O-SEQUENCES. Let M be an ideal in R that is generated by monomials. Then M is called a *lex-initial ideal*, or *segment ideal*, if each homogeneous component M_i has as basis the first $\dim_k M_i$ monomials of degree i, those largest in the lex order.[2] (Equivalently, M is an initial ideal in the graded lex order of [**CLO**, p.57].)

A sequence S of integers is called an *O-sequence* if there is a monomial ideal M such that $S = H(R/M)$.[3]

For example, $(x_1^2, x_1 x_2, x_1 x_3)$ is a lex-initial ideal.

We'll see shortly that, because of Macaulay's Theorem, in the defintion of O-sequence, M may be assumed to be lex-initial.

The following theorem was first proved by F. H. S. Macaulay. It was given a more elegant proof by F. Whipple, and was generalized by G. Clements and B. Lindström [**Mac3, Wh, ClL**]. For more details on the theorem, see the reference [**BruH**], which is excellent.

THEOREM C.2. MACAULAY'S THEOREM ON HILBERT FUNCTIONS. Let $H = H(I)$ be the Hilbert function of a graded ideal I in R; so $H_i = \dim_k I_i$. Then there is a unique lex-initial (segment) ideal $M = M_H$ such that $H = H(M)$.

COROLLARY C.3. Let $H = (1, h_1, h_2, \ldots)$ be a sequence of integers. Then the following statements are equivalent:

A. H is the Hilbert function of a quotient $S = R/I$ for some graded ideal I of R.

---

[2] The presentation in [**BruH**, pp.155–6] stresses the cobasis for I, so uses the reverse-lex order; their segment ideal is the last (smallest in reverse-lex order) t monomials of degree d. By replacing the variables $x_1, \ldots, x_r$ by $x_r, \ldots, x_1$ in the segment ideal of [**BruH**], we obtain our lex-initial ideal.

[3] The name O-sequence comes from *order ideal*, which is a subset M' of monomials closed under division. This is not an ideal in the usual sense, but rather M' is a basis of the graded complement of a monomial ideal M. Thus, an O-sequence gives the Hilbert function of the "order ideal" M', or, equivalently of the quotient R/M (see [**St3**, (2.1)],[**BruH**, (4.2.1)]).

B. H *is an O-sequence.*

C. *There is a lex-initial monomial ideal M such that $H = H(R/M)$.*

One significant consequence of the Macaulay Hilbert-Function Theorem is that there is a minimum possible growth of the Hilbert function $H(I)$, where I is a graded ideal of R. A more precise statement is given in the following corollary. To state it, set $r_i = \dim_k R_i$ and fix $0 < t \leq r_i$. Denote by $V = V(t, i, r)$ the vector space spanned by the first t monomials in R_i in the lex order, and by $(V(t, i, r))_{i+1}$ the degree-$(i+1)$ piece of the ideal generated by $V(t, i, r)$. Evidently,

$$(V(t, i, r))_{i+1} = R_1 \cdot V(t, i, r) = \langle \{x_1 v, \dots, x_r v\} \mid v \in V \rangle. \quad (C.1.1)$$

Also, if μ_t is the last monomial of $V(t, i, r)$, then

$$x_r \mu_t \text{ is the last monomial of } R_1 \cdot V(t, i, r). \quad (C.1.2)$$

COROLLARY C.4. MINIMAL GROWTH OF AN IDEAL. *Let I be a graded ideal of R, and set $t = \dim_k I_i = H(I)_i$. Then*

$$H(I)_{i+1} \geq \dim_k (R_1 \cdot V(t, i, r)).$$

EXAMPLE C.5. Suppose $r = 3$ and $H(I)_2 = 2$. Then $V(2, 2, 3)$ is equal to $< x_1^2, x_1 x_2 >$, and

$$R_1 \cdot V(2, 2, 3) = \langle x_1^3, x_1^2 x_2, x_1^2 x_3, x_1 x_2^2, x_1 x_2 x_3 \rangle. \quad (C.1.3)$$

So, by the corollary, $\dim_k I_3 \geq \dim_k R_1 \cdot V(2, 2, 3) = 5$.

If an ideal I' satisfies $H(I')_2 = 3$, then $H(I')_3 \geq 6$, since $V(2, 3, 3)$ is equal to $< x_1^3, x_1^2 x_2, x_1^2 x_3 >$, and $\dim R_1 V(2, 3, 3) = 6$.

The vector space $L = < x_1^2, x_1 x_2, x_2^2 >$ is spanned by the first three monomials of degree 2 in the reverse-lex order. Here $\dim R_1 \cdot L = 7$, not 6: the ideal $(L) = R \cdot L$ does not have minimal growth from degree 2 to degree 3. Thus, we cannot use the initial monomials in reverse-lex order to redefine the notion of segment ideal in Macaulay's Theorem.

We next consider the special case $r = 2$, which is used in the characterization of height-3 Gorenstein sequences; these are the Hilbert functions of graded Artin Gorenstein algebras $A = R/I$ where I is of height, or codimension, 3 (see [BE2, St1] and Theorem 5.25).

Let $H = \{h_i | i \geq 0\}$ be a sequence, and set

$$\nu(H) = \inf\{i \mid h_i < r_i\}.$$

Note that $\nu(H)$ is the order of any ideal I such that $H(R/I) = H$.

COROLLARY C.6. O-SEQUENCES WHEN $r = 2$. *If $h_i = i + 1$ for $i < \nu$ and if $h_{t+1} = 0$ for some t, then H is an O-sequence for $r = 2$ if and only if*

$$\nu = h_{\nu-1} \geq h_\nu \geq \ldots h_t \geq 0 \text{ and } h_i = 0 \text{ for } i \geq t + 1. \tag{C.1.4}$$

A graded ideal $I \subset R = k[x_1, x_2]$ satisfies the condition of maximum growth $H(R/I)_i = H(R/I)_{i+1} = s$ if and only if there exists an $h \in R_s$ such that $I_i = (h) \cap R_i$, and $I_{i+1} = (h) \cap R_{i+1}$. If also I is generated in degree i or less, then $H(R/I)_{i+k} = 0$ for $k \geq 0$.

PROOF. First, suppose H is an O-sequence and $H(R/I) = H$. If $i \geq \nu$, or equivalently, if $I_i \neq 0$, then by Corollary C.4, $\dim_k I_{i+1} > \dim_k I_i$. Since $\dim_k R_i = i + 1$, it follows that

$$h_i = \dim_k R_i - \dim_k I_i \geq \dim_k R_{i+1} - \dim_k I_{i+1} = h_{i+1}.$$

Condition (C.1.4) follows immediately.

Conversely, suppose that (C.1.4) holds. For $i \geq \nu$, form the vector subspace $V_i \subset R_i$ spanned by the monomials in the interval $[x_1^i, \ldots, x_1^{h_i} x_2^{i-h_i}]$. The smallest monomial of $R_1 V_i$ is $x_1^{h_i} x_2^{i+1-h_i}$, and it is greater than $x_1^{h_{i+1}} x_2^{i+1-h_{i+1}}$ because $h_i \geq h_{i+1}$. Set $I = \oplus_{i \geq \nu} V_i$. Then I is a lex-initial ideal such that $H(R/I) = H$. Thus H is an O-sequence.

The last statements constitute a particular case of Part ii of Proposition C.32 below. In this case, there is a simpler proof, found in [I1, p.56] and [Da, p.349]. For further generalizations to arbitrary r, see [Got4] and [BGM, pp. 219–221] and also Propositions C.33 and C.34. $\qquad \square$

The next Corollary generalizes the inequality in the first part of Corollary C.6 to arbitrary r.

COROLLARY C.7. *Let H be an O-sequence, i an integer. If $i \geq H_i$, then*

(i) $H_i \geq H_{i+1}$, *and*

(ii) H *is nonincreasing in degrees at least i.* $\tag{C.1.5}$

PROOF. Assume (i). Then $i + 1 \geq i \geq H_i \geq H_{i+1}$. So applying (i) to $i + 1$ yields $H_{i+1} \geq H_{i+2}$. Repeating yields (ii). Thus it remains to prove (i).

Let $s = H_i$. Since $i \geq s$, the last s monomials of degree i, in lex order, are

$$x_{r-1}^{s-1} x_r^{i-s+1}, \ldots, x_r^i. \tag{C.1.6}$$

Let $t = r_i - s$. Then the last (so t-th) monomial of the vector space $V(t, i, r)$ is $x_{r-1}^s x_r^{i-s}$, and the last monomial of $R_u \cdot V(t, i, r)$ is $x_{r-1}^s x_r^{i+u-s}$, which immediately precedes the last s monomials of R_{i+u}. It follows that $\mathrm{cod}(R_u \cdot V(t, i, r)) = s$ in R_{i+u}. Taking $u = 1$, we have by Theorem C.2 that $H_{i+1} \leq s$. This completes the proof of (i) and of the Corollary. □

C.2. Macaulay and Gotzmann polynomials

EXAMPLE C.8. MINIMAL GROWTH PERSISTS. Let $V = V(t, i, r)$ and $I = (V)$. So I is the ideal generated by the first t monomials of degree i, say $v_1, \ldots, v_t$. Then $R_1 \cdot V(t, i, r)$ is generated by the monomials in the set on the right in (C.1.1), and they clearly form an interval in the lex order, starting with x_1^{i+1}. So, $R_1 \cdot V(t, i, r)$ is equal to $V(t', i+1, r)$ for some t'.

Continuing by induction, we conclude that, for every $u \geq 1$,

$$R_u \cdot V(t, i, r) = V(t^{(u)}, i + u, r)$$

for some $t^{(u)}$. Thus, for $I = (V)$, the growth of the Hilbert function $H(I)$ at each step is always the minimum allowed by Corollary C.4.

It is not hard to show that the values of the Hilbert function $H(I)_z$ are given for $z \geq i$ by a polynomial $Q(t, i, r)$ in z with coefficients in the rational numbers $\mathbb{Q}$, and so the values of $H(R/I)_z$ too are given by a polynomial $P(t, i, r)$; see [**Got1**, (2.4)], [**BruH**, Corollary 4.2.9], and Remark C.11 below.

DEFINITION C.9. Given integers r, d, t with $0 < t \leq r_d$, we define polynomials $P(t, d, r)$ and $Q(t, d, r))$ in $\mathbb{Q}[z]$ by the formulas,

$$P(t, d, r)(z) = \dim_k \big(R_z / (R_{z-d} \cdot V(t, d, r)) \big),$$
$$Q(t, d, r)(z) = \dim_k \big(R_{z-d} \cdot V(t, d, r) \big).$$

We say that $P(z)$ and $Q(z)$ in $\mathbb{Q}[z]$ are the *Macaulay polynomial* and the *Gotzmann polynomial* of an r-variable lex-initial ideal if $P = P(t, d, r)$ and $Q = Q(t, d, r)$ for some d, t.

To describe these polynomials, we first determine the dimension of $V(t, d, r)$ in terms of the exponents in the t-th monomial of R_d in the lex order. We denote by R^i the polynomial ring $k[x_i, \ldots, x_r]$, so $R^1 = R$. Let t satisfy $0 < t < r_d$, and write the t-th monomial $\mu_{t,d}$ of R_d as

$$\mu_{t,d} = x_1^{a_0 - 1} x_2^{a_1 - a_0} \cdots x_k^{a_{k-1} - a_{k-2}} x_{k+1}^{a_k - a_{k-1} + 1} x_r^{d - a_k} \qquad \text{(C.2.1)}$$

for a unique sequence of integers a_i with

$$0 < a_0 \leq \ldots \leq a_k \leq d \text{ and } 0 \leq k \leq r - 2. \tag{C.2.2}$$

Then $V(t, d, r)$ satisfies

$$V(t, d, r) = x_1^{a_0} \cdot R_{d-a_0} \oplus x_1^{a_0-1} x_2^{a_1-a_0+1} \cdot R_{d-a_1}^2$$
$$\oplus x_1^{a_0-1} x_2^{a_1-a_0} \cdots x_{i-1}^{a_{i-2}-a_{i-1}} \cdot x_i^{a_{i-1}-a_{i-2}+1} \cdot R_{d-a_{i-1}}^i \oplus \cdots$$
$$\oplus x_1^{a_0-1} x_2^{a_1-a_0} \cdots x_{k+1}^{a_k-a_{k-1}+1} \cdot R_{d-a_k}^{k+1} \tag{C.2.3}$$

Since $\dim_k R_u^i = \binom{r-i+u}{u}$, we have proved the existence part of the following lemma.

LEMMA C.10. [**Got2**, (2.3)] *Suppose that d, r are fixed, and that t satisfies $0 < t < r_d$. Then t has a unique "Macaulay representation,"*

$$t = \binom{d-a_0+r-1}{r-1} + \cdots + \binom{d-a_k+r-1-k}{r-1-k}, \tag{C.2.4}$$

where the integers $a_0, \ldots a_k$ satisfy (C.2.2). The a_i may be determined from the t-th monomial $\mu_{t,d}$ by writing it in the form (C.2.1). Furthermore, if $Q(t, d, r) = Q(t', d', r)$, then $k = k'$ and $a_i = a_i', 0 \leq i \leq k$.

PROOF. Given a second representation of t determined by $d' \geq d$, repeated application of Observation (C.1.2) yields $\mu_t' = x_r^{d'-d} \mu_t$, whence $a_i' = a_i$. □

REMARK C.11. MACAULAY AND GOTZMANN POLYNOMIALS. What are the Macaulay and Gotzmann polynomials? Answers are given by Equations (C.2.6), (C.2.7) and (C.2.9) below. The preceding equations (C.2.2)–(C.2.6) are well known (see [**Got1, Sp, Got4**], which are based on [**Mac3, Wh**]).

First, we determine the Gotzmann polynomials. By a similar argument to that of Lemma C.10 ([**Got1**, (2.4)], [**Got2**, 2.3]), we see that if $i \geq d$, then

$$\dim(V(t, d, r))_i = \binom{i-a_0+r-1}{r-1} + \cdots + \binom{i-a_k+r-1-k}{r-1-k}. \tag{C.2.5}$$

Hence the polynomial $Q(t, d, r) \in \mathbb{Q}[z]$ is given by the formula,

$$Q(t, d, r)(z) = \binom{z-a_0+r-1}{r-1} + \cdots + \binom{z-a_k+r-1-k}{r-1-k}, \tag{C.2.6}$$

where the a_i satisfy (C.2.2); here we expand the binomial coefficients in (C.2.6) in the usual way. It follows from Lemma C.10, that the a_i are uniquely determined, given Q and r.

Note also that the right hand side of (C.2.6) depends only on the sequence $a_0, \ldots, a_k$ and r, but not on d and t; whence, so does $Q(t, d, r)$. In particular, we may take $d = a_k$: doing so is equivalent to assuming that $\mu_{t,d}$ is not divisible by x_r. Since $\dim_k R_z = \binom{r+z-1}{r-1}$, the Macaulay polynomial $P(t,i,r)$ is given by the formula,

$$P(t,d,r) = \binom{r+z-1}{r-1} - Q(t,d,r). \tag{C.2.7}$$

In the preceding discussion, we followed Gotzmann's presentation, so (C.2.7) is not the "Macaulay representation" of $P(t,i,r)$ that is found in [**BruH, Gr1, Gr2, BGM**]. We now explain the latter. Set $c = r_d - t$, and expand c in its Macaulay representation,

$$c = \binom{k(d)}{d} + \binom{k(d-1)}{d-1} + \cdots + \binom{k(1)}{1}$$

$$\text{where } k(d) > k(d-1) > \ldots > k(1) \geq 0. \tag{C.2.8}$$

Then we have

$$P(t,d,r) = \binom{z+k(d)-d}{k(d)-d} + \binom{z+k(d-1)-d}{k(d-1)-(d-1)} + \cdots + \binom{z+k(1)-d}{k(1)-1}. \tag{C.2.9}$$

In the above expressions, any terms at the end with $k(i) < i$ vanish, and so are suppressed. It follows from [**BruH**, p.158] that if the t-th monomial of R_d is written in the form,

$$m = x_{j(1)}x_{j(2)}\cdots x_{j(d)}, \text{ with } 1 \leq j(1) \leq \ldots \leq j(d), \tag{C.2.10}$$

then in (C.2.9) we have

$$k(i) = r + 1 - j(d+1-i) + i - 2. \tag{C.2.11}$$

The terms "Macaulay" and "Gotzmann" polynomials are not standardized in the literature. For further discussion see the references already mentioned, and as well [**BaS2, Bla, Gas1, Gas2**].

DEFINITION C.12. GOTZMANN NUMBER. Given a Gotzmann polynomial $Q = Q(t,i,r)$, set $\varphi_r(Q) = a_k$ where a_k is the number defined by (C.2.6). Given a Macaulay polynomial $P = P(t,i,r)$, so $Q = \binom{z+r-1}{r-1} - P$, set $\varphi_r(P) = \varphi_r(Q)$. In general, we will suppress the subscript, and write $\varphi(P)$ and $\varphi(Q)$.

It is not hard to see from (C.2.10) that $\varphi_r(P)$ is the number of terms in its Macaulay representation ([**Gr1**]).

EXAMPLE C.13. If $r = 4$, then the 5-th monomial of degree 3 is $x_1 x_2^2$. So by (C.2.1), we have $(a_0, a_1) = (2,3)$. By (C.2.10), we have $(k(1), k(2), k(3)) = (2,3,5)$; alternatively, the third Macaulay representation of $c = \dim_k R_3 - 5$ is $15 = \binom{5}{3} + \binom{3}{2} + \binom{2}{1}$, from which we can read off the k-sequence. Thus, we have

$$Q = Q(5,3,4) = \binom{z-2+3}{3} + \binom{z-3+2}{2} = \binom{z+1}{3} + \binom{z-1}{2},$$
$$P = P(5,3,4) = \binom{z+2}{2} + \binom{z}{1} + \binom{z-1}{1}.$$

The Gotzmann number $\varphi(P) = \varphi(Q)$ is $a_1 = 3$, which is also the number of terms in the Macaulay representation of P.

EXAMPLE C.14. Consider the Macaulay polynomial $P = s$ (constant), when $r \geq 2$ and $d \geq s$. Let $t = r_d - s$, and recall from the proof of Corollary C.7 that, in the vector space $V(t, d, r)$, the last, so t-th, monomial is $x_{r-1}^s x_r^{d-s}$ and that the last monomial of $R_i \cdot V(t, d, r)$ is $x_{r-1}^s x_r^{d+i-s}$, which immediately precedes the last s monomials of R_{d+i}. Thus, by definition, $\varphi(P) \leq s$; the opposite inequality can be seen directly by noting that, if $d' = s - 1$ and $t' = r_{s-1} - s$, then $\mathrm{cod}\, R_i \cdot V(t', d', r) = s + i$, not s.

In the expression (C.2.1), we have $a_{r-2} = s$, so $k = r - 2$. From this observation, or directly, we conclude that the Gotzmann number $\varphi(P) = s$. Also, if $r \geq 3$, then $a_0 = \ldots a_{r-3} = 1$.

Alternatively, the Macaulay expansion (C.2.9) for $P = s$ satisfies $k(i) = i$ for $1 \leq i \leq s$, and all other terms vanish. So $\varphi(P) = s$, the number of terms. If $d > s$, then the expression (C.2.11) also gives $k(i) = i - 1$ for $d \geq i > s$, but the terms for which $k(i) < i$ are omitted.

Let $I \subset R$ be a graded ideal, and M the lex-initial ideal having the same Hilbert function; the existence of M is assured by Macaulay's Theorem C.2. Let $\nu(I)$ be the order of I; it is the smallest integer ν such that $I_\nu \neq 0$. For $i \geq \nu(I)$, let μ_i be the last monomial in M_i in lex order, and set $\Upsilon = \{\mu_i | i \geq \nu(M)\}$. Pick simultaneously a degree $\varphi(I)$ and a monomial $\mu_{\varphi(I)} \in M$ as follows: let α_1 be the smallest power of x_1 appearing in any element of Υ; let α_2 be the smallest power of x_2 appearing among those elements having x_1 appear with power α_1; and so on; finally, define

$$\varphi(I) = \alpha_1 + \cdots + \alpha_r \text{ and } \mu_{\varphi(I)} = x_1^{\alpha_1} \cdots \cdot x_r^{\alpha_r}.$$

Choose t so that $\mu_{\varphi(I)}$ is the t-th monomial of degree $\varphi(I)$.

Macaulay's Theorem C.2 implies the following result.

COROLLARY C.15. *In degrees i at least $\varphi(I)$, the Hilbert function $H(R/I)$ is the following polynomial:*

$$H(R/I) = H(R/V(t, \varphi(I), r)) = P(t, \varphi(I), r). \tag{C.2.12}$$

PROOF. Since M is a graded-lex-initial ideal, $\mu_{i+1} \leq x_r \cdot \mu_i$ for $i \geq \nu(I)$. Hence the power of x_1 appearing in μ_i is nonincreasing with i, so attains the minimum value α_1, then is constant, say for $i \geq i_1$ with i_1 minimum. Similarly, for $i \geq i_1$, the power of x_2 appearing in μ_i stabilizes at its minimum value α_2, say for $i \geq i_2$ with i_2 minimum, and so on up to $r - 1$. Clearly, $i_{r-1} = \varphi(I)$. Furthermore, if $i \geq \varphi(I)$, then $\mu_{i+1} = x_r \mu_i$. Hence $M \cap m^{\varphi(I)} = (V(t, \varphi(I), r))$. Therefore, $H(R/I) = P(t, \varphi(I), r)$ for $i \geq \varphi(I)$.   □

There is the minimum possible growth of $H(I)$ from degree d to $d + 1$ if and only if, correspondingly, there is the maximum possible growth of $H(R/I)$. The latter condition is given a concise combinatorial treatment in [BruH, Theorem 4.2.10] or [St3].

As a matter of notation, if $c = H(R/I)_d$ has d-th Macaulay representation as in (C.2.8), and if $P(t, d, r)$ is defined from it as in (C.2.9), set $c^{(d)} = P(t, d, r)(d + 1)$.

COROLLARY C.16. *Under the above conditions,*

$$H(R/I)_{d+1} \geq c^{(d)}. \tag{C.2.13}$$

C.3. Gotzmann's Persistence Theorem and m-Regularity

We begin by stating G. Gotzmann's Persistence Theorem C.17. The persistence problem had been stated by D. Berman [Be] in the context of ideals in R; its answer is classical when $r = 2$ (see Corollary C.6 above). Related theorems were proved recently by A. Aramova, J. Herzog, and T. Hibi [AHH] and by V. Gasharov [Gas2].

For $u \geq 0$ and $0 \leq t \leq r_d$, define recursively $N(t, d, r, 0) = t$, and

$$N(t, d, r, u) = \dim_k R_1 \cdot V(N(t, d, r, u - 1), d, r)$$

It is clear that $N(t, d, r, u) = \dim_k R_u \cdot V(t, d, r) = Q(t, d, r)_{z=d+u}$; see Example C.8.

If $t < r_d$, then we have

$$\dim_k R_1 \cdot V(t + 1, d, r) > \dim_k R_1 \cdot V(t, d, r). \tag{C.3.1}$$

Indeed, by Example C.8 both spaces are spanned by monomials forming an interval in the lex order with initial monomial x_1^{d+1}. By (C.1.2) the smallest monomial of $R_1 \cdot V(t + 1, d, r)$ is smaller than the smallest monomial of $R_1 \cdot V(t, d, r)$. Hence (C.3.1) holds.

Note that (C.3.1) implies that $N(t, d, r, u)$ is strictly monotonically increasing as a function of t when $0 \leq t < r_d$.

Let A be an arbitrary Noetherian ring, and set $S = A[x_1, \ldots x_r]$. Given a graded S-module M, denote by M_d its piece of degree d.

THEOREM C.17. ([Got1, Satz, p.61]) PERSISTENCE FOR IDEALS OF MINIMAL GROWTH. *Let I be a homogeneous ideal of S, generated by I_d, and set $M = S/I$. If M_i is A-flat of rank $P(t, d, r)_i$ for $i = d, d+1$, then M_i is so for all $i \geq d$.*

In particular, take A to be a field k. Let $I \subset R$ be a graded ideal, and $d > 0$. Set $t = \dim_k I_d$, and assume $\dim_k I_{d+1} = N(t, d, r, 1)$, which is the condition of minimal growth. Consider $J = (I_d)$, the ideal generated by I_d. Then $\dim_k J_{d+u} = N(t, d, r, u)$ for all $u \geq 0$; that is, minimal growth persists. Such a vector space I_d having minimal growth in degree d+1 is called a "Gotzmann space" in [**BruH, HeP**].

In the special case where $A = k$ and I is monomial, the proof is somewhat easier than in the general case, and Gotzmann proved it in [**Got1**, (2.12)] using part of his version of the theory of Castelnuovo–Mumford m-regularity; see Section C.4 below. Gotzmann then derived the general case using Grothendieck's theory of the Hilbert scheme, and at the same time, he obtained a new description of the Hilbert scheme, which is discussed in the next section.

M. Green refined the proof of the Persistence Theorem in the special case $A = k$ ([**Gr1**], see also [**BruH**, Theorem 4.3.3, p.172, 1998 ed.]). Green's key new ingredient is his Theorem 1 on p.77, which gives a bound on the dimension of the quotient I_d/tI_{d-1} for a general element t of R_1. A generalization of M. Green's result to a generic homogeneous element of arbitrary degree when char $k = 0$ was shown by J. Herzog and D. Popescu; they apply this result to some conjectures in higher Castelnuovo theory and Cayley–Bacharach theory [**HeP**]. V. Gasharov extended the Herzog–Popescu result to char $k = p$ [**Gas3**]. In his 1996 notes, Green gives a slightly different proof of the special case using a "Crystallization Principle" based on D. Bayer and M. Stillman's criterion for m-regularity ([**BaS2**], [**Gr2**, p. 41,47]).

Let $Z \subset \mathbb{P}^{r-1}$ be the closed subscheme defined by the graded ideal $I_Z \subset R$. Recall that there exists a polynomial $P_Z \in \mathbb{Q}[z]$ such that for j large enough,

$$\dim_k (R/I_Z)_j = P_Z(j);$$

in other words, $P_Z(z)$ is equal to the Hilbert function of R/I_Z for $j \gg 0$. This polynomial $P_Z(z)$ is called the *Hilbert polynomial* of Z. Let $W \subset R_v$ and $u > 0$. Recall that, by definition,

$$W : R_u = \{h \in R_{v-u} \mid R_u h \subset W\}.$$

Recall also that the *ancestor ideal* $\overline{W_v}$ of W_v is defined by the formula,

$$\overline{W_v} = \sum_{1 \leq u \leq v} (W_v : R_u) + (W_v);$$

see Equation (2.3.3). Finally, given a graded ideal J of R, recall that its *saturation*, denoted $Sat(J)$, is defined by the formulas,

$$Sat(J) = \lim_{v \to \infty} \overline{J_v} = \{h \mid \exists u \geq 0 \text{ s.t. } R_u h \subset J\} \tag{C.3.2}$$

$$= \bigcup_{v \geq u \geq 0}(J_v : R_u).$$

COROLLARY C.18. *Consider the ideal* $J = (I_d)$ *introduced after Theorem C.17. Its saturation* $Sat(J)$ *is equal to the ancestor ideal* $\overline{J_d}$, *and*

$$Sat(J) \cap (x_1, \ldots, x_r)^d = J.$$

Furthermore, $P(t, d, r)$ *is the Hilbert polynomial of the subscheme* Z_J *of* $\mathbb{P}^{r-1}$ *defined by* J.

PROOF. Note that, for $u \geq 0$ and $i \geq d$, we have

$$\langle R_u \cdot J_i \rangle : R_u = J_i. \tag{C.3.3}$$

Indeed, by descending induction on u, it suffices to prove for every $i \geq d$ that $(J_{i+1} : R_1) = J_i$. Since $J = (I_d)$, the growth from J_i to J_{i+1} is always minimal by Theorem C.17. Set $t = \dim_k J_i$ and $t' = \dim_k(J_{i+1} : R_1)$. Suppose $t' > t$. Then by Corollary C.4,

$$\dim_k R_1(J_{i+1} : R_1) \geq \dim_k R_1 V(t', i, r) = N(t', i, r, 1).$$

By the monotonicity proved above,

$$N(t', i, r, 1) > N(t, i, r, 1) = \dim_k J_{i+1},$$

which is a contradiction. Thus (C.3.3) holds. Hence $h \in Sat(J)$ if and only if $h \in (J) + J_d : R_1 + \ldots$. Therefore, $Sat(J)$ is the ancestor ideal $\overline{J_d}$. Finally, it is immediate from the definition of $P(t, d, r)$ and the assumption on $J = (I_d)$ that $P(t, d, r)$ is the Hilbert polynomial of Z_J. □

It is not hard to show that, if minimal growth happens from degree d to d+u, it happens from d to d+1 (see [HeP, Lemma 4.6(ii)]). The following consequence of Macaulay's Theorem C.2 was proved by E. Sperner.

COROLLARY C.19. [Sp, p.161–163] *Let* P_Z *be the Hilbert polynomial of a closed subscheme* Z *of* $\mathbb{P}^{r-1}$. *Then there is an integer* $i > 0$ *such that, setting* $t = r_i - P_Z(i)$, *we have* $P_Z = P(t, i, r)$.

REMARK C.20. By Sperner's Corollary, the set of Hilbert polynomials of schemes in $\mathbb{P}^{r-1}$ is exactly the set of Macaulay polynomials of Remark C.11. This statement leaves open the question of recognizing which Hilbert polynomials occur for "good" schemes, for example Cohen–Macaulay schemes or Gorenstein schemes. For more, see [**BruH**, 1998 ed. §4.4].

Recall that a graded ideal I of R is called *m-regular* if

$$H^i(\mathbb{P}^{r-1}, \tilde{I}(m-i)) = 0 \text{ for } i > 0 \qquad (C.3.4)$$

where $\tilde{I}$ is the sheaf associated to I. Recall that the *Castelnuovo–Mumford regularity*, or *m-regularity*, of I is the integer $\sigma(I)$ defined by the formula,

$$\sigma(I) = \min\{m \mid I \text{ is } m\text{-regular}\}; \qquad (C.3.5)$$

see [**Mum**, p.99], and [**BaM**, §3]).

D. Mumford proved that, if I is m-regular, then it is t-regular for all $t \geq m$; moreover, its saturation is generated in degrees m and above by its piece of degree m.

PROPOSITION C.21. [**Mum**, p. 99] *If $\tilde{I}$ is m-regular and if $t \geq m$, then $H^0(\tilde{I}(t))$ is spanned by $H^0(\tilde{I}(m)) \otimes H^0(\mathcal{O}(t-m))$, and $\tilde{I}$ is t-regular, that is, $H^i(\tilde{I}(t-i)) = 0$ for $i > 0$.*

The next Proposition is well known, and follows from the definition of m-regularity; for a proof see [**Got1**, (1.2)].

PROPOSITION C.22. *Let $Z \subset \mathbb{P}^{r-1}$ be a closed subscheme, P its Hilbert polynomial, and $\mathcal{I}_Z$ its saturated ideal. If $\mathcal{I}_Z$ is m-regular, then $H(R/\mathcal{I}_Z)_i = P(i)$ for $i \geq m$.*

There has been much work on bounding the regularity degree of $\mathcal{I}_Z$ for subschemes $Z \subset \mathbb{P}^n$ in terms of, say, the degrees of generators of $\mathcal{I}_Z$. For arbitrary schemes, by an example of E. Mayr and A. Meyer the bounds must be doubly exponential, but for "good" schemes the bounds are often even linear (see [**MayM**]; also [**BaS1**] for an effective regularity criterion, [**BaM**] for discussion and further references, and also [**BaS3, HoaM, Mal1, Mal3, Mal4, MiV, Pa, SmSw, Cu**]).

Given $P(z) \in \mathbb{Q}[z]$, set

$$\sigma_r(P) = \inf\{m \mid \mathcal{I}_Z \text{ is } m\text{-regular for every } Z \subset \mathbb{P}^{r-1}$$

$$\text{with Hilbert polynomial } P\}.$$

We will sometimes suppress the subscript, and write $\sigma(P)$.

A. Grothendieck proved, using the finiteness theorem for Chow coordinates, that $\sigma_r(P) < \infty$ [**Gro1**, p.221-7], [**ChW**]. D. Mumford introduced the theory of m-regularity, and showed that, given r, there is a polynomial in the coefficients of P that gives an upper bound for $\sigma_r(P)$ [**Mum**, p. 101]. G. Gotzmann proved a significantly improved upper bound for $\sigma_r(P)$, showing that a subscheme Z of $\mathbb{P}^{r-1}$ having Hilbert polynomial P, is $\varphi_r(P)$-regular (see Definition C.12).

LEMMA C.23. GOTZMANN'S REGULARITY THEOREM[**Got1**, (2.9)]. *If P is the Hilbert polynomial of a subscheme of $\mathbb{P}^{r-1}$, then $\sigma(P) \leq \varphi(P)$.*

Although the following result is an immediate consequence, and was known to experts, it appears not to have been stated explicitly in this form in the literature. G. Gotzmann [priv. comm.] ascribes it to D. Bayer, who showed in [**Ba**] that the Castelnuovo–Mumford regularity $\sigma(I)$ attains its maximum $\sigma(P)$ at the lex-segment ideal, and in addition gave a simple combinatorial expression for $\sigma(P)$.[4] Our proof depends on noting that when $x_r \nmid \mu_{t,d}$ then the saturated lex-segment ideal determined by $V(t, d, r)$ has Castelnuovo–Mumford regularity degree d. A lex-segment ideal is special in other ways, being a smooth point of the Hilbert scheme $\mathbf{Hilb}^P(\mathbb{P}^{r-1})$, by a result of A. Reeves and M. Stillman [**ReeS**].

PROPOSITION C.24. *Let P be a Hilbert polynomial. Then*

$$\sigma(P) = \varphi(P).$$

PROOF. Set $d = \varphi(P)$ and $t = \binom{r+d-1}{r-1} - P(d)$. Then the t-th monomial $\mu_{t,d}$ of R_d has no x_r factor, and $P = P(t, d, r)$; see Remark C.11 just after (C.2.6). Set $s = \sigma(P)$. Then $s \leq d$ by Lemma C.23. Consider $I = Sat(V(t, d, r))$. Then $H(I)_d = t$ by Proposition C.22; hence, the last monomial of I_d is $\mu_{t,d}$. Suppose $s < d$. Then $I_d = R_1 I_{d-1}$ by Proposition C.21. Let μ be the last monomial of I_{d-1}. Then $x_r \mu$ is the last monomial of $R_1 I_{d-1}$, and $x_r \mu \neq \mu_{t,d}$. This is a contradiction. Hence, $s = d$, as asserted.   □

These regularity bounds are used, when constructing the Hilbert scheme, to show the finiteness of the construction. S. L. Kleiman's article [**Kle1**, Theorem 3] contains a generalization of Mumford's result on regularity; his SGA VI article [**Kle2**] develops the theory further

---

[4][**Ba**, §II.10.1, II.10.5], cited also in [**HaM**, p.8]. G. Gotzmann mentioned as completing the proof of equality, a result in D. Bayer and M. Stillman's article [**BaS2**, Proposition 2.9], (if char $k = 0$ and I is a Borel fixed monomial ideal generated in degrees $\leq m$ and having a generator of degree-m then $\sigma(I) = m$), and his [**Got2**, (2.3)] (see (C.2.5) above). See also M. DeMazure's prenotes [**DeM**].

and uses it to prove a number of finiteness theorems for the Picard scheme, which had been announced by A. Grothendieck.

C.4. The Hilbert scheme $\mathbf{Hilb}^P(\mathbb{P}^{r-1})$

We discuss Gotzmann's description of the Hilbert scheme, which he obtained along with his persistence theorem, on the basis of Grothendieck's construction of the Hilbert scheme, and of his own version of Mumford's theory of m-regularity [**Got1**, **Gro1**, **Mum**].

Recall that the k-rational points of the Hilbert scheme $\mathbf{Hilb}^P(\mathbb{P}^{r-1})$ represent the closed subschemes Z of $\mathbb{P}^{r-1}$ whose Hilbert polynomial is P. In other words, there is a point $p_Z \in \mathbf{Hilb}^P(\mathbb{P}^{r-1})$, whose residue class field is k, corresponding to a $Z \subset \mathbb{P}^{r-1}$ if and only if $\dim_k(R/\mathcal{I}_Z)_j = P(j)$ for $j \gg 0$, where $\mathcal{I}_Z$ denotes the saturated ideal of Z. More generally, the T-maps $T \to \mathbf{Hilb}^P(\mathbb{P}^{r-1})$ correspond bijectively to the T-flat closed subschemes of $\mathbb{P}^{r-1} \times T$ whose fibres have Hilbert polynomial P.

Consider the category $\mathcal{C}$ of Noetherian k-schemes, and its subcategory $\mathcal{C}_A$ of affine schemes. Every scheme X defines a contravariant functor on $\mathcal{C}$ to the category of sets; it associates to a scheme T the set $X(T)$ of maps $T \to X$. Moreover, any functor on $\mathcal{C}$ restricts to a functor on the subcategory $\mathcal{C}_A$; for convenience, let $X(A)$ stand for $X(\mathrm{Spec}(A))$. Thus there are maps of categories,

$$\mathcal{C} \to ((\text{functors on } \mathcal{C})) \to ((\text{functors on } \mathcal{C}_A)). \tag{C.4.1}$$

The composition is fully-faithful; that is, given two schemes, every map between their functors on $\mathcal{C}_A$ arises from a unique map between the schemes (see [**EiH**, Proposition IV-2]). In short, a scheme is determined by its functor. A functor is said to be *representable* by a scheme if the functor is isomorphic to the one that arises from the scheme.

EXAMPLE C.25. Given $1 \leq p < r_d$, recall that the *Grassmannian* $\mathbf{Grass}^p(R_d)$ is the projective variety that parameterizes the vector subspaces of R_d with codimension p. More precisely, $\mathbf{Grass}^p(R_d)$ represents the functor of summands of R_d with corank p. In other words, $\mathbf{Grass}^p(R_d)(A)$ consists of the A-summands of F of S_d with corank p, where $S_d = A \otimes_k R_d$. (Note that the following conditions on an A-submodule F of S_d are equivalent: (1) F is a direct summand; (2) S_d/F is projective; and (3) F is locally a direct summand. See [**Ei2**, pp. 615–16].) Letting $q = \dim_k R_d - p$, we denote $\mathbf{Grass}^p(R_d)$ also by $\mathbf{Grass}(q, R_d)$, which thus parametrizes subspaces of R_d having dimension q.

Grothendieck constructed the Hilbert scheme as a closed subscheme of a suitable Grassmannian. His main tool was his construction of the (relative) flattening stratification of a sheaf.

THEOREM C.26. [Gro1] *Given a polynomial P and integers r and d, if $d \geq \sigma_r(P)$, then the Grassmannian $\mathbf{Grass}^{P(d)}(R_d)$ contains a closed subscheme $\mathbf{Hilb}^P(\mathbb{P}^{r-1})$ that represents the functor of A-flat subschemes Z of $\mathbb{P}_A^{r-1}$ whose fibers have Hilbert polynomial P. A subscheme Z corresponds to the submodule $(\mathcal{I}_Z)_d$ of S_d.*

The k-rational points of $\mathbf{Hilb}^P(\mathbb{P}^{r-1})$ represent all the vector subspaces $V \subset R_d$ such that the graded ring $R/(V)$ has Hilbert polynomial P. More generally, the elements of $\mathbf{Hilb}^P(\mathbb{P}^{r-1})(A)$ represent all the direct summands F of S_d such that the graded ring $S/(F)$ is A-flat with Hilbert polynomial P.

To move on to Gotzmann's work, fix d and form the product,

$$\mathbf{G} = \mathbf{Grass}^{P(d)}(R_d) \times \mathbf{Grass}^{P(d+1)}(R_{d+1}).$$

On $\mathcal{C}_A$, define a subfunctor $\mathbb{W}(A)$ of $\mathbf{G}(A)$ by

$$\mathbb{W}(A) = \{(F, G) \in \mathbf{G}(A) \mid F \cdot S_1 = G\}. \qquad (\text{C.4.2})$$

REMARK C.27. G. Gotzmann assumed that $d \geq \varphi_r(P)$, and defined $\mathbb{W}(A)$ using the inclusion $F \cdot S_1 \subset G$ in place of the equation $F \cdot S_1 = G$ in (C.4.2). As he remarked, it is easy to show that, since $d \geq \varphi_r(P)$, the inclusion implies the equation [Got1, (3.1),(3.2)]. Here's how (priv. comm., Dec. 1998). Let M be a maximal ideal of A, and set $F' = F/MF$ and $G' = G/MG$. Then $R_1 F' \subset G'$ since $S_1 F \subset G$. Now, $\dim R_1 F' = \dim G'$ by Macaulay's Theorem. Hence $R_1 F' = G'$. So $S_1 F + MG = G$. Hence, by Nakayama's Lemma, the localizations at M of $S_1 F$ and G are equal. Since M is arbitrary, $S_1 F = G$.

Thus it is a strong condition to require $S_1 F$ to be contained in a direct summand G of S_{d+1} of the right rank. Furthermore, Gotzmann's approach leads to local equations for the Hilbert scheme as a subscheme of the product $\mathbf{G}$. Indeed, the condition that $S_1 F \subset G$ is equivalent to the condition that the map $S_1 F \to H$ vanish, where $H = S_{d+1}/G$. Thus the condition is realized by the subscheme whose ideal is the image of $S_1 F \otimes H^*$ in the structure sheaf.

In the proof of the next Proposition, we show that the inclusion condition may be replaced by two rank conditions. They provide an alternative approach, which leads to a proof of Proposition C.30.

The following result about $\mathbb{W}(A)$ is implicit in Grothendieck's construction of $\mathbf{Hilb}^P(\mathbb{P}^{r-1})$.

PROPOSITION C.28. *For any d, the subfunctor $\mathbb{W}(A)$ of $\mathbf{G}(A)$ is representable by a subscheme $\mathbf{W}$ of $\mathbf{G}$, and the first projection embeds $\mathbf{W}$ in $\mathbf{Grass}^{P(d)}$. If $d \geq \sigma_r(P)$, then $\mathbf{W}$ contains the image of $\mathbf{Hilb}^P(\mathbb{P}^{r-1})$ under the (closed) diagonal embedding; on the level of functors, an A-flat subscheme Z of $\mathbb{P}_A^{r-1}$ is carried to the pair $((\mathcal{I}_Z)_d, (\mathcal{I}_Z)_{d+1}) \in \mathbb{W}(A)$.*

PROOF. The question is local. So consider an affine open subscheme $\mathbf{U}$ of $\mathbf{Grass}^{P(d)}(R_d)(A)$, and let A be its coordinate ring. Let $F \in \mathbf{Grass}^{P(d)}(R_d)(A)$ be the "universal" submodule of S_d, which corresponds to the inclusion map of $\mathbf{U}$. For each variable x_i, consider the map $F \to S_{d+1}$ given by multiplication by x_i, and form the sum,

$$u \colon F^{\oplus r} \to S_{d+1}.$$

Then the image of u is simply $F \cdot S_1$.

Set $q = P(d+1)$ and $q^\vee = r_{d+1} - q$, and consider these two conditions:

$$\mathrm{rk}(u) \leq q^\vee \quad \text{and} \quad \mathrm{rk}(u) > q^\vee - 1. \tag{C.4.3}$$

The first condition is closed, and the second is open; formally, both are given by the appropriate ideals of minors. So together, they define a subscheme $\mathbf{V}$ of $\mathbf{U}$.

Let A' be an A-algebra, $(F', G') \in \mathbb{W}(A')$. Then $F' = A' \otimes F$; moreover, $A' \otimes u$ satisfies the conditions in (C.4.3) because this map's image is equal to G'. So the first projection is an injection,

$$\beta \colon \mathbb{W}(A') \hookrightarrow \mathbf{V}(A').$$

We'll show β is a bijection.

Let $F' \in \mathbf{V}(A')$. Then $F' = A' \otimes_A F$. Set $C' = \mathrm{Cok}(u) \otimes A'$. Then C' is a locally free A'-module of rank q; indeed, keeping in mind Fitting's lemma [Ei2, 20.4, p. 493], see [Gro1, pp. 221-15–16] or [Mum, 8-1⁰, pp. 55–57] or [Ei2, 20.8, p. 495]. Set $S' = A' \otimes_k R$ and $G' = F' \cdot S_1'$. Then G' is a direct summand of S_{d+1}' of corank q since $S_{d+1}'/G' = C'$ and C' is locally free of rank q.

By construction, $(F', G') \in \mathbb{W}(A')$. Hence β is a bijection, and so $\mathbf{V}$ represents $\mathbb{W}|U$. Now, $\mathbf{V}(A) \cong W(A) \subset \mathbf{G}(A)$; so the identity map of $\mathbf{V}$ lifts to a map $\mathbf{v} \colon \mathbf{V} \to \mathbf{G}$. Then $\mathbf{v}$ is a section of the first projection of $\mathbf{G}$; so $\mathbf{v}$ is an embedding. Thus the first assertion holds.

The second assertion follows directly from the theorem.   □

In general, the two subschemes of $\mathbf{G}$ are certainly not equal, as it takes more than two degrees to determine the Hilbert polynomial. However, if $d \geq \varphi_r(P)$, then they are equal! Gotzmann proved it in two steps. First, he proved that the embedding of $\mathbf{Hilb}^P(\mathbb{P}^{r-1})$ in $\mathbf{W}$ is open as well as closed. Then, inspired by Hartshorne's proof of the connectedness of the Hilbert scheme, he degenerated an arbitrary pair $(F, G) \in \mathbb{W}(A)$ into one in which F is spanned by monomials. So he could now apply the special case of his Persistence Theorem, which he proved for this purpose.

This part of the argument can be simplified by using M. Green's lovely proof, which immediately yields the general case of the Persistence Theorem over a field [**Gr1**]. Indeed, in Gotzmann's treatment, Corollary C.18 is derived from Theorem C.29. However, M. Green proved the Persistence Theorem and so its Corollary C.18, directly for arbitrary graded ideals of R ([**Gr1**, p.82], see also [**BruH**, 1998 ed. §4.3], [**Gr2**, p. 47]). Finally, it is immediate from Corollary C.18 that every geometric point of $\mathbf{W}$ lies in $\mathbf{Hilb}^P(\mathbb{P}^{r-1})$.

THEOREM C.29. GOTZMANN'S HILBERT SCHEME THEOREM ([**Got1**, Satz,§(3.4),(3.7)]). *If $d \geq \varphi_r(P)$, then the subschemes $\mathbf{Hilb}^P(\mathbb{P}^{r-1})$ and $\mathbf{W}$ of $\mathbf{G}$ are equal.*

In particular, Gotzmann's Theorem and the proof of Proposition C.28 imply that the first condition in (C.4.3), which is the condition of minimal growth, already alone suffices to define $\mathbf{Hilb}^P(\mathbb{P}^{r-1})$ as a subscheme of $\mathbf{Grass}^{P(d)}(R_d)$.[5] Hence we can describe this Hilbert scheme by local equations as follows.

Set $p = P(d)$ and $p^\vee = r_d - p$. Use the set of all the monomials as a basis of R_d, and associate to each r_d by $p^\vee$ matrix a subspace V of R_d of codimension p, namely, its column space. Choose a subset K of p monomials of degree d, and form the affine space $\mathbf{A}_K$ of those matrices with the $p^\vee$ by $p^\vee$ identity matrix as the submatrix corresponding to the monomials outside K. Then $\mathbf{A}_K$ is an affine coordinate chart of the Grassmanian $\mathbf{Grass}^p(R_d)$, and as K varies, these charts cover.

PROPOSITION C.30. *In the affine coordinate chart $\mathbf{A}_K$ of the Grassmannian $\mathbf{Grass}^{P(d)}(R_d)$, the Hilbert scheme $\mathbf{Hilb}^P(\mathbb{P}^{r-1})$ is defined by*

---

[5]That the determinantal conditions suffice to give $Hilb^P(\mathbb{P}^{r-1})$ as a reduced algebraic set is shown in D. Bayer's thesis, where it is also explicitly conjectured that they give the scheme structure [**Ba**, p. 134]. This stronger statement was later asserted, but not proved, in the first author's survey [**I5**, Theorem, p.312], in [**HaM**, p.9], and also in [**Cat**, p. 584].

the ideal of all the $(q^\vee + 1)$ by $(q^\vee + 1)$ minors of the r_{d+1} by $rp^\vee$ matrix whose columns are the vectors $x_i \cdot v_j \in R_{d+1}$ for $1 \leq i \leq r$ and $1 \leq j \leq p^\vee$.

REMARK. We stated Proposition C.30 in local terms since this form might be more useful for applications. It is equivalent to say that $\mathbf{Hilb}^P(\mathbb{P}^{r-1})$ is the determinantal scheme, in the sense of Definition 5.19 (the definition there is valid over an arbitrary base), associated to the following map of bundles:

$$\mu : \mathcal{S} \otimes \mathcal{E}_1 \to \mathcal{E}_{d+1}.$$

Here $\mathcal{E}_i$ is the trivial sheaf $R_i \otimes_k \mathcal{O}_{\mathbf{Grass}}$, and $\mathcal{S}$ is the tautological subsheaf of $\mathcal{E}_d$ on $\mathbf{Grass}^{P(d)}(R_d)$; also, μ is the induced multiplication map.

EXAMPLE C.31. If $r = 4$, $d = 3$, and $\mu_{t,d} = x_2^2 x_3$, then from (C.2.1), $a_0 = 1$, $a_1 = 3$, $a_2 = 3$. So

$$Q(t,3,4) = \binom{z-1+3}{3} + \binom{z-3+2}{2} + \binom{z-3+1}{1},$$

and $t = Q(t,3,4)_{z=3} = 12$; whereas, $c = 8 = \binom{4}{3} + \binom{3}{2} + \binom{1}{1}$, so

$$P = \binom{z+1}{1} + \binom{z}{1} + \binom{z-2}{1} = 3z - 1,$$

of regularity degree $\sigma(P) = 3$. Then $p = P(3) = 8$ and $q = P(4) = 11$; whereas, $p^\vee = Q(3) = 12$ and $q^\vee = Q(4) = 24$. The Hilbert scheme $\mathbf{Hilb}^P(\mathbb{P}^3)$ is defined locally on the affine coordinate charts of $Grass(12, R_3)$ by the vanishing of the 25×25 minors of the 35×80 matrix whose columns are $x_1 \cdot v_j \in R_4$, as in Corollary C.30.

When $r = 4$ and $P = 6z - 2$, the Hilbert polynomial of a canonical curve, then $\sigma(P) = 12$ since $P = \binom{z+1}{1} + \binom{z}{1} + \cdots \binom{z-4}{1} + 6$, so the Macaulay expansion of P will need 12 terms, implying $\sigma(P) = 12$. Here $t = r_{12} - P(12) = 386$: by writing 386 as in Lemma C.10 we have $Q = Q(386, 12, 4) = \binom{z-1+3}{3} + \binom{z-7+2}{2} + \binom{z-12+1}{1}$ and $\mu_{386,12} = x_2^6 x_3^6$. To obtain all components of $\mathbf{Hilb}^P(\mathbb{P}^3)$ one must work in degree 12, high compared to the Castelnuovo regularity 3 of a canonical curve ([**Cat**, (2.23)]).

D. Bayer studied the action of $Sl(r)$, and characterized the weight decompositions of Hilbert points with respect to a maximal torus [**Ba**]. This was further described by D. Bayer and I. Morrison ([**BaM**], see also [**HaM**, §4B]). Recently, R. Notari and M. L. Spreafico studied the stratification of $\mathbf{Hilb}^P(\mathbb{P}^{r-1})$ by the locally closed subschemes

$\mathbf{Hilb}^P(\mathbb{P}^{r-1})_M$ parametrizing those subschemes Z having a given monomial ideal M as initial ideal of $\mathcal{I}_Z$; they show that the strata are either affine schemes (when M is itself saturated), or locally closed subschemes of affine schemes, and they give equations for them [**NotSp**].

G. Gotzmann used his identification of the Hilbert scheme in further work, [**Got2, Got5**], the former on simply connected Hilbert schemes (see Proposition C.36 below, and also D. Mall's related [**Mal2**]). As well, F. Catanese used the Gotzmann Theorem to bound the complexity of the Hilbert scheme [**Cat**, p. 584].

There has been other recent work related to the connectedness or smoothness of the general Hilbert scheme: A. Reeves on "radius", K. Pardue and D. Mall concerning the connectedness of the Hilbert function strata $Hilb^H(\mathbb{P}^n)$ of the Hilbert scheme (generalizing G. Gotzmann's proof of connectedness when $n = 2$ in [**Got3**]), and J. Cheah's study of smooth nested punctual Hilbert schemes [**Che, Ree, ReeS, Pa, Mal4**].

For a survey of earlier work on the punctual Hilbert scheme to 1986 see [**I5**]. For further references, and a discussion of the construction of the Hilbert scheme, see the Lecture notes by E. Sernesi and by E. Sernesi and C. Ciliberto [**Ser, CiS**], and two Ergebnisse Volumes, the first by J. Kollár [**Ko**, §1.1], who also discusses the relation with the Chow variety, and the second by E. Viehweg, who also treats geometric invariant theory on the Hilbert scheme [**Vi**, Chapters 1,7].

C.5. Gorenstein sequences having a subsequence of maximal growth, and $\mathbf{Hilb}^P(\mathbb{P}^{r-1})$

Recall that a Gorenstein sequence T is a sequence that occurs as the Hilbert function of a (graded) Artin Gorenstein quotient R/I. So T is symmetric. One consequence of this symmetry and of the Persistence Theorem C.17 is that the growth from T_i to T_{i+1} rarely attains the maximum permitted by Corollary C.4 of Macaulay's Characterization Theorem C.2.

Indeed, set $t = r_i - T_i$. Then Corollary C.4 says that

$$T_{i+1} \leq r_{i+1} - N(t, i, r, 1).$$

When equality occurs, then I is rather special. For one thing, the Persistence Theorem implies that $Sat((I_i)) = Sat((I_{i+1}))$ and that the Hilbert polynomial of $R/(I_i)$ is $P(t, i, r)$. So far though, we have not used the symmetry of T. However, it restricts I further, as is illustrated by Proposition C.33 below.

A similar phenomenon was studied by A. Bigatti, A. Geramita, and J. Migliore [**BGM**] in a different context. They considered a subvariety X of $\mathbb{P}^{r-1}$ whose Artin minimal reduction $(R/(L))/\mathcal{I}_X) + (L))$, where L is a suitable vector space of linear forms, has a Hilbert function H with a constant subsequence $(H_d = s, H_{d+1} = s)$ where $s \leq d$. If X is a reduced (smooth) set of points satisfying a uniform position property, they show that then X must lie on a reduced irreducible curve of degree s [**BGM**, Theorems 4.7]. They also study maximum growth in a wider context.

We are particularly interested in the case that the Hilbert polynomial $P(t, i, r)$ is constant, and this case is treated in the following Proposition. The first two parts are immediate consequences of Gotzmann's work, and he treats Part (i) explicitly when $r = 3$ in his paper [**Got3**, p.544].

Recall that we denote by $\mathbf{Grass}^s(R_d)$ or $\mathbf{Grass}(s^\vee, R_d)$ the Grassmanian parametrizing s-dimensional quotients of R_d, or equivalently, $s^\vee$-dimensional subspaces, where $s^\vee = r_d - s$.

PROPOSITION C.32. REGULARITY FOR THE CONSTANT POLYNOMIAL

i. Let $P = s$ be the constant polynomial. Then $\sigma(P) = \varphi(P) = s$.

ii. Let I be a graded ideal of R such that, for some $d \geq s$,

$$\dim_k R_d/I_d = \dim_k R_{d+1}/I_{d+1} = s. \qquad (C.5.1)$$

Set $Z = \operatorname{Proj} R/(I_d)$. Then Z is a subscheme of $\mathbb{P}^{r-1}$ of dimension 0 and degree s. Furthermore, the saturated ideal $\mathcal{I}_Z$ of Z satisfies $(\mathcal{I}_Z)_i = (I_d)_i$ for $i \geq d$ and $(\mathcal{I}_Z)_{d+1} = I_{d+1}$.

iii. If $d \geq s$, then $\mathbf{Hilb}^s(\mathbb{P}^{r-1})$ is the subscheme of $\mathbf{Grass}^s(R_d)$ determined by the condition $\dim_k V \cdot R_1 = r_{d+1} - s$.

PROOF. Part (i) holds because $\varphi(P) = s$ by Example C.14 and because $\sigma(P) = \varphi(P)$ by Proposition C.24. Alternatively, $\sigma(P) = s$ by Theorem 1.69.

For Part (ii), first note that, since $d \geq s$, by Example C.14, there is minimal growth in size from I_d to I_{d+1} in the sense of Corollary C.4 of Macaulay's theorem C.2. So since $R_1 I_d \subset I_{d+1}$, we have $R_1 I_d = I_{d+1}$. Part (ii) then follows from Gotzmann's Persistence Theorem C.17 and its Corollary C.18 because $P(t, d, r) = s$ holds for $t = r_d - s$ in view of Example C.14.

Finally, Part (iii) follows from Part (i) and Proposition C.30. □

Recall that the *socle degree* of a Gorenstein sequence T is the largest integer j such that $T_j \neq 0$.

PROPOSITION C.33. *Let s, d, j be integers satisfying $0 < s \leq d < j$ and let $T = (T_i)$ be a Gorenstein sequence of socle degree j having a constant subsequence $(T_d = s,\ T_{d+1} = s)$. Then the following assertions hold.*

   i. *The sequence T is nonincreasing for $i \geq d$ and nondecreasing for $i \leq j - d$. If $d \leq j/2$, then T is nondecreasing for $i \leq d$, constant at s for $d \leq i \leq j - d$, and nonincreasing for $i \geq j - d$; in particular, then T is bounded above by s.*

   ii. *Let $f \in Gor(T)$. Set $I = \mathrm{Ann}(f)$ and $Z = \mathrm{Proj}\, R/(I_d)$. Then Z is a subscheme of $\mathbb{P}^{r-1}$ of dimension 0 and degree s, and is an annihilating scheme of f. If also $d \leq j/2$, then Z is a tight annihilating scheme of f (see Definition 5.1). Always, $I_i = (\mathcal{I}_Z)_i$ for $i \leq \max(d, j - d)$.*

PROOF. To prove (i), we won't use the hypothesis $T_d = T_{d+1}$. Indeed, Corollary C.7 implies that T is nonincreasing for $i \geq d$. Now, T is symmetric about $j/2$; hence, T is also nondecreasing for $i \leq j - d$. If $d \leq j/2$, then the assertions about T follow formally. Thus (i) is proved.

Consider (ii). Let I be a Gorenstein ideal with $H(R/I) = T$, and let $J = Sat((I_d))$. Since $s \leq d$, part (ii) of Proposition C.32 implies that $J_i = (I_d)_i$ and $H(R/J)_i = s$ for $i \geq d$. It also says that Z is of dimension 0 and degree s. Moreover, Corollary C.18 and Lemma 2.17 imply that $\mathcal{I}_Z \subset I$; so Z is an annihilating scheme of f. If $d \leq j/2$, then $\ell\mathrm{diff}(f) = s$ by Part (i); since $\deg(Z) = s$, therefore Z is a tight annihilating scheme by Definition 5.3. Since $J_d = I_d$, Lemma 2.17 also implies $J_i = I_i$ for $i \leq d$. If $d \leq j/2$, then also $H(R/J)_i = H(R/I)_i = s$ for $d \leq i \leq j - d$, implying that $J_i = I_i$ for $i \leq j - d$ in that case. $\square$

More generally, we have the following result with a similar proof.

PROPOSITION C.34. *Let P be an arbitrary Hilbert polynomial, let d, j be integers satisfying $\varphi(P) \leq d < j$, and let T be a Gorenstein sequence of socle degree j such that $T_d = P(d)$ and $T_{d+1} = P(d + 1)$. Then the following assertions hold.*

   i. *The sequence T satisfies the bound $T_i \leq P(i)$ for $i \geq d$. If $d \leq j/2$, then $T_u \leq P(j - u)$ for $u \leq d$.*

   ii. *Let $f \in Gor(T)$. Set $I = \mathrm{Ann}(f)$ and $Z = \mathrm{Proj}\, R/(I_d)$. Then Z is a subscheme of $\mathbb{P}^{r-1}$ with Hilbert polynomial P. Furthermore, we have $\mathcal{I}_Z \subset I$, with equality in degrees d and $d + 1$.*

EXAMPLE C.35. If $r \geq 3$ and if $T = (1, r, T_2, \ldots, r, 1)$ satisfies $j(T) = 9$ and $T_4 = 4$, then any $f \in Gor(T)$ has a degree-4 tight annihilating scheme $Z = \operatorname{Proj} R/(I_4)$ where $I = \operatorname{Ann}(f)$. Indeed, the symmetry of T implies that $T_5 = 4$; so the assertion follows from Part (ii) of Proposition C.33.

Next we state a result of G. Gotzmann, which describes the structure of subschemes of $\mathbb{P}^{r-1}$ with certain Hilbert polynomials $P(t, d, r)$. The condition that $k \leq r - 3$ is equivalent to the condition that, in the lex order, the t-th monomial μ_t of R_d is not divisible by x_{r-1}. We may also take d to be the minimum possible; that is, if $P(t, d, r) = P(t', d', r)$, then $d' \geq d$. It is equivalent to assume that μ_t is not divisible by x_r; see Remark C.11. The integers $a_0 \leq \ldots \leq a_k$ appear in Gotzmann's expansion of the polynomial $Q(t, d, r)$; see (C.2.2). Set $a_{-1} = 1$.

PROPOSITION C.36. ([**Got4**, Propositions 1,2], [**Got2**, Satz 1])[6] *Let $P = \binom{z+r-1}{r-1} - Q(t, d, r)$, and assume that the Gotzmann polynomial $Q(t, d, r)$ has $k \leq r - 3$, see (C.2.2). Then each saturated ideal I of R with Hilbert polynomial $P(R/I) = P$ has, after a suitable linear transformation of $\mathbb{P}^{r-1}$, a generating set of the form,*

$$(f_0 x_1, f_0 f_1 x_2, f_0 f_1 f_2 x_3, \ldots, f_0 \cdots f_{k-1} x_k, f_0 \cdots f_k), \qquad \text{(C.5.2)}$$

where $\deg f_i = a_i - a_{i-1}$ for $0 \leq i \leq k-1$ and $\deg f_k = a_k - a_{k-1} + 1$. Furthermore, the Hilbert scheme $\mathbf{Hilb}^P(\mathbb{P}^n)$ is irreducible and simply connected, and under the action on it of the upper triangular group, $Z_o = \operatorname{Proj} R/(V(t, d, r))$ is the only fixed point.

EXAMPLE C.37. The Hilbert function $T = (1, 3, 4, 5, 4, 3, 1)$ has maximum growth from degrees 2 to 3. So T corresponds to the graded lex-initial ideal $(V(2, 2, 3))$, whose Hilbert polynomial is $P(z) = z + 2$. Part (ii) of Proposition C.36 implies that, if $f \in Gor(T)$, then $I = \operatorname{Ann}(f)$ is identical in degrees 2 and 3 with the saturated ideal $\mathcal{I}_Z$ of a subscheme Z of $\mathbb{P}^2$ having Hilbert polynomial $P = z + 2$; namely, Z is either a line union a disjoint point or a line with an embedded point. It follows that $I_2 = h \cdot V$, where $h \in R_1$ and V is the two dimensional subspace of R_1 defining the point in $\mathbb{P}^2$. This is the Hilbert function T_2 of Example 3.6. Note that $Q(z) = \binom{z+2}{2} - P = \binom{z-1}{2} + \binom{z-1}{1}$, so $(a_0, a_1) = (2, 2)$. Although k is $r - 2$, not $r - 3$ as is assumed in Proposition C.36, nevertheless $\mathcal{I}_Z$ is of the form (C.5.2); indeed, $\mathcal{I}_Z = (hx, hy)$, with h, x and y each of degree one.

---

[6]There is a typographical error in the degree of f_k in Proposition 2 of [**Got4**]. Equation (C.5.2) agrees with the analogous Formula (2.3)(ii) of [**Got2**].

j	6	7	8	9	10
$\dim_k \mathcal{T}_F$	57	63	71	83	96
$\dim(C_H)$	50	60	71	83	96

TABLE C.1. Dimension of the tangent space $\mathcal{T}_F$ at a point of $Gor(T)$ with $T = Sym(H, j)$, and the corresponding value of $\dim(C_H)$; see Example C.38.

Recall that $Sym(H_Z, j)$ denotes the symmetrization of the sequence H_Z about $j/2$ (see p. 108).

EXAMPLE C.38. Let $P = P(5, 3, 4) = \binom{z+2}{2} + \binom{z}{1} + \binom{z-1}{1}$. Then the fifth monomial of R_3 is $x_1 x_2^2$, and Gotzmann's expansion has $a_0 = 2$ and $a_1 = 3$; see Example C.13. By Proposition C.36, any subscheme Z of $\mathbb{P}^3$ with Hilbert polynomial P satisfies $\mathcal{I}_Z = (fx, fg)$, where $\deg f = 1$ and $\deg x = 1$ and $\deg g = 2$. We claim that, if F is general enough in $(\mathcal{I}_Z)_6^{\perp}$, then the Gorenstein ideal $I = Ann(F)$ contains $\mathcal{I}_Z$; furthermore, then

$$H(R/I) = (1, 4, 9, 15, 9, 4, 1) = Sym(H_Z, 6),$$

and $I_i = (\mathcal{I}_Z)_i$ for $i = 2, 3$.

This claim was confirmed by a calculation using the symbolic algebra program "Macaulay" in the special case $Z_0 = Proj R/(V(5, 3, 4))$; the general case then follows by deforming from this special case. This construction determines a subfamily C_H of $Gor(T)$ where $T = H(R/I)$. The tangent space $\mathcal{T}_F$ is of dimension 57 at a general point F lying over Z_0. The construction can be compared with similar constructions in Chapter 6; the latter yielded irreducible subfamilies of certain $Gor(T)$, consisting of forms having tight annihilating schemes of dimension 0, of a certain kind (as smooth, or "compressed"). Here the "annihilating scheme" is of higher dimension, an idea introduced by M. Boij.

By choosing different values of j, we may construct in this way other examples of Hilbert functions $T = Sym(H_Z, j)$. For $j = 8, 9, 10$, the dimension of the tangent space $\mathcal{T}_F$ is equal to the dimension of the subfamily C_H; hence, C_H is an irreducible component of $Gor(T)$. By the argument in the proof of Theorem 6.27, which depends on the result of P. Maroscia (see [Mar, Theorem 1.8], and Theorem 5.21(A) above), we see that $Gor(T)$ has at least two irreducible components: a second component lies over a family of smooth 0-dimensional subschemes of $\mathbb{P}^3$. The first such example has $j = 8$ and $T =$

$(1, 4, 9, 15, 22, 15, 9, 4, 1)$; this example is similar to some examples that were constructed previously by M. Boij (see [Bo2] and Theorem 6.42).

Here we use the formula, $\dim C_H = P(j) + \dim Hilb^H(\mathbb{P}^3)$; it holds since F is chosen in $(\mathcal{I}_Z)_j{}^\perp$, which is a vector space of dimension $P(j)$. To see that $\dim Hilb^P(\mathbb{P}^3) = 11$, note that the choice of Z is equivalent to the choice of two linear forms f, x and of a degree-2 form g mod xR_1, up to constant multiples; hence, $\dim Hilb^P(\mathbb{P}^3) = 3 + 3 + 5$. To produce Table C.1, we used the symbolic algebra program "Macaulay" to calculate the dimension of the tangent space T_F at a point $F \in Gor(T)$ lying over Z_0.

ACKNOWLEDGMENT. The authors of the Appendix would like to acknowledge the helpful comments of G. Gotzmann and V. Kanev.

Author addresses for Appendix C:

Anthony Iarrobino
Department of Mathematics, 567 Lake Hall
Northeastern University
Boston, MA 02115
USA
e-mail address: iarrobin@neu.edu

Steven L. Kleiman
Department of Mathematics
Room 2-278, MIT
77 Mass Ave, Cambridge, MA 02139-4307
USA
e-mail address: kleiman@math.mit.edu

Examples of "Macaulay" Scripts

We used the D. Bayer, M. Stillman algebra program "Macaulay" [BaS1] to find the dimension of the tangent space to $Gor(T)$ at the point parametrizing a given degree-j homogeneous form f: by Theorem 3.9 this is $\dim_k(R_j/I_j^2)$ where $I = \text{Ann}(f)$. If $f \in PS(s,j;r)$, so $f = L_1^j + \cdots + L_s^j \in \mathcal{R}_j$, and $L_1, \ldots, L_s$ are general enough elements of $\mathcal{R}_1$, and $s \le r_{t-1}$, $t = \lfloor j/2 \rfloor$ then by Lemma 3.16 it suffices to calculate $\dim_k(R_j/(\mathcal{I}_Z^{(2)})_j)$ where Z is the vanishing ideal at the corresponding points of $\mathbb{P}^n$. The following calculation shows that the hyptheis "general enough" is necessary in Lemma 3.16. In practice, to find a Gorenstein ideal I we replace $\mathcal{R}$ by the divided power ring $\mathcal{D}$, find $f = L_1^{[j]} + \ldots + L_s^{[j]}$, then calculate the ideal $I = \text{Ann}(f)$ in the contraction action of R on $\mathcal{D}$.

Warning: Using the contraction action – <l_from_dual ... ... in "Macaulay" – with ordinary powers, will give wrong answers. See Appendix A.

Dimension of the tangent space at a form F with annihilating scheme $Z = 10$ points on a rational normal curve. We find $\dim_k T_F$, $F = L_1^7 + \cdots + L_{10}^7 \in \mathcal{R}_7$, where $\ell\text{diff}(F) = 10$, and F has a tight annihilating scheme Z consisting of 10 points on a RNC in $\mathbb{P}^4$.

```
<ring 5 v-z r
<ideal L v v+w+x+y+z v+2w+4x+8y+16z v+3w+9x+27y+81z
   v+5w+25x+125y+625z v+7w+49x+343y+2401z v-3w+9x-27y+81z
   v-2w+4x-8y+16z v-5w+25x-125y+625z v-7w+49x-343y+2401z
<pow_entry L 7 10 POWERS  [This script forms L_1^{[7]}, ..., L_{10}^{[7]}, where
```
$$L_2^{[7]} = v^7 + v^6 w + \cdots \ ].$$
```
random 10 1 M
mult POWERS M F   [F is a linear combination of L_1^{[7]}, ..., L_{10}^{[7]}]
<l_from_dual F I  [Annihilator in contraction action]
std I I
hilb I   Ans: 1 5 9 10 10 9 5 1   (This is T)
```

```
pow I 2 I2
std I2 I2
hilb I2   Ans: 1 5 15 35 50 51 50 44 26 13 9 5 1 = H(R/I²).
```

Thus, the dimension of the tangent space to $Gor(T)$ at F is $\dim_k T_F$ $= \dim_k(R_7/I_7^2) = 44$. Compare with Example 6.14, where $j = 6$ in place of $j = 7$ here, but the tangent space has the same dimension.

Hilbert function of the square of the vanishing ideal at Z. The scheme $Z \subset RNC \subset \mathbb{P}^4$ corresponds to the linear forms L above, and is defined by the ten points $Z = \{(1,0,0,0,0),(1,1,1,1,1),(1,2,4, 8,16),\ldots,(1,-7,49,-243,2401)\}$. By Lemma 6.1(c) Z is the unique tight annihilating scheme for F. We find $H(R/\mathcal{I}_Z^2)$.

```
transpose L T [needed because of how "coef" works]
coef T MONOM M
transpose M P [P is now a 5 by 10 matrix with rows 1 a a² a³ a⁴]
<points P IZ
std IZ IZ [vanishing ideal IZ = I_Z]
hilb IZ Ans: (1,4,4,1). [thus, H(R/I_Z) = (1,5,9,10,10,...) and
         τ(Z) = 3, so (Z,F) satisfy the hypotheses of Lemma 6.1(c)]
pow IZ 2 SQ
std SQ SQ
hilb SQ  Ans: 1 4 10 20 15 1 -1
```

Thus, $H(R/\mathcal{I}_Z^2) = (1,5,15,35,50,51,50,50,\ldots)$, satisfying $H(R/\mathcal{I}_Z^2) \geq H(R/I^2)$, a consequence of $I = \mathrm{Ann}(F) \subset \mathcal{I}_Z$. This calculation is the line $SQ(5)$ in Table 6.1.

Hilbert function for the symbolic square $\mathcal{I}_Z^{(2)}$

```
transpose L T
coef T MONOM M
transpose M P [P is now a 5 by 10 matrix with rows 1 a a² a³ a⁴]
<ideal W x2 x2 x2 x2 x2 x2 x2 x2 x2 x2 [to get weights all 2]
<points P J W
std J J [vanishing ideal J = I_Z^(2)]
hilb J Ans: 1 4 10 19 13 3 codimension = 4
```

Thus $H(R/\mathcal{I}_Z^{(2)}) = (1,5,15,34,47,50,50,\ldots)$. This calculation is the line $SYMSQ(5)$ of Table 6.1. See Example 6.17.

REMARK. Here $\dim_k(R/(\mathcal{I}_Z^{(2)})_7) = \dim_k(R/(\mathcal{I}_Z^2)_7) = 50 > \dim_k T_F = 44$. When the linear forms $L_1,\ldots,L_{10}$ determining the summands L_i^j of F are not sufficiently general – here they lie on a RNC – then the dimension of the tangent space is not computed by $\dim_k(R/(\mathcal{I}_Z^{(2)}))_7$, as Lemma 3.16 does not apply.

APPENDIX E

Concordance with the 1996 Version

We here give a comparison of the sections and chapters with those of the May, 1996 manuscript [**IK**]. We do this because the manuscript was circulated, and there are several references to it in the published literature. Several of these are to the result describing the tangent space to $\mathbf{Gor}(T)$, which is Theorem 3.9 here.

Almost all sections retained from the earlier version have many changes and improvements; often the theorems stated here are stronger. We also tightened language, for example distinguishing between "family" of varieties and "parameter space". Conversion to LaTeX allowed a clearer notation, for example $\mathbf{Gor}(T)$ for scheme and $Gor(T)$ for the related variety or algebraic set. Many of the changes are related to our desire to make the book more user-friendly, and to include more exposition and examples. This led to changes throughout, but is especially true of the added Introduction, Sections 1.2, 1.3, the two survey Sections 4.4, 6.4, the added Sections 5.2, 5.4, 5.6, much of Appendix B and the new Appendix C. We outline below the main changes.

The Introduction is new, an informal preparation for the book.

Chapter 1 is completely rewritten; and Sections 1.2, 1.3 are new. Section 1.2, although an introduction to the cases $j = 2, 3$ surveys some new results. Section 1.3 concerns the binary forms case $r = 2$ and Hankel matrices; it also assembles material not before collected in one place. Section 1.4 is updated from the old Detailed Summary.

Section 2.1 on the Waring problem is new, and contains a new result in char p, the proof of Theorem 1.61 for char $p \nmid j$. Sections 2.2 and 2.3 are largely from the old Chapter 1, but contain improvements, especially in char $k = p$.

Chapter 3 is the old Chapter 2, with a new result — Proposition 3.14 for corank one catalecticant varieties.

Sections 4.1, 4.2 are the old Sections 3A, 3B, with some improvements and updating.

Section 4.3 is the old Appendix A2. This extends several of the results of Section 4.2, but uses a different approach.

Section 4.4 describes recent developments concerning $Gor(T)$ when $r = 3$, some inspired by the 1996 version of this book. This is rather more narrow than the survey by the first author in [I9], written in 1997, but the narrow focus allows us to explain certain developments in more depth.

Chapter 5 is largely revised and rewritten from the old Section 4A (new 5.1), 4E (new 5.3) and 4F (new 5.7). The new Section 5.2 contains expository material on flat families of 0-dimensional schemes and limit ideals, and a determinantal construction of the postulation Hilbert scheme. Section 5.3 is greatly expanded, with added exposition and details of proofs as well as additional results. Section 5.4, a summary on power sums, and Section 5.5, an application, are new. Section 5.6 is also new; it collects results concerning lengths of a form and the closure of $PS(s, j; r)$, some not explicitly stated in the 1996 manuscript.

Chapter 6 is mostly from the old Sections 4B–4D, with some improvements. However, a new update Section 6.4 is added.

Chapter 7 is from the old Chapter 5, but with many revisions, and some improvements.

Chapter 8 is based on the old Chapter 6.

Chapter 9 on open problems has been extensively rewritten. A few of the problems mentioned as open in 1996 have since been solved! (See, for example, Sections 4.4, 6.4).

Appendix A corresponds to the old Appendix A3, but is completely rewritten and extended. Appendix B is new except for a part of Section B.2, which is from the old Appendix A1. The new Appendix C, by the first author and S. L. Kleiman, states basic results concerning extremal Hilbert functions, states Gotzmann's construction of the Hilbert scheme, and as well shows a determinantal variation of that construction, conjectured by D. Bayer; Section C.5 gives applications. Appendix D is the old Appendix A4.

There are rather more references. This is partly due to our decision to survey recent work in Sections 4.4 and 6.4, and to the new Appendix C; and partly due to our wish to give the reader the choice to look further. We have added A. M. S. Math Review numbers after several of the references, when we felt the review might be particularly useful. Because of the added introductory material, we decided to omit a List of Theorems. However, we added substantially to the indexes, and have for the reader's convenience included short definitions in the Index of Notation.

References

[ABW]   Akin K., Buchsbaum D., Weyman J.: *Schur functors and Schur complexes*, Adv. Math. **44** (1982), 207–270.

[Al]   Alexander J.: *Singularités imposable en position general à une hypersurface projective*, Compositio Math. **68**(1988), no. 3, 305–354.

[AlH1]   ———, Hirschowitz A.: *Un lemme d'Horace différential: application aux singularités hyperquartiques de $\mathbb{P}^5$*, J. Algebraic Geometry **1** (1992), no. 3, 411–426. MR 93e:14004.

[AlH2]   ———, ———: *La méthode d'Horace éclaté: application à l'interpolation en degré quatre*, Invent. Math. **107** (1992), no. 3, 585–602.

[AlH3]   ———, ———: *Polynomial interpolation in several variables*, J. Algebraic Geometry **4** (1995), 201–222. MR 96f:14065.

[AlH4]   ———, ———: *Generic hypersurface singularities*, Proc. Indian Acad. Sci. Math. Sci. **107** (1997), no. 2, 139–154.

[AnL]   Angéniol B., Lejeune-Jalabert M.: *Calcul différentiel et classes caractéristiques en géométrie algébrique*, Travaux en Cours (Works in Progress), **38**, Hermann, Paris, (1989).

[AHH]   Aramova A., Herzog J., Hibi T.: *Gotzmann theorems for exterior algebras and combinatorics*, J. Algebra **191** (1997), no.1, 174–211.

[ACGH]   Arbarello E., Cornalba M., Griffiths P. A., Harris J.: *Geometry of Algebraic Curves, Vol. I*, Grundlehren Math. Wiss. Bd. 267, Springer-Verlag, Berlin and New York (1985).

[AM]   Atiyah M., Macdonald I. G.: *Introduction to commutative algebra*, Addison-Wesley, Reading MA (1969).

[Bas]   Bass H.: *On the ubiquity of Gorenstein rings*, Math. Z. **82** (1963), 8–28.

[Ba]   Bayer D.: *The Division algorithm and the Hilbert scheme*, thesis, (1982) Harvard U., Cambridge.

[BaM]   ———, Mumford D.: *What can be computed in algebraic geometry?* Computational Algebraic Geometry and Commmutative Algebra (Cartona 1991), Cambridge Univ. Press, Cambridge, (1993) pp.1–48.

[BaMo]   ———, Morrison I.: *Standard bases and geometric invariant theory I. Initial ideal and state polytopes*, J. Symbolic Computation, **8** (1988), 209–217.

[BaS1]   ———, Stillman M.: *Macaulay, A system for computation in algebraic geometry and commutative algebra*, (1982-1990), Source and object code available for Unix and Macintosh computers from http://www.math.columbia.edu/ bayer/Macaulay.html.

[BaS2] ———, ———: *A criterion for detecting m-regularity*, Invent. Math. **87**(1987), no. 1,1–11.

[BaS3] ———, ———: *On the complexity of computing syzygies. Computational aspects of commutative algebra.* J. Symbolic Comput. **6** (1988), 135–147. MR 90g:68071.

[Be] Berman D.: *Simplicity of a vector space of forms: finiteness of the number of complete Hilbert functions*, J. Algebra **45** (1977), 52–57.

[BeI] Bernstein D., Iarrobino A.: *A nonunimodal graded Gorenstein Artin algebra in codimension five*, Comm. in Algebra **20** # 8 (1992), 2323–2336.

[BGM] Bigatti A., Geramita A., Migliore J.: *Geometric consequences of extremal behavior in a theorem of Macaulay*, Trans. Amer. Math. Soc. **346** (1994), no. 1, 203–235. MR95e:14040.

[Bla] Blancafort C.: *Hilbert functions of graded algebras over Artinian rings*, J. Pure Appl. Algebra **125** (1998), 55–78. MR 98m:13023.

[BofS] Boffi G., Sánchez R.: *On the resolutions of the powers of Pfaffian ideal*, J. Algebra **152** (1992), 463–491.

[Bo1] Boij M.: *Gorenstein Artin algebras and points in projective space*, Bull. London Math. Soc. **31** (1999), no. 1, 11–16.

[Bo2] ———: *Components of the space parametrizing graded Gorenstein Artin algebras with a given Hilbert function*, Pacific J. Math. **187** (1999), no. 1, 1–11.

[Bo3] ———: *Betti numbers of compressed level algebras*, J. Pure Appl. Algebra, **134** (1999), no. 2, 111–131.

[Bo4] ———: *Nonunimodal Betti numbers of Gorenstein algebras*, Math. Scand. (to appear).

[Bo5] ———: *Betti Number Strata of the Space of codimension three Gorenstein Artin algebras*, preprint, in preparation. Also, abstract of talk at Conference: "Homological methods in Commutative Algebra and Algebraic Geometry", May 24-June 13, 1996, at Genoa.

[BoL] ———, Laksov, D.: *Nonunimodality of graded Gorenstein Artin algebras*, Proc. A.M.S. **120** (1994), 1083–1092.

[Bor] Borel A.: *Linear Algebraic Groups*, Benjamin, New York (1969).

[Bou] Bourbaki N.: *Algèbre, Chap. I - IX*, Hermann, Masson, (1970–1980).

[Bre] Brennan J.: *Invariant theory in characteristic p: Hazlett's symbolic method for binary quantics*, Factorization in integral domains (Iowa City, IA, 1996), Lecture Notes in Pure and Appl. Math., **189** (1997) Dekker, New York, pp. 257–269.

[Bri] Briançon J.: *Description de* $\mathbf{Hilb}^n\mathbb{C}\{x, y\}$, Invent. Math. **41** (1977), 45–89.

[BrS] Brodman M., Sharp R.: *Local cohomology: an algebraic introduction with geometric applications*, Cambridge Studies in Advanced Mathematics, **60**, Cambridge University Press, Cambridge (1998).

[Bro] Bronowski J.: *The sums of powers as canonical expression*, Proc. Cambridge Philos. Soc. **29** (1933), 69–82.

[Bru] Bruns W.: *"Jede" endliche freie Auflösung ist freie Auflösung eines von drei Elementen erzeugten Ideals*, J. Algebra 39 (1976), no. 2, 429–439. MR 53 # 2925.

[BruH] ———, Herzog J.: *Cohen-Macaulay Rings*, Cambridge Studies in Advanced Mathematics # 39, Cambridge University Press, Cambridge, U.K., (1993); revised paperback edition, 1998.

[BruV] ———, Vetter U.: *Determinantal rings*, Lecture Notes in Math. # 1327, Springer-Verlag, Berlin and New York (1988).

[BE1] Buchsbaum D., Eisenbud D.: *What makes a complex exact?*, J. Algebra 25 (1973), 259–268.

[BE2] ———, ———: *Algebra structures for finite free resolutions, and some structure theorems for codimension three*, Amer. J. Math. 99 (1977), 447–485.

[BE3] ———, ———: *Gorenstein ideals of height 3*, Seminar D. Eisenbud/B. Singh/W. Vogel, Vol. 2, pp. 30–48, Teubner-Texte zur Math.,48, Teubner, Leipzig (1982).

[Bu1] Burch L.: *On ideals of finite homological dimension in local rings*, Proc. Camb. Phil. Soc. 64 (1968), 941–948.

[Bu2] ———: *A note on the homology of ideals generated by three elements in local rings*, Proc. Camb. Phil. Soc. 64 (1968), 949–952.

[Cam] Campanella G.: *Standard bases of perfect homogeneous polynomial ideals of height 2*, J. Algebra 101 (1986), 47–60.

[C-J] Catalano-Johnson M.: *The possible dimensions of the higher secant varieties*, Amer. J. Math 118 (1996), 355–361.

[Cat] Catanese F.: *Chow varieties, Hilbert schemes, and moduli spaces of surfaces of general type*, J. Algebraic Geom. 1 (1992), 561–595. MR 93j14005.

[Cha1] Chandler K.: *Regularity of the powers of an ideal*, Comm. in Algebra 25 (1997), 3773–3776.

[Cha2] ———: *Higher infinitesimal neighborhoods*, J. Algebra 205 (1998), 460–479.

[Cha3] ———: *A brief proof of the maximal rank theorem for generic double points in projective space*, preprint, 15p. (1998).

[Che] Cheah J.: *Cellular decomposition for nested Hilbert schemes of points*, Pacific J. Math. 198, no. 1, 39–90.

[ChiC] Chiantini L., Coppens M.: *Grassmannians of secant varieties*, preprint (1999).

[ChoI1] Cho Y., Iarrobino A.: *Inverse systems of zero-dimensional schemes in $\mathbb{P}^n$*, preprint, (1999).

[ChoI2] ———, ———: *Gorenstein Artin algebras arising from punctual schemes in projective space*, preprint (1999).

[ChoJ1] ———, Jung B.: *Dimension of the tangent space of $Gor(T)$*, The Curves Seminar at Queen's. Vol. XII (Kingston, ON, 1998), Queen's Papers in Pure and Appl. Math., 114, Queen's Univ., Kingston, ON, (1998), 29–41.

[ChoJ2] ———, ———: *The dimension of the determinantal scheme $V_s(t,t,2)$ of the catalecticant matrix*, to appear, Communications in Algebra.

[ChW] Chow W. L., van der Waerden B. L.: *Über zugeordnete Formen und algebraische Systeme von algebraischen Mannigfaltigkeiten (Zur Algebraischen Geometrie IX)*, Math. Ann. 113 (1937), 692-704.

[CiS]   Ciliberto C., Sernesi E.: *Families of varieties and the Hilbert scheme*, Lectures on Riemann surfaces (Trieste, 1987), World Sci. Publishing, Teaneck, NJ, (1989) pp. 428-499.

[Cle]   Clebsch A.: *Ueber Curven vierter Ordnung*, Journal für die Reine und Angewandte Mathematik (Crelle), Bd. **59** (1861), 125-145.

[ClL]   Clements G., Lindström L.: A generalization of a combinatorial theorem of Macaulay, J. Combinatorial Theory **7** (1969), 230-238.

[Co1]   Coble A.: *Associated sets of points*, Trans. Amer. Math. Soc. 24 (1922), 1-20.

[Co2]   Coble A.: *Algebraic geometry and theta functions*, Amer. Math. Soc. Coll. Publ. # 10, Providence R. I. (1929), (3rd ed. 1969).

[Con]   Conca A.: *Straightening law and powers of determinantal ideals of Hankel matrices*, Advances in Math. **138** (1998), 263-292.

[CoV1]   ———, Valla G.: *Hilbert functions of powers of ideals of low codimension*, Math. Z. 230 (1999), no. 4, 753-784.

[CoV2]   ———, ———: *Betti numbers and lifting of Gorenstein codimension three ideals*, to appear, Communications in Algebra.

[CLO]   Cox D., Little J., O'Shea D.: *Ideals, Varieties, and Algorithms*, 2nd. edition, (1997), Springer UTM, Springer, New York.

[Cu]   Cutkosky S. D.: *Irrational Asymptotic Behaviour of Castelnuovo-Mumford Regularity*, preprint, 10p., (1999), math.AG/9902037, available at http://xxx.lanl.gov/.

[Da]   Davis E. : *Complements to a paper of P. Dubreil*, Ricerche di Matematica **37** (1988), 347-357.

[DP]   De Concini C., Procesi C.: *A characteristic free approach to invariant theory*, Adv. Math. **21** (1976), 330-354.

[DeM]   DeMazure M.: *Results of R. Stanley and D. Bayer: Prenotes*, preprint (1984).

[DeP]   De Poi P.: *On higher secant varieties of rational normal scrolls*, Matematiche (Catania) **51** (1996), 3-21. MR 98k:14075

[DGM]   Diaz S., Geramita A., Migliore J.: *Resolutions of subsets of finite sets of points in projective space*, to appear, Comm. in Algebra.

[Di]   Diesel S. J.: *Some irreducibility and dimension theorems for families of height 3 Gorenstein algebras*, Pacific J. Math. **172** (1996), 365-397.

[Dix]   Dixon A.: *The canonical forms of the ternary sextic and quaternary quartic*, Proc. London Math. Soc. (2), **4** (1906), 223-227.

[DK]   Dolgachev I., Kanev V.: *Polar covariants of plane cubics and quartics*, Advances in Math. **98** (1993), 216-301. MR94g14029.

[DO]   Dolgachev I., Ortland D.: *Point sets in projective spaces and theta functions*, Asterisque t. **165** (1988).

[DuPW]   du Plessis A., Wall C. T. C.: *The geometry of topological stability*, London Math. Society Monographs. New series, **9**, Oxford Science Publ., Clarendon Press, Oxford Univ. Press, New York (1995). MR 97k:58024.

[EaN]   Eagon J., Northcott D. G.: *Ideals defined by matrices and certain complex associated with them*, Proc. Roy. Soc. London Ser. A **269** (1962), 188-204.

[EhR]   Ehrenborg R., Rota G.-C.: *Apolarity and canonical forms for homogeneous polynomials*, European J. of Combinatorics **14** (1993), 157-182.

[Ei1]   Eisenbud D.: *Linear sections of determinantal varieties*, Amer. J. Math. **110** (1988), 541–575.

[Ei2]   Eisenbud D.: *Commutative algebra with a view toward algebraic geometry*. Graduate Texts in Math. # 150, Springer-Verlag, Berlin and New York (1994).

[EiG]   ———, Goto S.: *Linear free resolutions and minimal multiplicity*, J. Algebra **88** (1984), 89–133.

[EiH]   ———, Harris J.: *Schemes: the language of algebraic geometry*, (1992), Wadsworth & Brooks/Cole, Belmont, California.

[EiL]   ———, Levine H.: *The degree of a C^∞ map germ*, Annals of Math. **106** (1977), 19–38.

[EiP1]   ———, Popescu S.: *Gale duality and free resolutions of ideals of points*, Invent. Math. **136** (1999), 419–449.

[EiP2]   ———, ———: *The projective geometry of the Gale transform*, to appear, Journal of Algebra, D. Buchsbaum volume.

[EliI]   Elias J., Iarrobino, A.: *The Hilbert function of a Cohen-Macaulay local algebra: extremal Gorenstein algebras*, J. Algebra **110** (1987), 344–356.

[El]   Ellingsrud G.: *Sur le schéma de Hilbert des variétés de codimension 2 dans $\mathbb{P}^e$ à cône de Cohen-Macaulay*, Ann. scient. Ec. Norm. Sup, 4e série, t.8 (1975), 423–432.

[ElS]   ———, Strømme S. A.: *Bott's formula and enumerative geometry*, Journal A.M.S., **9** (1996), 175–193.

[Ell]   E. B. Elliot, *An introduction to the algebra of quantics*, 2nd ed., Oxford Univ. Press, London and New York, (1913).

[Em]   Emsalem J.: *Géométrie des Points Epais*, Bull. Soc. Math. France **106** (1978), 399–416.

[EmI1]   ———, Iarrobino A.: *Some zero-dimensional generic singularities; finite algebras having small tangent space*, Compositio Math. **36** (1978), 145–188.

[EmI2]   ———, ———: *Inverse system of a symbolic power I*, J. Algebra. **174** (1995), 1080–1090.

[Fog]   Fogarty J.: *Algebraic families on an algebraic surface*, Amer. J. Math. **10** (1968), 511–521.

[Fr]   Fröberg R. : *An inequality for Hilbert series of graded algebras*, Math. Scandinavica **56** (1985), 117–144.

[FL]   ———, Laksov D.: *Compressed algebras*, Conf. on Complete Intersections in Acireale, (S.Greco and R. Strano, eds), Lecture Notes in Math. # 1092, Springer-Verlag, Berlin and New York, (1984), pp. 121–151.

[FW]   Fukui T., Weyman J.: *Cohen-Macaulay properties of Thom-Boardman strata I: Morin ideal*, to appear, Proc. London Math. Soc.

[Fu1]   Fulton W.: *Schubert varieties in flag bundles for the classical groups*, Proc. of the Hirzebruch 65th Birthday Conference on Algebraic Geometry (Ramat Gan, 1993), Israel Math. Conf. Proc. **9**, Bar-Ilan Univ., Ramat Gan, (1996) pp. 241–262.

[Fu2]   ———: *Determinantal formulas for orthogonal and symplectic degeneracy loci*, J. Differential Geom. **43** (1996), no. 2, 276–290. MR 98d:14004.

[FuH]   _____, Hansen J.: *A connectedness theorem for projective varieties, with application to intersections and singularities of mappings*, Ann. of Math. **110** (1979), 159–166.

[Gaf]   Gaffney T.: *Multiple points, chaining, and the Hilbert schemes*, Amer. J. Math. **110** (1988), 595–628.

[Gas1]   Gasharov V.: *Extremal properties of Hilbert functions*, Illinois J. Math. **41** (1997), 612–629.

[Gas2]   _____: *Green and Gotzmann theorems for polynomial rings with restricted powers of the variables*, J. Pure and Applied Algebra, **130** (1998), no.2, 113–118.

[Gas3]   _____: *Hilbert functions and homogeneous generic forms II*, Compositio Math., **116** (1999), no. 2, 167–172.

[G1]   Geramita A. V.: *Inverse systems of fat points: Waring's problem, secant varieties of Veronese varieties and parameter spaces for Gorenstein ideals*, The Curves Seminar at Queen's, Vol X, (Kingston, 1995), Queen's Papers in Pure and Appl. Math., **102** (1996), 1–114.

[G2]   _____: *Catalecticant varieties*, in Lecture notes in Pure and Applied Mathematics, **206** (1999), F. Van Ostaeyen, ed., Dekker, New York.

[GGP]   _____, Gimigliano A., Pitteloud Y.: *Graded Betti numbers of some embedded n-folds*, Math. Annalen, **301** (1995), 363–380.

[GGR]   _____, Gregory D., Roberts G.: *Monomial ideals and points in projective space*, J. Pure Appl. Algebra **40** (1986), 33–62.

[GHS1]   _____, Harima T., Shin Y. S.: *An alternative to the Hilbert function for the ideal of a finite set of points in* $\mathbb{P}^n$, The Curves Seminar at Queen's. Vol. XII (Kingston, 1998), Queen's Papers in Pure and Appl. Math., **114** (1998), 67–96.

[GHS2]   _____, _____, _____: *Extremal point sets and Gorenstein ideals*, The Curves Seminar at Queen's. Vol. XII (Kingston, 1998), Queen's Papers in Pure and Appl. Math., **114** (1998), 97–140. Also to appear, Advances in Math.

[GKS]   _____, Ko H. J., Shin Y. S.: *The Hilbert function and the minimal free resolution of some Gorenstein ideals of codimension 4*, Communications in Algebra, **26** (1998), 4285–4307.

[GM]   _____, Maroscia P.: *The ideal of forms vanishing at a finite set of points in* $\mathbb{P}^n$, J. Algebra, **90** (1984), 528–555.

[GMR]   _____, _____, Roberts G.: *The Hilbert function of a reduced k- algebra*, J. London Math. Soc. (2), **28** (1983), 443–452.

[GMi1]   _____, Migliore J.: *Hyperplane Sections of a Smooth Curve in* $\mathbb{P}^3$, Comm. Alg. **17** (1989), 3129-3164).

[GMi2]   _____, _____: *Reduced Gorenstein Codimension Three Subschemes of Projective Space*, Proc. Amer. Math. Soc. **125** (1997), 943–950.

[GO]   _____, Orecchia F.: *The Cohen-Macaulay type of s lines in* $\mathbb{A}^{n+1}$, J. Algebra **70** (1981), 116–140.

[GPS]   _____, Pucci M., Shin Y. S. : *Smooth points of* $\mathcal{G}or(T)$, J. of Pure and Applied Algebra, **122** (1997), 209–241. MR98m:13031

[GSc]   _____, Schenck H.: *Fat points, inverse systems, and piecewise polynomial functions*, J. Algebra 204 (1998), 116–128, MR 99d:13019.

[GS] _____, Shin Y. S. : *k-configurations in* $\mathbb{P}^3$ *all have extremal resolutions,* J. Algebra 213 (1999), 351–368.

[Gh] Gherardelli F.: *Observations about a class of determinantal varieties,* (Italian, English summary), Istit. Lombardo Accad Sci. Lett. Rend. A **130** (1996), 163-170. MR 99b:14052.

[Gor] Gorenstein D.: *An arithmetic theory of adjoint plane curves,* Trans. Amer. Math. Soc. **72**, 414–436.

[GoS] Gorodentsev A.L., Shapiro B. Z.: *On associated discriminants for polynomials in one variable* Beiträge Algebra Geom. **39** (1998), no. 1, 53–74.

[Göt] Göttsche L.: *Hilbert schemes of zero-dimensional subschemes of smooth varieties,* Lecture Notes in Math. # 1572, Springer-Verlag, Berlin and New York, (1994).

[Got1] Gotzmann G.: *Eine Bedingung für die Flachheit und das Hilbertpolynom eines graduierten Ringes,* Math. Z. **158** (1978), no. 1, 61–70. MR 58:641.

[Got2] _____: *Einige einfach-zusammenhängende Hilbertschemata,* Math. Zeit. **180** (1982), 291–305.

[Got3] _____: *A stratification of the Hilbert scheme of points in the projective plane,* Math. Z. **199** (1988), no. 4, 539–547. MR 89k:14003.

[Got4] _____: *Some irreducible Hilbert schemes,* Math. Z. **201** (1989), 13–17.

[Got5] _____: *Topologische Eigenschaften von Hilbertfunktion-Strata.* Habilitationsschrift, Westfälische Wilhelms-Universität Münster, (1993).

[GrY] Grace J. H., Young A.: *The algebra of invariants,* (1903), Cambridge University Press, Cambridge; reprint New York: Chelsea.

[Gra] Granger M.: *Géométrie des schémas de Hilbert ponctuels,* Mém. Soc. Math. de France (N.S.),**8** (1983).

[Gre] Green E. L.: *Complete intersections and Gorenstein ideals,* J. Algebra **52** (1978), 264–273.

[Gr1] Green M.: *Restrictions of linear series to hyperplanes, and some results of Macaulay and Gotzmann,* Algebraic Curves and Projective Geometry (Trento 1988), (E. Ballico and C. Ciliberto, eds.), Lecture Notes in Math. # 1389, Springer-Verlag, Berlin and New York, (1989) pp. 76–86.

[Gr2] _____: *Generic initial ideals,* Summer School in Commutative Algebra (Bellaterra, July 16-26, 1996), (J. Elias, J. M Giral, R. M. Miró-Roig, S. Zarzuela, organizers), vol. II, Centre de Recerca Matematica, Bellaterra, (1997), pp. 13–85; also in Six Lectures on Commutative Algebra, J. Elias, ed. Progress in Math. #166, Birkhäuser), (1998).

[GriH] Griffiths Ph., Harris J.: *Principles of algebraic geometry,* John Wiley and Sons, New York (1978).

[Gro1] Grothendieck A.: Techniques de construction et théorèmes d'existence en géometrie Algèbrique, Sem. Bourbaki # 221 (1961), or Fondements de la Géometrie Algèbrique, Sem. Bourbaki, 1957-1962, Secretariat Math. Paris (1962).

[Gro2] _____, _____: *Eléments de Géométrie Algébrique* IV_3, Publ. Math. I. H. E. S. **28** (1966).

[Gro3] _____, Dieudonné J.: *Eléments de Géométrie Algébrique I,* Springer-Verlag (1971).

326   REFERENCES

[GruP]   Gruson L., Peskine C.: *Courbes de l espace projectif: variétés de sécantes*, Enumerative Geometry and Classical Algebraic Geometry, (Le Barz P. and Hervier Y., eds.), Prog. Math Vol. 24, (1982), Birkhauser, Boston, pp. 1–31.

[Gu1]   Gundelfinger S.: *Zur Theorie der binären Formen*, Göttingen Nachr. **12** (1883)

[Gu2]   Gundelfinger S.: *Zur Theorie der binären Formen*, J. Reine Angew. Math. (Crelle J.), **100** (1886), 413–424.

[Hai]   Haiman M.: *(t, q)-Catalan Numbers and the Hilbert Scheme*, Selected papers in honor of Adriano Garsia (Taormina, 1994). Discrete Math. **193** (1998), 201–224.

[Hari1]   Harima T.: *Some examples of unimodal Gorenstein sequences*, J. Pure and Applied Algebra **103** (1995), 313–324.

[Hari2]   _____: *Characterization of Hilbert functions of Gorenstein Artin algebras with the weak Stanley property*, Proc. Amer. Math. Soc. **123** (1995), 3631-3638.

[Hari3]   _____: *A note on Artinian Gorenstein algebras of codimension three*, J. Pure and Applied Algebra, **135** (1999), 45–56.

[Ha]   Harris J.: *Algebraic Geometry, a First Course*, Graduate Texts in Math. # 133, Springer-Verlag, Berlin and New York (1992).

[HaM]   _____, Morrison I.: *Moduli of Curves*, Graduate Texts in Math. #187, Springer-Verlag, Berlin and New York (1998).

[HaT]   _____, Tu L.: *On symmetric and skew-symmetric determinantal varieties*, Topology **23** (1984), no. 1, 71–84.

[Har1]   Hartshorne R.: *Connectedness of the Hilbert scheme*, Publ. Math. de l'I.H.E.S. **29** (1966), 5–48.

[Har2]   _____: *Algebraic geometry*, Springer-Verlag, Berlin and New York, (1977).

[HasT]   Hassett B., Tschinkel Y.: *Geometry of equivariant compactifications of* $\mathbb{G}_a^n$, preprint, 22p. (1999), math.AG/9902073, available at xxx.lanl.gov.

[HeiR]   Heinig G. , Rost K.: *Algebraic methods for Toeplitz-like Matrices and Operators*, Operator Theory, Advances and Applications, # 13, Birkhäuser, Basel and Boston (1984).

[Hel]   Helmke U.: *Waring's problem for binary forms* J. Pure Appl. Algebra **80** (1992), no. 1, 29–45.

[He]   Herzog J.: *Ein Cohen-Macaulay Kriterium mit Anwendungen auf den Konormalen Modul und den Differential-modul*, Math Z. **163** (1978), no.2, 149–162. MR80a:13025.

[HeP]   _____, Popescu D.: *Hilbert functions and generic forms*, Compositio Math. **113** (1998), no. 1, 1–22.

[HeTV]   _____, Trung N. V., Valla G.: *On hyperplane sections of reduced irreducible varieties of low codimension*, J. Math. Kyoto Univ. **34** (1994), no. 1, 47–72.

[Hi1]   Hilbert D.: *Üeber die vollen Invariantensysteme*, Math. Ann. **42** (1890), 473–534.

[Hi2l]   _____: *Theory of algebraic invariants*, (Translated from the German and with a preface by Reinhard C. Laubenbacher. Edited and with an introduction by Bernd Sturmfels) Cambridge Univ. Press, Cambridge (1993). MR97j:01049

[HMP] Hinrichsen D., Manthey W., Prätzel-Wolters D.: *The Bruhat decomposition of finite Hankel matrices*, Systems and Control Letters **7** (1986), 173–182.

[Hir] Hirschowitz A.: *La methode d'Horace pour l'Interpolation à plusieurs variables*, Manuscr. Math. **50** (1985), 337–388.

[HRW] ———, Rahavandrainy O., Walter C.: *Quelques strates de Brill-Noether du schéma de Hilbert de* $\mathbb{P}^2$, C. R. Acad. Sci. Paris Sér. I Math. 319 (1994), no. 6, 589–594. MR 96b:14005.

[HoaM] Hoa Lê Tuân, Miyazaki C.: *Bounds on Castelnuovo-Mumford regularity for generalized Cohen-Macaulay graded rings*, Math. Ann. **301** (1995), no. 3, 587–598.

[HoE] Hochster M., Eagon John A.: *Cohen-Macaulay rings, invariant theory, and the generic perfection of determinantal loci* Amer. J. Math. **93** (1971), 1020–1058.

[HoL] ———, Laksov D.: *The linear syzygies of homogeneous forms*, Commun. Algebra **15** (1987), 227–239.

[Hum] Humphreys J. E.: *Linear algebraic groups* Springer-Verlag, Berlin and New York (1975).

[Hun1] Huneke C.: *Invariants of Liaison*, Algebraic Geometry (Ann Arbor, Michigan, 1981), Lecture Notes in Math. # 1008, Springer-Verlag, Berlin and New York (1983) pp. 65–74.

[Hun2] ———: *Numerical invariants of liaison classes*, Inv. Math. **75** (1984), 301–325.

[I1] Iarrobino A.: *Punctual Hilbert Schemes*, Mem. Amer. Math. Soc. **10** # 188 (1977), Amer. Math. Soc., Providence.

[I2] Iarrobino A.: *Deforming complete intersection Artin algebras. Appendix: Hilbert function of* $\mathbb{C}[x,y]$, Singularities, part I, (Arcata, Calif. 1981), (P. Orlik, ed.), Proc. Symp. Pure Math. # 40, Amer. Math. Soc., Providence, RI, (1983) pp. 593–608.

[I3] ———: *Compressed algebras: Artin algebras having given socle degrees and maximal length*, Trans. AMS **285** (1984), 337–378.

[I4] ———: *Compressed algebras and components of the punctual Hilbert scheme*, pp.146-185 in Algebraic Geometry, Sitges 1983, Lecture Notes in Math. vol 1124, Springer-Verlag (1985).

[I5] ———: *Hilbert scheme of points: Overview of last ten years*, in Algebraic Geometry, Bowdoin 1985, S. Bloch, editor, PSPM #46 Part 2, Amer. Math. Soc., Providence (1987) pp. 297-320.

[I6] ———: *Associated graded algebra of a Gorenstein Artin algebra*, Mem. Amer. Math. Soc. **107** (1994) # 514, Amer. Math. Soc.,Providence.

[I7] ———: *Inverse system of a symbolic power II: The Waring problem for forms*, J. Algebra **174** (1995), 1091–1110.

[I8] ———: *Inverse system of a symbolic power III: thin algebras and fat points*, Compositio Math. **108** (1997), 319–356.

[I9] ———: *Gorenstein Artin algebras, additive decompositions of forms and the punctual Hilbert scheme*, in Commutative Algebra, Algebraic

Geometry, and Computational Methods [Proceedings of Hanoi Conference in Commutative Algebra (1996)], D. Eisenbud, ed., Springer-Verlag (1999), 53–96.

[IK] _____, Kanev V.: *The length of a homogeneous form, determinantal varieties of catalecticants, and Gorenstein algebras*, preprint, 150 p., (1996) [preliminary version of this book].

[IY] _____, Yaméogo J.: *Graded ideals in k[x,y] and partitions,II: Ramification and a generalization of Schubert calculus*, preprint, 44p., alg-geom/9709021, available at xxx.lanl.gov

[Ik1] Ikeda, H.: *Results on Dilworth and Rees numbers of Artinian local rings*, Japan J. Math (N.S.) **22** (1996), 147–158.

[Ik2] _____: *The Dilworth and Ress numbers of one-dimensional modules*, Comm. Alg. **25** (1997), 2627–2633.

[IlR] Iliev A., Ranestad K.: *K3 surfaces of genus 8 and varieties of sums of powers of cubic fourfolds*, to appear in Trans. of Amer. Math. Soc. (1999).

[Io] Iohvidov I. S.: *Hankel and Toeplitz Matrices and Forms: algebraic theory*, transl. by Thijsse P. of 1974 Russian edition (Gohberg I., ed.), Birkhäuser, Boston, (1982).

[Is1] Iskovskih V. A.: *Fano 3-folds, I* (in Russian), Izv. Akad. Nauk SSSR Ser. Mat. **41** (1977), no. 3, 516–562; English translation in Math. USSR Izvestija **11** (1977) 485–527). MR 80c:14023a.

[Is2] _____: *Fano 3-folds, II* (in Russian), Izv. Akad. Nauk SSSR Ser. Mat. **42** (1978), no. 3, 506–549. MR 80c:14023b.

[JLP] Jozefiak T., Lascoux A., Pragacz P.: *Classes of determinantal varieties associated with symmetric and skew-symmetric matrices* (Russian), Izv. Akad. Nauk SSSR Ser. Mat. **45** (1981), no. 3, 662–673. MR 83h14044.

[JPW] Jozefiak T., Pragacz P., Weyman J.: *Resolutions of determinantal varieties*, Tableaux de Young et foncteurs de Schur en algèbre et géometrie (Torun, 1980), Asterisque **87–88** (1981), pp. 207–220.

[Ka] Kanev V.: *Chordal varieties of Veronese varieties and catalecticant matrices*, math/9804141 available at xxx.lanl.gov; to appear, Journal of Math. Sci. (New York) volume dedicated to Iskovskih's 60th birthday.

[Ke1] Kempf G.: *On the geometry of a theorem of Riemann*, Ann. of Math. (2) **98** (1973), 178–185.

[Ke2] _____: *On the collapsing of homogeneous bundles*, Invent. Math. **37** (1976), 229–239.

[KW] King A., Walter C. H.: *On Chow rings of fine moduli spaces of modules*, J. reine angew. Math. **461** (1995), 179–187.

[Kle1] Kleiman S.: *Toward a numerical theory of ampleness*, Annals of Math. **132** (1974), 163–176.

[Kle2] _____: *Les Théorèmes de Finitudes pour les Fonctuers de Picard*, Exposé XIII in Théorèmes des Intersections et théorème de Riemann-Roch (SGA VI), pp. 616-666, Lecture Notes in Math. # 225, Springer-Verlag, Berlin and New York, (1971) , pp. 616-666.

[KleU] _____, Ulrich B.: *Gorenstein algebras, symmetric matrices, self-linked ideals, and symbolic powers*, Trans. A.M.S. **349**, No. 12, (1997), 4973–5000.

[Kl.H]   Kleppe H.: *Deformations of schemes defined by vanishing of Pfaffians*, J. Algebra **53** (1978), no. 1, 84–92.

[KL]   _____, Laksov D.: *The algebraic structure and deformations of Pfaffian schemes*, J. Algebra **64** (1980), no.1, 167–189. MR 82b:14027.

[Kl1]   Kleppe J. O.: *Deformations of Graded Algebras*, Math. Scand. 45 (1979), 205–231.

[Kl2]   _____: *The smoothness and the dimension of PGOR(H) and of other strata of the punctual Hilbert scheme*, J. Algebra 200 (1998), 606–628.

[KlM-R]   _____, Miró-Roig R. M.: *The dimension of the Hilbert scheme of Gorenstein codimension 3 subschemes*, J. Pure and Applied Algebra, **127** (1998), no. 1, 73–82.

[KMMNP]   _____, Migliore J., Miró-Roig R. M., Nagel U., Peterson C.: *Gorenstein liaison, complete intersection liaison invariants, and unobstructedness*, preprint, (1998).

[Ko]   Kollár J.: *Rational curves on algebraic varieties*, Ergebnisse der Mathematik und ihrer Grenzgebiete 3 Folge, Band 32 Springer-Verlag, Berlin, Heidelberg (1996).

[Kr]   Kreuzer M.: *Some applications of the canonical module of a zero- dimensional scheme*, Zero-dimensional Schemes, F. Orecchia and L. Chiantini, (1994), Walter de Gruyter, Berlin, pp. 243–252.

[Ku1]   Kung J. P. S.: *Canonical forms for binary forms of even degree*, Invariant theory, (Koh S. S., ed.) Lecture Notes in Math., # 1278, Springer-Verlag, Berlin and New York, (1987) pp. 52–61.

[Ku2]   _____: *Gundelfinger's theorem on binary forms*, Stud. Appl. Math. **75** (1986), 163–170.

[Ku3]   _____: *Canonical forms of binary forms: variations on a theme of Sylvester*. Invariant theory and tableaux (Minneapolis, MN, 1988), IMA Vol. Math. Appl., **19** (1990), Springer, New York-Berlin, pp. 46–58.

[KuR]   _____, Rota G.-C.: *The invariant theory of binary forms*, Bull. Amer. Math. Soc. (N.S.) **10** (1983), 27–85.

[Kur]   Kurosh A.: *Course in advanced algebra*, Moscow (1965); translation "Higher algebra" by George Yankovsky, 1972, reprinted 1988, "Mir", Moscow.

[KusM]   Kustin A. R., Miller M.: *Classification of the Tor-algebras of codimension four Gorenstein local rings*, Math. Z. **190** (1985), no. 3, 341–355.

[KusU]   _____, Ulrich B.: *A family of complexes associated to an almost alternating map, with applications to residual intersections*, Mem. Amer. Math. Soc. **95**, # 461, (1992) Amer. Math. Soc., Providence.

[Kut]   Kutz R. E.: *Cohen-Macaulay rings and ideal theory in rings of invariants of algebraic groups*, Trans. A. M. S. **194** (1974), 115–129.

[Lak]   Laksov D.: *Indecomposability of restricted tangent bundles*, Tableaux de Young et foncteurs de Schur en algèbre et géometrie, Asterisque **87–88** (1981) 207–220.

[La1]   Lascoux A.: *Syzygies des variétés déterminantales*, Adv. Math. **30** (1978), 202–237.

[La2]   _____: *Forme canonique d'une forme binaire*, Invariant theory, (Koh S. S., ed.) Lecture Notes in Math., # 1278, Springer-Verlag, Berlin and New York (1987) 44–51. MR 89a:12005

[Las]   Lasker E.: *Zur Theorie der Kanonischen Formen.* Math. Annalen **58** (1904), 434–440.

[Lur]   Lüroth J.: *Einige Eigenschaften einer gewissen Gattung von Curven vierten Ordnung*, Math. Ann. **1** (1868), 38–53.

[Mac1]   Macaulay F. H. S.: *On a method for dealing with the intersections of two plane curves*, Trans. A. M. S. **5** (1904), 385–400.

[Mac2]   _____: *The Algebraic Theory of Modular Systems*, Cambridge Univ. Press, Cambridge, U. K. (1916); reprinted with a foreword by P. Roberts, Cambridge Univ. Press, London and New York (1994).

[Mac3]   _____: *Some properties of enumeration in the theory of modular systems*, Proc. London Math. Soc. **26** (1927), 531–555.

[Macd]   Macdonald I. G. : *Symmetric Functions*, Cambridge Univ. Press, Cambridge, U.K.

[Mal1]   Mall D.: *Betti numbers, Castelnuovo Mumford regularity, and generalisations of Macaulay's theorem* Comm. Algebra **25** (1997), no. 12, 3841–3852.

[Mal2]   _____: *Characterisations of lexicographic sets and simply-connected Hilbert schemes* Applied algebra, algebraic algorithms and error-correcting codes (Toulouse, 1997), Lecture Notes in Comput. Sci., # 1255 (1997) Springer, Berlin, pp. 221–236.

[Mal3]   _____: *Combinatorics of polynomial ideals, Grobner bases and Hilbert Schemes*, Habilitationsschrift (Jan, 1997), ETH Zurich.

[Mal4]   _____: *Connectedness of Hilbert function strata and other connectedness results*, to appear, J. of Pure and Applied Algebra.

[M-DP]   Martin-Deschamps M., Perrin D.: *Sur la classification des courbes gauches, I*, Asterisque, **184-185** (1990).

[Mat]   Matlis E.: *Injective modules over Noetherian rings*, Pacific J. Math.,**8** (1958), 511–528.

[Ma1]   Matsumura H.,*Commutative Algebra*:   W. A. Benjamin Co., New York,1970.

[Ma2]   _____: *Commutative Ring Theory*, Cambridge Studies in Adv. Math. # 8, Cambridge University Press, Cambridge, U. K., (1986).

[Mar]   Maroscia P.: *Some problems and results on finite sets of points in* $\mathbb{P}^n$, Open Problems in Algebraic Geometry, VIII, Proc. Conf. at Ravello, (C. Cilberto, F. Ghione, and F. Orecchia, eds.) Lecture Notes in Math. # 997, Springer-Verlag, Berlin and New York, (1983), pp. 290–314.

[MayM]   Mayr E., Meyer A.: *The complexity of the word problem for comutative semigroups and polynomial ideals*, Adv. in Math. **46** (1982), 305–329.

[Mig]   Migliore J.: *Introduction to Liaison Theory and Deficiency Modules*, Progress in Math. # 165, Birkhaäuser, Boston, (1998).

[MiV]   Miyazaki C., Vogel W.: *Bounds on cohomology and Castelnuovo-Mumford regularity*, J. Algebra **185** (1996), 626–642. MR 98g:13016.

[MoSi]   Morales M., Simis A.: *Symbolic powers of monomial curves in* $\mathbb{P}^3$ *lying on a quadric surface*, Comm. Algebra **20** (1992), 1109–1121. MR 93c:13005.

[Muk]   Mukai S.:   *Fano 3-folds*, Complex projective geometry, (Trieste, 1989/Bergen, 1989), London Math. Soc. Lect. Notes Ser. # 179,. Cambridge University Press, Cambridge, (1992), pp. 255–263.

[Mum]   Mumford D.: *Lectures on curves on an algebraic surface*, Ann. Math
            Studies # 59, Princeton Univ. Press, Princeton (1966).

[Nak1]   Nakajima H.: *Heisenberg algebras and Hilbert schemes of points on pro-
            jective surfaces*, Ann. of Math. (2) **145** (1997), no. 2, 379–388.

[Nak2]   ———: *Lectures on Hilbert scheme of points on projective surfaces*,
            to appear, A.M.S. University Lecture Notes # 18, Amer. Math. Soc.,
            Providence.

[No]   Northcott D. G.: *Injective envelopes and inverse polynomials.*, J. Lon-
            don Math. Soc. (2), **8** (1974), 290–296.

[NoR]   ———, Rees D.: *Principal systems*, Quart. J. Math. Oxford (2), **8**
            (1957), 119–127.

[NotSp]   Notari R., Spreafico M. L.: *A stratification of the Hilbert scheme by ini-
            tial ideals, and applications*, preprint (1999), submitted to Manuscripta
            Math.

[Or]   Orecchia F. : *One-dimensional local rings with reduced associated graded
            rings, and their Hilbert functions*, Manuscripta Math. **32** (1980), no. 3-4,
            391–405.

[Pa]   Pardue K.: *Deformations of graded modules and connected loci on the
            Hilbert scheme*, The Curves Seminar at Queen's, Vol. XI (Kingston, ON,
            1997), Queen's Papers in Pure and Appl. Math., 105, Queen's Univ.,
            Kingston, ON, (1997), pp. 131–149.

[Po]   Porras O. : *Rank varieties and their desingularizations*, J. Algebra **186**
            (1996), 677–723.

[Por]   Porter R.: *The handbook of Latex and AMS-Latex Samples*, preprint
            (1997).

[Puc]   Pucci M.: *The Veronese variety and catalecticant matrices*, J. Algebra
            202 (1998), 72–95.

[Pup]   Puppe V.: *Simply connected manifolds without S^1 symmetry*, Algebraic
            Topology and Transformation Groups (Gottingen, 1987), Lecture Notes
            in Math. # 1361, Springer-Verlag, Berlin and New York (1988) pp.
            261–268.

[RaZ]   Ragusa A., Zappalà G.: *Properties of 3-codimensional Gorenstein
            schemes*, preprint 15 p., (1999).

[Rah]   Rahavandrainy O.: *Quelques composantes des strates de Brill-Noether
            de $Hilb^N(P^2)$*, C. R. Acad. Sci. Paris Sér. I Math. 322 (1996), no. 5,
            455–460. MR 97b:14006.

[RS]   Ranestad K., Schreyer F.-O.: *Varieties of sums of powers*, preprint, 33
            p. (1998), math.AG/9801110 available at xxx.lanl.gov.

[Rat]   Rathmann J.: *The uniform position principle for curves in characteris-
            tic p*, Math. Ann. **276** (1987), 565–579.

[Ree]   Reeves A.: *The radius of the Hilbert scheme*, J. Alg. Geom. 4 (1995),
            639–657.

[ReeS]   ———, Stillman M.: *Smoothness of the lexicographic point*, J. Alg.
            Geom. **6** (1997), 235–246.

[Rei]   Reichstein B.: *On expressing a cubic form as a sum of cubes of linear
            forms*, Linear Algebra Appl. 86 (1987), 91–122.

[ReiR]   ———, Reichstein Z.: *Surfaces parametrizing Waring presentation of
            smooth plane cubics*, Michigan Math. J. **40** (1993), 95–118.

[Rey]   Reye T.: *Erweiterung der Polarentheorie algebraischer Flächen*, J. Reine
           Angew. Math. **78** (1874), 97–122.

[Rez1]   Reznick B.: *Sums of even powers of real linear forms*, Memoirs of Amer.
           Math. Soc. **96** # 463, Amer. Math. Soc., Providence (1992).

[Rez2]   ———: *Homogeneous polynomial solutions to constant coefficient
           PDE's*, Adv. Math. **117** (1996), no. 2, 179–192.

[Ri]   Richmond H. W.: *On Canonical forms*, Quart. J. Pure Appl. Math. **33**
           (1904), 331–340.

[Ro]   Room T. G.: *Geometry of Determinantal Loci*, Cambridge Univ. Press,
           Cambridge, U. K., (1938).

[S-D]   Saint-Donat B.: *Sur les equations définissant une courbe algébrique*,
           C.R. Acad. Sci. Paris A **274** (1972), 324–327.

[Sa]   Sauer T.: *Smoothing projectively Cohen-Macaulay Space Curves*, Math.
           Ann **272** (1985), 83–90.

[ScSt]   Scheja G. , Storch U.: *Über Spurfunktionen bei vollständigen Durch-
           schnitten*, J. Reine Anew. Math. **278** (1975), 174–190.

[Sc1]   Scorza G.: *Sopra la theoria delle figure polari delle curve piane dei 4°
           ordine*, Ann. di Mat. (3) **2** (1899), 155–202; reprinted in his Opere Scelte,
           Vol.1. Edizione Cremonese, 1960, pp. 24–72.

[Sc2]   ———: *Un nuovo therema sopra le quatiche piane generali*, Math. Ann.
           52 (1899), 457–461; reprinted in his Opere Scelte, Vol.1., Edizione Cre-
           monese, 1960, pp. 73–77.

[Sch1]   Schenzel P.: *Über die freien Auflösungen extremaler Cohen-Macaulay
           ringe*, J. Algebra **64** (1980), 93–101.

[Sch2]   ———: *Castelnuovo's index of regularity and reduction numbers*, Top-
           ics in algebra, Part 2 (Warsaw, 1988), Banach Center Publ., 26, Part 2
           (1990), PWN, Warsaw, pp. 201–208.

[Schr]   Schreyer F-O.: *Geometry and algebra of Fano 3-folds of genus 12*, (Ab-
           stract), Syzygies and Geometry (Conference at Northeastern Univ.),
           Math Dept, Northeastern Univ., Boston (1995) pp.101–104.

[Sek]   Sekiguchi H.: *The upper bound of the Dilworth number and the Rees
           number of Noetherian local rings with a Hilbert function*, Adv. in Math.
           124 (1996), 197–206.

[Ser]   Sernesi E.: *Topics on families of projective schemes* Queen's Papers in
           Pure and Applied Mathematics, 73. Queen's University, Kingston, Ont.,
           (1986), viii+203 pp. MR 88b:14006.

[Sha]   Shafarevich I.: *Basic algebraic geometry*, Springer- Verlag, Berlin and
           New York, (1972).

[Sho]   Shokurov V. V.: *The Noether-Enriques theorem on canonical curves* (in
           Russian), Mat. Sbornik **86** (1971), 361–403; English transl. in Math.
           USSR Sbornik **15** (1971), 361–403).

[SiV]   Simis A., Vasconcelos W.: *The syzygies of the conormal bundle*, Amer.
           J. Math. **103** (1981), 203–224.

[SmSw]   Smith K., Swanson I.: *Linear bounds on growth of associated primes*
           Comm. Algebra **25** (1997), 3071–3079.

[Sp]   Sperner E.: *Über einem kombinatorischen Satz von Macaulay und seine
           Anwendungen auf die Theorie der Polynomideale*, Abh. math. Sem.
           Univ. Hamburg **7** (1930), 149–163.

[Sr1]   Srinivasan H.: *Decomposition formulas for Pfaffians*, J. Algebra **163** (1994), 312–334.

[Sr2]   _____: *Algebra structures for graded free resolutions*, Commutative algebra: syzygies, multiplicities, and birational algebra (South Hadley, MA, 1992), Contemp. Math., **159** (1994), Amer. Math. Soc., Providence, RI, pp. 357–365. MR96e:13011.

[Sr3]   _____: *A grade five Gorenstein algebra with no minimal algebra resolutions*, J. Algebra **179** (1996), no. 2, 362–379. MR96j:13012.

[St1]   Stanley R.: *Hilbert functions of graded algebras* Advances in Math. **28** (1978), 57–83.

[St2]   _____: *Weyl Groups, the hard Lefschetz theorem and the Sperner property*, S.I.A.M. J. Algebra, Disc. Math., **1**, No. 2, (1980), 168–184.

[St3]   _____: *Combinatorics and Commutative Algebra*, Progress in Math. # 41, Birkhäuser, Boston (1981); 2nd edition, 1996.

[Stu]   Sturmfels B.: *Four counterexamples in combinatorial algebraic geometry*, to appear, Journal of Algebra, D. Buchsbaum volume.

[Sy1]   Sylvester J. J.: *An essay on canonical forms, supplement to a sketch of a memoir on elimination* George Bell, Fleet Street, 1851; reprinted in his Collected Math. Papers, Vol. I, Paper 34.

[Sy2]   _____: *On a remarkable discovery in the theory of canonical forms and of hyperdeterminants* Philos. Mag. (4) **2** (1851), 391–410; reprinted in his Collected Mathematical Papers, Vol. I, Paper 41).

[Sy3]   _____: *Collected Mathematical Papers*, Cambridge: Cambridge Univ. Press (1904–1912).

[Sy4]   _____: *Sur une extension d'un théorème de Clebsch relatif aux courbes de quatrième degré*, Comptes Rendus de l'Acad. de Sciences **102** (1886), 1532–1534, reprinted in his Collected Math. Works, IV, pp. 527–530.

[Te]   Teissier B.: *Sur une inégalité à la Minkowski pour les multiplicités*, Annals of Math., **106** (1977), 38–44.

[Ter1]   Terracini A.: *Sulle V_k per cui la variet'a degli $S_h(h + 1)$-seganti ha dimensione minore dell'ordinario*, Rend. Circ. Mat. Palermo **31** (1911), 527–530.

[Ter2]   Terracini A.: *Sulla rappresentazione delle coppie di forme ternarie mediante somme di potenze di forme lineari*, Annali di Mat. Pura Appl. Serie III, **24** (1915), 1–10.

[TrV]   Trung N. V., Valla G.: *The Cohen-Macaulay type of points in generic position*, J. Algebra **125** (1989), no. 1, 110–119.

[Tu]   Tu L.: *The conectedness of symmetric and skew-symmetric degeneracy loci: even ranks*, Trans. Amer. Math. Soc. **313** (1989), no. 1, 381–392.

[Vi]   Viehweg E.: *Quasi-projective moduli for polarized manifolds*, Ergebnisse der Mathematik und ihrer Grenzgebiete 3 Folge, Band 30 Springer-Verlag, Berlin, Heidelberg (1995).

[VP]   Vinberg E. B., Popov V. L.: *Invariant theory*, (Russian), Algebraic geometry - 4, Itogi Nauki i Tekhniki: Sovr. Probl. Mat., # 55, VINITI, Moscow, (1989), pp. 137–314; English transl. in Encyclopaedia Math. Sci., Springer-Verlag.

[Wak]   Wakeford E. K.: *On canonical forms*, Proc. London Math. Soc. (2) **19** (1920), 403–410.

[Wall]   Wall C. T. C.: *Determination of the semi-nice dimensions*, Math. Proc. Camb. Philos. Soc. **81** (1977), 351–364.

[Wa1]   Watanabe J.: *A note on Gorenstein rings of embedding codimension three*, Nagoya Math. J. **50** (1973), 227-232.

[Wa2]   _____: *The Dilworth number of Artinian rings and finite posets with rank function* Commutative Algebra and Combinatorics, Advanced Studies in Pure Math. Vol. 11, Kinokuniya Co./North Holland, Amsterdam, (1987) pp. 303–312.

[Wa3]   _____: *The Dilworth number of Artin Gorenstein rings* Adv. Math. **76** (1989), no. 2, 194–199. MR90j:13023.

[Wa4]   _____: *Hankel Matrices and Hankel Ideals* (preprint,1985); 2nd rev. in Queen's Papers in Pure and Applied Mathematics 102, Curves Seminar at Queen's University, vol. 10, (A. Geramita, ed.), (1996), pp. 351–363; 3rd. rev. Proc. School Sci. Tokai Univ. **32** (1997), 11–21. MR 97g:13022,98c:13020.

[Weyl]   Weyl H.: *The classical groups*, 2nd ed., Princeton Univ. Press, Princeton, (1946).

[We1]   Weyman J.: *Resolutions of the exterior and symmetric powers of a module*, J. Algebra **58** (1979), 333–341.

[We2]   _____: *On the equations of conjugacy classes of nilpotent matrices*, Invent. Math. **98** (1989), 229–245.

[We3]   _____: *On the Hilbert function of multiplicity ideals*, J. Algebra **161** (1993), 358–369.

[We4]   _____: *Gordan ideals in the theory of binary forms*, J. Algebra **161** (1993), 370–391.

[Wh]   Whipple F.: *On a theorem due to F. S. Macaulay*, J. London Math. Soc. **28** (1928), 431–437.

[Ya1]   Yameogo J.: *Décomposition cellulaire de variétés paramétrant des idéaux homogènes de $\mathbb{C}[[x,y]]$. Incidence des cellules I*, Compositio Math. **90** (1994), 81–98.

[Ya2]   _____: *Décomposition cellulaire de variétés paramétrant des idéaux homogènes de $\mathbb{C}[[x,y]]$. Incidence des cellules II*, J. reine angew. Math. bf 450 (1994), 123–137.

Index

Index of Names

Index of Notation

Printing: Weihert-Druck GmbH, Darmstadt
Binding: Buchbinderei Schäffer, Grünstadt

Lecture Notes in Mathematics

For information about Vols. 1–1530
please contact your bookseller or Springer-Verlag

4. Lecture Notes are printed by photo-offset from the master-copy delivered in camera-ready form by the authors. Springer-Verlag provides technical instructions for the preparation of manuscripts. Macro packages in T_EX, L^AT_EX2e, $L^AT_EX2.09$ are available from Springer's web-pages at

http://www.springer.de/math/authors/b-tex.html.

Careful preparation of the manuscripts will help keep production time short and ensure satisfactory appearance of the finished book.

The actual production of a Lecture Notes volume takes approximately 12 weeks.

5. Authors receive a total of 50 free copies of their volume, but no royalties. They are entitled to a discount of 33.3 % on the price of Springer books purchase for their personal use, if ordering directly from Springer-Verlag.

Commitment to publish is made by letter of intent rather than by signing a formal contract. Springer-Verlag secures the copyright for each volume. Authors are free to reuse material contained in their LNM volumes in later publications: A brief written (or e-mail) request for formal permission is sufficient.

Addresses:

Professor F. Takens, Mathematisch Instituut,
Rijksuniversiteit Groningen, Postbus 800,
9700 AV Groningen, The Netherlands
E-mail: F.Takens@math.rug.nl

Professor B. Teissier, DMI, École Normale Supérieure
45, rue d'Ulm,
F-7500 Paris, France
E-mail: Teissier@ens.fr

Springer-Verlag, Mathematics Editorial, Tiergartenstr. 17,
D-69121 Heidelberg, Germany,
Tel.: *49 (6221) 487-701
Fax: *49 (6221) 487-355
E-mail: lnm@Springer.de